TETRAHEDRON ORGANIC CH
Series Editors: J E Baldwin, F

VOLUME 21

Biodiversity and Natural

Product Diversity

Related Pergamon Titles of Interest

BOOKS

Tetrahedron Organic Chemistry Series:
CARRUTHERS: Cycloaddition Reactions in Organic Synthesis
CLARIDGE: High-Resolution NMR Techniques in Organic Chemistry
FINET: Ligand Coupling Reactions with Heteroatomic Compounds
GAWLEY & AUBÉ: Principles of Asymmetric Synthesis
HASSNER & STUMER: Organic Syntheses Based on Name Reactions and
Unnamed Reactions
McKILLOP: Advanced Problems in Organic Reaction Mechanisms
OBRECHT & VILLALGORDO: Solid-Supported Combinatorial and Parallel
Synthesis of Small-Molecular-Weight Compound Libraries
PERLMUTTER: Conjugate Addition Reactions in Organic Synthesis
SESSLER & WEGHORN: Expanded, Contracted & Isomeric Porphyrins
TANG & LEVY: Chemistry of C-Glycosides
WONG & WHITESIDES: Enzymes in Synthetic Organic Chemistry
LI & GRIBBLE: Palladium in Heterocyclic Chemistry

RAHMAN: Studies in Natural Products Chemistry *(series)*

JOURNALS

BIOORGANIC & MEDICINAL CHEMISTRY
BIOORGANIC & MEDICINAL CHEMISTRY LETTERS
TETRAHEDRON
TETRAHEDRON LETTERS
TETRAHEDRON: ASYMMETRY

*Full details of all Elsevier Science publications are available on www.elsevier.com or from your nearest
Elsevier Science office*

Biodiversity and Natural Product Diversity

FRANCESCO PIETRA

Centro Linceo Interdisciplinare "Beniamino Segre", Roma, Italy /
Università di Trento, Trento, Italy

2002

PERGAMON

An imprint of Elsevier Science

Amsterdam – Boston – London – New York – Oxford – Paris
San Diego – San Francisco – Singapore – Sydney – Tokyo

ELSEVIER SCIENCE Ltd
The Boulevard, Langford Lane
Kidlington, Oxford OX5 1GB, UK

First edition 2002

Library of Congress Cataloging in Publication Data
A catalog record from the Library of Congress has been applied for.

British Library Cataloguing in Publication Data
A catalogue record from the British Library has been applied for.

ISBN: 0-08-043707-9 (Hardbound)
ISBN: 0-08-043706-0 (Paperback)

Printed and bound in Great Britain by
CPI Antony Rowe, Chippenham and Eastbourne

To Alessandro Ballio,
who made this work possible

Table of Contents

Preface

This book is designed to provide an all encompassing vision of the diversity of natural products in the perspective of biodiversity. Both living organisms and fossil remains are taken into account, without any bias for either the sea, land, or extreme environments. Understandably, however, this is not intended to be a comprehensive portrait of natural product diversity, which would be beyond my resources and would demand a multi-volume treatment.

My aim was to focus the attention on representative examples - drawing from both my own experiences and the most recent literature - in a balanced coverage of the various niches, considering the local-geographic rather than the global-taxonomic distribution of natural products. To this concern, the attention is focused on the molecular skeletons (grouped in charts for classes of metabolites for each local-geographic environment) as differences in the coding genes are expected to be larger in directing the molecular framework than its detail, i.e. the functional groups. In this perspective, the molecular diversity is semi-quantitatively evaluated by a molecular complexity index. It was encouraging, during this compilation, to see similar strategies applied to the conservation of endangered taxa (Rodríguez 2000) and the comparison of natural products with synthetic compounds (Henkel 1999).

Then, the natural product diversity is examined at the functional level, that is signaling and defensive agents, and how the wild diversity is exploited, modified through biotechnological techniques, and recreated or imitated via total synthesis. To cope with these aspects, the full molecular details are considered. In this analysis, the value of the natural product as a drug is kept within the limits imposed by the advancement in the knowledge of the molecular structure of the targets, which is opening the way to the rational design of drugs.

Next, threatening of natural product diversity is examined, either by human activity or mass mortality during the geological eras. Molecular skeletons are shown, for the same reasons above about the natural product diversity from extant organisms. Management and conservation of natural products are also briefly discussed.

The major difficulties encountered in gathering the data were for microorganisms on land: papers from the golden period of the antibiotics contain scarce geographic information and this problem has continued to the present with work in industry.

On the taxonomic/phylogenetic side, not only the genus and species are given for the eukaryotes, but also the order, which is most significant with respect to the natural product distribution in living organisms. An exception is made for land plants, keeping with tradition to give also the family. Changes in the way living things are taxonomically described are kept to a minimum, preferring to refer to widely accepted positions and names. This has found recently an authoritative support (Lewin 2001).

The terminology "higher organisms" or "lower organisms" is used freely in this book for conciseness, without any implication with respect to the rank of the organisms along the biological scale. In particular, "higher" is not intended to mean more evolved, since, provided that the phylogeny is cleaned from uncertainties (Huelsenbeck 2000) and the evolution is evaluated for all its possible aspects, the extent of the evolution is related to the age of the group and not to its position along the tree of life.

What is also avoided is any attempt at a revival of the past, lost pristine conditions, since the increased energy demand (which is at the basis of the loss of natural product diversity) was unavoidable with the advances in technology, the prolonged life span in rich countries, and uncontrolled human population growth in the developing world. I believe that the only hope for saving the Earth's remaining living resources rests largely in new technologies for food, clean energy, and material production.

With the exception of a few landmark papers, such as Robinson's total synthesis of tropinone (Robinson 1917) and Blumer's essay on fossil pigments (Blumer 1965), for reasons of space the most recent references are given, preferably review papers; only the first author is given in the text. This allows us easily to trace back all previous work, although regrettably the original discoveries are not appropriately acknowledged. To leave space for the titles of the papers, which indicate to the reader the importance of reading beyond this book, a smaller font is used for the references.

Examples in the tables, and trade names of compounds, were chosen without any commitment or preference for any particular commercial source. Trade names are intended to be registered trade names. The data provided in this book are in good faith, but I make no warranty, expressed or implied, nor assume any legal liability or responsibility for any purpose for which these data are used.

I welcome criticism, corrections, and suggestions about gaps or redundancy in coverage that will help to make a second edition of this book even more useful.

Francesco Pietra
Rome, October 2001

Definitions of abbreviations for the charts and tables

Actinom. = Actinomycetales

Ala = alanine

alkal. = alkaloid(s)

Ang. = Angiospermae or angiosperm(s)

Amphib. = Amphibia or amphibian(s)

Arach. = Arachnida

Arg = arginine

Ascid. = Ascidiacea

Ascom. = Ascomycotina

Asn = asparagine

Asp = aspartic acid

Aster. = Asterozoa

Axinel. = Axinellida

AY = Ayensu 1979

Bact. = bacteria

Basidiomyc. = Basidiomycotina

BC = British Columbia

BR = Bruneton 1995

Bryoph. = Bryophyta

Bryoz. = Bryozoa

C = Chart

Calc. = Calcarea

carboh. = carbohydrate(s)

Chloroph. = Chlorophyta

Chrysoph. = Chrysophyta

Cilioph. = Ciliophora

Cnid. = Cnidaria

cosmop. = cosmopolitan

Crust. = Crustacea

Cyanobact. = cyanobacteria

cyclopept. = cyclopeptide(s)

Cys = cysteine

Demosp. = Demospongiae

Dendroc. = Dendroceratida

Dendroph. = Dendrophylliidae

Deuterom. = Deuteromycotina

Diatom. = Diatomae

Dictyoc. = Dictyoceratida

Dinofl. = Dinoflagellatea

diterp. = diterpene(s) or diterpenoid(s)

Echin. = Echinodermata

Eumyc. = Eumycota

freshw. = freshwater

Glu = glutamic acid

pGlu = pyroglutamic acid = 5-oxoproline

Gly = glycine

GBR = Great Barrier Reef

Gymn. = Gymnospermae or gymnosperm(s)

Halichon. = Halichondrida

Hemich. = Hemichordata

Hexactin. = Hexactinellida

Hist = histidine

Holoth. = Holothuroidea

Homoscl. = Homoscleromorpha

Hyp = hydroxyproline

Ins. = Insecta

invertebr. = invertebrate(s)

Ile = isoleucine

isopr. = isoprenoid(s) or isoprene

Leu = leucine

Lys = lysine

Mamm. = Mammalia

meroditerp. = meroditerpene(s)

Met = methionine

MI = The Merck Index 12:2 (1998)

Moll. = Mollusca

monoterp. =   monoterpene(s) or
                      monoterpenoid(s)

Myxobact. = Myxobacteria

NC = New Caledonia

Nemert. = Nemertea

Nepheliosp. = Nepheliospongida

Nudibr. = Nudibranchia

NZ = New Zealand

Okin. = Okinawa

Opisthobr. = Opisthobranchia

orphan = orphan molecule or molecular skeleton (the producing organism is not known)

Osteich. = Osteichthyes

pantrop. = pantropical

pept. = peptide(s) or cyclopeptide(s), depsipeptide(s), lipopeptide(s)

Phaeoph. = Phaeophyceae

Phe = phenylalanine

PNG = Papua New Guinea

Poecil. = Poecilosclerida

polyket. = polyketides

polysacch. = polysaccharide(s)

Porif. = Porifera (sponges)

Prol = proline

Rhodoph. = Rhodophyceae

Ser = serine

sesquiterp. = sesquiterpene(s)

shikim. = shikimate(s)

T = table

thiopept. = thiopeptide(s)

Thr = threonine

triterp. = triterpene(s)

Trp = tryptophan

Turbell. = Turbellaria

Tyr = tyrosine

Ulvoph. = Ulvophyta

undeterm. = undetermined

Val = valine

Vertebr. = Vertebrata

Part I. The concept of biodiversity

blown by all the winds that pass
and wet with all the showers
Louis Stevenson

Chapter 1. Defining biodiversity

Biodiversity is difficult both to define and measure. Central to the issue is the concept of species, which has proved most elusive to generations of biologists. Darwin himself was never convinced of the prevailing positions at his time and debates about the concept of species have continued to the present and will continue apace.

Although the reader is referred to the biological literature for a thorough illustration of these concepts, a survey of the prevalent positions may serve as an introduction and a basis to our intents. Mayr's concept of biological species as "a group of actually or potentially interbreeding natural populations that are reproductively isolated from other such groups" (Mayr 1963, 1969) is much encompassing, although Mallet's criticism that "Reproductive isolation is a kind of mystical definition..." may be defended. Other definitions of species, such as ecological species (for groups that occupy different ecological niches), evolutionary species (defining a species as a single lineage of ancestral-descendant populations that remain separate from other such lineages in an evolutionary tree), and geno-species (to account for asexual bacterial strains that allow interchange of genetic material) also serve the scope.

In the light that a concept of species that satisfactorily encompasses all organisms can hardly be found, we are now in a position to examine the prevalent definitions of biodiversity.

1.1 Biodiversity at species level

Biodiversity is commonly understood as the number of different species in a given ecosystem. By doing so, the huge number of insects (estimated from two million to ten million species, Speight 1999) gives the wrong impression that biodiversity on land is far greater than in the oceans. Actually, we will see that the species is a misleading basis on which to compare biodiversity on land and in the sea.

Choosing species as a measure of biodiversity is also far from straightforward. The concept of species is poorly defined with microorganisms, particularly for non-sexual or unculturable ones. When sibling and cryptic species are also accounted for, the difficulty in making an overall estimate of the biodiversity increases, particularly for the sea, where sibling species are plentiful. Accounting for sibling species, a fourfold increase of biodiversity in the sea is expected (Knowlton 1993). Endosymbionts pose even more difficult problems, to the point that there is no agreement whether they should be taken into account in any evaluation of biodiversity.

In any event, the mere inclusion of microorganisms in the species to be enumerated makes a catalogue of the existing species far from being an affordable enterprise. Dealing with viruses is especially difficult: although they do not live as isolated entities, they are found widespreadly in the oceans, possibly involved in regulating the biodiversity (Fuhrman 1999).

The microorganisms deserve much attention, representing the bulk of phylogenetic diversity, thus contributing enormously to the biological evolution. The high turnover makes also the microorganism an attractive entity for the laboratory search of correlations between species diversity and their productivity, i.e. the "rate of production of organic matter by a community" (Kassen 2000).

1.2 Biodiversity at higher taxonomic levels

Taking higher taxonomic levels as an estimate of biodiversity (May 1994), more phyla are found in the oceans than on land. Also, with the only exception of the arthropods, phyla with representatives both on land and in the sea are more speciose in the latter.

Of the thirty-three known phyla of extant animals, only one, Onychophora, is exclusive of land, while as many as twenty-one phyla are exclusive of the sea. The latter comprise the brachiopods (lamp shells, about 300 species), chaetognates (arrow worms, about 50 species), ctenophores (comb jellies, over 80 species), echinoderms (about 600 species distributed in the crinoids, i.e. sea lilies, and feather stars, the asterozoans, i.e. starfishes, brittle stars and basket stars, and the echinozoans, i.e. sea urchins, sand dollars and heart urchins), echiurians (spoon worms, about 300 species), gnathostomulids (about 100 species), hemichordates (nearly 100 species, distributed in the enteropneusts - acorn worms - pterobranchs, and planktosphaerids), kinorhynchs (about 100 species), loricifers, phoronids (horsheshoe worms, about 15 species), placozoans (2 species), pogonophors (about 100 species), priapulids (4 species), and rotifers (more than 1500 species).

The conclusion is that the taxonomic diversity is larger in the sea than on land.

1.3 Biodiversity at genetic level

Biodiversity can also be considered within the species or even within the population. This is called genetic diversity, signifying that each individual has its own genetic make-up. This allows the species to adapt to environmental changes and furnishes the seed for speciation, by which biological evolution occurs.

Even this definition of biodiversity is not free of pitfalls, however, and it may open the door to ontology. To add to this concern, arbuscular mycorrhizal fungi in the order Glomales are a particularly vexing problem. The mycorrhyzae are asexual organisms, which poses problems in defining biodiversity; they also show different rDNA sequences on different nuclei, which are many, enclosed in a single spore of these coenocytic organisms (Hijri 1999; Hosny 1999). It is perhaps this high genetic diversity that has granted success to these organisms for more than 400 million years.

1.4 Biodiversity at ecosystem level

Biodiversity can also be examined at ecosystem level. This is of special concern for ecology, because, although attention to ecological problems is increasing, the choice of the appropriate scale to look at the events is problematic (Levin 1999). The spatial and temporal scales needed are often of such large size to make any reliable observation difficult. This is especially true for the sea, due to the lack of barriers which, therefore, makes any subdivision into ecosystems difficult. This is why any quantitative evaluation of biodiversity at ecosystem level is far from being an easy task.

Chapter 2. The course of biodiversity

The discovery of transposons has opened new vistas on the evolutionary events: evolutionary changes, once thought to be small and rare, have been recognized to be large and frequent. Genetically unstable bacterial populations adapt more rapidly than stable populations (Radman 2001). Marine snails in the genus *Conus*, being unusually able to mutate through their third exon, are another good example. Proneness to mutation gives flexibility to the array of toxic peptides produced by these mollusks, which accounts for their extraordinary success, testified by the existence of about 500 species. Adaptation serves to maintain the efficiency of the peptidic venom for the snails' harpoons, used in capturing other invertebrates and even fish.

Other examples of easy mutation are found with flower plants in the genus *Ipomoea*, to which morning glory belongs. They are characterized by a reshaping of the genome by transposons. Equally flexible are the ciliates: in the formation of a new macronucleus, the DNA portion comprised between the coding regions is removed, bringing about order in the region.

Admittedly, bacteria, *Conus*, *Ipomea*, and the ciliates have unusually high mutation ability, but examples of natural genetic engineering are encountered often, even with the vertebrates. The history of the acquisition of the immune system is illustrative. Present in the invertebrates as merely an innate immunity by the way of phagocytic cells, the adaptive immune system of vertebrates resulted from the sudden introduction of a transposon just where a remote forerunner of the antibody genes resides. The antibody gene complex could evolve this way, giving rise to B and T lymphocytes. This occurred with sharks, probably when they first appeared in the Ordovician era.

Biodiversity may also be fostered locally without any genomic rearrangement. A redistribution of species suffices, in the concept of the refugee. This may be a physical shelter, such as a foreign empty shell for the hermit crab, or a chemical weapon, such as the distasteful metabolites of certain seaweeds that provide a refugee for grazeable seaweeds. Other examples of correlation between the diversity of symbionts and the hosts are provided by mycorrhizal fungi in symbiosis with terrestrial plants (van der Heijden 1998). Leaf-cutting ants in coevolution with plants, and antibiotic-producing actinomycetales (Currie 1999), are examples of associated biodiversity (Swift 1996).

All these represent evolutionary adaptations to the changing environment, in the survival of the fittest (Brookfield 2001), which was at the basis of Darwin's thoughts and has never been disproved. Biodiversity is always in a fluctuating condition. Becoming affirmed is not a condition for species to escape evolution. Natural laws impose a finite length of time to any species, which is more important than a finite length of time imposed to individuals (Sgró 1999). Then, the course of the evolution brings the species to extinction, providing space and resources to new species.

Unstable and evanescent analogical (no gene) life of primordial times (Woolfson 2000) is difficult to imagine in the perspective of secondary metabolites. Digital (DNA) life has more concrete foundations, dating to at least 3,450 My ago, according to cyanobacterial remains (Mojzsis 1996). The first eukaryotes are latecomers, appeared in the fossil record 2,500 My ago (Schopf 1993), and probably inherited homologues both of tubulin and actin from bacteria (van den Ent 2001).

These chronologies are represented in Fig. 2.I, in the perspective of a sea vs land occupancy and mass extinctions. It is seen that the cycads and the ginkgophytes appeared during the end-Permian, following the disappearance of the pteridosperms, which had predominated in the carboniferous. This does not mean that the follower descends necessarily from the former, since the pteridosperms, rather than the cycads and ginkgos, are considered the ancestors of the angiosperms. The ginkgos and the cycads simply replaced the pteridosperms. The same occurred following the end-Cretaceous mass

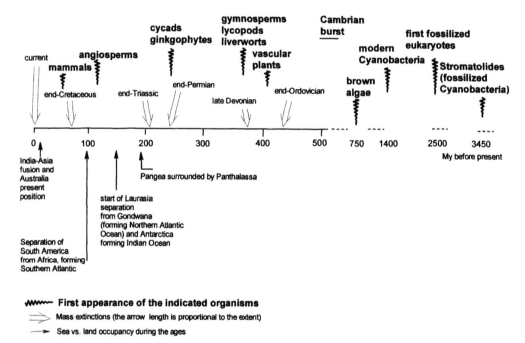

Figure 2.I. First appearance of organisms in the perspective of mass extinctions and sea vs land occupancy

extinction, when the mammals took the place left vacantly by the reptiles, paving the way to the appearance of the humans.

Biodiversity may also be fostered when the ecosystem is unstable, like a pond subjected to alternating periods of dryness and wetness, populated by plants giving C_3 or C_4 metabolites, or shifting to this biosynthetic mode according to the availability of CO_2 and the need of saving water (Keeley 1998).

Ecological factors may also determine both the phenotype and speciation (Tregenza 1999). At least at the phenotype level, biodiversity may also be fostered by predation; this can be seen in the framework of the evolution of signals, such as with electric fish (Stoddard, P.K. 1999). Threatening of biodiversity, and the resistance opposed by the ecosystems, are examined in Part VI.

Part II. The relationship between biodiversity and natural product diversity

The genetic message, the programme of the present-day organism, resembles a text without an author, that a proof-reader has been correcting for more than two billion years ...

François Jacob

Chapter 3. Taxonomy, phylogeny, and natural products

Relationships between the nature of secondary metabolites and the taxonomy of the organismic source have often been sought. This is particularly true in botany, where chemotaxonomy has a long tradition, now supplemented by genomic analysis (Grayner 1999). Since taxonomy and phylogeny go hand in hand, exclusive metabolites may have phylogenetic significance.

These ideas have been expanded in an ecological and evolutionary perspective for land plants (Gottlieb 1998). In the animal kingdom, cladograms based on the distribution of natural products have been set up for demosponges (Andersen 1996).

Any relationship between secondary metabolites and the taxonomy of the organismic source is subject to rapid revision, however. The analytical techniques for the isolation of natural products present in tiny amounts in intricate mixtures, and the spectral techniques for structure elucidation, have greatly advanced in the last three decades. This has two major consequences: first, that compounds or compound classes previously thought to be exclusive of certain taxa are more and more frequently found in phylogenetically and ecologically distant organisms (Pietra 1995), and second, that the occurrence of secondary metabolites in families of compounds, differing either in the carbon skeleton or merely in the stereochemistry, has become the norm. Curiously, families of compounds are reported from time to time as a novel observation, in connection with libraries of compounds from combinatorial synthesis or biosynthesis (Brady 2001).

The characterization of the metabolites also deserves the greatest care in order that chemotaxonomy remains a valid science. In particular, a close scrutiny of stereochemical relationships in a biosynthetic cascade may allow us to relate the metabolite distribution to lineages; this is a task that genetics cannot yet perform (Guella 1997A).

Oddly enough, while ecology receives increasing attention, alpha taxonomy, that is the search and description of new species, has been neglected (Winston 2000). Granting agencies and their professional referees have considered alpha taxonomy a jump into the past, which undermines the study of both biodiversity and natural product diversity, if it is true that only 10% of extant species has been described. This attitude is now changing; signs appear of a renewed interest in alpha taxonomy, relying on bioinformatics (Bisby 2000; Edwards 2000A). Integration of taxonomic, ecological, and biogeographical data from various databases existing around the world is the aim of recent projects (Species 2000).

With the coming of programs for personal computers for the treatment of molecular data, taxonomy has largely become digital (DNA) phylogeny. A link with natural products has also been promised (Bisby 2000), but how it could be realized is unclear: large databases of natural products exist indeed but, because of the economic interests involved, the access is very expensive, in contrast with free access to taxonomic data. Even without budgetary restrictions, it would be an enormous enterprise - perhaps beyond the human resources - establishing the organismic origin for the huge number of natural products that have been reported. One should have to rely on poor descriptions in chemical journals, often without the support of a qualified taxonomist. For streptomycin, a famous antibiotic substance, I found no way to trace back where the productive actinomycete, *Streptomyces griseus*, came from, in spite of gracious help from the specific learned institutions. I could only guess, from the original literature, that the "heavily manured field soil... or the throat of a chicken", from which the actinomycete was isolated (Schatz 1944), was in a temperate area, probably the US.

Lower organisms that face a strong interspecific competition are the best organismic sources. However, no steady relationship has ever been found between the taxonomic rank and the secondary metabolic productivity. Bacteria in a single genus, *Streptomyces*, have emerged as rich sources of antibiotics. Most microorganisms are not known to produce any unusual metabolite. This is counterbalanced, with productive strains, by the occurrence of secondary metabolites in families, where the members may differ merely in the functional groups, or are more distantly interrelated along a biogenetic cascade, involving rearrangement of the carbon backbone.

Biosynthetic studies of triterpenoids of different skeletal type from homogeneously the same species of plant, carried out with chimeric enzymes, revealed that great chemical diversity may stem from tiny differences in the genes coding for the relevant multifunctional enzymes (Chapter 13.1). A few changes in the amino acids at the active site may result in metabolites differing in the carbon skeleton. This suggests that plants have evolved chimeric enzymes acting as multifunctional triterpene synthases (Chapter 14.1 and Chart 14.1: Kushiro 1999).

Secondary metabolites that may be taken as biomarkers, and sequences of rRNA, concur in viewing separate groups within both the marine ciliates (Pietra 1997) and the archaeans. For the latter, however, identity of lipid biomarkers was assumed without paying attention to chirality (Hinrichs 1999); if diastereomers are involved, different enzymes for the different archaean groups are implied, while genomic differences may have passed unnoticed because of insufficient resolution.

Chapter 4. The problem of unculturable species

A systematic evaluation of the biological position of living things started with Linnaeus in the 18th century, albeit limited, with few exceptions, to conspicuous species. This lopsided bias has continued down to our days, judging from the meager number of taxonomists devoted to microorganisms, estimated to a mere 2-3% of the whole community of taxonomists (May 1994).

Things worsen if unculturable microorganisms are considered. They are commonly believed to represent 99.9% of the whole microbial community, in contrast with the view that "the global richness of free-living microbial species is moderate" (Finlay 1999). Between the two positions, cautious estimates have been made as to the number of new species of microorganisms that remain to be discovered (May 1994).

In any event, referring to "unculturable species" may be misleading. Microorganisms often resist culture because of inappropriate media and conditions. Overcoming these problems may be far from a simple task. It may require lengthy and systematic trials, such as for the dinoflagellate *Gambierdiscus toxicus*, involved in ciguatera seafood poisoning. Years elapsed until the culture of this protist succeeded on a natural substrate (Satake 1997).

Admittedly, strict symbiotic and parasitic species that have long adapted to the host, renouncing to a part of their genome to rely largely on the genomic resources of the host, are candidate unculturable species. Biotechnological techniques may help avoiding the problem of the culture, however, as discussed in Chapter 14.1, lending more importance to secondary metabolites in the evolution of biodiversity.

At any event, microorganisms remain far less known than conspicuous species. Microorganisms are subject to high dispersal, as in the recently discovered gene flow for planktonic foraminifers of Arctic and Antarctic waters (Darling 2000). The composition of the soil may be also relevant (Yaalon 2000). This is why making an estimate of the diversity of natural products is difficult.

Chapter 5. Natural product diversity: at which rank?

5.1 The molecular rank

In recent years, we have all watched or heard the increasing popularization of the concept of biodiversity, with all the consequences. Things worsen if the diversity of natural products is considered, which is not surprising because the language of chemistry is not widely spoken. To the layman, chemistry evokes such catastrophes as Bhopal and Seveso, which have more to do with misusing chemistry than chemistry as a science. The layman is also seldom prone to acknowledge that man (and domesticated animals) live longer and better thanks to the drugs developed by the chemist. Even at the learned level, the language of chemistry is often mishandled, or bypassed, as if chemistry were a specialist side branch and not a science at the core of life.

Popularizing chemistry finds many obstacles. The systematization of the names of substances is particularly threatening (Bensaude-Vincent 2001), in a new kind of Esperanto that no one can master at the speed of the speech. Luckily, trivial names for natural products have never been dismissed. Admittedly, digitalization requires systematization, but natural product retrieval is best done through the very language of chemistry, that is the structural formula.

At any event, defining and measuring the diversity of natural products remains a difficult problem, at least as defining and measuring biodiversity. If the different molecules present in a given ecosystem are taken as a measure of natural products diversity, the land seems to be far richer than the sea. This reflects the huge variety of secondary metabolites produced by vascular plants on a restricted number of themes, which were isolated during more than 150 years of studies.

Giving less importance to functional groups and focusing on the molecular skeleton, we can get a perspective of natural products diversity more closely to the genome. The reason is that differences in the coding genes are larger in directing the construction of the molecular framework than its details.

Approaching natural product diversity from the molecular skeleton gives special importance to genes that are specific of the taxon, neglecting those that control the primary biochemical processes, which are common to all organisms and thus do not contribute to the diversity of natural products. Sometimes, however, it is difficult and arbitrary distinguishing secondary metabolites from primary metabolites. Glyceryl ethers of the cell walls of the archaeans (Chapter 10) are a case in point. Therefore, "natural product" stands here for any well defined organic molecule found in nature, not limited to the secondary metabolite of the chemical tradition.

Special problems are posed by compounds that have many points of attachment in forming new bonds, such as the carbohydrates, which show the highest structural diversity for a single class of compounds. This fulfils the requirement of recognition processes (Davis 2000) through glycosaminoglycans placed at the cell surface and extracellular matrix. The structural differences in the disaccharide repeating units derive mostly from different sulfation at three hydroxyl groups and one amino group (Venkataraman 1999; Gabius 2000). Glycosides, such as the glycosphingolipids

(Vankar 2000), behave similarly. These functional groups do not affect the planar skeleton, which is our basis of judgment of the diversity of natural products.

Comparing the different molecular frameworks of natural products with one another poses several problems. The method of choice depends on the intents, which, for an approach to nature as in this book, differ from synthetic strategies (Chanon 1998). First, the drawing of the molecular skeletons is curtailed to the essential, reducing all carbon-carbon bonds to single bonds and disregarding the stereochemistry, except crossed bonds for reasons of clarity. This means that stereochemical series, such as the *ent*-forms of terpenoids, do not show up.

Second, compounds of a different biogenesis are grouped together under the same skeletal type, such as marine and terrestrial quinolizidine alkaloids, although the latter only are known to derive from lysine. This position was dictated by the scarce knowledge of biosynthetic pathways for metabolites of recent isolation, in particular marine metabolites. I made an exception for documented instances of historical and traditional importance, such as the tryptamine-secologanin and phenylalanine-secologanin alkaloids (Charts 6.1.1.A, 6.1.2.A and 6.1.3.A), where the different biogenetic moieties are shown.

Third, heteroatoms are only shown if present at heterocyclic positions or making an intrinsic part of functional groups needed for a realistic representation of the metabolite, such as the carbonyl group of macrolides.

Since the build up of a more complicated molecular framework requires a more elaborated enzyme system, a semi-quantitative evaluation of the molecular diversity can be made by adding a term for the molecular complexity. Although a mathematical topological index (Bertz 1981, 1993) could be used in this regard, an empirical counting metric is sufficient. This is provided by the complexity metric S and the size metric H (Whitlock 1998), making things much easier. S is the sum of a constant times the number of rings, plus a constant times the number of unsaturations, plus a constant times the number of heteroatoms, plus finally a constant times the number of chiral centers. H is the number of bonds in the molecule (one to three from a single to a triple bond). Thus, H is related to the molecular size. Although the stereochemistry is disregarded, chiral centers are counted for the metric S. Both S and H can be secured instantly from a program that runs on the simplest personal computer.

Reducing all carbon-carbon bonds to single bonds may create unrealistic chiral centers, increasing artificially the complexity value, particularly when aromatic rings are present. I made due account of this artifact in drawing conclusions as to the skeletal complexity, particularly on comparing aromatic compounds with other classes of compounds. The problem becomes less serious on comparing alkaloids from different ecosystems, since aromatic (indole) alkaloids are ubiquitous. To avoid creating unrealistically high complexity values, large polyphenols may be drawn as aromatic compounds. On the other hand, aromatic rings are not counted for the complexity metric S (Whitlock 1998). Worse, no adequate treatment of the functional complexity is yet available (Chanon 1998).

5.2 The taxonomic and ecological rank

Once it is agreed that natural product diversity can be best evaluated from the molecular skeleton of the metabolites, the choice is between a global-taxonomic and a local-geographic distribution. At a time that declining of natural product diversity has taken a non linear trend, the choice of a local-geographic distribution can be better defended for the same reasons as the distribution of endangered taxa (Rodríguez 2000). Accordingly, representative molecular skeletons are grouped per biogenetic class for each ecosystem, alongside the productive taxa (Chapters 6-10).

In dealing with ecosystems we are forced to arbitrary choices because, unlike taxonomy, the organization of the ecosystems and biomes is not a science. Ecological subdivisions are made according to one's particular need (Odum 1983). To fit our needs, the distribution of living organisms and ecosystems should be such as to flow smoothly into a description of the distribution of natural products. The traditional subdivision of land into biomes fits these purposes, whereas no such general commodious subdivision has ever been proposed for the oceans, seen as a continuum. Certain marine ecosystems warrant separate attention, however, in particular the coral reefs, where the natural product diversity is far larger than for any other ecosystem. Because of all these problems, a simplified presentation of ecosystems, biomes, and biodiversity was adopted here. Throughout, east and west are arbitrary, with Europe as the dividing line.

Not only rare species, but also widely distributed species, either cultivated or of recent introduction into regions of compatible climates and resources, may give unusual secondary metabolites. In any case, a valid taxonomic assignment is a prerequisite to any reliable estimate of the distribution of natural products. Major problems to this concern are posed by microorganisms, where even the definition of species is often questionable. This is particularly vexing, since many microorganisms are of old age and wide adaptability to the most diverse conditions, occupying all niches. Currents in the sea and rivers across the continents, and winds in the atmosphere, favor the spreading of microorganisms. On land, particularly in the soil, the microorganism diversity increases in the order archaeans, algae, protozoans, fungi, and bacteria; probably the same order holds for the sea.

The organismic sources of the metabolites are reported in the charts at the lowest possible taxonomic level (species, genus, family, order, phylum), as required by the distribution of the metabolites at issue. With terrestrial plants the tradition of giving also the family was always followed, although a broader exploration of plants, and the advent of high-sensitivity analytical systems, have rendered the family of scarce taxonomic significance. With few exceptions of highly specific metabolites, the order is the taxonomic rank most representative of the distribution of natural products.

Whenever possible, the fundamental binomial Linnaeus Latin names and the authorities of the taxonomic assignments are given. Widely understood, vulgarized taxonomic terms, are also used, such as angiosperms and gymnosperms. Occasionally, the common names for terrestrial plants are also used, without full commitment, however, since they have no meaning outside the

English-speaking world, in particular the German-speaking world, which has his own *ad hoc* vocabulary.

Part III. Natural product diversity at ecosystem level

And the gold of that land is good:
there is bdellium and the onyx stone.
Genesis 2.12

Chapter 6. Terrestrial and freshwater biomes

It was concluded in Chapter 5.2 that the contrasting needs of detail and conciseness in describing the local-geographic distribution of natural products can be best met by subdividing land into biomes. Traditionally, these comprise the tropical rain forest, grassland and savanna, scrub and deciduous forest, temperate grassland, deciduous forest, chaparral, and taiga and tundra (Fig. 6.I).

This scheme adapts nicely to endemic species and, to a certain extent, also to species that have been introduced from a long time. In contrast, cultivated and recently introduced species can hardly be dealt with satisfactorily by any kind of ecosystem subdivision.

For each biome, the metabolites are grouped per chemical class in graphical charts, in the order alkaloids, peptides (including also depsipeptides, lipopeptides, and glycopeptides), polypeptides, isoprenoids, fatty acids and polyketides, shikimates, and carbohydrates. Attention is also paid to

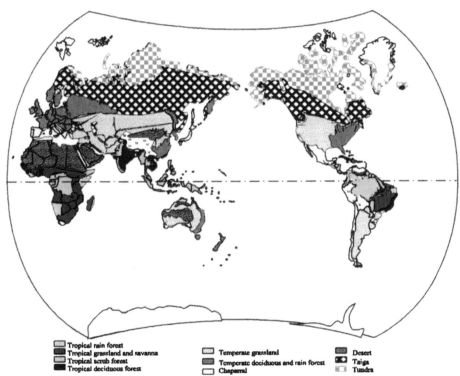

Tropical rain forest
Tropical grassland and savanna
Tropical scrub forest
Tropical deciduous forest

Temperate grassland
Temperate deciduous and rain forest
Chaparral

Desert
Taiga
Tundra

Figure 6.I. Terrestrial and freshwater biomes

biome-specific proteins. To some extent, these groups correspond to biogenetic classes.

The different classes of molecular skeletons are grouped into different boxes in the charts; each skeleton is accompanied by the value of the complexity metric, S, the specific complexity metric, S/H, and, in boldface characters, by a high-rank taxonomic assignment of the organismic source.

Correspondingly, the molecular-skeleton class is indicated in boldface characters in the captions, with a further subdivision into families of skeletons, marked with a line underneath. This allows us a quick view as to the taxonomic distribution of the natural products, which is complementary to their ecological distribution.

Peak values and average values for S and S/H for the various biogenetic classes of metabolites follow the chart title; they are discussed in Chapter 11.

6.1 Tropical rain forest, grassland and savanna, scrub and deciduous forest

The tropics and subtropics are comprised within 30^0 of latitude on both sides of the equator. Various ecosystems are represented: the tropical rain forest (Jacobs 1981), grassland and savanna, scrub and deciduous forest, and arid land (AY 1979) (Fig. 6.I).

Tropical rain forests occupy wide areas of Central and South America, politically belonging to Guatemala, Honduras, Nicaragua, Costa Rica, Panama, non-mountain portions of Peru, Ecuador, Colombia, Venezuela, Guyana, Surinam, French Guiana, and the whole northern Brazil. The latter represents 40% of the world's remaining tropical rain forest. The African tropical rain forest occupies more restricted area in Liberia, the Ivory coast, Cameroon, Gabon, Congo, the Central Africa Republic, and northern Zaire. Eastern Madagascar is also a region of rain forests. In the East, rain forests are found in southwestern India, Bangladesh, Burma, Malaysia, the Philippines, Indonesia, and Melanesia, up to the Solomon islands (Mueller-Dombois 1999).

Of the 250,000 species of known plants, about 170,000 grow in the tropical rain forest, more than half native to the Americas. Many of these, mostly broad-leaved evergreens, large ferns, lianas, and epiphytic orchids, have been transplanted to other tropical regions or have adapted to indoor culture. Tropical rain forests in the Indo-Pacific are dominated by resinous trees in the Dipterocarpaceae, which make a sheltering canopy. The African tropical rain forest is the poorest in species.

Under these competitive conditions for space and resources, tropical plants have evolved a variety of signaling secondary metabolites. Some serve to attract pollinator insects, other ones as a defense against grazers and parasites. This makes the tropical land a center of high natural product diversity.

The tropical ecosystem called grassland and savanna is found in the southern American continent (particularly northern Colombia, Bolivia, and central Brazil) and most extensively in Africa (from just below the Sahara to the Cape Province). In the East it is limited to northern Australia.

The tropical scrub forest is met in the Americas (southwestern Mexico, western Bolivia, with a foot into Argentina, and easternmost Brazil, where it does not reach the coast), Africa (Ethiopia, Somalia, a part of Angola, and Namibia), western Madagascar, and the East (the whole Yemen, and parts of Afghanistan, Pakistan, and India).

Ample zones of the Americas are covered by the tropical deciduous forest (Caribbean, particularly San Salvador and Cuba, western Ecuador and easternmost Brazil, reaching the Uruguay border) and the East (as the main vegetation of India, Thailand, Laos, Vietnam, the southernmost China, and northeastern Australia).

Arid land (southwestern US, northern Mexico, Sahara regions, Saudi Arabia, and central Australia) is the world of the Cactaceae (AY 1979). The highly differentiated zonation of tropical mountain regions escapes these schemes.

In the tropical biomes, plants and insects contribute the largest number of species and are also responsible for the bulk of secondary metabolites. The diversity of tropical microorganisms is much less known.

Characteristic skeletons of natural products from these areas are grouped in charts for the Americas, Africa, and the East, according to the strategy set forth in Chapters 5.1 and 5.2. Because of our focus on the native origin of the species, the charts reflect a distribution of metabolites prevalent in the past, before the massive globalization brought about by modern man. However, the situation can be updated through data provided in the same charts for cultivated species and in Chapter 16.2.2 for introduced species. Both cultivated and alien species are the cause of decrease in natural product diversity, second only to the genetic modification and extinction of species.

Examination of the charts in Chapters 6.1.1 to 6.1.3 shows that the American and eastern tropical areas contribute more that the African tropics to the diversity of natural products. In particular the African tropics are poor in terpenoids, except secologanin, which is involved on a world basis in the formation of mixed-biogenesis alkaloids.

Secologanin does not know frontiers, but the biogenetic routes to which it takes part are different for the three main tropical areas. Tropical America is characterized by cinchona and ipecac alkaloids (Chart 6.1.1.A), tropical Africa by ibogaine/tabernanthine and voacamine alkaloids (Chart 6.1.2.A), and tropical East by corynanthe, yohimbe, catharantus, vindoline/vinblastine, and strychnos alkaloids (Chart 6.1.3.A). This reflects the different genera of plants evolved in the three tropical biomes, as indicated in these charts. From the pharmaceutical point of view, it is the East that has given more, and the culture of *Cinchona* spp. for quinine and *Vinca rosea* for vindoline/vinblastine are rare examples of medicinal plants developed as major crops, owing to their great germ plasm value.

The American tropics are also peculiar for a variety of steroids and alkaloidal steroids from the amphibians (Chart 6.1.1.I/FA/PO and 6.1.1.A).

Values of the skeletal complexity metric S for these alkaloids are small, in the range 36-49. In contrast, S/H values are relatively high, in particular for strychnos alkaloids (0.88, Chart 6.1.3.A); this reflects a high molecular complexity that stems from the many cycles and chirality centers, in spite of the low molecular weights.

Unusual proteins are also characteristic of the biome: free-sulfydryl type for both tropical Americas and the East, and taste-modifying type, like miraculin, monellins, and thaumatin, for tropical Africa.

6.1.1 American tropical and subtropical land

Chart 6.1.1.A Alkaloids (skeletons) and proteins from organisms of the American tropical and subtropical land (*S*max=54, av=25; *S/H*max=0.63, av=0.48).

Alkal.: amine: polyamine (spiders, Arach., from Nicaragua, Costa Rica, and Peru: Braekman 1998).; azoxy: cycasin (*Cycas* spp. seeds, Pinophyta, Gymn. from Guam: BR, MI); diaza-adamantane: acosmine (*Acosmium panamense* (Benth.) Yakovlev, Fabaceae, Ang., also introduced to Central Africa: Nuzillard 1999); indolizidine: pumiliotoxin B (Panamanian *Dendrobates pumilio*, Dendrobatidae frog: Daly 2000); 251D, 251H and 341A (Ecuadorian *Epipedobates tricolor* (Boulenger, 1899), Dendrobatidae frog, Amphib.: Daly 2000); isoquinoline: tubocurarine (bark of *Chondrodendron tomentosum* R. & P., Menispermaceae, Ang.: MI); *Erythrina* (*Erythrina* spp., pantrop. Fabales, Ang.: MI); piperidine, pyrrolidine, and pyrrolizidine: epibatidine (orphan from Ecuadorian frog *Epipedobates tricolor* (Boulenger, 1899), Dendrobatidae, Amphib.: Daly 2000); histrionicotoxin (orphan from *Dendrobates histrionicus* Dendrobatidae frogs, Amphib.: Daly 2000); tropane (cocaine from *Erythroxylum coca* Lam., Erythroxylaceae, Ang. from Peru mountains: MI); betalains: (*Opuntia* spp., native to American arid land, but also transplanted, and other Cactaceae, Ang.: AY); pyrrolizidine: (pantropical *Solenopsis* spp., fire ants, Hymenoptera, and *Glomeris* and *Polyzonium* spp., millipedes, Diplopoda, Ins.: Braekman 1998); steroidal: batrachotoxins (dietary orphans from Colombian *Phyllobates aurotaenia*, Dendrobatidae frog, Amphib.: Daly 2000); solasodine-like (*Solanum* spp., Solanales, Ang. from Mexico: AY); tomatidine (*Lycopersicon esculentum* Mill., Ang. from S America, cultivated and genetically engineered: MI); tryptophan/phenylalanine-secologanin: *Cinchona* alkal.: cinchonidine (*Cinchona* spp. Rubiales, Ang. from S America: BR; cultivated in Indonesia and Zaire for quinidine and in Tanzania, Burundi, India, Kenya, Guatemala, Peru, Ecuador, Bolivia, Rwanda, Sri Lanka, Colombia, and Costa Rica for quinine); ipecac (*Uragoga ipecacuanha* (Brot.) Baill, Rubiales, Ang. from Nicaragua: MI).

Proteins: free-sulfydryl: papain and chymopapain (papaya, *Carica papaya* L., Violales, Ang. native to S America, cultivated in Zaire, Uganda, Tanzania, and the East) and bromelain (pineapple, *Ananas comosum*, Bromeliales, Ang. native to S America, cultivated in Africa); NO-releasing heme: nitrophorin (blood-sucking Ins. *Rhodnius prolixux*: Walker 1999).

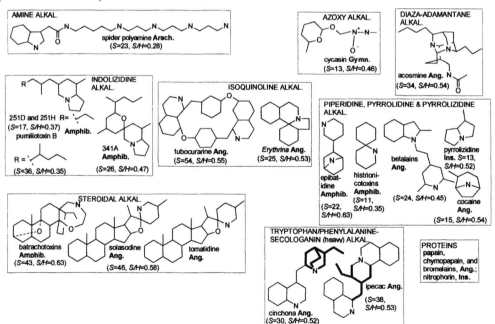

Chart 6.1.1.I/FA/PO Isoprenoids and polyketides (skeletons) from organisms of the American tropical and subtropical land (I: Smax=40, av=24; S/Hmax=0.56, av=0.41. FA/PO: S=22; S/H=0.18).

Sesquit.: <u>cadinanes</u>: gossypol (*Gossypium* spp., Malvales, and other widely cultivated Ang.: MI).

Diterp.: <u>cyclized cembranes</u>: kempane, trinervitanes, secotrinervitanes, rippertane, longipane (pantrop. termite soldiers, Nasutitermitinae, Ins.: Wahlberg 1992B); <u>ingenane</u>: ingenol esters *abeo*-related to phorbol esters (*Euphorbia* spp., Euphorbiales, Ang. from arid Peru: AY); <u>3,8-cyclodolabellane</u>: unidentif. endophytic fungus (isolated from *Daphnopsis americana*, Ang. from Costa Rica: Brady 2001); <u>stemodanes</u>: maritimol (Caribbean *Stemodia maritima* L., Scrophulariaceae, Ang.: Toró 2000).

Steroids: <u>norcholestanes</u>: diosgenin (*Agave* spp., Ang. from arid Mexico: AY).

Tetraterp.: <u>linear</u>: lycopene (*Lycopersicon esculentum* Mill., Ang., also cultivated and genetically engineered: MI); Cp473 (*Curtobacterium flaccumfaciens* pvar. *poinsettiae* pathogen Bact. of *Poinsettia* spp., Euphorbiaceae, Ang: Häberli 2000).

Polyket.: <u>polyenes</u>: amphotericins (*Streptomyces* spp. Actinom. Bact. from soil in Venezuela: MI); filipin (*Streptomyces filipinensis* from Philippine soil: MI); natamycin (*Streptomyces natalensis* from S Africa soil; candicidin from *Streptomyces griseus*: MI).

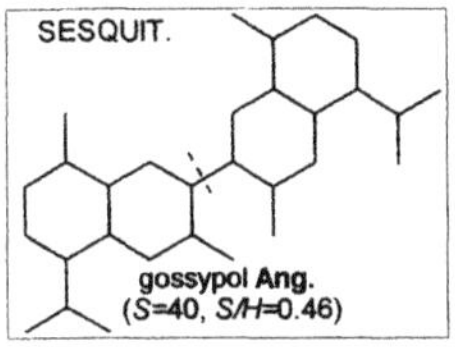

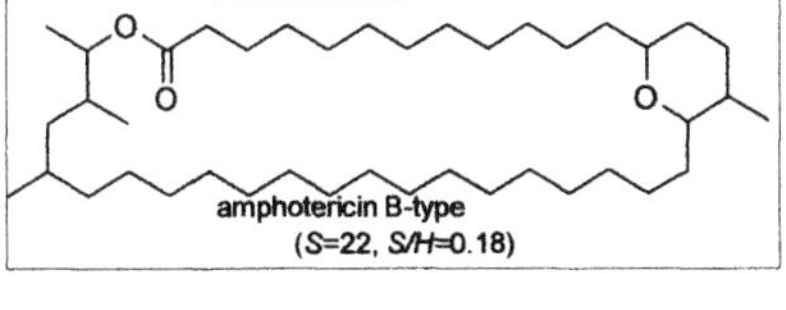

SESQUIT.

gossypol Ang.
(S=40, S/H=0.46)

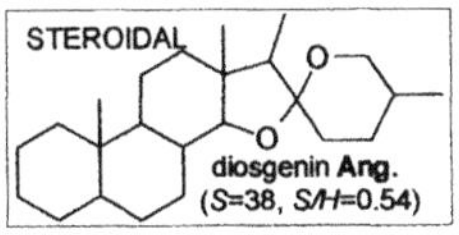

STEROIDAL

diosgenin Ang.
(S=38, S/H=0.54)

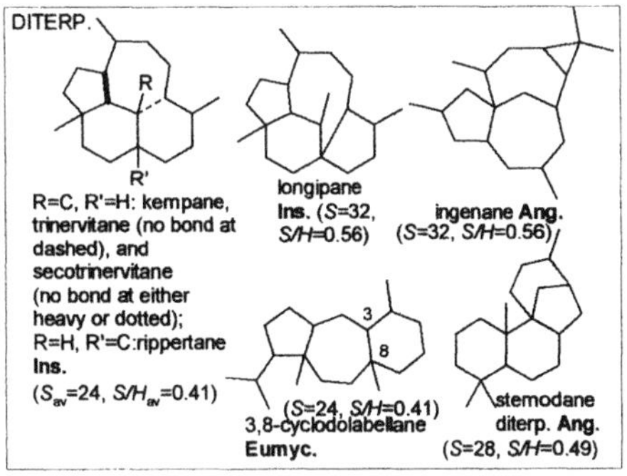

DITERP.

R=C, R'=H: kempane, trinervitane (no bond at dashed), and secotrinervitane (no bond at either heavy or dotted); R=H, R'=C:rippertane **Ins.** (S_{av}=24, S/H_{av}=0.41)

longipane **Ins.** (S=32, S/H=0.56)

ingenane Ang. (S=32, S/H=0.56)

(S=24, S/H=0.41) 3,8-cyclodolabellane **Eumyc.**

stemodane diterp. Ang. (S=28, S/H=0.49)

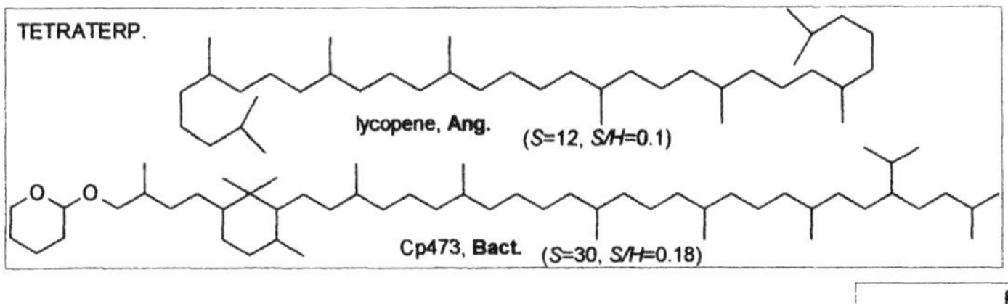

TETRATERP.

lycopene, **Ang.** (S=12, S/H=0.1)

Cp473, **Bact.** (S=30, S/H=0.18)

POLYKETIDES

amphotericin B-type
(S=22, S/H=0.18)

6.1.2 African tropical and subtropical land

Chart 6.1.2.A Alkaloids, amino acids, peptides (skeletons), and proteins from organisms of the African tropical and subtropical land (A: $Smax$=78, av=39; S/Hmax=0.67, av=0.6. P: S=70, S/H=0.38).

Alkal.: amine: polyamine (spiders from E and S Africa, C 6.1.1.A: Braekman 1998); indole: reserpine (*Rauwolfia vomitoria* Afzel., Gentianales, Ang. from rain forest: MI; AY); pantrop. *Erythrina* (C 6.1.1.A); schizozygane (*Schizozygia caffaeoides*, Ang. : Hubbs 1999); lupanine: acosmine (*Acosmium panamense*, Fabaceae, Ang. from Congo: Nuzillard 1999); pyrrolizidine: pantrop. Ins. and diplopods (see C 6.1.1.A); tryptophan-secologanin: ibogaine and tabernanthine (*Tabernanthe iboga* Baill., Gentianales, Ang. from rain forests: BR); dimeric, voacamine (*Voacanga africana* Stapf., Apocynaceae, Ang.); steroidal (conessine-like from *Holarrhena* spp., Gentianales, Ang. from arid Africa: MI; AY).

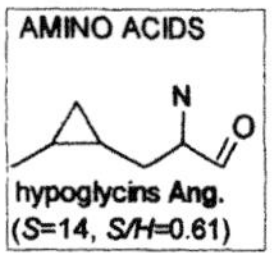

Amino acids: hypoglycins (*Blighia sapida* Kon., Sapindales, Ang., also introduced to the Caribbean: BR).

Pept.: sanglifehrin A (depsipept.) (*Streptomyces* sp., Actinom. Bact. from Malawi soil: Sanglier 1999).

Taste-modifying proteins: miraculins (red berries of *Synsepalum dulcificum* (Schum.) Daniell, Sapotaceae, Ang. from W Africa, MW 44,000); monellins (*Dioscoreophyllum cuminsii* Diels, Ranunculales, Ang. from rain forests, 91 amino acids, MW 10,700: BR); thaumatins (*Thaumatococcus danielli* Benth., Zingiberales, Ang. from rain forests, MW 22,000: BR).

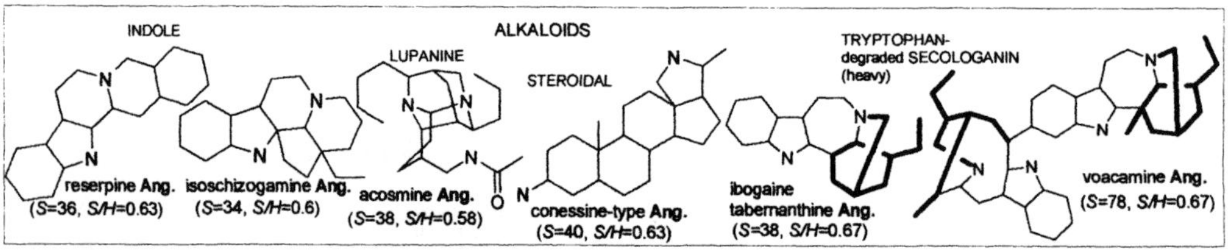

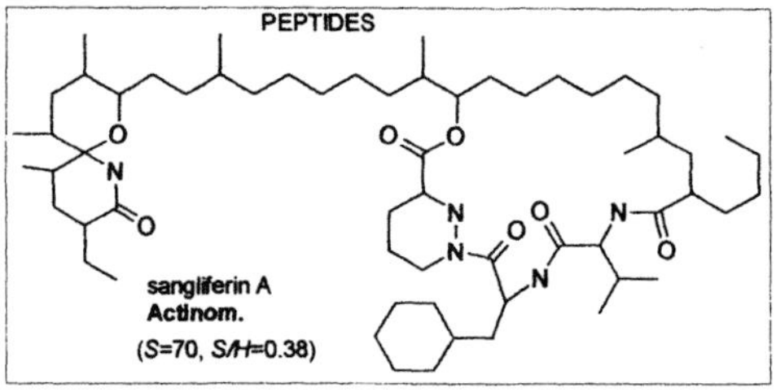

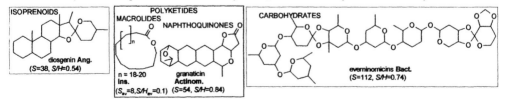

Chart 6.1.2.I/FA/PO/C Isoprenoids, polyketides, and carbohydrates (skeletons) from organisms of the African tropical and subtropical land (I: S=38; S/H=0.54. FA/PO: $Smax$=46, av=21; S/H max=0.87, av=0.35. C: S=112; S/H=0.74).

Isopr.: DITERP.: cyclized cembranes: (pantrop., C 6.1.1.I); STEROIDS: norcholestanes: diosgenin (*Dioscorea* spp.: MI and other Liliales, Ang.:AY).

Polyket.: macrolides: 27-29 membered: (African termite *Pseudacanthotermes spiniger* Plasman 1999); naphthoquinones: granaticin (*Streptomyces olivaceus*, Actinom., Bact. from soil in W Africa: MI); orthosomicins: everninomicin (*Micromonospora carbonacea* var. *africana*, Bact. from Kenya: Nicolaou 1999)

6.1.3 Eastern tropical and subtropical land

Chart 6.1.3.A/P Alkaloids and peptides (skeletons) and proteins from organisms of the eastern tropical and subtropical land (A: Smax=68, av=27; S/Hmax=0.88, av=0.58. P: Smax=84, av=72; S/Hmax=0.51, av=0.47).
Alkal.: polyamine: (spiders from eastern and southern Australia, C 6.1.1.A, C 6.1.2A: Braekman 1998); indole: reserpine and rescinnamine (pantrop., C 6.1.2.A) and ajmalicine (*Rauwolfia serpentina* L. Benth., Gentianales, Ang. from India: MI; AY); isoquinoline: pantrop. *Erythrina* (C 6.1.1.A).; piperidine, pyrrolidine, pyrrolizidine, and quinazoline: piperine and clavicine (*Piper* spp., Piperaceae, Ang. from Philippines, India, Indochina: MI); pyrrolizidines (pantrop., C 6.1.1.A); tetraponerines (*Tetraponera* sp., fire ants, Formicidae, Ins. from PNG: Braekman 1998); vasicine (*Adhatoda vasica* Nees, Acanthaceae, Ang. from E India: MI); steroidal: conessine-like (*Holarrhena* spp., Gentianales, Ang. from India: MI; AY); solasodine-like (*Solanum* spp., Solanales, Ang. from India: AY); tryptophan-secologanin: cinchona (C 6.1.1.A), native to S America, cultivated in Indonesia; yohimbe and corynanthe (*Rauwolfia serpentina* L. Benth., Gentianales and *Corynanthe yohimbe* K. Schum., Rubiales, Ang. from S. Asia: BR); strychnos (*Strychnos nux-vomica* L., Gentianales, Ang. from S Asia: BR); lochnericine, vindoline, vinblastine and biogenetic precursor, catharanthine (Madagascar periwinkle, *Vinca rosea* L. [= *Catharanthus roseus* G. Don.], Apocynaceae, Ang., also cultivated: MI).
Pept.: verucopeptin (*Actinomadura verrucosospora*, Bact. from the Philippines: Sugawara 1993); vancomycin (*Streptomyces orientalis*, Actinom., Bact. from Borneo soil: MI); caerulein-like (from Australian tree frog, *Litoria splendida*, and *Hyla caerulea*, Amphib. from W Australia: Wabnitz 1999).
Proteins: free-sulfydryl: ficin (MW 23,88-25,000, from *Ficus* spp., Moraceae, Ang.: MI).

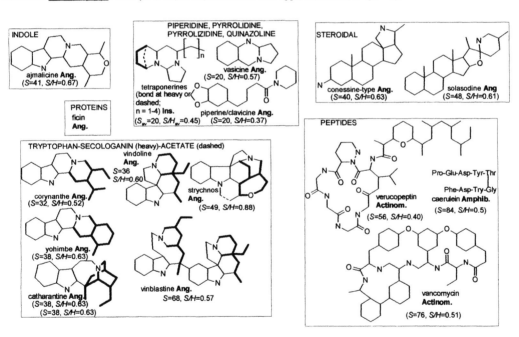

Chart 6.1.3.I Isoprenoids (skeletons) from organisms of eastern tropical and subtropical land (*S*max=63, av=48; *S/H*max=0.83, av=0.48).

Diterp.: cyclized cembranes: kempane, trinervitanes, secotrinervitanes, rippertane, and longipane (pantrop. termite soldiers, C 6.1.1.I/FA/PO); cyclopropabenzazulene: prostratin (*Homalanthus nutans* (Forster) Pax, from Samoa: Gustafson 1992, and phorbols from *Croton tiglium* L., both Euphorbiaceae, Ang.); homo-diterp.: unusually chlorinated (South African *Chromodoris hamiltoni*, Nudibr.: Pika 1995); ingenane: ingenol esters, *abeo*-related to phorbol esters, from *Euphorbia* spp., Euphorbiales, Ang. from arid Burma and India:AY); daphnane: close to tigliane and phorbols (Euphorbiaceae, Thymelaeaceae, Ang. and extended daphnane from *Trigonostemon reidioides* Craib, Euphorbiaceae from Thailand: Jayasuriya 2000); Diels-Alder dimers: maytenone (*Maytenus dispersus*, Celestraceae, Ang. from Australia but found also in plants from tropical S America: Alvarenga 2000).

Sesterterp.: retigeranic acid-like: retigeranic acid (*Lobaria retigera* (Bory) Trevis), 1869, Eumyc. from India: Turner 1983); stellatic acid-like (*Aspergillus stellatus* (Curzi) 1934, Eumyc. from Pakistan: Turner 1983).

Triterp.: azadirachtin-like tetranortriterp.: azadirachtin (neem tree, *Azadirachta indica* A. Juss., Meliaceae, Ang.: MI); quassinoids: (biosynthetically related to triterp.) including seco- and further degraded forms (*Quassia amara* L., Sapindales, Ang. from Jamaica quassin: DSII, and other metabolites: Connolly 1997).

Mixed biogenesis: sesquiterp.-shikim.: fissistigmatins (*Fissistigma bracteolatum* Chatt., Annonaceae, Ang. from N Vietnam: Porzel 2000).

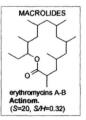

Chart 6.1.3.FA/PO Fatty acids and polyketides (skeletons) from organisms of eastern tropical and subtropical land (*S*max=38, av=24; *S/H*max=0.64, av=0.41).

Mono-bicyclic: long-chain: (*Achyranthes aspera*, Amaranthaceae, Ang. from India: Misra 1993); citrinin: (Australian plant *Crotolaria crispata* and *Aspergillus, Clavariopsis, Penicillium* spp., Eumyc.: MI).

Polycyclic: endriandric acid A-like: (*Endiandra introrsa* C.T. White, Lauraceae, Ang. from Australian forests: MI).

Macrolides: erythromycin-like: erythromycin (*Streptomyces erythreus*, Actinom., Bact. from the Philippine soil and picromycins, from temperate soil: MI).

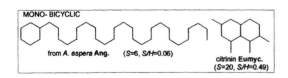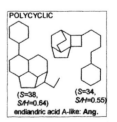

6.2 Temperate grassland, deciduous forest, and chaparral

In the northern hemisphere, temperate land is comprised from 30^0 N to approximately the Arctic circle. In the southern hemisphere, it is limited to the southernmost parts of Australia and America. Everywhere it is dominated, in decreasing extent, by grassland, deciduous and rain forests, and chaparral (Fig. 6.I).

Grassland occupies vast areas of the western and central parts of North America, central and southern Russia, northern China, Argentina, and the inside southern Australia.

Temperate deciduous and rain forest - which was an immense continuum before man-made deforestation - is only conserved in eastern US, central Europe, the British Isles, southern Scandinavia, eastern China, Korea, and the coastal parts of the Kamchatka Peninsula.

The chaparral is typical of the Mediterranean coasts, except Libya and Egypt. It is also found in California, southern Australia, and around Cape Town. The latter isolated spot supports many micro-habitats, highly significant for plant biodiversity.

Lakes, rivers, swamps, and marshes - common in temperate areas - contribute little to the diversity of natural products. Abundant dull-green grass and dull-colored fish and mollusks characterize lakes and rivers, in contrasts with the vivid colors of tropical fish and seaweeds. Haplosclerid sponges are occasionally abundant in freshwater, but their secondary metabolism is limited to demospongic acids (Dembisky 1994), in contrast with the variety of metabolites from marine sponges in the same order. Where not for cyanobacteria (which are as rich of unusual metabolites as the marine strains), tropical amphibians, and aquatic fungi, freshwater ecosystems would have passed unnoticed in this book.

Overall, however, the immensity of temperate land corresponds to a most various secondary metabolic production, different from that of tropical land. The most renowned alkaloids belong to the morphine class (Chart 6.2.A1), and, in combination with isoprenoids, to the ergot and triterpene classes (Chart 6.2.A2). Prominent in the peptides are the cyclosporins (the first of which was isolated from a fungus collected in Norway), streptogramins, and β-lactams (Chart 6.2.P). The isoprenoids are represented by pyrethrin monoterpenes, cedrane sesquiterpenes, ginkgolide and taxane diterpenes, ophiobolane sesterterpenes, and arborane and amyrin-like triterpenes (Chart 6.2.I). In the polyketides, epothilones, recently discovered from Myxobacteria, and the long known rapamycin, are two prominent classes of macrolides (Chart 6.2.FA/PO/C).

Chart 6.2.A1 Alkaloids (skeletons) from organisms of temperate land (including C 6.2.A2: *S*max=49, av=33; *S/H*max=0.74, av=0.55)

Alkal.: <u>hydroxamate</u>: trichostatin (*Streptomyces hygroscopicus*, Actinom., Bact.: MI); <u>indole</u>: madindolines (*Streptomyces nitrosporeus* from soil sample in Wisconsin US: Sunazuka 2000); paraherquamides (VM55599 from *Penicillium* sp., Eumyc. from soil in Turkey: Stocking 2000); <u>isoquinoline</u>: morphine-like (*Papaveum somniferum L.,* Papaverales, Ang. from Asia and Asia Minor: AY; cultivated in Turkey, India, Burma, Thailand also for codeine, papaverine, and noscapine); aporphines (boldine, bulbocapine, isothebaine, glaucine from Papaverales and Laurales: MI); degraded aporphines (aristolochic acid from Aristolochiales, Ang.: MI); berberine-bridge (protopine, berberine, coptisine from *Papaver dubium* L. from Asia and Europe:AY); <u>isoxazole</u>: muscimol (widely distributed *Amanita muscaria* L. (Fr.), Basidiomyc., Eumyc.: MI); <u>perhydroazaazulene</u>: *Stemona* alkal. (stemoamide, stenine, croomine, tuberostemonine, stemonine from *Stemona tuberosa* and *Croomiacea* spp.: Jacobi 2000, and isotemofoline from *Stemona japonica*: Kende 1999, Stemonaceae, Asian Ang.); <u>piperidine and pyrrolidine</u>: gelsemine (jasmine, *Gelsemium sempervirens* (L.) Ait., Loganiaceae, Ang. from Asia and southern US: MI); lincomycin (*Streptomyces lincolnensis* var. *lincolnensis*, Actinom., Bact.: MI); tropane (Brassicaceae, Convolvulaceae, Erythroxylaceae, Euphorbiaceae, Olacaceae, Proteaceae, Rhizophoraceae, Solanaceae, notably among the latter *Atropa belladonna*, cultivated in US, Europe, India, China for atropine, originating from central and S Europe, *Datura matel*, cultivated in Asia for scopolamine, and *Hyoscyamus niger*, cultivated in temperate areas for hyosciamine: BR, MI); *Lycopodium* spp. (lycopodine, cernuine, huperzine from cosmop. Lycopodiaceae, Lycopodiophyta: MI, Kozikowski 1999; Herbert 1989); <u>tyrosine</u>: phenanthroindolizidines: tylocrebrine, tylophorine (Asclepiadaceae, Gentianales, and Moraceae, Urticales, Ang): MI)

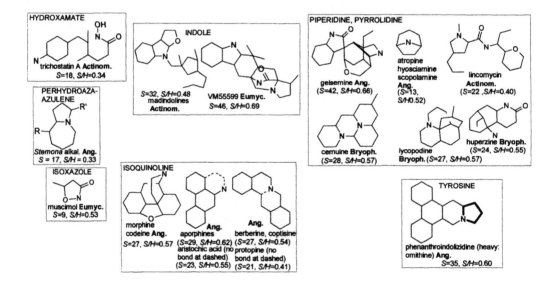

Chart 6.2.A2 Isoprene alkaloids (skeletons) from organisms of temperate land (for *S* and *H*, see C 6.2.A1).
Isopr. alkal.: *Aconitum* diterp.: (*Aconitum*n spp. and *Delfinium* spp. Ranunculales, Ang., widely occurring: BR, AY); triterp. and steroidal: Buxaceae-class (Euphorbiales, Ang.); solanidine and jervine (Solanaceae and Liliaceae, resp., Ang.: BR); samandarine and samandenone (European salamanders, Amphib.: Braekman 1998); *Daphniphyllum*: daphnezomines (*Daphniphyllum* spp., Daphniphyllales, Ang. from Japan: Morita 1999); tryptophan-dimethylallyl pyrophosphate: Ergot (*Clavices purpurea* (Fr.) Tul. (1883), parasitic Ascom., Eumyc., growing on rye and other grains: Lohmeyer 1997); cyclopiazonic acids (*Penicillium cyclopium* Westling (1911) and *Aspergillus flavus* Link ex Fries, Deuterom., Eumyc., growing on ground-nuts: Herbert 1989); tryptophan-secologanin: vincamine (Old World periwinkle, *Vinca minor* L., Gentianales, Ang.: BR); quinolizidine: veratridine (European while hellebore, *Veratrum album* L., and *Schoenocaulon officinale* (Schlecht. & Cham.), Liliaceae, Ang.: MI).

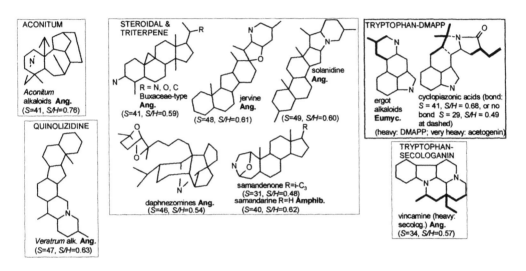

Chart 6.2.P Peptides (skeletons) from organisms of temperate land (Smax=137, av=52; S/Hmax=0.65, av=0.49).

Cyclosporins: cyclosporin A (*Tolypocladium inflatum* W. Gams (1971), Eumyc. from Wisconsin soil: Kleinkauf 1997); reutericyclins (lipopept.): (*Lactobacillus reuteri* Kandler *et al.* 1982, lactic acid Bact.: Höltzel 2000); group-A streptogramins (depsipetides): virginiamycin M_1 = streptogramin A (*Streptomyces virginiae* and *S. graminofaciens*), griseoviridin (*S. griseus*): MI; valinomycin (depsipept.): (*Streptomyces fulvissimus*, Actinom., Bact.: MI); waltherines: (*Waltheria douradinha*, Malvales, Ang. from S America: Morel 1999); β-lactams: penicillins, penem alkal. (*Penicillium* and *Cephalosporium* spp., Deuterom., Eumyc.; cephalosporins: (*Cephalosporium, Nocardia,* Eumyc., and *Streptomyces* spp., Actinom., Bact.: Martin 1997); amanitin (*Amanitas phalloides* (Fr.), Basidiom., Eumyc.: MI); alliins: allylcysteine sulfoxide (European garlic, *Allium sativum* Linn., and Asiatic onion, *Allium cepa* Linn., Liliales, Ang., also cultivated: BR); thiopept. antibiotics: thiostrepton and promothiocin A (bacterial strains, characterized by tri- or tetrasubstituted pyridine and dehydro amino acids, in addition to thiazole and oxazole rings as in marine-derived analogues: Bagley 2000)

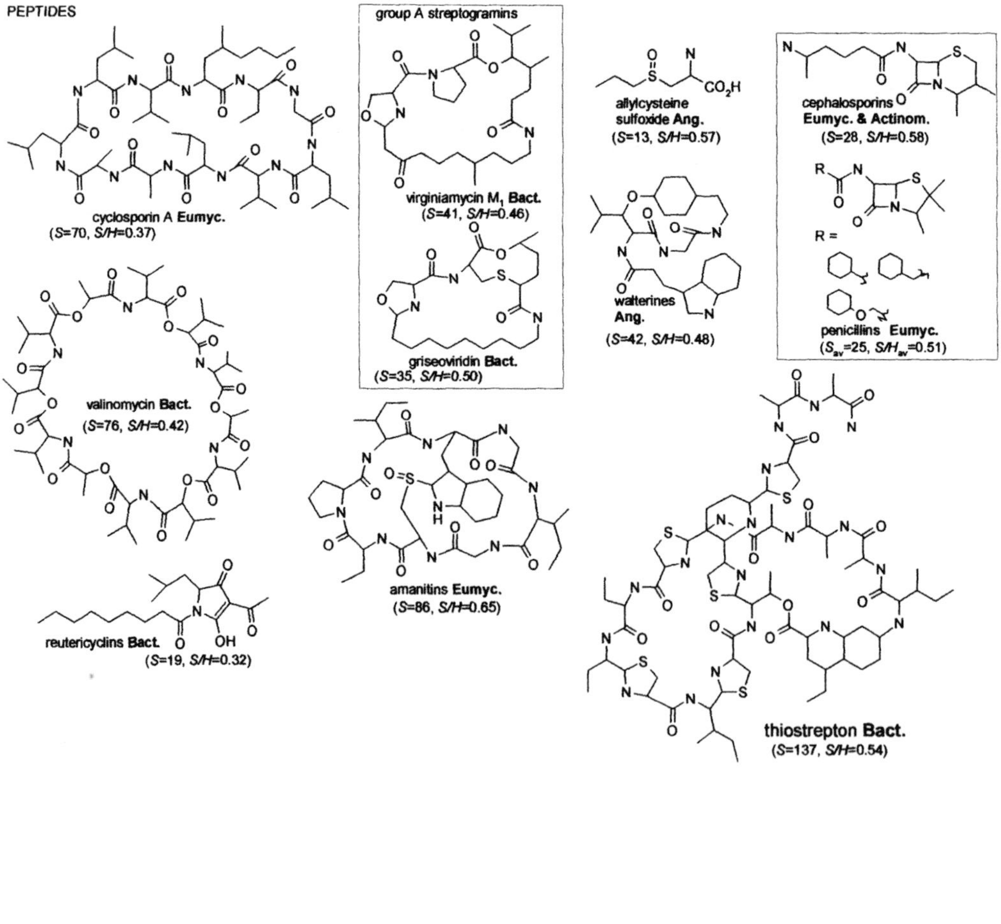

PEPTIDES

cyclosporin A Eumyc.
(S=70, S/H=0.37)

valinomycin Bact.
(S=76, S/H=0.42)

reutericyclins Bact.
(S=19, S/H=0.32)

group A streptogramins

virginiamycin M₁ Bact.
(S=41, S/H=0.46)

griseoviridin Bact.
(S=35, S/H=0.50)

amanitins Eumyc.
(S=86, S/H=0.65)

allylcysteine sulfoxide Ang.
(S=13, S/H=0.57)

waltherines Ang.
(S=42, S/H=0.48)

cephalosporins Eumyc. & Actinom.
(S=28, S/H=0.58)

R =

penicillins Eumyc.
(S_{av}=25, S/H_{av}=0.51)

thiostrepton Bact.
(S=137, S/H=0.54)

Chart 6.2.I Isoprenoids (skeletons) from organisms of temperate land (Smax=55, av=28; S/Hmax=0.93, av=0.45).

Monoterp.: pyrethrin-like: (*Tanacetum cinerarifolium,* Asterales, Ang.: Br)

Sesquiterp.: cedrane and duprezianane: (*Juniperus* spp., Cupressaceae, Gymn.: Barrero 2000); rearr. dumortanes: (*Dumortiera* spp., Marchantiales liverworth, Bryoph. from Argentina: Asakawa 1995); pseudoguaiane: (*Ambrosia maritima* L., Compositae, Ang. from Mediterranean and Africa:AY).

Sesquiterp. dimers: elemane/eudesmane : macrophyllidimer A and dimeric elemane: (*Inula macrophylla* from Uzbekistan: Su 2000).

Diterp.: cyclized cembranes: basmane and virgane (*Nicotiana* spp., Solanales, Ang. of American origin; cultivated in temperate and subtropical regions throughout the world: Walberg 1992B); labdane-derived: aphidicolane and rosenonolactone (Eumyc.: Turner 1983); pleuromutilane: pleuromutilin (*Pleurotus* and European *Clitopilus* spp., Basidiom., Eumyc.: Erkel 1997); ginkgolides: ginkgolide B (*Ginkgo biloba* L., Ginkgoaceae, Gymn., Xitianmu Mountain, China: MI); grayanotoxane: grayanotoxins I and II (*Rhododendron, Kalmia, Leucothoe,* Ericales, Ang.: BR; MI); pimarane(iso): 15(13→12)-*abeo*-isopimarane (*Orthosiphon aristatus,* Lamiaceae, Ang. from Japan: Shibuya 1999); taxane: (*Taxus* spp., Gymn. from Pacific and Europe and *Taxomyces andreanae* Strobel, A. Stierle, D. Stierle & W.M. Hess (1993), Eumyc., endophytic in Pacific Yew: Stierle 1993).

Sesterterp.: ophiobolanes (= prenylfusicoccanes): (*Helminthosporium* and *Drechslera* spp., Eumyc.: Turner 1983); terpestacin and fusoproliferin-like: (*Artrinium* sp. and maize pathogen, *Fusarium proliferatum* (Matsush.) Nirenberg ex Gerlach & Nirenberg (1976), Eumyc., resp.: Hanson 1996; astellatol-like: (*Aspergillus stellatus* Curzi (1934) Eumyc.: Hanson 1996).

Triterp.: β-amyrin-like: glycyrrhizin (Mediterranean *Glycyrrhiza glabra* L., Fabales, Ang., widely cultivated: MI); arborane: (Gramineae, Ang. from Japan and other temp. regions: Ohmoto 1969); 14(13→12)*abeoαH*-serratane: piceanols (*Picea jezoensis* subsp. *hondoensis* (Mayr) P.A. Schmidt, Gymn.: Tanaka 1999); salvadione-like: non squalenoids, presumably originating from geranylpyrophosphate addition to icetexone-like precursor (*Salvia bucharica* M. Pop., Lamiaceae, Ang.: Ahmad 1999).

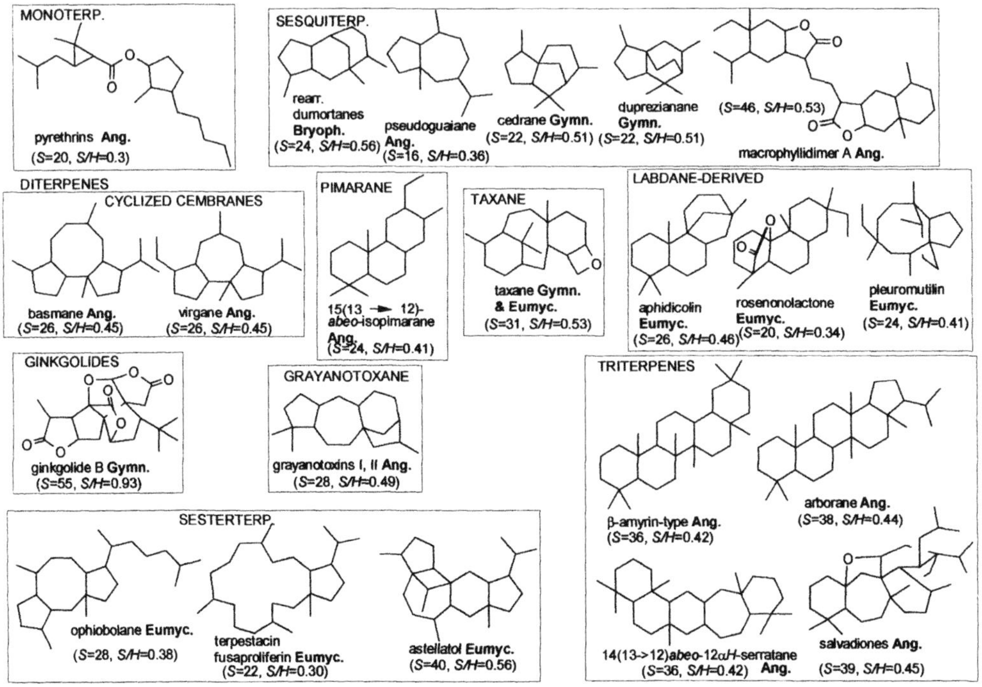

Chart 6.2.FA/PO/C Fatty acids, polyketides, and glycosides (skeletons) from organisms of temperate land (Smax=49, av=29; S/Hmax=0.77, av=0.42).

Non-macrocyclic: citrinin-like: (*Aspergillus, Clavariopsis, Penicillium* spp., Deuterom., Eumyc: MI);

mevinic acids: mevastatin and lovastatin (*Penicillium, Aspergillus, Monascus, Agomycetes* spp., Deuterom., Eumyc.: Endo 1997); squalestatins: squalestatin S1 = zaragozic acid A (*Phoma* sp., Eumyc. from Portugal: Endo 1997 and CP (nonandride) molecules: Dabrah 1997); tetracyclines: *Streptomyces* spp., Actinom., Bact.: MI); phomoidride = CP molecule-like: (unidentified phomaoid fungus from twigs of *Juniperus ashei* from shrub forest in Dripping Springs, Texas: Dabrah 1997, which embody a (not shown here) bridgehead double bond at the limit of Bredt's rule); sterigmatocystin-like: (*Aspergillus* and *Chaetomium* spp., Deuterom., Eumyc., in Europe and other temperate land, carcinogenic: Frisvad 1998); podophyllotoxin-like: (N American *Podophyllum peltatum* L., Podophyllaceae, Ang.: MI)

Macrolides and macrolactams: epothilones: (*Sorangium cellulosum*, Myxobact.: MI); oligomycins: (*Streptomyces* sp., Actinom., Bact.: MI); rapamycins: (*Streptomyces hygroscopicus* from Easter Is: MI); tacrolimus (*Streptomyces tsukubaensis*: MI); lactimidomycins: (*Streptomyces amphibiosporus* from Japan: Sugawara 1992).

Other macrocyclic polyket.: paracyclophanes: cylindrocyclophanes (*Cylindrospermum licheniforme* Kützing, Nostocales, from American Type Culture Collection, and nostocyclophanes from *Nostoc linckia* (Roth) Bornet ex Bornet & Flahault, from Univ. of Texas Culture Collection, freshwater Cyanobact.: Hoye 2000).

Aminoglycosides: streptomycin (first isolated from *Streptomyces griseus* (Krainsky) Waksman et Henrici, Bact. isolated from unspecified manured field soil: Schatz 1944; MI)

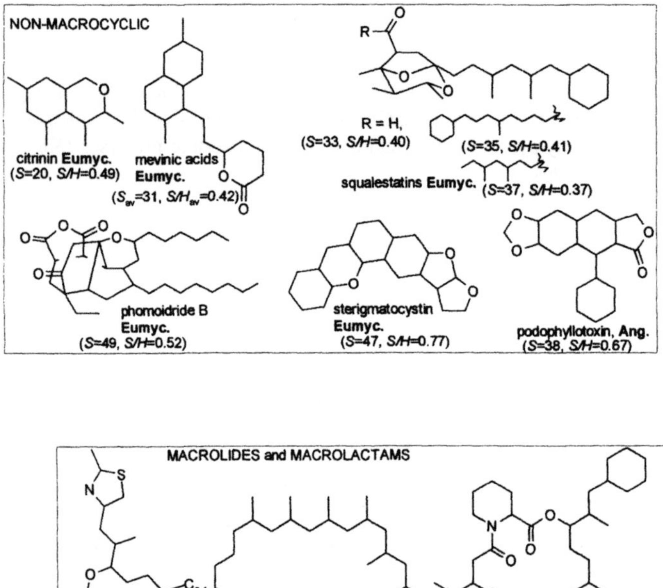

NON-MACROCYCLIC

citrinin Eumyc. (S=20, S/H=0.49)
mevinic acids Eumyc. (S$_{av}$=31, S/H$_{av}$=0.42)
squalestatins Eumyc. R = H, (S=33, S/H=0.40) (S=35, S/H=0.41) (S=37, S/H=0.37)
phomoidride B Eumyc. (S=49, S/H=0.52)
sterigmatocystin Eumyc. (S=47, S/H=0.77)
podophyllotoxin, Ang. (S=38, S/H=0.67)

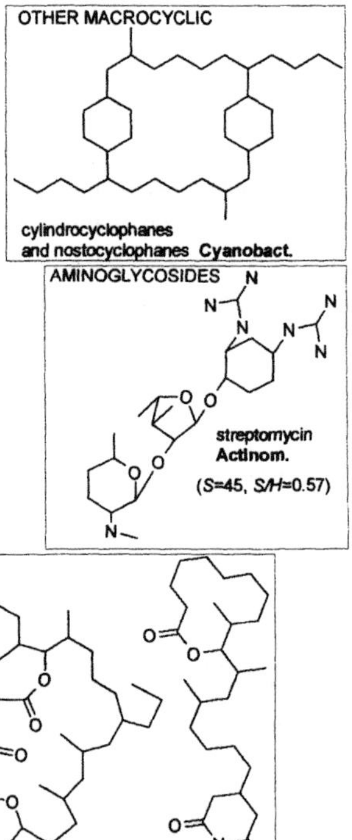

OTHER MACROCYCLIC

cylindrocyclophanes and nostocyclophanes Cyanobact.

AMINOGLYCOSIDES

streptomycin Actinom. (S=45, S/H=0.57)

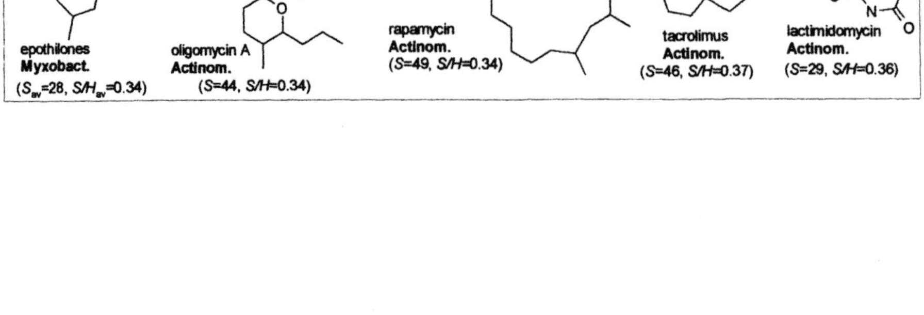

MACROLIDES and MACROLACTAMS

epothilones Myxobact. (S$_{av}$=28, S/H$_{av}$=0.34)
oligomycin A Actinom. (S=44, S/H=0.34)
rapamycin Actinom. (S=49, S/H=0.34)
tacrolimus Actinom. (S=46, S/H=0.37)
lactimidomycin Actinom. (S=29, S/H=0.36)

6.3 The taiga and the tundra

The northern conifer forest (taiga) extends at high latitudes, above the temperate deciduous and rain forest and grassland (Fig. 6.I). It is dominated by spruces, *Picea* spp., and firs, *Abies* spp. It is an immense continuum, only interrupted by the Atlantic Ocean, from Alaska to above the Kamchatka Peninsula, comprising also islands, like Iceland.

Representative metabolites from the taiga are volatile lower terpenes and abietane diterpenes, produced by the gymnosperms (Chart 6.3).

The tundra extends from the northern border of the taiga to the Arctic circle. It is a treeless ground with frozen subsoil that only allows mosses and lichens growing abundantly. Typical lichen metabolites are in the orcein-class of dyes (Chart 6.3).

Chart 6.3 Alkaloids, isoprenoids, fatty acids/polyketides, and shikimates (skeletons) from organisms of the taiga and the tundra (A: *S*max=34, av=18; *S/H*max=0.56, av=0.32. I: *S*max=24, av=13; *S/H*max=0.48, av=0.4. FA/PO: *S*max=12, av=8; *S/H*max=0.71, av=0.38. S: *S*=0.57; *S/H*=0.57).

Alkal.: azoxy: elaiomycin (*Streptomyces hepaticus*, Actinom., Bact. from Ontario: MI); orcein-class: orcein, constituent of litmus dye (*Roccella tinctoria* DC. (1805), *Lecanora tartarea* (L.) Ach., lichens from Scandinavian tundra, and also lichens from temperate and tropical land: MI).

Isopr.: MONOTERP.: pinane: α- and β-pinene and menthane: α- and β-phellandrene (*Abies balsamea* (L.) Mill., Pinaceae, Gymn. of N conifer forest: MI); bornane, menthane, pinane: pinene, limonene, and bornyl acetate (*Abies alba* Mill., Pinaceae, Gymn. of N conifer forest); bornane, pinane, camphane, menthane: pinene, phellandrene, limonene, camphene, bornyl acetate (*Abies sibirica* Ledeb., Pinaceae, Gymn. of Siberian conifer forest: MI). DITERP.: abietane: abietic acid (*Abies balsamea* (L.) Mill., Pinaceae, Gymn. from northern conifer forest, from which retene was formed in fossilized wood, or obtained by hydrogenation; 18-norabietane also found in decayed wood: MI).

Shikim.: picein-like: (*Picea* spp., Pinaceae, Gymn. from N conifer forest; also from willow bark: MI).

Fatty acid/polyket.: agaricic-acid-like: (*Fomes laricis* (Jacq.), Basidiomyc., Eumyc. from European and Russian northern conifer forest: MI); maltol-like: maltol (γ-pyrone from Pinaceae, Gymn. from the conifer forest: MI).

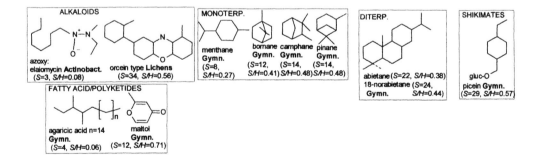

Chapter 7. The oceans

The oceans cover about 71% or the Earth's crust, or 80% when the southern hemisphere alone is considered. Nonetheless, free oceanic waters are largely a void of life, except planktonic small organisms.

Hydrothermal vents and cold seeps, in deep zones of the oceans, are rich in microorganisms, mollusks, and worms, but without other forms of life that characterize the coasts (in particular photosynthetic organisms and their symbionts) and seamounts.

Seamounts are areas of endemism, richer than the hydrothermal vents and cold seeps. A recent study, which required extensive sampling in difficult areas, represents a change in the approach to marine biodiversity, with a focus on the species (Richer de Forges 2000). Moreover, contrary to the past, this study was not limited to organisms of commercial interest.

Neglected for a long time, a census of marine life is also being supported on a wide front in a planned ten-year intervention through the National Oceanographic Partnership Program. What remains unclear, is how and how far marine microbial life will be evaluated in these programs, which is a pervasive difficulty that dates to linnaean times and is still largely unsolved.

7.1 Zonation of the seas and oceans

Although the oceans are physically uninterrupted, a geographical subdivision into a few major basins is traditional and useful: the Pacific, Atlantic, Indian, and Arctic and Antarctic Oceans, in the given order of decreasing size and depth. The Pacific Ocean is connected to shallow seas: the South China Sea, East China Sea, Sea of Japan, Sea of Okhotsk, Coral Sea, Tasman Sea, Bering Sea, Gulf of Alaska, and Gulf of California. The Indian Ocean also has many branches: the Red Sea, the Arabian Sea, the Persian Gulf, and the Bay of Bengal.

These basins were formed from the uninterrupted Panthalassa that surrounded the sole existing continent, Pangea. First, about 180 My ago, Laurasia started to separate from both Gondwana, forming northern Atlantic Ocean, and Antarctica (Fig. 7.1.I). Later, about 90 My ago, South America separated from Africa, forming the southern Atlantic Ocean. Finally, about 15 My ago, India became fused to Asia, while Australia moved to its present position. This helps understanding certain latitudinal trends of biodiversity and natural product diversity in the sea, in particular their decrease on moving eastward from the Great Barrier Reef toward the Tuamotu archipelago.

The biological zonation and the water temperature, which is related to the latitude, are also important in determining the distribution of natural products. The intertidal and subtidal zones, down to 200 m depth, correspond to the continental shelf, where most seaweeds and productive invertebrates grow, especially in tropical and subtropical areas. These roughly correspond to the solar tropics and subtropics, besides a limited coastal zone in the eastern Pacific with unusually warm waters for this latitude.

In coral reefs, high natural product diversity stems from the many sponges, cnidarians, corals, and ascidians. Temperate waters are poorer in species. *Alcyonium coralloides*, which is one of the few Mediterranean soft corals, is an example: the richness of complex terpenoids yielded by this soft coral is comparable to that of tropical alcyonaceans (D'Ambrosio 1990), but the latter comprise many species, each one with its own genes for the secondary metabolism.

Focusing on the distribution of marine invertebrates, the seas can be divided into eleven zones (Fig. 7.1.II, George 1979). Four of these are exclusively coastal zones: the Mediterranean (the geographical Mediterranean basin and the coastal areas and islands from northwestern France to the middle of Angola in the Atlantic), Caribbean (from South Carolina to São Paulo in Brazil, including the geographical Caribbean and the Gulf of Mexico), Panamanian (coastal zones from southern

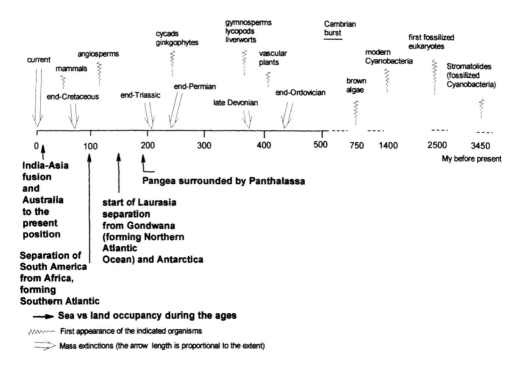

Figure 7.1.I Sea vs land occupancy in the perspective of the first appearance of organisms and mass extinctions

California to central Peru, comprising the Galápagos archipelago), and Zealandic (from the northeastern plateau of New Zealand to the plateau off the middle of Western Australia, including Tasmania and the Great Australian Bight).

Both coastal areas and open oceans are found in the immense Indo-Pacific and Atlantic zones. Moving eastward, the Indo-Pacific comprises a sector from southern Japan to Natal, up to the Tuamotu and Hawaii archipelagos. The North Pacific and South Pacific are further subdivisions of the Pacific Ocean. Further zonation defines the North Atlantic (including the coastal areas from North Carolina to southern Greenland in the west, and a sector from off the coasts of Senegal to Novaya

Zemlya in the East) and the South Atlantic (embracing the Atlantic Ocean and southern Indian Ocean).

The Antarctic area at the extreme South and the Arctic area at the extreme North complete the classical zonation of the oceans (George 1979). I have added a further commodious zonation, the Internal Seas.

Originally conceived for the distribution of marine invertebrates (George 1979), these zones account nicely for the distribution of the seaweeds too. The distribution of marine microorganisms

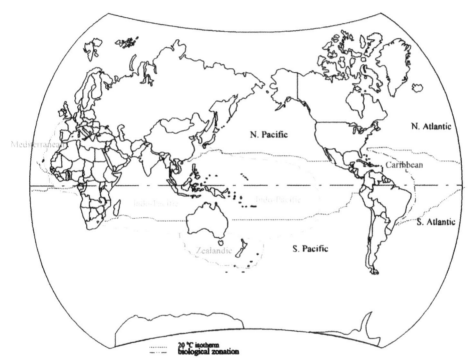

Figure 7.1.II. Biological zonation of the oceans

remains uncertain because of scarce *ad hoc* studies. However, the myriads of phenomena of association and parasitism typical of marine life suggest that areas of high macro-organismic biodiversity are also areas of high micro-organismic biodiversity. Anyway, microorganisms have a wider range of occupancy than macroorganisms, comprising extreme conditions that the latter cannot endure. This receives specific attention in Chapter 10.

Our perusal of the marine zones begins from those of highest biodiversity, the Indo-Pacific and the Caribbean, rich of coral reefs. Fringing reefs, barrier reefs, and atolls are the main forms of coral reefs (Fig. 7.1.III). The longest fringing reef - in its simplest form of a fringe along the coast - is found in the Red Sea. The largest barrier reefs, including lagoons and coral knolls, are found, in the order, in northeastern Australia and New Caledonia, the latter comprising the largest lagoon. Most atolls, with their ring of reefs, lagoon, and patch reefs (not indicated in Fig. 7.1.III for simplicity), are encountered in the Indian and western Pacific Oceans.

In coral reefs, the density of animal life, stemming not only from reef-building scleractinian corals and calcareous algae but also from non-reef-building soft corals, ascidians, bryozoans, soft seaweeds,

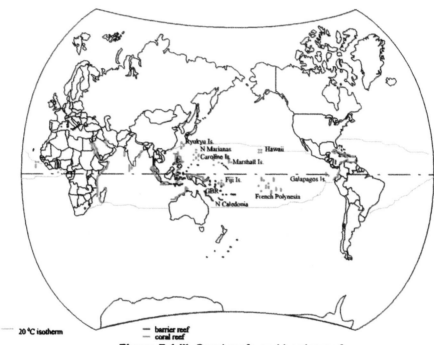

Figure 7.1.III. Coral reefs and barrier reefs

and other invertebrates, is unsurpassed even by the tropical rain forest (May 1994). Circumstantial evidence suggests that also during the geological times the tropics have contributed more than temperate and cold areas to the biological evolution (Jablonski 1993). Coral reefs have strict requirements for growth: average water temperatures between 20 and 30 $^\circ$C (Fig. 7.1.III), direct light for the symbiotic photosynthetic zooxanthellae, normally high salinity, and unpolluted waters. Because of these requirements, coral reefs can best grow between the tropics, away from large river effluents. Coral reefs that have adapted outside these ideal limits are an exception, such as in the Persian Gulf, which is subjected to a wider temperature excursion.

Sampling species on coral reefs has required the most various techniques. These comprise, in the order of increasing difficulty and expenses, hand at low tide, snorkeling in shallow waters, ordinary scuba diving down to 40-50 m depth, and deep-water scuba diving down to 150 m depth with special gas mixtures and equipments. In certain areas, particularly of the Indian Ocean, diving requires physical protection from sharks. Deep-water dredging, carried out in New Caledonian seamounts, and sampling by manned submarines in the Bahamas, have revealed many undescribed demosponges and echinoderms that have greatly contributed to our knowledge of the biology and natural product chemistry of coral reef areas.

Deep-sea has a different meaning to different scientists. For the marine ecologist, in a broad perspective, deep-waters begin from the upper abyssal zone, that is 1,000 m in depth (George 1981; Vernberg 1981). The marine natural product chemist refers rather to the zones of productive organisms and the most common techniques of sampling: deep-sea organisms are those thriving at depths below normal scuba diving. This is the position taken in this book, fixing arbitrarily the limit of shallow waters to 80 m depth.

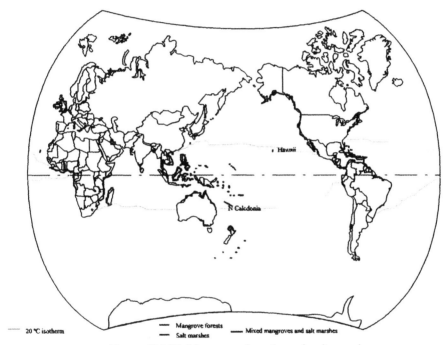

Figure 7.1.IV. Mangrove forests and salt marshes

Tropical and subtropical coasts are also areas of mangrove forests (Fig. 7.1.IV). Although often mixed with coral reefs, the mangroves can also grow in waters that are too cold, or not salted enough, for reef building corals, such as along the coasts of the Gulf of Guinea. Because of wide homoeostatic properties, mangrove forests fringe 60-75% of the tropical shores, extending also to the subtropics, often as imported species, like in the Hawaiian archipelago. The greatest variety of mangroves is found in the western Indo-Pacific, whereas in the western Atlantic their typical distribution, from the outer seaward edge toward inland, is limited to species in the genera *Rhizophora* (red mangrove), *Avicennia* (black mangrove), and *Laguncularia* (white mangrove).

In the monsoon season the salinity in mangrove forests may drop to very low levels, limiting the variety of marine life to halotolerant species, in particular lignicolous fungi. Besides metabolites from fungi and the mangroves themselves (Miles 1991), secondary metabolites in the mangrove areas have been isolated from a few species of red seaweeds, ascidians, and demosponges. Although modest in natural product diversity, mangrove forests play important functions in the coastal defense against sea level rise, for fish breeding, and as an important resource of sea food and charcoal for local populations.

Salt marshes, mixed with the mangroves in certain warm regions, account for the formation of 10% of the atmospheric methyl bromide and methyl chloride. These two gases react with stratospheric ozone (Rhew 2000).

7.2 Indo-Pacific

That the Indo-Pacific (Fig. 7.1.II) is an area of high natural product diversity was first revealed by the studies carried out in the 1970's at the Roche Research Institute on the Great Barrier Reef (Baker 1980). Studies worldwide have since confirmed this richness, unsurpassed by any other area, either in the sea or on land. The variety of small peptides from the Indo-Pacific is impressive, characterized, particularly with the ascidians, by the condensation of the cysteine NH_2 and SH groups with the COOH group of an adjacent amino acid. Thiazole and thiazoline rings are formed, which alternate with the normal peptide bonds formed by the other amino acids (Chart 7.2.P1-3). Correspondingly, threonine forms methyloxazoline moieties (Chart 7.2.A, 7.2.P1, 7.2.P2). These condensations impose a drastic conformational bias to peptides (Abbenante 1996).

3-Alkylpyridinium salts are intermediates in the formation of macrocyclic alkaloids isolated from demosponges (Chart 7.2.A). Their far-reaching significance is discussed in Chapter 7.6. Many bis-indole, β-carboline, and pyridoacridine alkaloids were found in sponges, ascidians, and bacteria (Chart 7.2.A). Guanidine C_{11} alkaloids, isolated from Indo-Pacific sponges (Chart 7.2.A), are less characteristic, being also found in Caribbean sponges (Chart 7.3.A/P).

The variety of monoterpenoids is modest, limited to ochtodane monoterpenes from red seaweeds (Chart 7.2.I1), which are also found in specimens from the Galápagos Islands (Chart 7.4.P/I/PO). Many exclusive sesquiterpenes, sesterterpenes (Chart 7.2.I1), triterpenes, sterols, and polyprenyls (Chart 7.2.I2) were found. The new skeleton vannusane, from a strain of *Euplotes vannus* from the Celebes Sea, may be seen as a most unusual triterpene: its C_{30} isoprenoid backbone bears no obvious relationships with squalene but it may arise along a profound deviation from the squalene pathway (Guella 1999A). However, sesquiterpenes subsequently isolated from this ciliate appear as vannusane halves, suggesting a possible alternative biogenetic rationalization of vannusanes from the coupling of two *ad hoc* sesquiterpene moieties (Guella 2001B). A biosynthetic study is needed to resolve the puzzle.

The discovery of polyketides that, like maitotoxin and palytoxin, have extraordinarily high molecular weight was startling. Although their complexity metric S is also high, the ratio to the size metric, S/H, has modest values. This reflects the quasi-repetitive sequence of ethereal rings of these metabolites (Chart 7.2.FA/PO1). An ample variety of structures has been found with the amphidinolides, macrolides isolated from dinoflagellates in a single genus, *Amphidinium*. The bryostatins from bryozoans, and the superstolides and sphinxolides from sponges, add to the variety of macrolides (Chart 7.2.PO2). Macrolactams from cyanobacteria and macrocarbocycles from ascidians complete this unique assortment of metabolites.

Chart 7.2.A Alkaloids (skeletons) from the Indo-Pacific (*S*max=49, av=26; *S/H*max=0.80, av=0.50).
Alkal.: amine: (*Aplidium* sp., Polyclinidae, Ascid. from GBR: Carroll 1993); benzonaphthyridine: aaptamines
(Hadromerida, Porif. from Okin. and GBR: Shen 1999); guanidine: sceptrins (pantrop. Agelasida and Nepheliosp.,
Porif.: Walker 1981), ptilomycalins (Asteroidea, Echin. from NC and Poecil., Porif.: Palagiano 1995); corallistine
(*Corallistes fulvodesmus* Lévi & Lévi, Lithistida, Porif. from NC, depth 500m: Debitus 1989); phloeodictynes
(*Phloeodictyon* sp., Nepheliosp., Porif. from NC, depth 235m: Kourany-Lefoll 1994); indole: trisindoline and
vibridole (*Vibrio* spp., Bact., from Okin. and the Red Sea: Kobayashi 1994; Bell 1994); eudistomins (Ascid. from
GBR and Caribbean: Davis 1998); didemnimides (*Didemnum conchyliatum* (Sluiter, 1898), Didemnidae, Ascid.
from mangroves in Bahama: Vervoort 1999; bis-β-carboline, Ascid. from GBR: Kearns 1995); zyzzin (*Zyzza
massalis* (Dendy), Poecil., Porif. from NC, depth 235m: Mancini 1994A); nortopsentin D (Halichon. and Axinel.,
Porif. from NC, depth 300m: Mancini 1996); gelliusines (Haplosclerida, Porif. from NC, depth 300m: Bifulco
1995A); oxaquinolizidine: xestospongins from *Xestospongia* spp., Nepheliosp., Porif.: Baldwin 1998; oxazole and
isoxazole: bengazoles (Choristida, Porif. from W Pacific and the Red Sea: Fernández 1999); pyrinodermins
(*Amphimedon* sp., Haplosclerida, Porif. from Okin.: Tsuda 1999A); macrocyclic: motuporamines (*Xestospongia
exigua* (Kirkpatrick) Nepheliosp., Porif. from PNG: Williams 1998); xestospongins (Nepheliosp. and Haplosclerida,
Porif. from SW Pacific: Hoye 1994); halicyclamines and haliclonacyclamines (*Haliclona* sp., Haplosclerida, Porif.
from PNG and GBR: Clark 1998); keramaphidin B (Haplosclerida from Okin. and Nepheliosp., Porif. from PNG:
Kobayashi 1996B); manzamines (Demosp., Porif. from W Pacific: Ohtani 1995); allopurine: akalone (*Agrobacterium aurantiacum*, Bact. from Okin.: Izumida 1997; pyridoacridine: Porif., Ascid. from Okin. and W
Pacific and, less commonly, Caribbean (de Guzman 1999); pyrrolidine and piperidine: agelastatins (*Agelas
dendromorpha* Lévi, Agelasida, Porif. from NC, depth 260m: D'Ambrosio 1994).

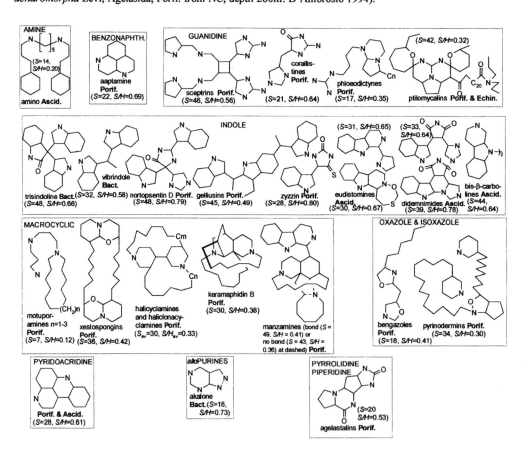

Chart 7.2.P1 Cyclohexa- and cycloheptapetides, and 'linear' peptides (skeletons) from the Indo-Pacific (including C 7.2.P2 and 7.2.P3: Smax=101, av=55; S/Hmax=0.67, av=0.46).

Non-macrocyclic pept.: milnamide-like: milnamide A (*Auletta* sp. Axinel., Porif. from PNG: Crews 1994 and hemiasterlins from *Hemiasterella* sp., Hadromerida, from S Africa and *Cymbastela* sp., Axinel., Porif., from PNG: Coleman 1995); majusculamide D and microcolins: (lipopept. from *Lyngbya majuscula* Gomont, Cyanobact.: Moore 1988; Nagle 1996; C 7.3A/P); lyngbyapeptin: (*Lyngbya* spp, Cyanobact. from PNG and Guam: Klein 1999; Luesch 2000); dysinin, dysidenin, dysideathiazoles, dysideapyrrolidones: (Indo-Pacific *Dysidea* spp. Dysideidae, Porif.: Dunlop 1982; Unson 1993); dysibetaines and dysiherbaines: (*Dysidea herbacea* (Keller) from Yao Is Micronesia: Sakai 1999); eusynstyelamides: (*Eusynstyela misakiensis*, Styelidae, Ascid. from the Philippines: Swersey 1994).

Cyclohexapept.: bistratamides: (*Lissoclinum* spp., Didemnidae, Ascid. from GBR and the Philippines: Foster, 1992); comoramides: (*Didemnum molle*, Didemnidae from Comoro Is: Rudi 1998); (hexa)patellins: (*Lissoclinum* sp. from Fiji Is.: Carroll 1996).

Cycloheptapept.: mayotamides: A and B (*Didemnum molle* (Herdman, 1886), Didemnidae, Ascid. from Comoro Is: Rudi 1998); mollamides: (*D. molle*, Didemnidae, Ascid. from GBR: Carroll 1994); trunkamides: trunkamide A (*Lissoclinum* sp., Didemnidae, Ascid. from GBR: Wipf 2000A); lissoclinamides: 7 (*Lissoclinum patella* Gottshaldt, Didemnidae, Ascid. from GBR: Boden 2000); pseudoaxinellins: (*Pseudoaxinella massa*, Axinel., Porif. from PNG: Kong 1992); (hepta)phakellistatins: phakellistatin 1 (pantrop. Axinel., Porif.: Pettit 1994; Herald 1997).

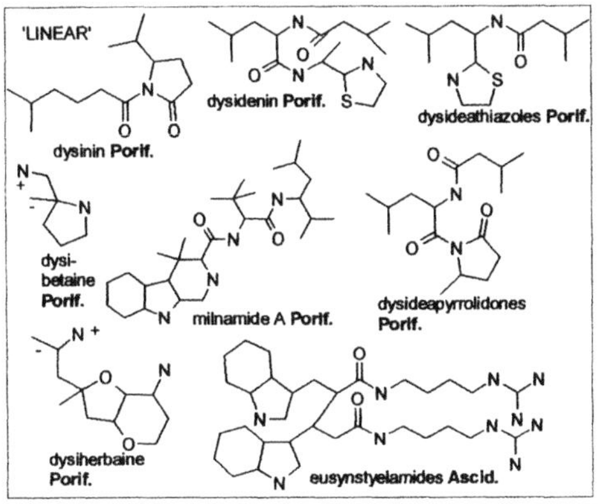

'LINEAR'
dysinin Porif.
dysidenin Porif.
dysideathiazoles Porif.
dysibetaine Porif.
milnamide A Porif.
dysideapyrrolidones Porif.
dysiherbaine Porif.
eusynstyelamides Ascid.

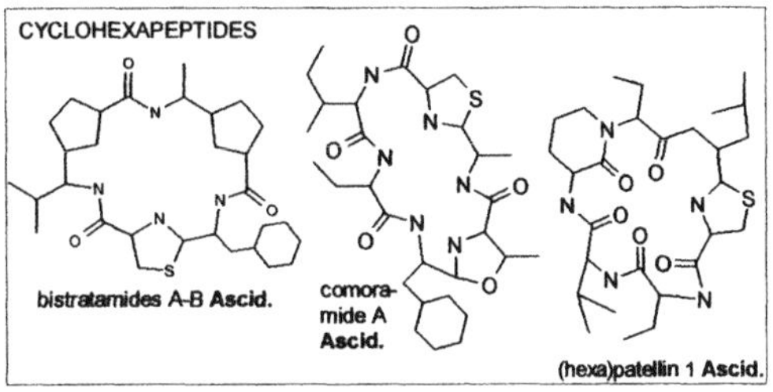

CYCLOHEXAPEPTIDES
bistratamides A-B Ascid.
comoramide A Ascid.
(hexa)patellin 1 Ascid.

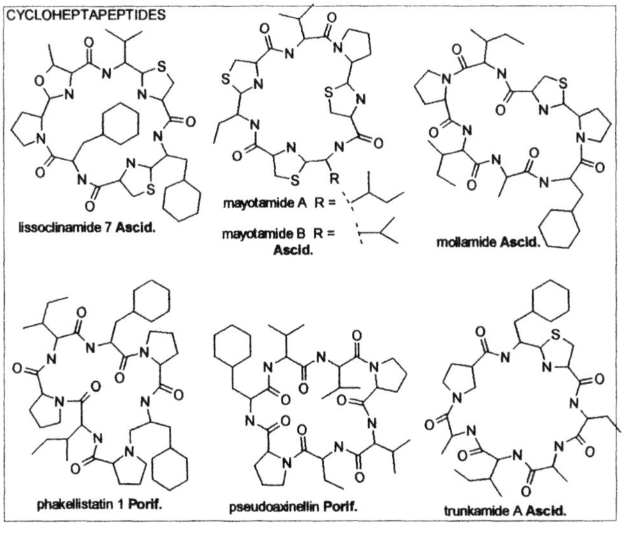

CYCLOHEPTAPEPTIDES
lissoclinamide 7 Ascid.
mayotamide A R =
mayotamide B R = Ascid.
mollamide Ascid.
phakellistatin 1 Porif.
pseudoaxinellin Porif.
trunkamide A Ascid.

Chart 7.2.P2 Cyclocta- and cyclodecapeptides (skeletons) from the Indo-Pacific (Smax=101, av=79; S/Hmax=0.67, av=0.55)

Cyclooctapept.: tawicyclamides: (*Lissoclinum patella* from Philippines: McDonald 1992); ulithiacyclamides: (*L. patella* from Palau: Wipf 2000A); (octa)patellins: 3 (*Lissoclinum* sp. from Fiji: Carroll 1996); ascidiacyclamides: (unidentif. colonial Ascid. from GBR: Boden 2000) ; patellamide D (*L. patella* from GBR: Wipf 2000A) .

Cyclodecapept.: loloatins: (*Bacillus* sp., Bact. from PNG: Gerard 1999); phakellistatins: 8 (*Phakellia* spp. Axinel. and *Stylotella aurantium* Kelly-Borges and Bergquist, Halichon. Porif. from W Pacific: Delbert 1997) .

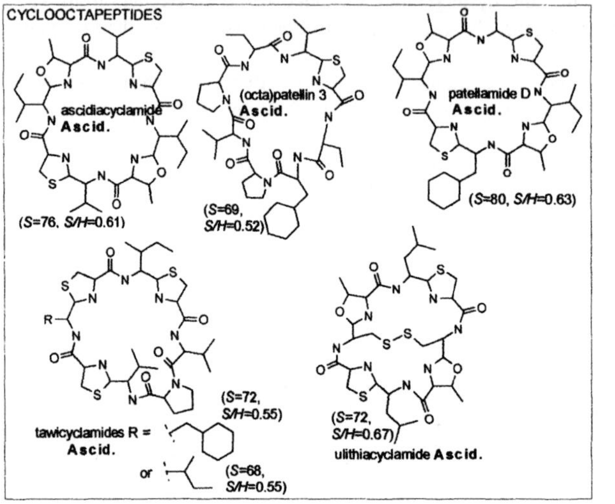

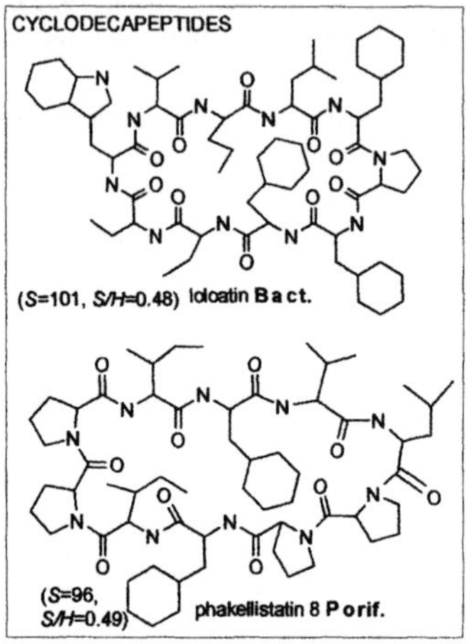

CYCLOOCTAPEPTIDES

ascidiacyclamide Ascid.
(S=76, S/H=0.61)

(octa)patellin 3 Ascid.
(S=69, S/H=0.52)

patellamide D Ascid.
(S=80, S/H=0.63)

tawicyclamides R =
Ascid.
(S=72, S/H=0.55)
or
(S=68, S/H=0.55)

ulithiacyclamide Ascid.
(S=72, S/H=0.67)

CYCLODECAPEPTIDES

(S=101, S/H=0.48) loloatin B act.

(S=96, S/H=0.49) phakellistatin 8 Porif.

Chart 7.2.P3 Other peptides (skeletons) from the Indo-Pacific (*S*max=78, av=42; *S/H*max=0.48, av=0.32)
Aplysamines and hemibastadins: (Verongida, Porif. from Hawaii and PNG: Pettit, 1996); bastadins: (*Ianthella* spp.
and *Psammaplysilla* spp., Verongida, Porif. from W Pacific: Franklin 1996); caprolactins: (unident. Bact. from Hawaii,
depth 5,065m: Davidson 1993 and Choristida, Porif. from W Pacific and Red Sea coral reefs, resp.: Fernández 1999);
jaspamides: (*Jaspis* spp., Choristida from PNG and *Hemiasterella* sp., Hadromerida, Porif., from S Africa; related
to geodiamolides from distant *Geodia* and *Cymbastela*, Porif.: Talpir 1994); hemiasterlins: (*Hemiasterella minor*
(Kirkpatrick) from S Africa: Talpir 1994); lyngbyapeptin A: *Lyngbya bouillonii* from PNG and *Lyngbya majuscula*
from Guam: Luesch 2000; majusculamides: depsipept. majusculamide C and lipopept. majusculamide D (*Lyngbya*
majuscula Gomont, Cyanobact. from Eniwetok atoll: Moore 1988); microsclerodermins: (*Microscleroderma* spp.,
Lithistida, Porif., from deep sea in NC and shallow waters in the Philippines and Palau: Qureshi 2000); microcolin-like
: microcolins A-B, ypaoamide (lipopept.) (*Lyngbya majuscula*, Cyanobact. from Guam: Nagle 1996).

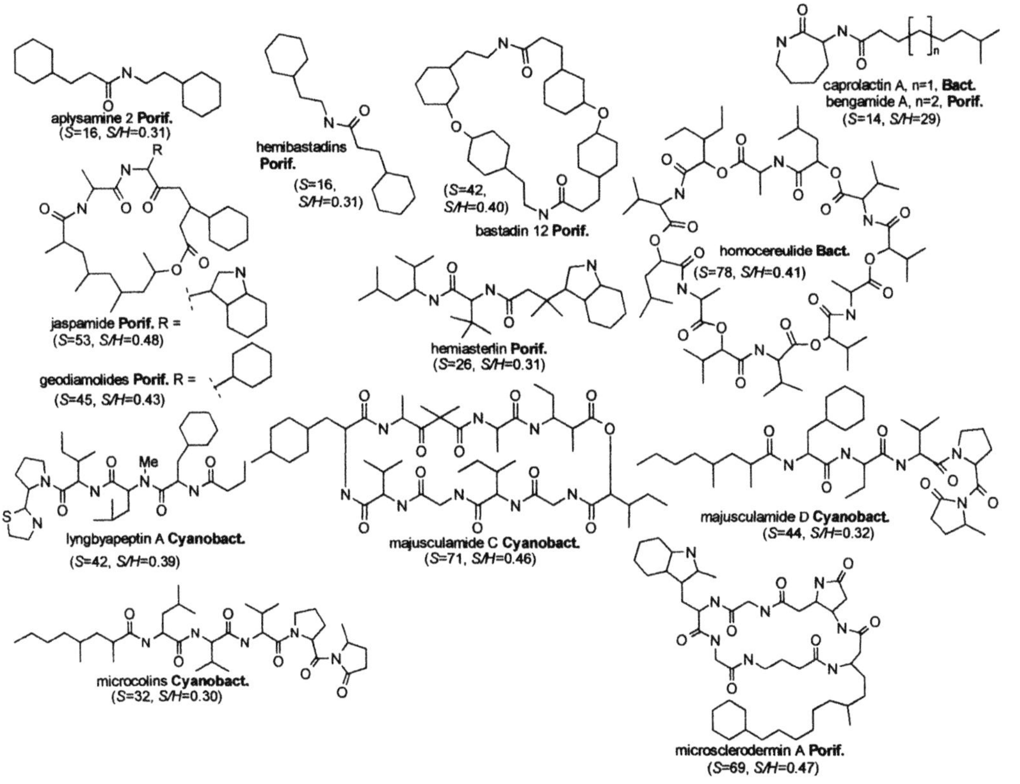

aplysamine 2 **Porif.**
(*S*=16, *S/H*=0.31)

hemibastadins
Porif.
(*S*=16,
S/H=0.31)

(*S*=42,
S/H=0.40)
bastadin 12 **Porif.**

caprolactin A, n=1, **Bact.**
bengamide A, n=2, **Porif.**
(*S*=14, *S/H*=29)

homocereulide **Bact.**
(*S*=78, *S/H*=0.41)

jaspamide **Porif.** R =
(*S*=53, *S/H*=0.48)

geodiamolides **Porif.** R =
(*S*=45, *S/H*=0.43)

hemiasterlin **Porif.**
(*S*=26, *S/H*=0.31)

lyngbyapeptin A **Cyanobact.**
(*S*=42, *S/H*=0.39)

majusculamide C **Cyanobact.**
(*S*=71, *S/H*=0.46)

majusculamide D **Cyanobact.**
(*S*=44, *S/H*=0.32)

microcolins **Cyanobact.**
(*S*=32, *S/H*=0.30)

microsclerodermin A **Porif.**
(*S*=69, *S/H*=0.47)

Chart 7.2.I1 Mono-, sesqui-, di-, and sesterterpenes (skeletons) from the Indo-Pacific (Smax=36, av=22; S/Hmax=0.63, av=0.38)

Monoterp.: <u>ochtodane</u>: (*Chondrococcus* (= *Portieria*) *hornemanni* from Tahiti and W Pacific., Gigartinales, Rhodoph.: Crews 1984).

Sesquit.: <u>spirodysane</u>: (*Dysidea* spp., Dysideidae, Porif. from GBR and Bermuda: Flowers 1998); <u>kelsoane</u>: (*Cymbastela hooperi*, Axinel., Porif. from GBR: König 1997); <u>trachyopsane</u>: (*Axinyssa aplysinoides* Dendy, 1922, Halichon., Porif. from Palau Is.: Patil 1997); <u>sinularane</u>: (Alcyonacea, Cnid. from Indonesia: Beechan 1978); <u>precapnellane and capnellane</u>: (Alcyonacea, Cnid. from Indonesia and S China: Morris 1998); <u>xenianovane</u>: (*Xenia* cf. *novae-britanniae*, Alcyonacea, Cnid. from GBR: Bowden 1987); <u>9,10-friedochamigrane</u>: (*Aplysia* spp. from Madagascar and Okin., and *Laurencia* sp. from Caribbean: Fedorov 2000).

Diterp.: <u>adociane</u>: (Haplosclerida, Porif. from GBR). <u>secoxenicane</u>: alcyonolide (Alcyonacea, Cnid. from Okin.; Kobayashi 1981).

Sesterterp.: <u>ircianane</u>: (*Ircinia wistarii* Bergquist, Dictyocer., Porif. from GBR: Coll 1997);

<u>dysideapalaunic acid</u>: (*Dysidea* sp., Dictyocer., Porif. from Palau: Hagiwara 1991); <u>lintenone-like</u>: (Dictyocer., Porif. from Bahama: Hanson 1996); <u>bilosespenes</u>: (*Dysidea cinerea* (Keller), Dictyocer., Porif. from Dalak archip.: Rudi 1999A); <u>diprenylfarnesylferane</u>: (manoalide and neomanoalide from Dictyocer., Porif. from Hawaii and W Pacific: Hanson 1996 and nor-sesterpenes from SW Pacific, including also prenylfarnesylferanes: D'Ambrosio 1998); <u>22(23→18)*abeo*-diprenylfarnesylferane</u>: (luffariellins from *Megalopastas* sp., Dendroc. from Enewetak atoll, and *Luffariella variabilis* (Polejaff), Dictyocer., Porif. from Palau, W Pacific: Butler 1992B).

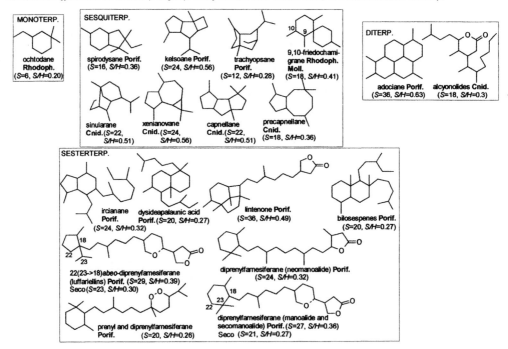

Chart 7.2.I2 Triterpenes, steroids, and polyprenyls (skeletons) from the Indo-Pacific ($Smax = 49$, av $= 33$; $S/Hmax = 0.52$, av 0.38).

Triterp.: <u>squalene derived</u>: sipholane and neviotane (*Siphonochalina siphonella* (Lévi, 1965), Haplosclerida, Porif. from the Red Sea: Ohtani 1991; puosane (puoside A-E from *Asteropus* sp., Choristida, Porif. from Truk lagoon, W Pacific: Ksebati 1988); yardenone, abudinol, and muzitone (*Ptilocaulis spiculifer*, Axinel., Porif. from the Dahlak Is.: Rudi 1999B); <u>non-squalene</u>: vannusane (*Euplotes vannus* (Müller, 1786), Hypotrichia, Ciliph. from Celebes Sea: Guella 1999A).

Steroids: <u>24-alkylcholestanes</u>: 24-*i*-Pr (*Pseudaxinyssa* spp., Axinel., Porif. from various areas: Zimmerman 1989); <u>9,11-secocholestane</u>: herbasterol (*Dysidea herbacea* (Keller), Dysideidae, Porif. from GBR: Capon 1985).

Polyprenyls: <u>toxiusane, toxicane, and shaagrockane</u>: (the ring separated by dashed lines represents the quinonoid portion) (*Toxiclona toxius*, Haplosclerida, Porif. from the Red Sea: Isaacs 1993); <u>mokupalide-like</u>: (*Megalopastas* sp., Dendroc., Porif. from Enewetak atoll, W Pacific: Yunker 1978).

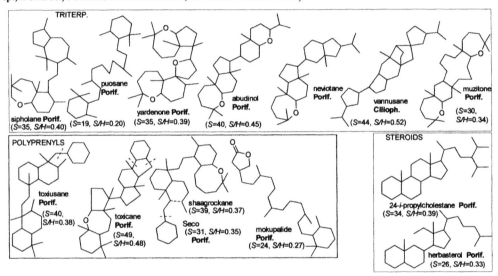

Chart 7.2.FA/PO1 Fatty acids and polyether polyketides (skeletons) from the Indo-Pacific (including C 7.2.FA/PO2: *S*max=296, av=85; *S/H*max=0.67, av=0.48).

Fatty acids: amphimic acids: (*Amphimedon* sp., Haplosclerida, Porif. from GBR: Nemoto 1997).

Polyket.: didemnenones: (*Didemnum voelzkowski* (Savigny, 1816), Didemnidae, Ascid. from Fiji; nakienones (*Synechocystis* sp., Chroococcales, Cyanobact. from Okin.: Nagle 1995); iantherans: (*Ianthella* sp., Verongida, Porif. from GBR: Okamoto 1999); helicascolides: (*Helicascus kanaloanus* Kohlmeyer, Ascom., Eumyc. from mangroves in Hawaii: Poch 1989).

Polyethers: ciguatoxins: (*Gambierdiscus toxicus* Adachi et Fukuio, 1979, Dinofl., and ciguateric fish from tropical Pacifi and Caribbean: Satake 1997); palytoxins: (*Palythoa* spp., Zoanthinaria, Cnid. from Hawaii and Okin., and free-swimming *Ostreopsis siamensis* Schmidt 1901, Dinofl., and crabs and fish from Okin.: Kahn 2001); maitotoxin: (*Gambierdiscus toxicus*, Dinofl. from French Polynesia: Murata 2000).

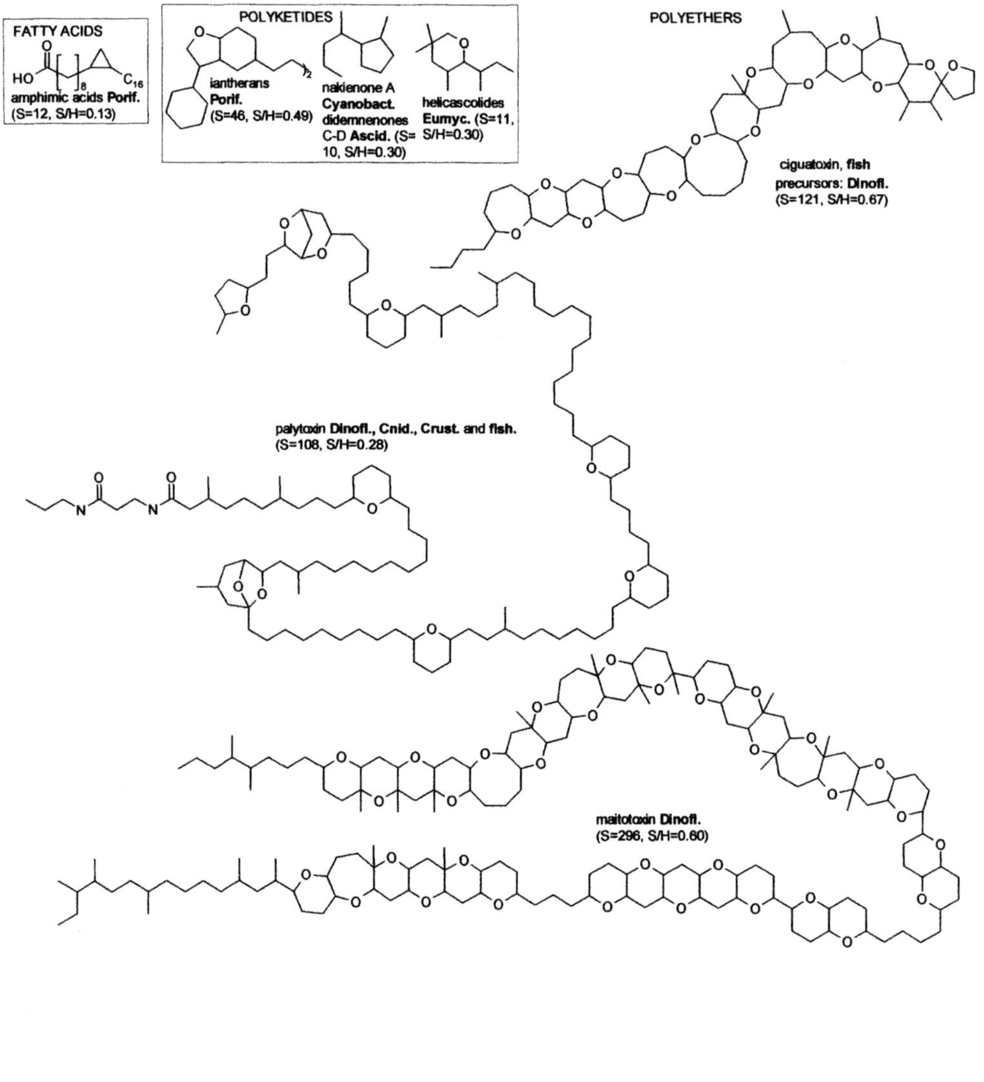

Chart 7.2.FA/PO2 Macrolides, macrolide-macrolactams, and macrocarbocyclic polyketides (skeletons) from the Indo-Pacific (Smax = 56, av = 32; S/Hmax = 0.49, av 0.34)

Macrolides: amphidinolides: A, B, C, D, E, K, J, N, R, T (*Amphidinium* spp., Dinofl., symbiotic or free-swimming, from Okin. or the Caribbean, resp.: Tsuda 2000); aplysiatoxins: (Nostocales, Cyanobact. from Marshall Is and *Stylocheilus longicauda* (Quoy and Gaimard, 1824), Anaspidea., Moll. and *Gracilaria coronopifolia* J. Agardh, Gigartinales, Rhodoph. from Hawaii: Nagai 1997); bryostatins: bryostatin 1 (*Bugula neritina* (L.), Cheilostomata, Bryoz. from Caribbean: Pettit 1991; superstolides: (*Neosiphonia superstes* Sollas, Lithistida, Porif. from NC, depth 500m: D'Auria, 1994); sphinxolides: (Nudibr., Moll. from Hawaii and deep-sea lithistids, Porif. from NC: Carbonelli 1999).

Macrolide-macrolactams: madangolide and laingolide A-like: (*Lyngbya bouillonii* Hoffman and Demoulin, Nostocales, Cyabact. from PNG: Klein 1999).

Macrocarbocyclic: longithorone A: (*Aplidium longithorax* Monniot, 1987, Aplousobranchia, Ascid. from Palau: Fu 1997 and GBR: Davis 1999B).

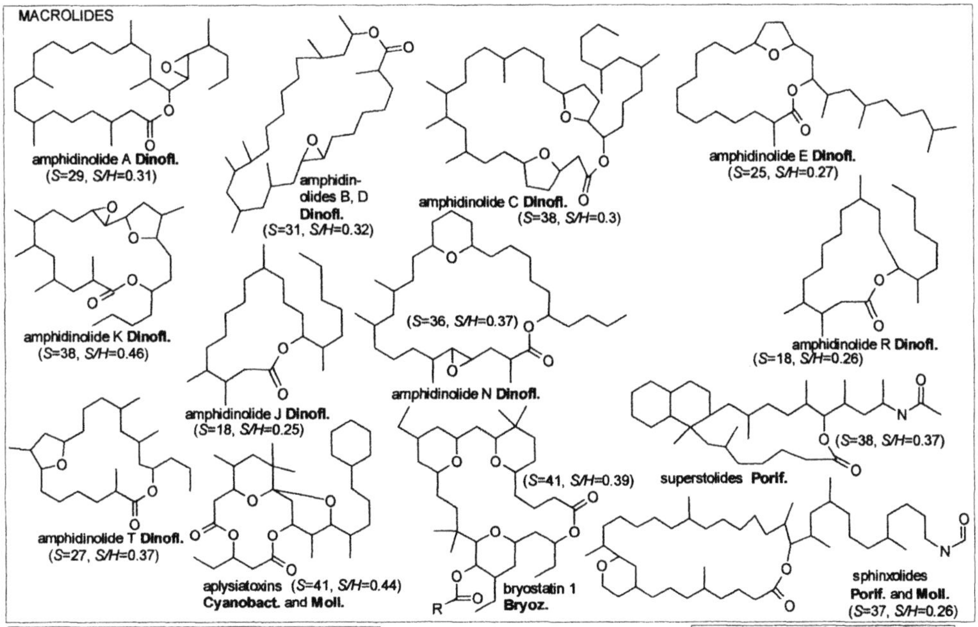

MACROLIDES

amphidinolide A **Dinofl.**
(S=29, S/H=0.31)

amphidin-olides B, D **Dinofl.**
(S=31, S/H=0.32)

amphidinolide C **Dinofl.**
(S=38, S/H=0.3)

amphidinolide E **Dinofl.**
(S=25, S/H=0.27)

amphidinolide K **Dinofl.**
(S=38, S/H=0.46)

(S=36, S/H=0.37)

amphidinolide R **Dinofl.**
(S=18, S/H=0.26)

amphidinolide J **Dinofl.**
(S=18, S/H=0.25)

amphidinolide N **Dinofl.**

(S=38, S/H=0.37)

amphidinolide T **Dinofl.**
(S=27, S/H=0.37)

(S=41, S/H=0.39) superstolides **Porif.**

aplysiatoxins (S=41, S/H=0.44)
Cyanobact. and **Moll.**

bryostatin 1 R **Bryoz.**

sphinxolides **Porif.** and **Moll.**
(S=37, S/H=0.26)

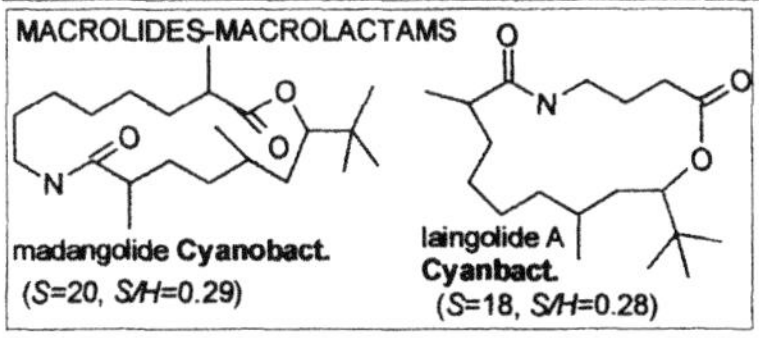

MACROLIDES-MACROLACTAMS

madangolide **Cyanobact.**
(S=20, S/H=0.29)

laingolide A **Cyanbact.**
(S=18, S/H=0.28)

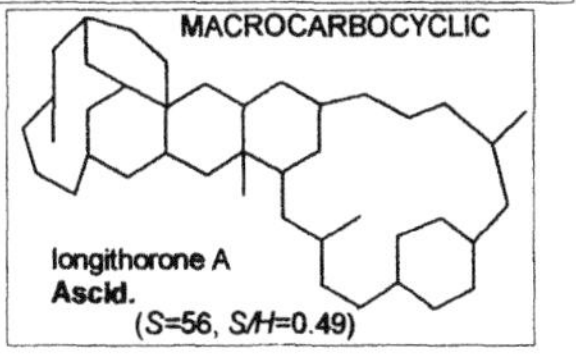

MACROCARBOCYCLIC

longithorone A **Ascid.**
(S=56, S/H=0.49)

7.3 Caribbean

Composed of beautiful coral reefs, the Caribbean Sea and the Gulf of Mexico have provided unique marine natural products. However, the active area of the Caribbean is small compared to the Indo-Pacific: the Brazilian coast is made inhospitable to coral reefs because of the fresh waters brought in by the Amazon river. Therefore, the natural product diversity of the Caribbean is second to the Indo-Pacific.

Perhaps the most unusual marine metabolites from the Caribbean are complex polyketides, called brevetoxins, produced by a toxic dinoflagellate, *Gymnodinium breve* (Chart 7.3.FA/PO).

The Caribbeans are also the realms of shallow water gorgonians, which have afforded new erythrane, pseudopterane, and sandresolide diterpenes (Chart 7.3.I).

Unique alkaloids and peptides were also isolated from Caribbean species (Chart 7.3.A/P), along with guanidine C_{11} alkaloids similar to those from the Indo-Pacific.

Chart 7.3.A/P Alkaloids and peptides (skeletons) from the Caribbean (A: *S*max=42, av=34; *S/H*max=0.66, av=0.5. P: *S*max=79, av=41; *S/H*max=0.5, av=0.4).
Alkal.: guanidine: ptilocaulin: (Poecil., Porif.: Harbour 1981); sceptrin (pantrop. Agelasida and Nepheliosp., Porif.: Walker 1981), ptilomycalin A (Poecil., Porif.: Palagiano 1995); indole: discodermindoles (Lithistida, Porif. from Bahama, depth 185m: Sun 1991); eudistomins (*Eudistoma olivaceum* Polycitoridae, Ascid. from mangroves: Rinehart 1988); nortopsentin (Halichon. and Axinel., Porif. from Bahama, depth 460m: Mancini 1996); sesquiterp. alk.: stachybotrins (*Stachybotrys* sp., Deuterom., Eumyc. from brackish w. in Florida: Xu 1992).
Pept.: barbamide and (lipopept.) microcolins and antillatoxin: (*Lyngbya majuscula* Gomont, Cyanobact.: Yokokawa 2000); salinamides: (depsipept. from *Streptomyces* sp. Actinom., Bact. from jellyfish *Cassiopeia xamachana* from Florida keys: Trischman 1994A).

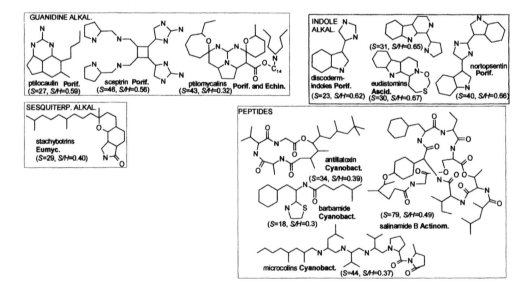

Chart 7.3.I Isoprenoids (skeletons) from the Caribbean (Smax=36, av=22; S/Hmax=0.45, av=0.32).

Monoterp.: ochtodane: (*Ochtodes* spp., Gigartinales, Rhodoph.: Crews 1984).

Diterp.: rearr. cembranes: erythrane (*Erythropodium caribeorum* Durchassaing & Michelotti, Gorgonacea, Cnid. from Caribbean: Wahlberg 1992B); pseudopterane (*Pseudopterogorgia acerosa* (Pallas), Gorgonacea, Cnid. from Caribbean: Wahlberg 1992A); sandresolides (*Pseudopterogorgia elisabetae* (Bayer), Gorgonacea, Cnid. from Caribbean: Rodríguez 1999A).

Sesterterp.: dysidiolide-like: (*Dysidea etheria* de Laubenfels, Dysideidae from Caribbean: Gunasekera 1996A).

Steroid: 9,11-secogorgostanes: (*Pseudopterogorgia americana* from Caribbean: Musmar 1983); secodinosterols: (*Pseudopterogorgia americana*, Gorgonacea, Cnid. from Caribbean: Miller 1995).

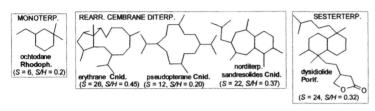

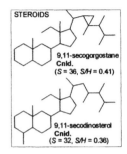

Chart 7.3.FA/PO Fatty acids and polyketides (skeletons) from the Caribbean (Smax=92, av=40; S/Hmax=0.45, av=0.44).

Simple polyket.: didemnenones: (*Trididemnum* cf. *cyanophorum*, Didemnidae, Ascid. from Bahama: Lindquist 1988); diosphenols (*Ulosa* sp., Halichon., Porif.: Wratten 1978).

Macrolide polyether: bryostatins: (*Bugula neritina* (L.), Cheilostomata, Bryoz.: Pettit 1991); caribenolides: (*Amphidinium* sp., free Dinofl.: Bauer 1995).

Polyether: hemibrevetoxins: (*Gymnodinium breve* Davis [= *Ptychodiscus brevis* Davis], free Dinofl.: Prasad 1989); brevetoxin-A-class: (*Gymnodinium breve*, free Dinofl.: Rein 1994).

7.4 Panamanian

The Panamanian marine area has afforded unique microbial metabolites. Unusual polyene polyketides, the macrolactins, were isolated from a deep-sea bacterium in culture. Unfortunately the strain has been lost, or has lost viability (Rychnovsky 1992). Unique cyclic heptapeptides, the cyclomarins, were obtained from the culture of an actinomycete, *Streptomyces* sp., isolated from sediments. The uniqueness of these metabolites contradicts the common assumption of scarce boundaries posed to microbial species in the sea.

Dolabellanes and spatanes have been isolated from Panamanian brown seaweeds, but these diterpene skeletons are also known from related species from other areas, the first ones from the Mediterranean and the second ones from Australia, Sri Lanka, and India.

Chart 7.4.P/I/PO Peptides, isoprenoids, and polyketides (skeletons) from the Panamanian (P: *S*=72; *S/H*=0.42. I: *S*max=26, av=17; *S/H* max=0.45, av=0.3. FA/PO: *S*=10; *S/H*=0.14).
Pept.: cyclomarins: (*Streptomyces* sp., Actinom., Bact. from S Californian sediments: Renner 1999).
Isopr.: MONOTERP.: ochtodane: (*Ochtodes crockeri* Setchell and Gardner, Gigartinales, Rhodoph. from Galápagos Is.: Crews 1984). DITERP.: dolabellane: (*Glossophora galapagensis* Taylor, Dictyotales, Phaeoph.: Sun 1979); spatane: (*Spatoglossum* spp., Dictyotales, Phaeoph.: Gerwick 1983).
Macrolides: macrolactins: (Bact. from US NE Pacific, depth 980 m: Rychnovsky 1992).

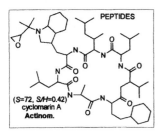

PEPTIDES

(S=72, S/H=0.42)
cyclomarin A
Actinom.

MONOTERP.

ochtodane
Rhodoph.
(S=6, S/H=0.2)

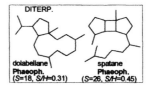

DITERP.

dolabellane
Phaeoph.
(S=18, S/H=0.31)

spatane
Phaeoph.
(S=26, S/H=0.45)

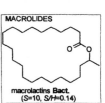

MACROLIDES

macrolactins Bact.
(S=10, S/H=0.14)

7.5 Mediterranean

The rogiolenynes are unique C-branched C_{15} polyketides isolated from the red seaweed *Laurencia microcladia* from Il Rogiolo, along the coast of Tuscany. Incisterol is an unusual degraded sterol isolated from the halichondrid sponge *Dictyonella incisa* from the coast of Liguria (Chart 7.5.A/I/PO).

The above strain of *L. microcladia* is biochemically different from all other seaweeds in nominally the same genus that have been examined so far. It is also amazing that, after a decade since their first report, the rogiolenynes remain the sole examples of C-branched C_{15} metabolites, in spite of the huge variety of compounds in this class, isolated on a world basis from seaweeds in the genus *Laurencia* and opisthobranch mollusks that feed on them.

The Mediterranean has also given peculiar triterpenes of squalene origin and an unusual cyclized cembranoid, coralloidolide C (Chart 7.5.A/I/PO). The latter resembles diterpenoids from tropical octocorals, indicating that these cnidarians, on migration to temperate waters, have conserved genes for secondary metabolites.

Chart 7.5.A/I/PO Alkaloids, isoprenoids, and polyketides (skeletons) from the Mediterranean (A: *S*=43; *S/H*=0.67. I: *S*max=38, av=35; *S/H*max=0.43, av=0.4. P: *S*=11; *S/H*=0.24).
Alkal.: indole: hamacanthins (Poecil., Porif. from Madeira Is., depth 548m: referred to in Capon 1998).
Isopr.: DITERP.: 3,7-cyclized cembrane: coralloidolide C (*Alcyonium* [= *Parerythropodium*] *coralloides* Pallas, 1766, Alcyonacea, Cnid. from East Pyrenean waters: D'Ambrosio 1990); TRITERP.: squalene derived: testudinariol (*Pleurobranchus testudinarius*, Notaspidea, Moll. from the Thyrrenian Sea: Spinella 1997); raspacionins (*Raspaciona aculeata*, Axinel., Porif. from eastern Pyrenean waters: Cimino 1993); STEROIDS: AB-nor: incisterol (*Dictyonella incisa*, Halichon., Porif. from the Ligurian Sea: Ciminiello 1990).
Polyket.: branched C_{15}: rogiolenynes (*Laurencia microcladia* of Il Rogiolo, Ligurian Sea: Guella 1992).

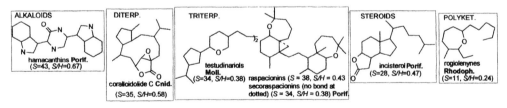

7.6 North Pacific

One may wonder whether the many metabolites known from the North Pacific connote a particular richness of this area or it was the commitment of Japanese chemists that left nothing hidden. On the other hand, this is a question that always arises when the natural product diversity is evaluated at local-geographic level.

Two groups of organisms from the North Pacific have been particularly well explored, microorganisms (actinomycetes, dinoflagellates, and filamentous fungi) and shallow water demosponges, mainly in the orders Lithistida, Choristida, Haplosclerida, and Nepheliospongida.

Bacterial metabolites include novel alkaloids, peptides, and macrocyclic lactams. The most unusual metabolites from dinoflagellate are polyethers, both linear and macrolides. The latter were also found abundantly in brown seaweeds (Chart 7.6.FA/PO). Filamentous fungi have afforded new isoindole and quinazoline alkaloids, as well as new-skeleton steroids (Chart 7.6.A/P/I).

3-Alkylamines, such as the haliclamines, isolated from the haplosclerid sponge *Haliclona* sp. (Chart 7.6.A/P/I), and cyclostellettamine, isolated from the choristid sponge *Stelletta maxima* from the same North Pacific area (Fusetani 1994), have been suggested as intermediates in the biosynthesis of complex macrocyclic alkaloids (Baldwin 1999). The latter were found not only in tropical Indo-Pacific nepheliospongid sponges (*inter alia*, motuporamines, xestospongins, keramaphidins, and manzamines, the latter also made by haplosclerid sponges, which also produce haliclyclamines and haliclonacyclamines, Chart 7.2.A) but also in a Mediterranean haplosclerid sponge, *Reniera* sp. (misenine, Guo 1998B).

3-Alkylpyridines supposedly lie also along another biosynthetic route, affording polymeric pyridinium compounds in marine haliclonid sponges: the Indo-Pacific *Callyspongia fibrosa* (Davies-Coleman 1993), the Caribbean *Amphimedon compressa* (Albrizio 1995), and the Mediterranean *Reniera sarai* (Sepčič 1997).

It is amazing that genes for the elaboration of 3-alkylamines are present in sponges belonging to such different ecosystems and orders as the Haplosclerida, Choristida, Nepheliospongida, and Haliclonida. This suggests a close phylogenetic relationship between these sponges, with genes that arose before their ecological separation.

Sponges from Japanese waters have also contributed mixed polyketides, like rubroside A and auranthoside A, tetramic acid glycosides (Chart 7.6.FA/PO), and metabolites reported in other parts of this book, such as nazumamide A (a tetrapeptide from a lithistid sponge, used as a laboratory tool, Table 13.5.II), calyculin A (a long-chain acetogenin, frequent target of total synthesis, Table 15.I, afforded also by other demosponges from different areas, Chart 8.2.A, including deep-water species, Table 9.III), cylindramide (a mixed-biogenesis macrolactam with analogues in terrestrial organisms, Chart 8.3.FA/PO), and mycalolide (an oxazole-bearing sponge macrolide, a target of total synthesis Table 15.I).

Chart 7.6.A/P/I Alkaloids, apoproteins, and isoprenoids (skeletons) from the North Pacific (A: *S*max=56, av=37; *S/H*max=0.9, av=0.56. P: *S*=16; *S/H*=0.22. I: *S*max=52, av=29; *S/H*max=0.42, av=034).

Alkal.: indole: neosurugatoxin and prosurugatoxin (*Babylonia japonica*, Moll. and symbiotic Actinom., Bact., from Japan: Kosuge 1985); isoindole: penochalasins (*Penicillium* sp., Eumyc. from *Enteromorpha intestinalis*, Ulvoph. from Japan: Numata 1996); quinazoline: fumiquinazolines (*Aspergillus fumigatus* Fres, Deuterom., Eumyc. from Japanes fish: Wang 2000); isopr.: altemicidin: (*Streptomyces sioyaensis*, Actinom., Bact. from Japan: Takahashi 1989); macrocyclic: haliclamine A and B (*Haliclona* sp., Haplosclerida, Porif. from the Uwa Sea, Japan: Fusetani 1989).

Lipopept.: korormicin: amino acid-fatty acid (*Pseudoalteromonas* sp., Bact.: Uehara 1999; Kobayashi 2000).

Apoproteins: green fluorescent protein: Bublitz 1998, and aequorin chromophores: Head 2000 (Scyphozoa, Cnid. from NW Pacific).

Isopr.: DITERP.: 4(3→2)*abeo*-reiswigane: mugipolasol (*Epipolasis* sp. of Japan: Umeyama 1998); rearr. cembrane: gersemolide (*Gersemia rubiformis* (Pallas), Alcyonacea, Cnid. from BC: Williams 1987); TRITERP.: xestovanins and isoxestovanins: (*Xestospongia vanilla* de Laubenfels, Nepheliosp., Porif. from BC: Morris 1991); STEROIDS: rearr. ergostane: gymnasterone, and 13(14→8)-*abeo*-ergostane,, dankasterones (*Gymnasclella dankaliensis* (Castellani), Deuterom., Eumyc. and *Halichondria japonica* Halichon., Porif. from Japan: Amagata 1999); 9,10-secocholestane: calicoferols (*Calicogorgia* sp., Gorgonacea, Cnid. from Japan, Ochi 1991); 13,17-secocholestane: isogosterones (*Dendronephthya* sp., Alcyonacea, Cnid. from Japan); mixed biogenesis: diaululusterol (*Diaulula sandiegensis* Nudibr., Moll. from BC: Kubanek 1999).

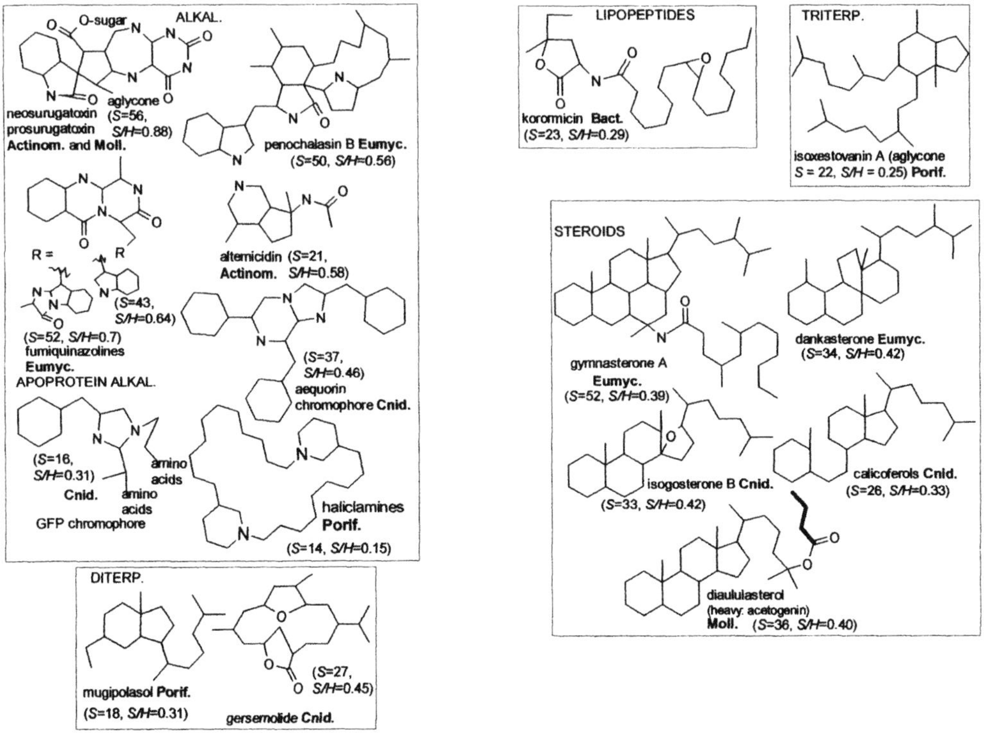

Chart 7.6.FA/PO Polyketides (skeletons) from the North Pacific (*S*max=68, av=32; *S/H*max=0.52, av=0.32).

Polyket.: <u>acarbocyclic</u>: amphidinols (*Amphidinium klebsii*, Dinofl. from Japan: Murata 1999); rubrosides and related aurantosides (*Siliquariaspongia japonica*, Lithistida, Porif. from off Hachijo-jima Is., Japan: Sata 1999); <u>macrolides</u>: ecklonialactones (Laminariales, Phaeoph. from Oregon and Japan: Todd 1994); goniodomins (*Goniodoma pseudogonyaulax*, Dinofl. from Japan: Murakami 1998); <u>macrolactams</u>: aburatubolactam A (*Streptomyces* sp., Actinom., Bact. from Moll. in Japan: Bae 1996); halichomycin (*Streptomyces hygroscopicus*, Actinom., Bact., from fish, Japan: Takahashi 1994).

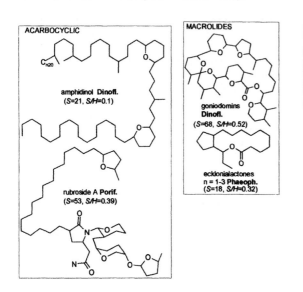

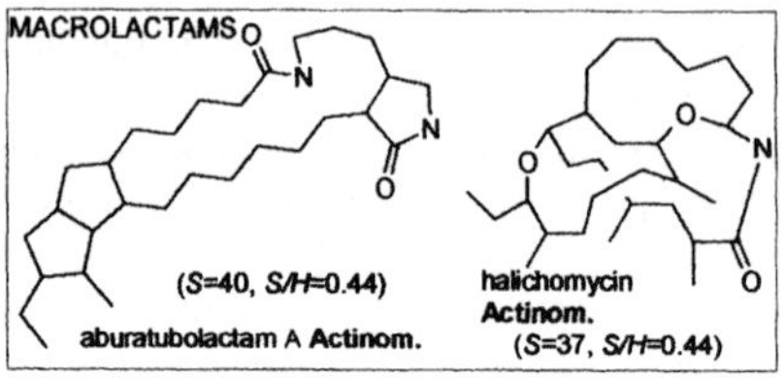

7.7 South Pacific

The South Pacific has contributed modestly to the diversity of natural products, although the Chilean coasts remain to be adequately explored. Worth noticing are diterpenes of cyclolabdane class and C_{15} oxocane acetogenins (Chart 7.7.I/FA/PO).

Chart 7.7.I/FA/PO Isoprenoids and fatty acid derivatives from the South Pacific (I: *S*=38; *S/H*=0.49. FA/PO: *S*=9; *S/H*=0.2).

Isopr.: DITERP.: <u>cyclolabdane-class</u>: epitaondiol (*Stypopodium flabelliforme* Weber van Bosse, Dictyotales, Phaeoph. from Easter Is: Sánchez-Ferrando 1995).

Fatty acid derivs.: <u>C_{15}-class</u>: oxocanes (*Laurencia claviformis* Börgensen 1924, Rhodoph., from Easter Is.: San-Martín 1997).

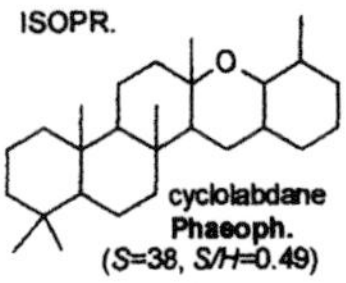

ISOPR.

cyclolabdane
Phaeoph.
(*S*=38, *S/H*=0.49)

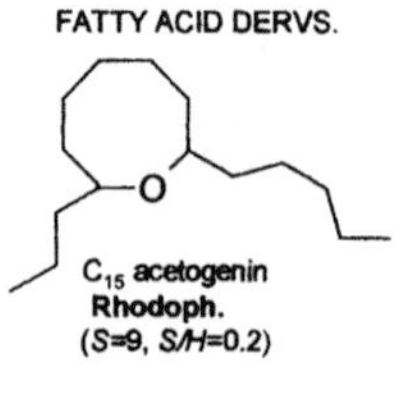

FATTY ACID DERVS.

C_{15} acetogenin
Rhodoph.
(*S*=9, *S/H*=0.2)

7.8 North Atlantic

The northeastern Atlantic includes the largest basin of brackish waters, the Baltic Sea. It is of recent origin, in the present form from the early Middle Age (Rheinheimer 1998). Being largely closed, and with a cycle of water exchange of more than thirty years, it is a particularly fragile ecosystem. It was only with the Sailing Olympics of 1972 that a sewage treatment plant was built for Kiel, a city facing the Baltic Sea with 200,000 inhabitants. Worse, damages by the Swedish industry will last long: chlorinated compounds from pulp bleaching, brought in by freshwater effluents, have made the Baltic Sea a most polluted area, where fishing boats had to drop anchor in the harbor. No unusual secondary metabolite is known from these polluted areas.

Other parts of the northeastern Atlantic are more productive: indole alkaloids isolated from bryozoans, seco-steroids from a gorgonian, *Gersemia fruticosa*, from the White Sea, toxic peptides from brackish-water cyanobacteria, and macrolides from dinoflagellates (Chart 7.8.A/P/PO).

Much practical interest has arisen from steroidal alkaloids, called squalamines, from sharks of the western Atlantic coast (Chart 7.8.A/P/PO). Although unexceptional in structure, these steroids, as a rare example from higher organisms, are precursors or models for new drugs (Tables 13.3 and 15.I).

Chart 7.8.A/P/PO Alkaloids, peptides, isoprenoids, and polyketides (skeletons) from the North Atlantic. (A: S max=39, av=36; S/H max=0.61, av=0.56. P: S=75; S/H=0.42. I: S max=36, av=29; S/H max=0.36, av=0.34. FA/PO: S max=25, av=18; S/H max=0.42, av=0.29).
Alkal.: indole: securamines and securines (*seco*-securamines) (*Securiflustra securifrons* (Pallas), Bryoz. from North Sea: Rahbæk 1997).
Pept.: nostocyclins: (*Nostoc* sp., Cyanobact. from brackish w. NE England: Kaya 1996).
Isopr.: TRITERP.: cucurbitacin-like: (*Adalaria loveni*, Nudibr., Moll. from the Norwegian Sea: Graziani 1995). STEROIDS: 9,11-secocholestanes: (*Gersemia fruticosa* Sars, 1860, Alcyonacea, Cnid. from White Sea: Koljak 1998); mixed biogenesis: squalamines (*Squalus acanthias,* Chondrich. from northern US Atlantic: Moore 1993).
Polyket.: bacillariolides: (*Pseudo-nitzschia multiseries*, Diatom. from NW Atlantic: Zheng 1997);
macrolides: hoffmanniolides (*Prorocentrum hoffmannianum* Faust, Dinofl. from NW Atlantic: Hu 1999).

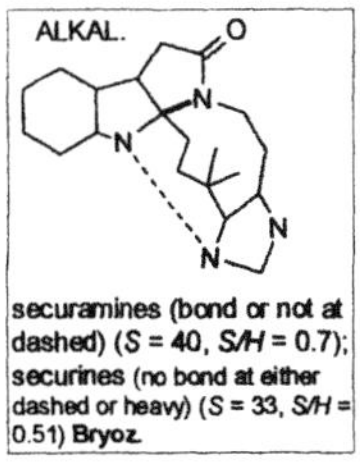

ALKAL.
securamines (bond or not at dashed) (S = 40, S/H = 0.7); securines (no bond at either dashed or heavy) (S = 33, S/H = 0.51) Bryoz.

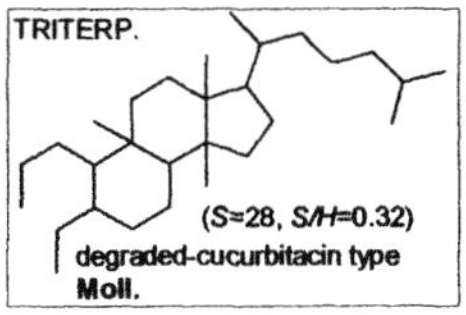

TRITERP.
(S=28, S/H=0.32)
degraded-cucurbitacin type
Moll.

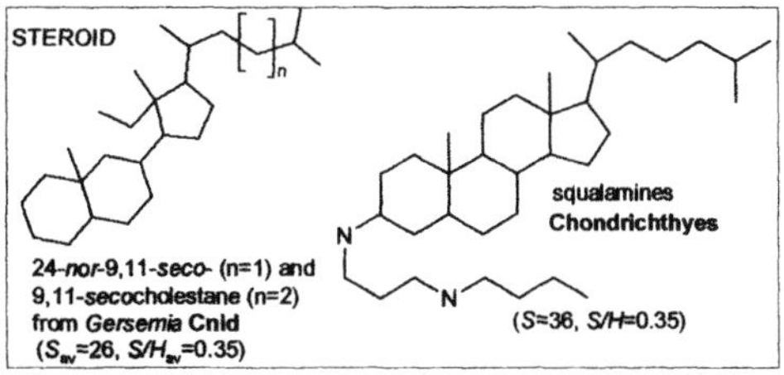

STEROID
24-*nor*-9,11-*seco*- (n=1) and 9,11-secocholestane (n=2) from *Gersemia* Cnid
(S_{av}=26, S/H_{av}=0.35)
squalamines
Chondrichthyes
(S=36, S/H=0.35)

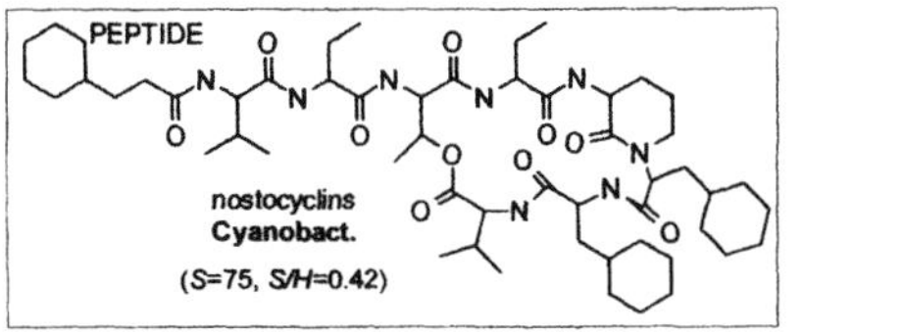

PEPTIDE
nostocyclins
Cyanobact.
(S=75, S/H=0.42)

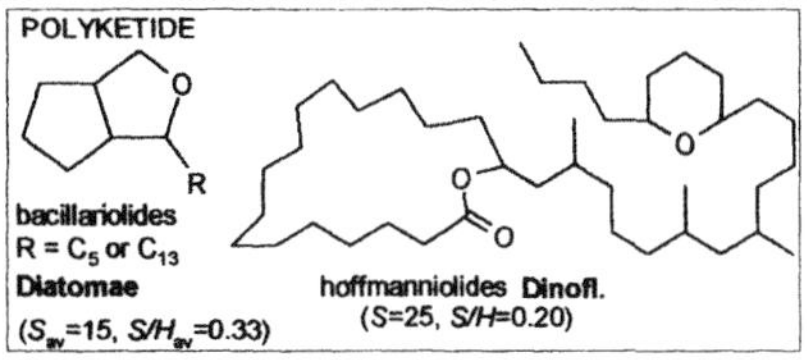

POLYKETIDE
bacillariolides
R = C_5 or C_{13}
Diatomae
(S_{av}=15, S/H_{av}=0.33)
hoffmanniolides Dinofl.
(S=25, S/H=0.20)

7.9 South Atlantic

Like the South Pacific, the South Atlantic also scarcely contributed to the diversity of natural products. Pregnane steroids from ascidians of the Argentine coast are the most unusual metabolites.

Squalene-derived triterpenes, sodwanones and *seco*sodwanones, isolated from the axinellid sponge *Axinella weltneri*, may be added if Sodwana Bay, at the border between South Atlantic and the Indo-Pacific, is considered in the South Atlantic.

Chart 7.9.I Isoprenoids (skeletons) from the South Atlantic (*S*max=37, av=30; *S*/*H*max=0.45, av=0.4).
Triterp.: squalene derived: sodwanones and *seco*sodwanones (*Axinella weltneri*, Axinel., Porif. from S Africa: Rudi 1999B).
Steroids: pregnanes: 20-Me-, 20-Et- 20-*i*Bu- (*Polizoa opunzia*, Ascid. from S Argentina: Palermo 1996).

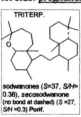

TRITERP.

sodwanones (*S*=37, *S*/*H*= 0.38), secosodwanone (no bond at dashed) (*S* =27, *S*/*H* =0.3) Porif.

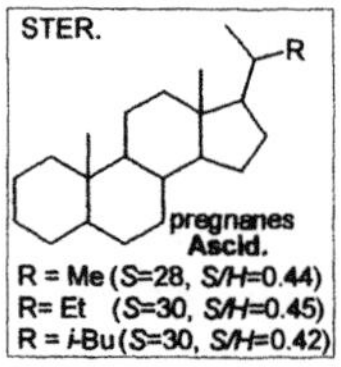

STER.

pregnanes Ascid.
R = Me (*S*=28, *S*/*H*=0.44)
R= Et (*S*=30, *S*/*H*=0.45)
R = *i*-Bu (*S*=30, *S*/*H*=0.42)

7.10 Zealandic

The Zealandic marine area has been extensively explored, affording several unusual metabolites (Chart 7.10.A/PO). Alkaloids bearing a macrocarbocycle are produced by dinoflagellates. Red seaweeds give macrocyclic ethers. Demosponges are the source of both pateamine A, a strongly

Chart 7.10.A/FA/PO Alkaloids, isoprenoids, and polyketides (skeletons) from the Zealandic A: (*S*max=43, av=35; *S*/*H*max=0.55, av=0.5. I: *S*=22; *S*/*H*=0.37. FA/PO: *S*max=31, av=23; *S*/*H*max=0.34, av=0.29).
Alkal.: piperidine: gymnodimines (*Gymnodinium* sp., Dinofl. from NZ: Steward 1997).
Isopr.: SESTERTERP. and NORSESTERT: luffarins (*Luffariella geometrica* Kirkpatric, Dictyocer., Porif., from S. Australia, depth 350 m: Butler 1992B); trunculins: (*Latrunculia* spp., Hadromerida, Porif. from S. Australia: Ovenden 1998).
Polyket.: mixed biogenesis: macrolides (pateamine A from *Mycale* sp., Poecil., Porif. from NZ: Northcote 1991); macrocyclic ethers: γ-pyrones (*Phacelocarpus* sp., Gigartinales, Rhodoph. from SE Australia and Tasmania: Murray 1995).

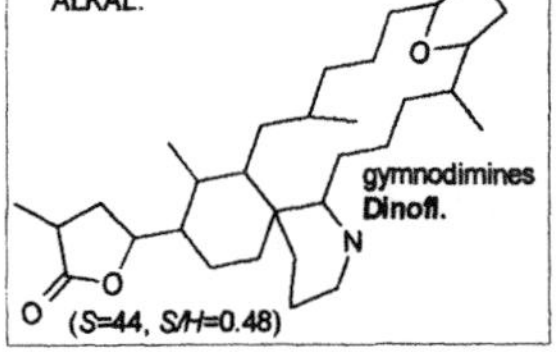

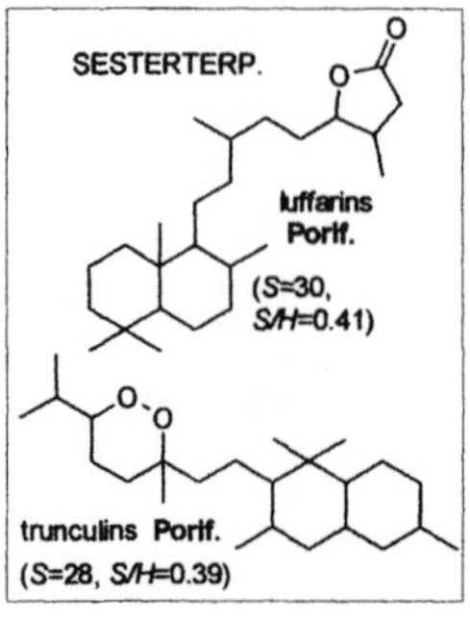

ALKAL.

gymnodimines Dinofl.
(*S*=44, *S*/*H*=0.48)

SESTERTERP.

luffarins Porif.
(*S*=30, *S*/*H*=0.41)

trunculins Porif.
(*S*=28, *S*/*H*=0.39)

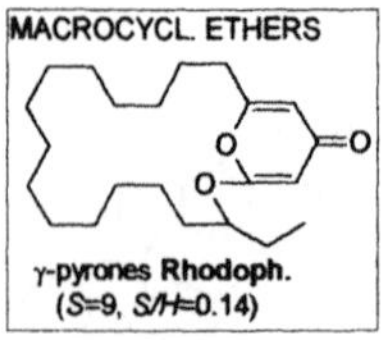

MACROCYCL. ETHERS

γ-pyrones Rhodoph.
(*S*=9, *S*/*H*=0.14)

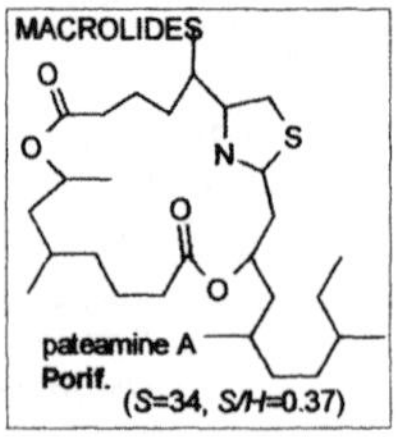

MACROLIDES

pateamine A Porif.
(*S*=34, *S*/*H*=0.37)

bioactive thiazole-containing macrolide, and new sesterterpenes and norsesterterpenes.

7.11 Arctic

Scymnol, a polyhydroxylated cholestane sterol of the bile alcohol class, isolated from the Arctic shark, *Scymnus borealis*, is one of the few new metabolites described from Arctic species. It is quite similar to ciprinol, however, isolated from the freshwater carp, *Cyprinus carpio* (Scheuer 1973).

As a metabolite of whales, which swim frequently in Arctic waters, the triterpene ambrein (Chart 8.2.I; Table 13.4) may be included here.

7.12 Antarctic

The Antarctic marine area has been widely explored thanks to various international projects, mostly geological but also involving the collection of living organisms. From these studies it has been concluded that competition between marine species in the Antarctic compares well to that in warm waters (McClintock 1997). This contrasts with the meager variety of unusual metabolites isolated from Antarctic marine organisms, limited to pyrimidine-indole alkaloids from ascidians and poecilosclerid sponges and suberitane sesterterpenes from hadromerid sponges (Chart 7.12.A/I) (for a complete list, see McClintock 1997).

These contrasting observations may perhaps be reconciled by admitting that warm-water marine organisms have adapted to the Antarctic. This has received circumstantial evidence from the finding of the same terpenoids in ciliate protists, *Euplotes* spp., from the Red Sea and the Antarctic. These

Chart 7.12.A/I Alkaloids and isoprenoids (skeletons) from the Antarctic (A: $S=23$; $S/H=0.57$. I: $S=30$; $S/H=0.42$).
Alkal.: indole: meridianins (*Aplidium meridianum* (Sluiter, 1906), Polyclinidae, Ascid. from Atlantic Antarctica, depth 100m: Franco 1998); psammopemmins (*Psammopemma* sp., Poecil. from Indian Antarctica, depth 258m: Butler 1992A).
Isopr.: SESTERTERP.: suberitane: (*Suberites* sp., Hadromerida, Porif. from Antarctica: Shin 1995).

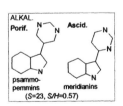

were originally thought to be different species, but at a closer examination, morphological strict similarities showed up, pointing to the same species (Pietra 1997).

7.13 Internal seas

The largest internal basins are the Black Sea and the Caspian Sea. The salinity of surface waters in the Black Sea is about 17‰ (half that of the Mediterranean), while in the Caspian Sea it is very low, except in the enclosed Zaliv Kara-Begaz, on the eastern side, where the salinity raises to 18.3‰.

Natural products from the Caspian Sea are limited to trivial fatty acids from red algae in the family Rhodomelaceae (Kurbanov 1997).

The diversity of natural products of the Black Sea is also modest, including unexceptional metabolites from red seaweeds (Güven 1970; Hamberg 1992) and brown seaweeds (Milkova 1997), fatty acids (Christie 1994) and sterols from sponges and other invertebrates (Milkova 1997), and common terpenoids from bryozoans (Hadjieva 1987).

This small cup of natural products is threatened: the Black Sea is a most polluted area, where fish catch has declined dramatically.

Chapter 8. The widespread distribution of natural products

8.1 The widespread distribution of natural products on land

Views on the distribution of secondary metabolites are changing. Once thought to be strictly taxon-specific, the same families of secondary metabolites, if not just the same compounds, are more and more frequently encountered in taxonomically, phylogenetically, and ecologically distant organisms (Chapter 3). It is the advancement in analytical technology that allows us to detect compounds that have escaped attention in the past, hidden in the bulk extracts or present in trace amounts.

The simplest, and probably most common, reason for the wide occurrence of a metabolite is the wide dispersal of the producing organism. This happens frequently with microbial products because of scarce barriers to the spreading of the microorganisms. Examples from bacteria and filamentous fungi are alkaloids and peptides (Chart 8.1.A/P), polyketides (Chart 8.1.FA/PO), and many carbohydrates (Chart 8.1.C). A special case is poly-(R)-3-hydroxybutanoic acid, which exists both as a high molecular weight ($n = 10^4$) microbial storage material, or a complexing oligomer ($n = 150$) (Chart 8.1.FA/PO). The latter, which is present in tiny amounts in all organisms, represents the fifth class of biomacromolecules (Seebach 2001), besides polynucleotides, polypeptides, proteins, starch (Conde-Petit 2001), and polysaccharides. Also alkaloids and peptides from plants and insects have wide distribution, which may be due to the ample adaptability of these associated organisms (Chart 8.1.A/P).

Wide distribution also occurs in metabolites that play the same role in phylogenetically and ecologically distant organisms, like peptides of the endothelin class (Chart 8.1.A/P), defensive monoterpenoids and steroids, the gibberellins (diterpenoid hormones) (Chart 8.1.I), and the alkaloid melatonin.

Dietary metabolites of insects (originary from plants) and opisthobranch mollusks (originary from cyanobacteria, seaweeds, sponges, bryozoans, and ascidians), are treated here as products from *de novo* synthesis. Co-evolution of insects with plants, and opisthobranchs with sponges, bryozoans, and ascidians, justifies this assumption. This poses no major problems in the interpretation of natural product diversity since our focus is on the ecological distribution of natural products and the co-evolving species make part of the same ecosystem. Migratory species do not fit this scheme, but, as far as it is known, phenomena of co-evolution involving secondary metabolites for migratory species are uncommon.

The presence of dietary metabolites connotes a wider distribution of natural products, at the detriment of their diversity. This is but a particular case of the general tenet that the wider the distribution, the narrower the diversity of natural products. The matter is complex, however, because the same molecule may arise from different biosynthetic pathways, implying different genes. To this category belong metabolites that play different functions in phylogenetically distant organisms, like the prostaglandins (Pietra 1995), although a firm conclusion about the biosynthetic pathways is still far to come (Varvas 1994; Hess 1999).

Wide occurrence of certain metabolites may be rationalized on purely chemical grounds. Thus, the chemical propensity to polycyclic folding in unsaturated precursors rationalizes the wide occurrence of dammarane, lupane, ursane, and cucurbitane triterpenes (Chart 8.1.I) (Pietra 1995). Widely occurring polyphenols of similar structure reflect the few different combination modes of the phenolic units (Chart 8.1.S). In all these examples, the trend of the natural product diversity diverges from that of biodiversity; the problem defies the geneticist, while the organic chemist is left alone to solve the matter.

Representative examples of alkaloids, isoprenoids, acetogenins, shikimates, and carbohydrates of wide distribution on land are shown in the following charts, where the various situations considered above are exemplified.

Chart 8.1.A/P Alkaloids, amino acids, and peptides (skeletons), and proteins widespread on land (A: Smax=48, av=28; S/Hmax=0.61, av=0.49. P: Smax=130, av=66; S/Hmax=0.43, av=0.39).

Alkal.: pyrrolidine and piperidine: azaphenalene, dimeric-azaph., and homotropane (ladybirds, Coccinellidae, tropical and temperate Coleoptera, Ins.: Braekman 1998; Lebrun 1999); tenuazonic-acid-like: (*Alternaria tenuis* Nees (1822), *Pyricularia oryzae* Cavara (1891), *Phoma sorghina* (Sacc.) Boerema, Dorenb. & Kesteren (1973), Deuterom., Eumyc.: MI); pyrrolizidine: (tropical and temperate Asteraceae and Boraginaceae, such as *Heliotropium* and *Trichodesma* spp., Ang.: AY; BR); phenethylisoquinoline: colchicines (temperate and subtropical *Colchicum* spp. and pre-desertic Sahara *Androcymbium gramineum* Macbridge, Liliaceae, Ang.:AY; MI); indole: melatonin (Amphib. and Mamm. skin: MI); isoindole: cytochalasins (*Aspergillus, Chaetomium, Chalara, Phoma,* and *Phomopsis* spp., Eumyc. growing on various crops in temperate, subtropical, and tropical regions: Numata 1996); diterp.-indole: paspalinine (*Penicillium* spp., Deuterom,, Eumyc.; Nicolaou 2000A); glycosidic: neomycin (*Streptomyces fradiae*: MI).

Amino acids: *N*-methylproline-class: stachydrine (Capparaceae from warm and arid regions, and other Ang.: MI).

Pept.: endothelins/sarafotoxins: (Mamm. endothelial cells/venom of burrowing asp *Atractaspis engaddensis,* Serpentes, Reptilia: Lee 1988); microcystins: (*Mycrocystis, Nodularia, Nostoc, Oscillatoria* spp., Cyanobact. of worldwide distribution: Kiviranta 1992); two-fold symmetric bicyclic octadepsipept.: echinomycin, triostin, BE-22179, and thiocoralline (T 15.I) (Bact.: Boger 2000).

Proteins: cecropins (induced from immune response in *Hyalophora cecropia* L., Lepidoptera, and other Ins. from various regions: MI).

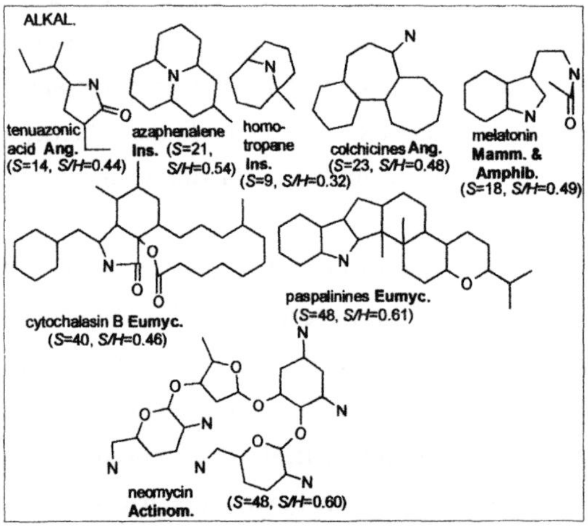

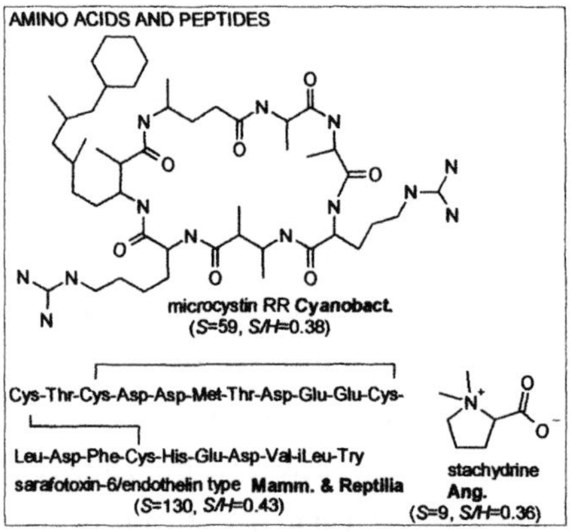

Chart 8.1.I Isoprenoids (skeletons) widespread on land (*S*max=45, av=29; *S*/*H*max=0.60, av=0.47).
Monoterp.: <u>menthane</u>: (Compositae, Ang. from temeperate and subtropical, not forested, areas and Bryoph.: AY);
thujane (Ang.: AY); <u>bornane</u>: camphor and borneol (tropical and subtropical Lauraceae and Mediterranean
Labiatae, Ang.:AY); <u>pinane</u>: (Ang. and Gymn.: MI; AY); <u>iridoids</u>: (Ang. and Argentine ant, *Iridomyrmex humilis*
Mayr., Ins.: BR; MI).
Sesquiterp.: <u>dimeric guaianes</u>: absinthin (Compositae: AY); longifolane (Bryoph. and Himalayan *Pinus longifolia*
Roxb., Gymn.: MI).
Diterp.: <u>cyclopropanebenzazulene</u>: phorbols (pantrop. and temperate Euphorbiaceae, Ang.: BR); tigliane (mostly
tropical and subtropical Euphorbiaceae, Ang.: MI); <u>kaurane</u>: (Ang.: McMillan 1997); <u>gibberellane</u>: (*Gibberella
fujikuroi* (Sawada) Wollenw. (1931), Ascom., Eumyc. and higher plants, ferns and algae: Rademacher 1997).
Triterp.: <u>dammarane/euphane</u>: (Ang: Connolly 1997); <u>lupane and ursane</u>: (Ang.: BR; Connolly 1997); <u>amyrin-
like</u>: (Compositae from temperate and subtropical, not forested, areas: AY, and *Glycyrrhiza glabra* Fabales, widely
cultivated for for glycyrrhizin: MI); <u>cucurbitane</u>: (Cucurbiitaceae, Ang.: MI; AY); <u>tetranortriterp</u>: (Asian neem
tree, *Azadirachta indica* A. Juss., and limonoids from European *Citrus* and other Ang.: BR; Connolly 1997);
<u>fusidane</u>: fusidic acid and helvolic acid (*Aspergillus* spp. and *Cephalosporium* spp., Eumyc., the latter also for
cephalosporin P$_1$: Erkel 1997).
Steroids: <u>withanolides</u>: (Ang. from Europe to India: BR); <u>cardenolides</u>: (*Strophanthus* spp. from Africa,
Madagascar and Indonesia, *Calotropis gigantea*, from the East, *Calotropis procera* Dryand. from Sahara and E
Africa, *Periploca* spp., Saharo-Mediterranean, Apocynaceae, and *Digitalis* from Europe and *Digitalis* spp. Ang.
cultivated in India for digitoxin: AY); <u>oestrones</u>: (Mamm.: MI and palms, Arecales, Ang.: AY).

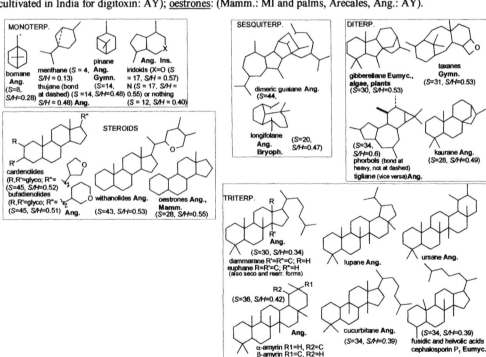

Chart 8.1.FA/PO Fatty acids and polyketides (skeletons) widespread on land (*S*max=72, av=32; *S/H*max=0.96, av=0.43).

Acarbocyclic: giganins: (Annonaceae, Ang. of Peru and N America: O'Hagan 1995); fumonisins: (*Fusarium* spp., Deuterom., Eumyc. on crops: MI); monensin-class: (*Streptomyces cinnamonensis* Okami, 1952, Actinom., Bact.: MI); mycophenolic acid-like: (*Penicillium* spp., Deuterom., Eumyc.: MI); patulin-like: (*Penicillium, Aspergillus, Gymnoascus* spp., Deuterom., Eumyc. on decaying fruits: MI); biopolymers: poly(*R*)-3-hydroxybutanoic acid (microbial storage polymer n=10⁴, Bact., or complexing n=150, present in all organisms: Seebach 2001).

Cyclohexane containing: "terminal ring"-class (*Achyranthes aspera* L., Amaranthaceae, Ang. from India and thermoacidophilic Bact: O'Hagan 1995; strobilurins: (*Strobilurus* spp., from tropical and temperate western countries, *Oudemansiella* spp., and *Agaricus, Mycena,* and *Crepidotus* spp., Basidiomyc., Eumyc.: Anke 1997; aflatoxins: (*Aspergillus flavus* Link (1832), *Aspergillus parasiticus* Speare (1912), Deuterom., Eumyc. on crops: MI); ochratoxins: (*Aspergillus* and *Penicillium* spp., Deuterom., on crops: MI); tremorgenic mycotoxins: janthitrem (*Penicillium janthinellum* Biourge (1923), Deuterom. on ryegrass in NZ, and nodulisporic acid from endophytic *Nodulisporium*, Eumyc. from Costa Rica: Magnus 1999).

Macrolides: tylosin-class: tylosin, carbomycins, leucomycins, midekamycins, miokamycins, rokitomycins (*Streptomyces* spp., Actinom., Bact.: MI); methymycin-class: methymycin and neomethymycin (*Streptomyces* spp.: MI); zearalenone-class: (*Fusarium* spp. and *Gibberella zeae* (Schwein.) Petch (1936), Eumyc.: Anke 1997); brefeldin-class: (*Penicillium* spp., Deuterom., Eumyc.: MI); polyether-class: nonactin (*Streptomyces* spp. MI and avermectins from *Streptomyces avermitilis*: MI); octalin-fused: nargenicin (*Nocardia argentinensis* ATCC, Bact.: Cane 1993); dimeric: SCH351448 (*Micromonospora* sp., Bact.: Hegde 2000).

Polycyclic: anthracyclines: (*Streptomyces* spp. ATTC, Actinom., Bact.: MI); tetracyclines: (*Streptomyces* spp. ATTC, Actinom., Bact.: MI); angucyclines: (*Streptomyces* spp. ATCC Bact.: MI); tetracenomycins: (*Streptomyces* spp. ATCC: MI); bisorbicillinoids: bisorbicillinol (T 15.I), bisobutenolide, and trichodimerol from *Trichoderma* sp., Eumyc.: Nicolaou 2000B).

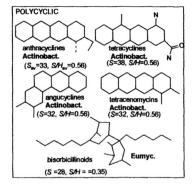

Chart 8.1.S Shikimates (skeletons) widespread on land (Smax=57, av=43; S/Hmax=0.48, av=0.39).
Tannins: lignans: epigallocathechin gallate (Asian black tea, *Camellia sinensis* (L.) Kuntze, Theaceae, Ang.: MI); procyanidins and prodelphinidins (widespread in plants); soluble hydrolyzable tannins: gallotannin (myrobalan, East Indian trees, *Terminalia* spp., oak, *Quercus* spp., *Lithocarpus* spp., and sumac, *Rhus* spp.); ellagitannin (Myrtaceae).

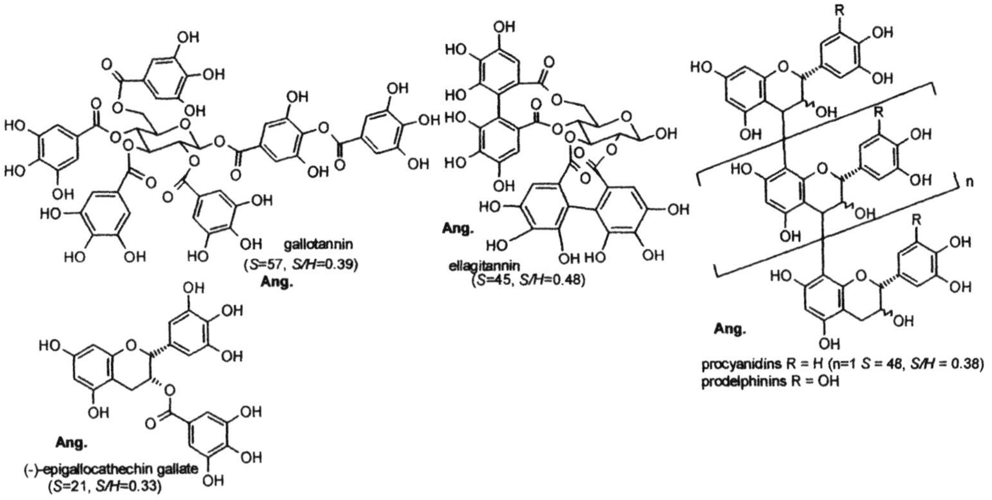

gallotannin
(S=57, S/H=0.39)
Ang.

ellagitannin
(S=45, S/H=0.48)
Ang.

procyanidins R = H (n=1 S = 48, S/H = 0.38)
prodelphinins R = OH

Ang.

(-)-epigallocathechin gallate
(S=21, S/H=0.33)

Chart 8.1.C Carbohydrates and glycosides (skeletons) widespread on land (Smax=115, S/Hmax=0.73, av=0.53).
Carboh.: orthosomycins: everninomycin (*Micromonospora carbonacea* from Kenya) and avilamycin (*Streptomyces viridichromogens*, Actinom., Bact.: MI).
Glycosides: cyanogenic: prunasin (Ang.: BR); linamarin (Euphorbiaceae, Ang., notably found in *Cassava* starch: BR); amygdalin (seeds of Rosaceae, Ang.: BR).

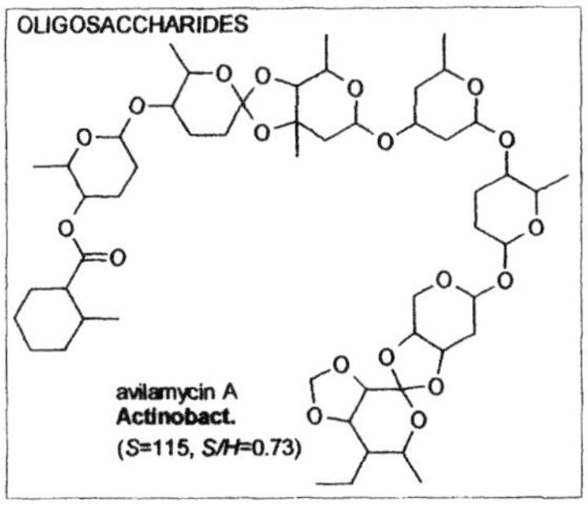

OLIGOSACCHARIDES

avilamycin A
Actinobact.
(S=115, S/H=0.73)

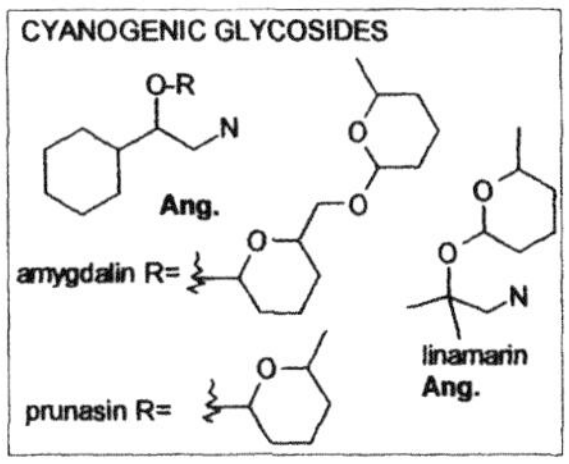

CYANOGENIC GLYCOSIDES

Ang.

amygdalin R=

prunasin R=

linamarin
Ang.

8.2 The widespread distribution of natural products in the oceans

The widespread distribution of certain marine metabolites in the oceans can be explained on similar lines as for common metabolites on land: either they play a key role in taxonomically, phylogenetically, and ecologically distant organisms, or the producing organisms are highly dispersed.

Prodigiosin, a pigment in many marine bacteria (Chart 8.2.A), and alterobactin and anguibactin, iron-sequestering peptides that determine the aggressiveness of certain marine bacteria, belong to the first category (Chart 8.2.P). The C_{15} acetogenins of cosmopolitan red seaweeds in the genus *Laurencia* are in the second category (Chart 8.2.FA/PO).

Both causes may be advocated for the widespread distribution of the silaffins. These polypeptides, as the name implies, show high affinity for silica, which is required to build up the skeleton of diatoms (Chart 8.2.P). Another example in this group is provided by certain tetrapyrroles that act as luciferins in marine dinoflagellates and other organisms (Chart 8.2.A; Tables 9.I and 13.5.II).

Purely chemical reasons, such as few possible combination modes of the building elements, may explain the wide diffusion of the phlorotannins in the sea (Chart 8.2.S), like the tannins on land (Chart 8.1.S). Certain polysaccharides, not shown in the charts, also have wide occurrence, produced by cosmopolitan seaweeds: alginic acids (brown algae in the genera *Fucus, Macrocystis, Laminaria*), carragenans (red seaweeds in the genus *Chondrus*), agar (red seaweeds in the genus *Gelidium*), furcellaran (read seaweeds in the genus *Furcellaria*), and fucans (various brown algae).

Chemical bias toward certain molecular arrangements may also determine the commonness of certain metabolites (Pietra 1995), such as the aplysinopsins (indole alkaloids of many sponges, cnidarians, and mollusks, Chart 8.2.A), bis-indole and guanidine alkaloids (Chart 8.2.A), isoprenoids (Pietra 1995, Chart 8.2.I), and polypropionates of mollusks, sponges, and fungi (Chart 8.2.FA/PO).

The toxic dinoflagellates responsible for the production of polyethers of the okadaic acid and prymnesin classes have recently occupied many new niches, which explains the increasingly frequent encounter of these tumor promoters and toxins (Chart 8.2.FA/PO).

Other examples in the following charts fit these general schemes. The different areas of origin are reported alongside the metabolites.

Chart 8.2.A Alkaloids (skeletons) widespread in the sea (Smax=52, av=32; S/Hmax=0.63, av=0.48).
Alkal.: pyridoacridine: (pantrop. Porif., Ascid., and cosmop. Actiniaria, Cnid.: de Guzman 1999); pyridinium oligomers and polymers: (Haplosclerida, Porif. from Micronesia and the Adriatic Sea: Sepčić 1997); polypyrrolic: prodigiosin, cycloprodigiosin (*Vibrio gazogenes* (Harwood *et al.* 1980) Baumann 1981, *Alteromonas* and *Beneckea* spp., including mutants, Bact.: Gerber 1983; Imamura 1994); magnesidins: from degradation of prodigiosin (Bact.: Imamura 1994); porphyrin-derivative: Dinofl. and Crust. luciferins (Dinofl. and Euphausiacea, Crust.: Nakamura 1989); guanidine: naamidines, leucettamines, clathridines (Calcinea, Calc., Porif.: Mancini, 1995), pseudozoanthoxanthins (cosmop. Zoanthinaria, Cnid.: Schwartz 1979); indole: aplysinopsins (pantrop. Dictyocer. and Choristida, Porif. and assoc. Nudibr., Moll., and pantrop. and temperate Dendroph., Cnid.: Guella 1989A); topsentins (cosmop. Porif.: Morris 1990); trikentrins and herbindoles (Axinel., Porif. from GBR and W Australia, resp.; Herb 1990); grossularines (rare α-carbolines from *Dendrodoa grossularia* Beneden, 1846, Ascid. from the English Channel and analogues from tropical W Pacific *Polycarpa aurata* (Quoy and Gaimard, 1834), Ascid.: Abas 1996); bis-indole: dragmacidin (Porif., depths 90-150m: Wright 1992); isoxazole: calyculins and clavosides (*Discodermia calyx*, Lithistida, from Japan and *Myriastra clavosa* (Ridley 1884), Choristida, Porif. from Palau: Fu 1998A).

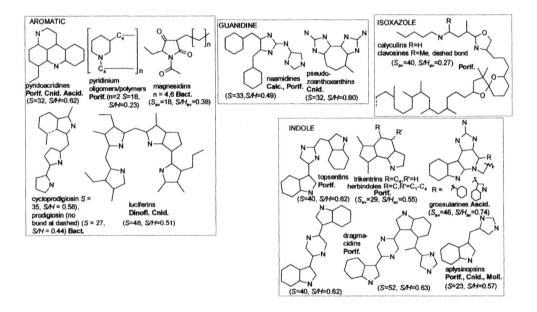

Chart 8.2.P Amino acid derivatives and peptides (skeletons) widespread in the sea (*S*max=72, av=43; *S/H*max=0.54, av=0.38).

Aminoacid derivs: malyngamides: (*Lyngbya majuscula* Gomont, Cyanobact. from Curaçao, Caribbean: Wu 1997 and the Côte d'Azur, Mediterr.: Mesguiche 1999).

Pept.: alterobactins (siderophores): (cosmop. *Alteromonas luteoviolacea* (*ex* Gauthier 1976) Gauthier 1982, Bact.: Reid 1993); anguibactin (siderophore): (*Vibrio anguillarum* Bergeman 1909, Bact.: Jalal 1989); didemnins: dehydrodidemin B (*Aplidium albicans* (Milne-Edwards, 1841), Polyclinidae; didemnin B *Trididemnum cereum* (Giard, 1872), Didemnidae, Ascid. from the Caribbean); dolastatins: dolastatin 10 (Anaspidea, Moll. from Mauritius Is. and Japan and Cyanobact. from W Pacific: Harrigan, 1999); orbiculamide A and discobahamins: (*Theonella* sp. from shallow Japanes waters and *Discodermia* sp., Lithistida, Porif., from Bahama depth 180m: Gunasekera 1994A); theonellapeptolid ID: (Lithistida from Okin. and Choristida, Porif., from NZ: Tsuda 1999B).

Polypept.: silaffins: (cosmop. Diatom.; Kröger 1999).

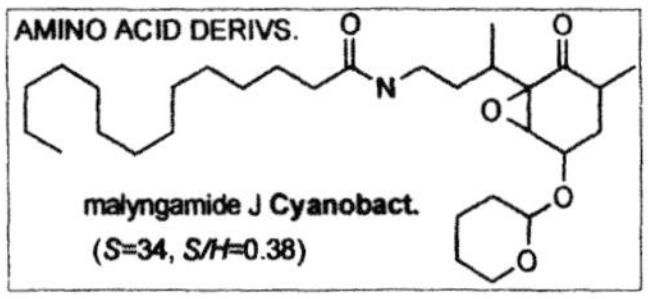

AMINO ACID DERIVS.

malyngamide J Cyanobact.
(*S*=34, *S/H*=0.38)

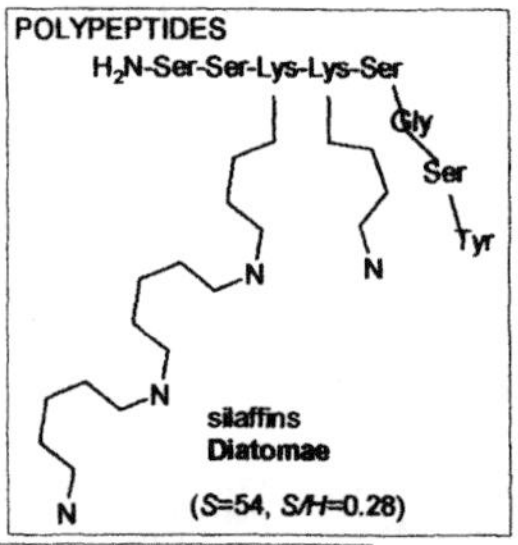

POLYPEPTIDES
H₂N-Ser-Ser-Lys-Lys-Ser

silaffins
Diatomae
(*S*=54, *S/H*=0.28)

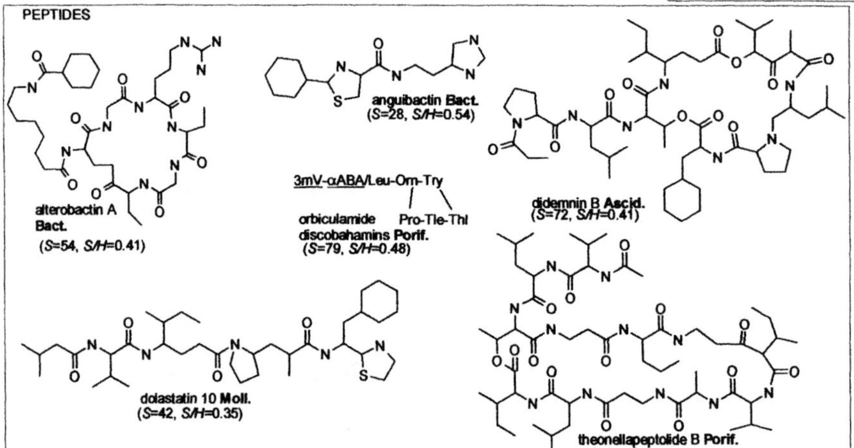

PEPTIDES

alterobactin A Bact.
(*S*=54, *S/H*=0.41)

anguibactin Bact.
(*S*=28, *S/H*=0.54)

3mV-αABA/Leu-Orn-Try
orbiculamide   Pro-Tle-Thl
discobahamins Porif.
(*S*=79, *S/H*=0.48)

didemnin B Ascid.
(*S*=72, *S/H*=0.41)

dolastatin 10 Moll.
(*S*=42, *S/H*=0.35)

theonellapeptolide B Porif.
(*S*=71, *S/H*=0.36)

Chart 8.2.I Isoprenoids (skeletons) widespread in the sea (Smax=42, av =21; S/Hmax=0.51, av=0.33).

Monoterp.: <u>mertensene:</u> (Gigartinales, Rhodoph. from GBR and Japan: Watanabe, K. 1989).

Sesquit.: <u>pupukeanane-type:</u> (Halichon., Porif., and assoc. Nudibr., Moll., from Hawaii, SW Pacific, and Japan: Okino 1996); <u>nakafuran-8 and nakafuran-9-like:</u> (pantrop. *Dysidea* spp., Dysideidae, Porif., and assoc. Nudibr., Moll.: Aiello 1996); <u>strongilane (rearr. drimane):</u> (cosmop. Demosp., Porif. and assoc. Nudibr., Moll.: Wright 1991; Bourguet-Kondracki 1992); <u>furodysane:</u> (pantrop. and Mediterranean *Dysidea* spp., Dysideidae, Porif. and assoc. Nudibr.: Searle 1994A); <u>cycloperforane and perforane:</u> (*Laurencia* spp., Rhodoph. from GBR and Canary Is: Coll 1989); <u>reiswigane:</u> (*Epipolasis* spp., Choristida, Porif. from Caribbean and Japan: Umeyama 1998).

Diterp.: <u>C-cyclized cembrane:</u> eunicellane (eunicellins, sarcodictyins, eleutherobins, valdivones from cosmop. Alcyonacea, Cnid.: Mancini 1999A); briarane (Gorgonacea, Pennatulacea and assoc. Nudibr., rarely Alcyonacea and Stolonifera, Cnid.: Sheu 1999); <u>spongiane</u> (Dictyocer. and Dendroc., Porif. and assoc. Nudibr., Moll.: Li 1999); <u>xenicane:</u> (Dictyotales, Phaeoph., and Pennatulacea, Cnid.: Miyaoka 1999); <u>pseudopterane:</u> (*Pseudopterogorgia* spp. from Caribbean and *Gersemia rubiformis* from BC, Alcyonacea, Cnid.: Williams 1987); <u>gersolane:</u> (*Gersemia rubiformis* from BC, Alcyonacea, Cnid.: Williams 1987).

Sesterterp.: <u>scalarane, homo- and bishomoscalarane:</u> (Dictyocer., Porif., and assoc. Nudibr., Moll.: Hanson 1996); <u>degraded C_{21}</u> (furanic from Dictyocer., Porif., and assoc. Nudibr.: Hanson 1996); <u>prenyllabdane:</u> (Halichon., Porif.: Hanson 1996).

Triterp.: <u>limatulane:</u> limatulone (Archeogastropoda, Moll. from intertidal California: Albizati 1985); <u>branched C_{25} and C_{30}</u> (Diatom., also found in sediments, C 16.1: Belt 2001); <u>ambrein:</u> ambergris (sperm whale, *Physeter macrocephalus* Linnaeus, 1758, Odontoceta, Mamm. from all oceans: MI).

Steroids: <u>amphisterane:</u> (*Amphidinium* spp., Dinofl., Euglenida, and and Chrysoph.: Shimizu 1996); <u>aplysterane:</u> (cosmop. Verongida, Porif., and tropical *Codium decorticatum* (Montagne), Caulerpales, Ulvoph.: Djerassi 1991); <u>gorgosterane</u> (Zooxanthellae, Dinofl. and invertebr. hosts: Giner 1991).

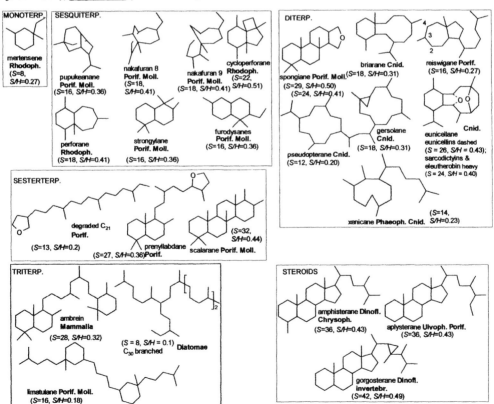

Chart 8.2.FA/PO Fatty acids and polyketides (skeletons) widespread in the sea (*S*max=154, av=29; *S/H*max=0.81, av=0.38).

C₁₁-class: ectocarpene-dictyotene, multifidene, hormosirene, and linear classs (Phaeoph.: Oldham 1996).
C₁₅-class: linear, tetrahydrofuranic, oxepane, oxocane, oxonane, and macrocyclic (cosmop. *Laurencia* spp., Rhodoph., and anaspidean grazers, Moll.: Guella 1999B, and rarely *Dasyphila plumariodes* Yendo, Ceramiales from the GBR: de Nys 1993).
Cyclopropane: gonyauline (Dinofl.: Roenneberg 1991).
Linear, furanone, 4-pyrone, hemiacetal-class polypropionates: (Opisthobr. and Pulmonata, Moll., and Homoscl., Porif.: Davies-Coleman 1998).
Macrolides: halichondrin B, isohomohalichondrin B, norhalichondrins (*Halichondria melanodocia* de Laubenfels, 1936, Halichon. from Micronesia and Japan, *Phakellia* sp. Axinel. from Japan, Demosp.: Pettit 1993); swinholides (*Theonella* spp., Lithistida, Porif., and ass. Bact. from W Pacific and Red Sea and assoc. *Chromodoris* spp., Nudibr., Moll. also from S Africa: Pika 1995).
Polyethers: <u>brevetoxin-B-class</u>: (Gymnodiniales, Dinofl. from Florida and Bivalvia feeding filters from NZ: Murata 1998); okadaic-acid class (Demosp., Porif. and assoc. Prorocentrales, Dinofl.: Konoki 1999); <u>prymnesins</u>: (*Prymnesium parvum* Carter 1967, Prymnesiophyta, cosmop. in brackish waters: Murata 2000).
2-Pyrone-class polypropionates: (Opisthobr. and Pulmonata, Moll., and Eumyc.: Davies-Coleman 1998).

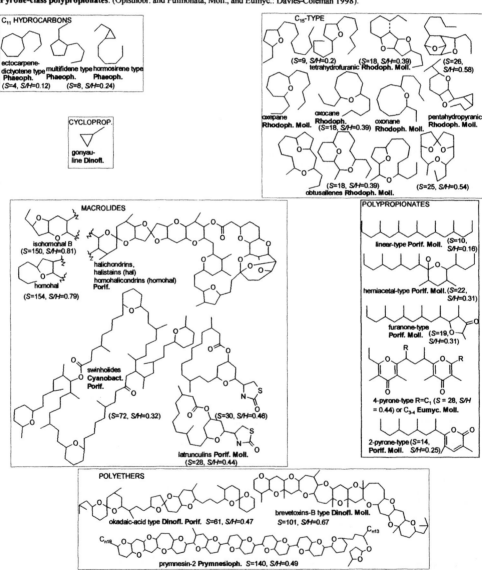

Chart 8.2.S Shikimates (skeletons) widespread in the sea (Smax=14, av=8; S/Hmax=0.16, av=0.08).
Phlorotannins: phloroglucinol derived: hydroxyhexaphloretol and terfucopentaphlorethol (*Carpophyllum maschalocarpum* Phaeoph. from NZ: Glombitza 1991); carrying dioxin-like units: 6,6'-bieckol (*Ecklonia kurome* Okamura from Japan: Fukuyama 1989). No sugar is involved in the polymerization, unlike with terrestrial tannins.

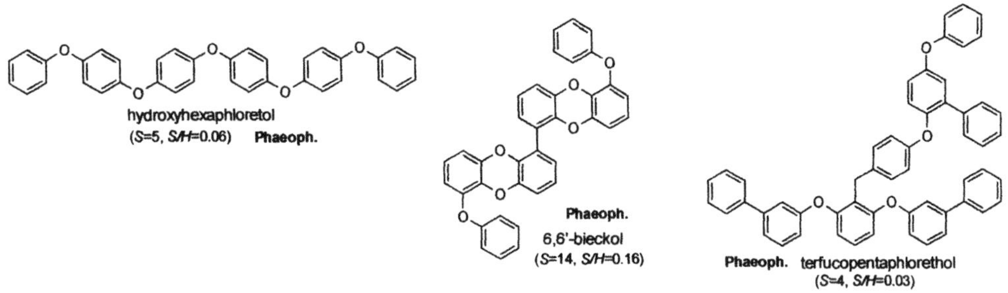

hydroxyhexaphloretol
(S=5, S/H=0.06)   **Phaeoph.**

Phaeoph.
6,6'-bieckol
(S=14, S/H=0.16)

Phaeoph. terfucopentaphlorethol
(S=4, S/H=0.03)

8.3 Found both on land and in the sea

It was only recently that the presence and significance of the same, or structurally closely related, natural products on land and in the sea has received specific attention (Pietra 1995). Three main conclusions were arrived at: (*a*) the highest frequency of structural similarities for secondary metabolites occurs with terpenoids from anthozoans and land plants; (*b*) biosynthetic routes became adapted to the function of the metabolite, such as a defensive one that must persist, or a hormonal one that needs to be rapidly switched off; so, the synthesis of the same compound in phylogenetically close organisms should occur by the same pathway, while different biosynthetic routes should be used by phylogenetically distant organisms, and (*c*) the contribution of marine bacteria to the genes for secondary metabolites in eukaryotes was scarce (Pietra 1995).

In case (*a*) above, the isotopic composition for the same metabolites in phylogenetically distant organisms may be different, because the isotopic fractionation depends on the particular biosynthetic pathways (Pietra 1995). This is at the basis of mass and NMR spectrometric control of adulterated food.

A major diversification of the enzymes of the early cnidarians during Cambrian times was advocated for case (*b*) above. Judging from fossil molecules, isoprenoids in a wide variety of structures were invented at those early times (Chapter 16.1). Later, this capability was transferred to the plants, where further diversification occurred. Symbionts may have taken part to these affairs, although disentangling their contribution would be arduous since the integrated biont may have formed a new whole, the holobiont (Pietra 1995).

The list of natural product similarities from land and the sea is immense, with metabolites of all chemical classes. On land these mostly derive from bacteria, cyanobacteria, plants, fungi, insects, amphibians, and mammals; in the sea, besides bacteria and cyanobacteria, the sponges, cnidarians, bryozoans, mollusks, ascidians, and sharks produce most. The ascidians are exclusively marine, and the sponges, cnidarians, and bryozoans are productive only in the sea. Dietary metabolites are also included, such as plant metabolites found in insects and metabolites from cyanobacteria, seaweeds, sponges, cnidarians, bryozoans, and ascidians found in opisthobranch mollusks.

The alkaloids range from the ubiquitous indole class to rare examples, such as benzylisoquinoline alkaloids in starfish. This is shown in Chart 8.3.A, which serves as preliminary documentation, until progress in biosynthetic studies allows us more definite conclusions.

Mycosporine and diketopiperazine types of peptides are typical of fungi but are also found in other taxa, on land and in the sea. Their wide occurrence may be explained by the chemical propensity of amino acids to couple, forming these chemically stable rings (Chart 8.3.P). Similarity of metabolites from terrestrial and marine fungi (Chart 8.3.P) is no surprise in view of the commonness of marine-adapted fungi.

The teleocidins from terrestrial actinomycetales and the lyngbyatoxins from marine cyanobacteria (Chart 8.3.P) are peptide alkaloids well known for their tumor promoting ability.

The isoprenoids contribute most to the list of structural similarities in the sea and on land. They range from common classes in both ecosystems, such as drimane sesquiterpenes, to rare classes in the sea, such as the trichothecenes (Chart 8.3.I1). The similarity in marine and terrestrial polyether triterpenes (Chart 8.3.I2) may be seen as convergence toward chemically favored structures, starting from squalene as a biosynthetic precursor. Similar conclusions may apply to polycyclic triterpenes.

In the fatty acid derivatives (Chart 8.3.FA/PO), the identity of prostaglandins from gorgonians and alcyonaceans in the sea, and mammals on land, has stimulated much experimentation (Varvas 1994) and computer calculations (Hess 1999). Although it is expected that biosynthetic routes in the sea differ from land because of the large phylogenetic distances of the organisms and the different functions of the metabolites at issue (Pietra 1995), details about the biosynthetic pathways are still to come (Varvas 1994).

In contrast, the biosynthesis of aplasmomycin in the sea and boromycin on land are expected to follow much the same pathways since actinomycetales are involved in both cases (Chart 8.3.FA/PO). Macrolactams are common metabolites in actinomycetales (Chart 8.3.FA/PO), which, as symbionts, may account for the presence of these metabolites also in marine demosponges.

Shikimates typical of terrestrial and marine angiosperms were rarely found in seaweeds (Chart 8.3.S). Finally, examples of terrestrial and marine glycosides are presented in Chart 8.3.C.

Chart 8.3.A Alkaloids (skeletons) typical of both marine and land organisms (*S*max=75, av=20; *S/H*max=0.91, av=0.56).

<u>Anabasine class</u>: Mar.: *Paranemertes peregrina* Coe, Nemert.: Kem 1971; Land: *Nicotiana glauca* Graham, Solanaceae, Ang.: MI.

<u>Indolizidine</u>: Mar.: protoplasm fusion with Bact.: Yamashita 1985; *Stelletta* spp. Choristida, Porif.: Matsunaga 1999; Ascid.: Raub 1992; Land.: orphans isolated from Ecuadorian *Epipedobates tricolor* (Boulenger, 1899), Dendrobatidae frog, Amphib., and fire ants, Ins.: Daly 2000; Convolvulaceae, Fabaceae, and Moraceae, Ang.: BR. <u>Benzylisoquinoline</u>: Mar.: rare in *Dermasterias imbricata* Grube, Aster., Echin. from BC: Burgoyne 1990; Land: Ang.: MI. <u>Isoquinoline</u>: Mar. invertebr.; Land: anhalamine from arid land and other Ang.:AY. <u>Guanidine</u>: tetrodotoxins (Mar.: Bact., Dinofl., Moll., Crust., Nemert., Osteich. puffers/parrotfish; Land: newts, Amphib.: Pietra 1995, Kodama 1996); saxitoxins (Mar.: Gonyaulacales, Dinofl. originary from NW Atlantic and symbiotic Bact., now worldwide: Committee O.R. 1999; Crust.: Scheuer 1996); Land: *Cylindrospermopsis raciborrskii* (Woloszynska) Cyanobact. from Brazilian freshwater: Lagos 1999). <u>Indole</u>: Mar.: lamellarins (pantrop. Ascid. and Mesogastropoda, Moll.; temperate Porif.: Davis, 1999A; Land: tropical Formicidae, ants: Braekman 1998); β-Carbolines (Mar.: *Pseudodistoma arborescens* Millar, Ascid. from NC; Land Zygophyllaceae, Ang.: Chbani 1993); carbazoles (Mar.: hyellazole and pyrrolo[2,3-*c*]carbazoles from tropical Pacific Cyanobact. and Japanese Dictyocer. Porif., and, also prenylated, Ascid., Bact.; Land: Ang., Bact: Sato 1993; Ihara 1997); physostigmines (Mar.: NE Atlantic *Flustra foliacea* (L.), Bryoz. and NW Pacific *Ciona savignyi*, Ascid.: Tsukamoto 1993; Land Calabar beans, Ang. from Nigeria and Gabon: BR); tryptamine-class (Mar.: Ascid. and various other marine invertebr.: Searle 1994B and Basidiomyc., Eumyc. and Ang.: MI); violacein-class (Mar.: *Alteromonas luteoviolacea* from Mediterranean Sea and *Chromobacterium* sp. from North Sea cod fish: Laatsch 1984; Land: *Chromobacterium violaceum*: Ballantine 1958); staurosporin-class (Mar.: *Eudistoma* spp., Ascid. and its predator *Pseuceros* sp., Turbell., and *Coriocella* spp., prosobranch Moll.: Cantrell 1999; Land: *Nocardiopsis* sp. and *Arcyria denudata*, Eumyc.: Schupp 1999); ergoline-class (Mar.: *Eudistoma* sp., Ascid.: Makarieva 1999; Land: Eumyc. and Ang.: BR). <u>Naphthyridinomycin-class</u>: Mar.: bioxalomycins from *Streptomyces* sp., Actinom., Bact.; Land naphthyridinomycins from *Streptomyces lusitanus*: Bernan 1997. <u>Phenazine</u>: phanazines (Mar.: pelagiomicins from tropical *Pelagiobacter variabilis*.Bact.; 5,10-dihydrophenocmycins from Caribbean *Streptomyces* sp.,Actinom.,Bact.:Pathirana 1992 and undeterm. Bact.: Bernan 1997); phenazine L-quinovose esters (Mar.: *Streptomyces* sp., from Californian ocean; Land: phencomycins in land *Streptomyces* spp.: Pathirana 1992). <u>Phenethylamine</u>: *S*-ring fused (Mar.: cosmop. Ascid. Makarieva 1995A; Land: mescaline from arid land peyote, Ang.: BR; ephedrine from Chinese *Ephedra* spp., Ang.: MI; epinephrine, norepinephrine, and dopamine in Mamm. and marine Cnid.: Pani 1994); chloramphenicol (Mar *Lunatia heros*, Moll. Land: *Streptomyces venezuaale*, Actinom., Bact. from soil: MI). <u>Pteridins</u>: Mar.: Diatom., Cnid. (Dendroph.), Porif. (Lithistida); Land: Ins.: Guerriero 1993). <u>Purine</u>: Mar.: caffeine in Mediterranean *Paramuricea chamaeleon* Koch, Gorgonacea, Cnid.: Imre 1987. Land.: caffeine in tropical coffee, maté leaves, theine and theobromine in tea leaves and cola nuts Ang.: BR. <u>Pyrazine</u>: Mar.: Hemich., Crust., Scyphozoa, Stolonifera, Cnid., Halichon., Porif.; Land.: tropical ants, Ins.: Braekman 1998. <u>Pyrrolidine and piperidine</u>: Mar.: *Telesto* sp., Telestacea, Cnid. from GBR: Bowden 1984; Ascid. Kiguchi 1997; Dinofl. and Porif.: Andersen 1996; Land: tropical fire ants, Ins., and Ang.: Braekman 1998. <u>Pyrroloiminoquinone</u>: *Damiria* alkal. (Mar.: Poecil. from Palau and deep-sea Porif. from Bahama: Stierle 1991); discorhabdins (Porif. from SE deep and shallow Indian and SW Pacific: D'Ambrosio, 1996A); makaluvamine A (*Zyzzya* cf. *marsailis*, Poecil., Porif. from Fiji Is. Land: *Didymium bahiense*, Mycetozoa from Japan: Ishibashi 2001 and Eumyc. and toads, Amphib.: Baumann, 1993). <u>Quinolizidine</u>: Mar.: Gorgonacea, Cnid.: Espada 1993, Porif.: Guo 1998A, Archaeogastropoda, Moll.: Martin 1986, Ascid.: Raub 1992, not derived from lysine; Land: Fabaceae, Ang., all from lysine: BR; AY). <u>Quinazoline</u>: Mar.: *Hincksinoflustra denticulata* Bryoz.: Blackman 1987; Land: *Adhatoda vasica* Nees, Acanthaceae, Ang. from India: BR. <u>Sesbanimide class</u>: Mar.: *Agrobacterium* sp., Bact., symbiont of Caribbean and Mediterranean Ascid.; Land: Leguminosae, Ang.: Acebal 1998). <u>Tetrahydroisoquinoline</u>: Mar.: saframycin-class, such as ecteinascidins from *Ecteinascidia turbinata* Herdman, 1880, Phlebobranchia, Ascid. from Caribbean, *Reniera* sp., Haplosclerida, Porif. from Caribbean and Palau, W Pacific; Land: saframycins and safracins from *Streptomyces lavendulae*, Actinom., Bact. from soil in Japan: Seaman 1998.

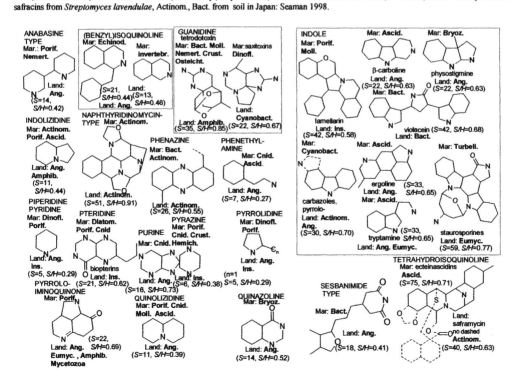

Chart 8.3.P Amino acids, peptides (skeletons), and proteins typical of both marine and land organisms (*S*max=69, av=37; *S/H*max=0.63, av=0.45).

Amino acids: γ-amino: (Mar.: calcareous Rhodoph.: Morse 1979; Land: neurotransmitter in Mamm.: MI); mycosporines: (Mar.: reef-buiding Scleract., Cnid. and cosmop. Holoth., Echin., Zoanthinaria, Dictyocer., Porif; Land: Eumyc.: Wu Won 1997 and freshwater Copepod, Crust., and Cyanobact.: Sommaruga 1999).

Pept.: andrimid-like (lipopept.): (Mar: andrimid and moiramides from Alaskan marine *Pseudomonas fluorescens* Migula 1895; Land: *Enterobacter* sp. and intracellular Bact. symbiont of brown planthopper from Thailand: Needham 1994); viscosin-like (depsilipopept.): (Mar. BC and Land: *Pseudomonas* spp., Bact.: Gerard 1997); surfactin-like: (Mar.: halobacillin from *Bacillus* sp., Bact. from Gulf of California, depth 124m: Trischman 1994B, related bacircines from *Bacillus* sp. from *Ircinia* sp., Porif. from GBR: Kalinovskaya 1995; related kailuins from unident. Bact. from Kailua Beach in Hawaii: Harrigan 1997; Land: iturins from land *Bacillus* spp.: Trischman 1994B); brevianamide-like: prenylated diketopiperazines (Mar: tryprostatins from *Aspergillus fumigatus* from estuarine sediment in Japan; Land: gypsetin from *Nocardia gypsea* var. *incurvata* and brevianamides from *Aspergillus ustus* and *Pecillium brevicompactum*: Schkeyryantz 1999); diketopiperazines: (Mar.: Bact: Cronan 1998, Eumyc.: Shigemori 1998; Porif.: Fu 1998B; Land: verticillins from Eumyc.: MI); eledoisin-like: (Mar.: *Ozaena* spp., Cephalopoda, Moll.; Land: tachykinins, e.g. substance P, have similar structure and physiological properties: MI); lyngbyatoxins/teleocidin-like: (Mar.: *Lyngbya majuscula* (Dillwyn) Harvey [= *L. majuscula* Gomont], Cyanobact. from Hawaii; Land: Bact.: Aimi 1990); methanofurans: widespread methanogenic Archaea thriving under extreme and normal conditions: White1988.

Proteins: papain/cathepsin-L protease family (Mar: spicules of *Tethya aurantia*, Porif. from California: Shimizu 1998; Land: kidney of Mamm. and tropical fruits of Ang.: MI).

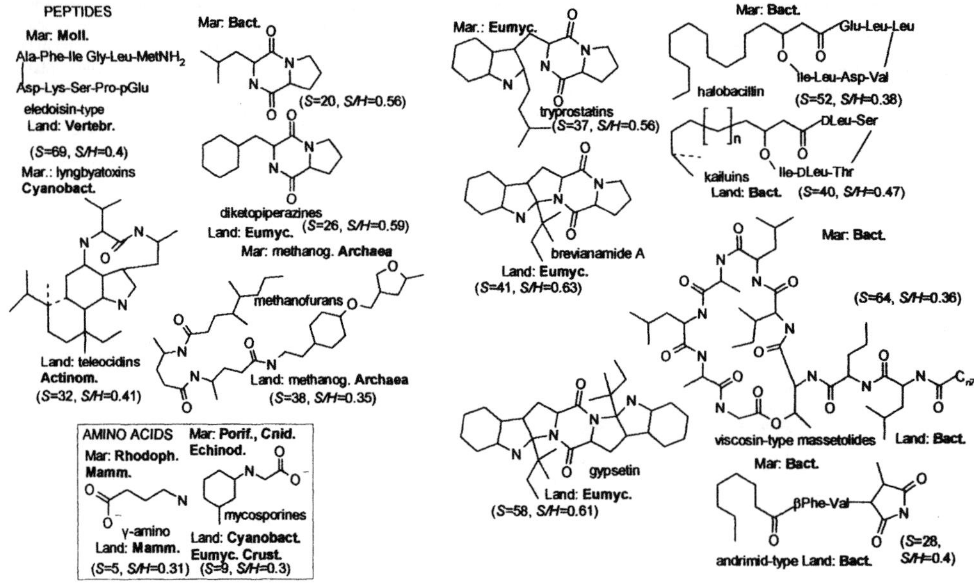

Chart 8.3.I1 Mono-, sesqui-, and diterpenes (skeletons) typical of both marine and land organisms (ref to Bryoph.: Asakawa 1995) (including C 8.3.I2: *S*max=46, av=24; *S/H*max=0.57, av=0.34).

Monoterp.: citronellol, *cis*- and *trans*-citral, neral, nerol, geranial, geraniol (Mar.: *Flustra foliacea* (L.), Cheilostomata, Bryoz. from North Sea: Christophersen 1985; Land: Bryoph., Ang.: MI, Eumyc: Turner 1983); linalool (Mar.: Cyanobact.; Land: mites, Arach.: Oldham 1996).

Sesquiterp.: africanane: Mar.: Alcyonacea, Cnid. from GBR and W Indian Oc.: Ramesh 1999 and *Lippia integrifolia*, Ang. from N Argentina: Catalan 1992 and Bryoph.); aristolane: Mar.: *Laurencia* spp., Rhodoph., Gorgonacea and Porif.: Jurek 1993; Land.: Bryoph.; aromadendrane: Mar.: cosmop. GBR Axinel., Porif. and *Laurencia* spp., Rhodoph.; Land Ang.: Pietra 1995; bisabolane: Mar.: cosmop. Porif., Alcyonacea, *Laurencia* spp., Rhodoph. Pietra 1995; Land: Bryoph.; brasilenane: Mar.: *Laurencia* spp.: Caccamese 1990; Land: Bryoph.; cadinane: Mar.: cosmop. GBR Axinel., Porif. and *Laurencia* spp., Rhodoph and Land Ang.: Pietra 1995; copaane: Mar.: Dictyotales, Phaeoph.: Amico 1979; Land: Bryoph.; calamenane: Mar.: Alcyonacea, Cnid.: Bowden 1986; Land Bryoph.; caryophyllane: Mar.: Alcyonacea: König 1993; Land: Bryoph.; chamigranes and cuparanes: Mar.: cosmop. *Laurencia* spp., Rhodoph., *Spongia zimocca* of Il Rogiolo: Guella 1992; Land: Bryoph.; rearr.: Land: Ang: Kouno 1999); drimane: Mar.: African elephant, Mamm.: Goodwin 1999; R=phenolic in Dictyocer., Porif., and assoc. Nudibr., Moll. from tropical W Pacific; Land: Ang. Pietra 1995 and Bryoph. and polygodial in Ang. and marine Moll.: Mori 2000; eremophilane: Mar.: Alcyonacea and Eumyc.: Jurek 1993; Land: Bryoph.; eudesmane: Mar.: *Laurencia* spp., Rhodoph., *Stylotella* sp. Porif., Alcyonacea, Cnid.: Matthée 1998; Land: Compositae, Ang.:AY); farnesane: Mar.: Dictyocer., Porif. and assoc. Nudibr., Moll.; Land: Ins.: Fontana 1998; germacrane: Mar.: Alcyon., Cnid.,, Phaeoph., Rhodoph.; Land: Bryoph.: Pietra 1995; guaiane: Mar.: Halichon., Porif. from Palau; Land: Ang.: Pietra 1995; humulane: Mar.: *Laurencia* spp., Rhodoph.: Takeda 1990; Land: Bryoph., Ang., including prenylhumulanes: Kubo 1999; lemnalane (= nardosinane): Mar.: pantrop. Alcyonacea and Gorgonacea, Cnid.; Land: Ang.: Su 1993; maaliane: Mar.: cosmop. GBR Axinel., Porif. and *Laurencia* spp., Rhodoph.; Land: Ang.: Pietra 1995; monocyclofarnesane and calamenene: Mar.: Alcyonacea, Cnid. from GBR, Bowden 1986; Land: Bryoph.; norguaiane: Mar.: chamazulene from Gorgonacea, Cnid.: Seo 1996; Land: Compositae, Ang.:AY; oppositane: Mar.: S Californian *Laurencia* spp., Rhodoph. and Mediterranean Axinel., Porif.: Wratten 1977; Land: Bryoph.; pacifigorgiane: Mar.: Gorgonacea, Cnid.: Izac 1982; Land: Bryoph.; penlane: Mar.: *Dysidea fragilis* (Montagu, 1818), Dictyocer., Porif.; Land: Ang.: Guella 1985; picrotoxininane: Mar.: *Spirastrella incostans* Dendy, Hadromerida, Porif. from Indian coast of the Bay of Bengal; Land: seeds of local Menispermaceae, Ang.: Sarma 1987; trichothecane: Mar.: rare from *Acremonium neo-caledoniae* Roquebert et Dupont, Deuterom. Eumyc. from NC lagoon; Land: *Fusarium, Calonectria, Cylindrocarpon, Gibberella, Hypocrea, Myrothecium, Spicelium, Stachybotrys, Trichothecium*, and *Verticimonosporium* spp., Eumyc.: Laurent 2000.

Diterp.: (iso)pimarane-like: Mar.: *Spongia zimocca*, Schmidt 1862, Dictyocer., Porif. from Ligurian Sea and *Laurencia* spp.: Guella 1992; Land: Eumyc.: Turner 1983, Bryoph., and Ang.: MI; serrulatane and amphilectane: Mar.: alkal. of *Pseudopterogorgia elisabethae* (Bayer) from Caribbean: Rodríguez 1999B; Land: *Capraria biflora*, Scrophulariaceae Ang. from Brazil: Comin 1963; fusicoccane: Mar. *Dictyota dichotoma*, Phaeoph. from Japan: Enoki 1983, Segawa 1987; Land: *Fusicoccum amygdali* Del., Eumyc.: Ballio 1968.

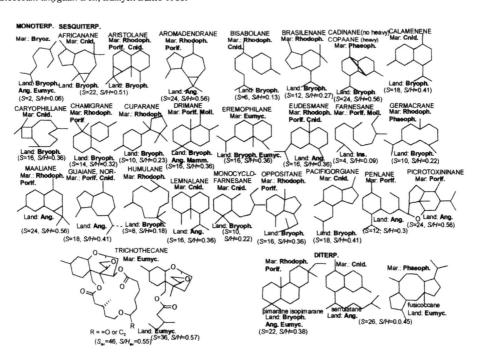

Chart 8.3.12 Sesterterpenes, triterpenes, steroids, tetraterpenes, and polyprenyls (skeletons) typical of both marine and land organisms (for *S* and *H* see C 8.3.11).

Triterp.: polyether: Mar.: thyrsiferol, venustatriol, and teurilene from *Laurencia* spp., Rhodoph.; Land: eurylene from *Eurycoma longifolia* Ang.: Suzuki 1987 and glabrescol from *Spathelia glabrescens*, Ang.: Bellenie 2001; friedelane: Mar.: Ulvoph.; Land: Ang.: DSII; oleanane: Mar.: Scleractinia, Cnid.: Sanduja 1984, Alcyonacea, Cnid.: Connolly 1997; Land: Ang.: DSII; taraxerane: Mar.: tropical Caulerpales, Ulvoph.: Santos 1971; Land: Ang.: DSII; AY; malabaricane: Mar.: tropical Choristida, Porif.: Kobayashi 1996A; Land: S Asian land *Ailanthus malabarica*, Simaroubaceae, Ang.: DSII; hopane: Mar.: Scuticociliatia, Ciliph.: Harvey 1997; Land: Bact., ferns Filicophyta: Rohmer 1989, Eumyc.: Connolly 1997; cycloartane: Mar.: rare nor-cycloartanes from tropical *Tydemania expeditionitis* Weber van Bosse, Caulerpales, Ulvoph.: Govindan 1994; Land: Ang.: Connolly 1997; AY; lanostane: Mar.: Echin. and, particularly nor-cycloartanes, from Demosp.: Antonov 1998; Land: animals: MI.

Steroids: 24-ethylcholestane: Mar.: poriferasterol, clionasterol, chondrillasterol, sitosterol in Porif.; Land: plants an toads: BR.

Tetraterp.: carotenoids/xanthophylls: peridinin in marine and freshw. Dinofl., except of genus *Gymnodinium*: Shimizu 1996.

Polyprenyls: vitamin K group (plants and Bact.) and vitamin E (Ang.) and ubiquinones (mitochondria).

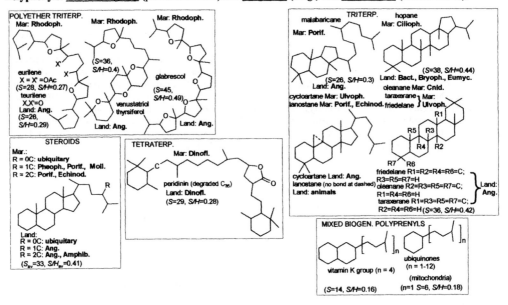

Chart 8.3.FA/PO Fatty acids and polyketides (skeletons) typical of both marine and land organisms (*S*max=69, av=25; *S/H*max=0.58, av=0.4).

C_6-C_n class: Mar.: aplysillin-A like: Mar.: Phaeoph. and deep-sea verongids, Porif., from Bahama; Land: Eumyc.: Gulavita 1995; biphenyls: Mar.: rare from Phaeoph. and *Axinella polycapella*, Porif. from Caribbean; Land: amarogentin, honokiol, magnolol, Ang.: Wratten 1981; griseofulvin-like: Mar: thelepin from *Thelepus setosus* (Quatrefages, 1865), Annelida from Hawaii: Higa 1974; Land: griseofulfin from *Aspergillus, Khushia, Nigrospoa, Penicillium*, and other Eumyc.: Erkel 1997.

Ether/polyether: pseudomonic acid like: Mar. and Land *Pseudomonas* and *Alteromonas*, Bact., in marine strains the acids are esterified by either anhydroornithine or pyrrothine, with further oxidation: Stierle 1992); pederin-like (Mar.: mycalamides from *Mycale* sp., Poecil. Porif. from NZ and onnamide A from *Theonella* sp., Lithistida, Porif. from Okin.; Land: pederin from land blister beetle *Paederus fuscipes* Curt., Ins.: Thompson 1994).

Macrolides: boromycin-like: Mar.: aplasmomycin from *Streptomyces griseus*, Actinom., Bact. from Japan sea mud; Land: boromycin from African soil *Streptomyces antibioticus*, Actinom., Bact.: Stout 1991; antimycin-like: Mar.: and Land *Streptomyces* spp.: Imamura 1993.

Mixed-biogenesis macrolactams: Mar.: discodermide, cylindramide and geodin, non-homogeneous family of ornithine/polyket.-chain derivs. from tropical, Japanese and S Australian Lithistida, Halichon., and Choristida Porif.: Capon 1999 and alteramide A from marine *Alteromonas* sp.: Shigemori 1992; Land: ikarugamycin from *Streptomyces phaechromogenes*, Actinom., Bact.: Capon 1999.

Polyacetylene alcohols: Mar.: cosmop. Nepheliosp., Haplosclerida, and Lithistida, Porif.; Land: Umbelliferae, Ang.: Ohta 1999, Fu 1999: non-terminal C-branched class in Porif.: Guerriero 1998.

Prostanoids: prostaglandin-class: Mar.: *Gracilaria* spp., Rhodoph., Alcyonacea, Gorgonacea, Cnid.; Land.: plants, Mamm.: Hamberg 1992; MI; Varvas 1994; clavulone/chlorovulone-class: Mar.: Stolonifera and Telestacea, Cnid.: Iwashima 1999; macrolide-class: Mar.: Nudibr., Moll.: Cimino 1991.

Sphingolipids: Mar.: Bact., Porif. (non-terminal C-branched class: Mancini 1994B); Land: Mamm.: Kobayashi 1995.

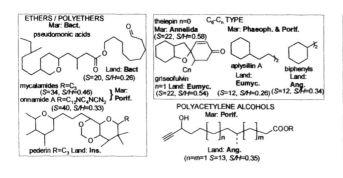

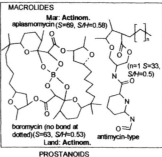

Chart 8.3.S Shikimates typical of both marine and land organisms (*S*max=56, av=22; *S/H*max=0.78, av=0.49).

Phenylpropanoids: <u>coumarins</u>: Mar.: 3,4,6-trihydroxycoumarin rare in *Dasycladus vermicularis* and other members of the Dasycladaceae, Dasycladales, Ulvoph.: Menzel 1983; Land: Asterales, Fabales, Apiales, and Sapindales, Ang.: AY; BR; *Amtherapeutic* spp., cultivated in Asia and Mediterranean regions for therapeutic xanthotoxin: MI; <u>flavonoids and chalcones</u>: Mar.: rare flavonoids in *Streptomyces* sp., Actinom., Bact.: Jiang 1997 and seagrass, Ang., and Rhodoph.: Wang 1998; Land: ubiquitary in Ang.: AY and also found in mosses, Bryoph.: Basile 1999); <u>isoflavonoids</u>: Mar.: rare in *Nerita albicilla*. Moll.: Sanduja 1985; Land: ubiquitary in Ang.:AY; <u>phenylpropanoids</u>: (Mar.: *Caulerpa* spp., Caulerpales, Ulvoph.: Mancini 1998 and as components of peptides from *Cliona* spp., Hadromerida, Porif.: Palermo 1998; Land: building block for lignin in Ang. on land and in the sea.

Quinones: <u>anthraquinones</u>: Mar.: Bact.: Schumaker 1995, Echin. and other marine invertebr.: Chang 1998; Land: Liliaceae, Ang., especially in tropical and temperate *Aloe* spp., and dimeric in Caesalpiniaceae, especially *Cassia* spp. from arid tropics: AY; *Cassia* spp. are cultivated in India and Egypt for danthron as a laxative; <u>phenanthroperylenequinones</u>: Mar.: living-fossil and fossil crinoids, Echin. (likely polyket.); Land: *Hypericum*, Theales, Ang.: Pietra 1995).

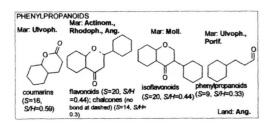

Chart 8.3.C Carbohydrates and glycosides (skeletons) found in marine and land organisms (*S*max=113, av=77; *S/H*max=0.62, av=0.5).

Carboh.: <u>ascorbic acid like</u>: Mar. seaweeds: Munda 1987, and Vertebr., where the sea lamprey is the most ancient lineage: Moreau 1998; Land: Vertebr. and ubiquitous in Ang. fruits: BR.

Steroidal saponins: land Ang., such as digitonin from *Digitalis purpurea* L., Scrophulariaceae: BR, and acanthaglycoside A from *Acantaster planci*, Aster., Echin.; also gorgonians and sponges: Gulavita 1994.

Triterp. saponins: examples: escin, a mixture of triterp. glycosides from *Aesculum hippocastanum* L., Hippocastanaceae: BR and bivittoside A from *Bohadschia bivittata*, Holoth., Echin.: Yayli 1999.

Chapter 9. Terrestrial vs marine natural product diversity

Life in the oceans differs in most respects from life on land. The most remarkable observation is that the oceans are poorer in species but richer in phyla with respect to land (Chapter 1.2). This finds no easy answer, however. Perhaps the reason for the larger number of species on land lies in a wider diversification of habitat and climate than in the sea, which required a great deal of adaptation, accompanied by speciation. In these affairs, coevolution of plants with insects was a major trigger of biodiversity.

It is more difficult to find a palatable answer to why more phyla are found in the sea than on land. None of the various hypotheses that have been offered has received overwhelming evidence. The view that has received most support is that life begun in the oceans, then the occupation of land started but some species found no adaptation and had to remain in the oceans or disappeared (May 1994). Whales are an exception, descending from land-dwelling mysterious artiodactyl or ungulate mammals (Luo 2000).

In any event, when higher taxa are examined, biodiversity is greater in the sea than on land (May 1994). This finds a parallel in a greater natural product diversity in the sea than on land at the molecular skeleton level (Part III). The reverse is true when biodiversity is considered at the species level and the actual metabolites are taken as a measure of natural product diversity (Part IV).

In the oceans it is the large (whales) that feeds on the small (zooplankton), whereas on land it is mostly the small (insects) that graze on the large (plants). Grazing in the sea differs from land, however, because seaweeds are mostly soft, whereas the large plants on land are mostly wooden and only in part exposed. These are particular aspects of a complicated relationship between body size and biodiversity. Generally, the distribution of body size is skewed, resulting in more small species than large species (Siemann 1996; Schmid 2000). An additional dimension must be taken into account if the internal biological scaling is considered (West 1999). Even this is not the whole story, however, because Siemann's relationship breaks down for very small organisms, possibly because they have a wide geographical distribution. This adds to the difficulty of relating biodiversity to the number of species, both in the sea and on land, because the very small is difficult to count and no quantitative relationship has ever been found to correlate the small to the large.

The huge variety of invertebrates thriving on coral reefs has no counterpart anywhere, even in tropical rain forests, where a large variety of taxonomic groups of animals is found with the arthropods only (May 1994). To strengthen the point, examination of DNA has revealed that genetically effective populations are smaller in the sea than on land (Ormund 1997). This finds analogy in the restricted distribution of certain marine natural products. An illustrative example is offered by a small family of C-branched C_{15} acetogenins called rogiolenynes, which were isolated from the red seaweed *Laurencia microcladia* from Il Rogiolo, a small spot along the coast of Tuscany, south of Livorno (Guella 1991). Elsewhere, the morphologically identical alga produces quite different metabolites, with strict geographical localization (Guella 2001A).

The relative contribution to the cup of natural products by marine and terrestrial organisms is outlined in Table 9.I. Column entries are the phylum, class or order, number of species, biogenetic class of metabolites produced, their bioactivity level, and a qualitative indication as to the average molecular complexity (Whitlock 1998). The latter property receives closer attention in Table 9.II for specific molecular skeletons. These data warrant several conclusions. First, unusual secondary metabolites on land derive mostly from green plants and arthropods, while in the sea are the algae, sponges, cnidarians, bryozoans, and ascidians that give most. This is true no matter if the molecular skeleton, or the actual metabolite, or even the bioactivity, is examined.

On land, the most ancient green plants (horsetail, ferns, lycopod, and gymnosperms) are peculiar for gammacerane and hopane triterpenes, besides a variety of lower terpenes. Green plants of more recent radiation - the angiosperms - have evolved characteristic alkaloids of the benzyl phenylethyl, benzyltetraisoquinoline, indolizinoquinoline, phenethylamine, phenethylisoquinoline, and terpenoid (secologanin) classes, besides iridoid monoterpenes and oleane triterpenes. Land-dwelling arthropods give a variety of toxic peptides, alkaloids, and diterpenes, whereas the productivity of marine arthropods is limited to new luciferins (Table 9.I).

In the sea, the brown algae give unusual diterpenes, the red seaweeds polyhalogenated mono- and higher terpenes and C_{15} acetogenins, and the green seaweeds cyclopeptides and masked-aldehyde sesquiterpenes (Table 9.I). Freshwater algae are fewer and, as far as it is known, unproductive.

Marine sponges have evolved two main groups of alkaloids that descend from either 3-alkylpyridines or halotyrosines. Other alkaloids from sponges are in the classes azacyclodecane, β-carboline, C_{11} or dimeric C_{11} bromopyrrole, (bis)imidazole, (bis)indole, imidazolyl pyrroloazepine, pyridoacridines, indolizidine, pyrroloiminoquinone, and (oxa)quinolizidine. Other unique products from marine sponges are sesterterpenes and their C_{21} degraded products, macrolides, polycyclic (hydro)quinones, polyethers, (epidioxy)polypropionates, and polyacetylenes that differ from those of land plants. In contrast, freshwater sponges are only known for very long chain fatty acids (demospongic acids), in common with marine sponges. The calcareous sponges - which inhabit shallow marine waters - have evolved unique Zn(II)-bound imidazole alkaloids and unusual macrolides, although I suspect that the latter derived from microbial species occasionally present.

In the cnidarians, the true medusae (Cubozoa) contain unusual toxic polypeptides in their nematocysts. The anthozoans are characteristic for the production of lower terpenoids similar to those from terrestrial plants (Pietra 1995). Bryozoans, in analogy with the sponges, are only productive in the sea, giving unusual metabolites, in particular novel macrolides (Table 9.I).

In the mollusks, only marine species are known for unusual metabolites. However, while polypropionates of marine pulmonates have *de novo* origin (Table 9.I), the secondary metabolites isolated from opisthobranch mollusks derive mostly from dietary cyanobacteria, seaweeds, and invertebrates.

At higher levels of the tree of life, the hemichordates are peculiar for pyrazine steroidal alkaloids. In the chordates are the ascidians that give most: alkaloids that are reminiscent of those isolated from both marine demosponges and terrestrial plants (Table 9.I). The most unusual metabolites from vertebrates are strongly bioactive steroidal alkaloids (squalamines) isolated from sharks. On land, the secondary metabolism of vertebrates is limited to scents.

Land vs sea microbial differentiation is generally scarce for the bacteria (which give the same kind of macrolides and peptides), archaeans (the same isoprenyl glyceryl ethers), and filamentous fungi (mostly the same metabolites). As to the archaeans, however, rich populations of mesopelagic species have recently been discovered (Karner 2001), and a high diversity has emerged at rDNA level (Moon-van der Staay 2001). If amenable to large cultures, these archaeans, lacking the environmental protection of those living under extreme conditions (Chapter 10), are potential sources of bioactive metabolites.

Unusual metabolites from dinoflagellates and ciliates are only known for marine species. However, because of culture problems, ciliates from land have not yet been adequately screened (Pietra 1997).

From the viewpoint of the diver, the most important characteristic of the sea is the increase of hydrostatic pressure by as much as 1 atm every 10 m of depth. Whether this has a bearing on the secondary metabolism cannot find a direct answer. First, the changing distribution of living organisms with the depth should be examined; this depends also on the temperature, light, and dioxygen and food availability, which generally decrease with the depth (Vernberg 1981). In particular, the algae and the corals that depend on photosynthetic symbionts are not found at depths greater than 100-200 m, depending on the transparency of the waters, which is higher in plankton-poor tropical regions.

Secondary metabolites from deep-sea organisms are a recent notion. Representative examples are shown in the charts of Chapter 7 and, on a broader basis, in Table 9.III. In the latter, the depth range, species, metabolites, and their bioactivity or technological use are indicated for the same zones as in Chapter 7. This affords an integrated perspective of the ecological and taxonomic distribution of the productive organisms and their secondary metabolites at the various depths.

For deep waters in the Indo-Pacific area, much was revealed by dredging campaigns carried out by geologists, marine biologists, and natural product chemists of the French ORSTOM (now IRD) in the economic zone of New Caledonia. Rich of seamounts, this area revealed the metabolic production of demosponges and echinoderms in the depth range 200-700 m, including living fossil lithistid sponges and crinoids. A manned submersible - formerly used for commercial sampling of corals - was used at the University of Hawaii to secure the blue pigments of gorgonians exclusive of those deep waters. In the Caribbean, a similar, more extensive, campaign was carried out by the Harbor Branch organization in the search for bioactive metabolites from demosponges in the depth range 80-400 m. Commercial dredging allowed to explore the Zealandic area at depths 80-400 m (mostly sponges), and the Mediterranean area in the depth range 80-300m, sporadically to 1,500 m (pennatulacean and scleractinian corals and brachiopods). The Antarctic and pre-Antarctic areas were sampled by dredging and beam-trawling, sponsored by international agencies (Chapter 7.12). Sampling of bacteria and fungi in the North Pacific relied much on sea bottom drilling.

Poly-indole alkaloids are a common theme of metabolites from deep-water sponges, particularly in the Caribbean, New Caledonia, and the Zealandic area (Table 9.III). Deep waters in the Caribbean have given pyrroloiminoquinone alkaloids from poecilosclerid sponges and pyridoacridine alkaloids from choristid sponges (Table 9.III). Both classes of alkaloids are also well known from shallow water organisms, however.

Macrolides from lithistid sponges and long-chain guanidine alkaloids (phloeodictyins) from nepheliospongid sponges are the most unusual metabolites from New Caledonian deep waters (Table 9.III).

Sampling in Mediterranean deep waters has revealed the first secondary metabolites from brachiopods and unique 10-hydroxyeicosa- and docosapolyenoic acids from scleractinian corals. That these fatty acids have also been found in a deep-water pre-Antarctic antipatharian coral (Table 9.III) must be attributed to convergence. Deep-water Mediterranean medusae and a lithistid sponge from New Caledonia, *Corallistes* sp., contain huge amounts of free porphyrins.

In summary, as far as it can be presently understood, there is little peculiarity as to the secondary metabolites of deep-water marine organisms, except, perhaps, the free porphyrins of medusae and sponges; an adaptive role of sensors in the absence of light may be suggested for them. The secondary metabolism - mostly unexceptional in the global perspective of marine organisms - seems to be more related to the range of the species - a composite of many factors - than to any effect clearly attributable to the hydrostatic pressure.

Table 9.I. Contribution to natural product diversity by land-dwelling vs marine organisms

Phylum or division[a]	Class or order	Nr of species on land/sea	Biogenetic classes of natural products[b]	Bioac-tivity level	Mol. com-plexity
Bacteria		overall > $4 \cdot 10^3$: Groom. 1992	Scarce diversification between land and sea. ALKAL.: indole, phenazine, pyperazine, pyrrole, polypyrrole. PEPT.: various classes, including siderophores. POLYKET.: lactones, macrolides, quinones. CARBOH.: (amino)glycosides, terpenoids	high	high
Archaea		large/very large: DeLong 1994; Karner 2001	MIXED BIOGENESIS: isoprenyl glyceryl ethers	low	medium
Phaeoph.	<Phaeo-ph.> (brown algae)	few/>1,500: May 1994	**Mar.:** ALKAL.: rare, of flavin-like. ISOPR.: Sesquiterp.: cadinane, calamelene, germacrane, mentane, muurolane. Diterp.: acarbocyclic, crenulatane, dolabellane, dolastane, merosesquiterp., (seco)spatane. FATTY ACID DERIVS. and POLYKET.: C_{11} hydrocarbons and *O/S* derivs. SHIKIM.: phlorotannins. CARBOH.: alginates, As-ribofuranosides, fucan, laminaran	**Mar.:** high	**Mar.:** high

Table 9.I. Contribution to natural product diversity by land-dwelling vs marine organisms

Phylum or division[a]	Class or order	Nr of species on land/sea	Biogenetic classes of natural products[b]	Bioactivity level	Mol. complexity
\<Rhodoph.\>		few/more than green plus brown: May 1994	Mar.: ALKAL.: (brominated) indole, pyrrole, and pept. ISOPR.: <u>Monoterp.</u>: polyhalogenated mertensene, ochtodane, and open chain; <u>Sesquiterp.</u>: brasilenane, (seco)chamigrane, cuparane, guimarane, perforane, and poitane; <u>Diterp.</u>: clerodane, parguerane, obtusane, and sphaerococcane; <u>Triterp.</u>: polyether. FATTY ACID DERIVS.: polyhalogenated C_{15}. CARBOH.: carrageenans	**Mar.:** high	**Mar.:** high
\<Ulvoph.\>		80/1,000: May 1994	Mar.: ALKAL.: indole. PEPT.: cyclopept. (*Bryopsis* spp.). ISOPR.: <u>Sesquiterp.</u>: brominated from *Neomeris* spp., otherwise aldehyde and masked aldehyde; <u>Diterp.</u>: aldehyde and masked aldehyde; <u>Triterp.</u>: friedelane and norcycloartane. SHIKIM.: brominated diphenylmethanes	**Mar.:** high	**Mar.:** high
Equiseto- phyta (horsetail)		22/0: Goom. 1992	FATTY ACID DERIVS.: polycarboxylic acids	low	low
Filicophyta (ferns)		10,000/0: Goom. 1992	ISOPR.: <u>triterp.</u>: gammacerane and hopane. POLYKET.: glycosidic. SHIKIM.: flavonoids and chromenes.	high	high
Lycopodio- phyta		590/0: Goom. 1992	ALKAL.: pyridine	low	low
Pinophyta (Gymn.)		600/0: Goom. 1992	ALKAL.: azoxy and phenethylamine. ISOPR.: mono-, di-, sesqui-, sester-, and triterp. SHIKIM.: glycosidic, phenylpropanoids, flavonoids	high	high
\<Magnolio- phyta (Ang.)\>		250,000/50: Goom. 1992	Land: ALKAL.: benzyl phenylethyl, benzyltetraisoquinoline, indolizinoquinoline, phenethylamine, phenethylisoquinoline, terpenoid (secologanin). ISOPR.: <u>Monoterp.</u>: iridoids; <u>Triterp.</u>: oleane	high	high
			Mar.: ISOPR.: <u>Diterp.</u>: cleisthanthene and clerodane. SHIKIM.: cinnamic acids.	**Mar.:** low	**Mar.:** low
Fungi		overall $1.5\,10^6$	Scarce diversification between land and sea.	high	high
Labyrin- thomorpha	**Thrausto- chytriales**	0/30	Mar.: FATTY ACIDS: docosahexaenoic acid. MIX. BIOGEN.: glycosphingolipids: Jenkins 1999	medium	low
Euglenozoa		$< 10^3/10^2$: May 1994	Mar.: ISOPR.: steroids, carotenoids: Liaaen-Jensen 1990	low	low
\<Chrypto- phyta\>		overall 12 genera	Mar.: SHIKIM.: styryl chromones	high	low

Table 9.I. Contribution to natural product diversity by land-dwelling vs marine organisms

Phylum or division[a]	Class or order	Nr of species on land/sea	Biogenetic classes of natural products[b]	Bioactivity level	Mol. complexity
<Chrysoph.>		400/few	**Mar.**: polychlorinated C_{24} enol sulfates	low	low
Diatom.		overall 5,600: CE 2001	**Mar.**: FATTY ACID DERIVS.: bacillariolides (C 7.8A/P/PO)	low	low
Dinozoa	<Dinofl.>	overall 1,500; few toxic (marine)	Land: dinosterols, dinocarotenoids	medium	medium
			Mar.: ALKAL.: β-carboline, guanidine, piperideine, and tetrapyrrole luciferins. ISOPR.: Steroids: 4-α-methyl. FATTY ACIDS: docosahexaenoic acid. POLYKET.: macrolides, polyhydroxy polyenes, and (high MW) polyethers	**Mar.**: high	**Mar.**: high
Cilioph.	**Hymenos tomatida**	overall > 4,000: Finlay 1999	**Mar.**: ISOPR.: Triterp.: gammacerane. POLYKET.: sulfated long-chain amides.	low	low
	<Hetero-trichia>		POLYKET.: bisanthrone pigments.	high	medium
	<Spiro-thrichia>		**Mar.**: ALKAL.: sulfated long-side-chain pyrrole alkal. ISOPR.: Sesquiterp.; Diterp.; Triterp.: in all three classes with aldehyde and masked aldehyde functional groups	**Mar.**: high	**Mar.**: high
<Porif.>	<Demo-sp.>	<100/1,000; George 1979, May 1994	Land: demospongic acids	low	low
			Mar.: ALKAL.: 3-alkylpyridine, azirine, azacyclodecanes, β-carboline, C_{11} or C_{22} bromopyrrole, guanidine, halotyrosine, (bis)imidazole, (bis)indole, imidazolyl pyrroloazepine, indolizidine, pept. alkal., pyridazine, (hydro)pyrimidine, macrocyclic and oligomeric pyridine or pyridinium, naphthyridine, oxazole, pteridine, pyridoacridine, pyrrole, pyrroloiminoquinone, (oxa)quinolizidine, terpenoid indolizidine. ISOPR.: Monoterp.: rare; Sesquiterp.:, Diterp., Sesterterp.: also degraded C_{21}; Triterp.. POLYKET.: macrolides, polyacetylenes, polycyclic (hydro)quinones, polyethers, (epidioxy)polypropionates.	high	high
	Calc.	0/50: George 1979	**Mar.**: ALKAL.: imidazole alkal. and Zn(II) complexes. POLYKET.: macrolides, long-chain β-amino alcohols	high	high
	Hexactin.	0/450: George 1979	**Mar.**: ISOPR.: Steroids: the same as in Demosp.	low	low

Table 9.I. Contribution to natural product diversity by land-dwelling vs marine organisms

Phylum or division[a]	Class or order	Nr of species on land/sea	Biogenetic classes of natural products[b]	Bioactivity level	Mol. complexity
	Cubozoa	0/16: George 1979	Mar.: PROTEINS : toxic	high	medium
<Cnidaria>	<Hydrozoa>	few/3,000: George 1979; May 1994	Mar.: ALKAL.: dihydropyrazine-like chromophores of bioluminescent proteins. POLYKET.: anthracenone polyphenols	Land: low; Mar.: medium	Land: low; Mar.: medium
	Anthozoa	0/6500: George 1979, May 1994	Mar.: ALKAL.: guanidine, imidazole, macrolide; di- and tri-terpene. ISOPR.: Sesquiterp.; Diterp.; Triterp. both exclusive and typical of terrestrial plants; Steroids: polyhydroxylated	high	high
<Moll.>	**Opisthobr.**		Mostly dietary metabolites from Cyanobact., seaweeds, sponges, cnidarians, bryozoans, and ascidians	high	high
	other Moll.	overall 5,000: Groom. 1992, May 1994	Mar.: ALKAL.: polycyclic aromatic and unique azaspiro. PEPT./PROTEINS: bioadhesive proteins; toxic pept. ISOPR.: Diterp.: cembrane, labdane; Triterp.: squalene-derived. POLYKET.: polypropionates. CARBOH.: various metabolites	Land: low; Mar.: high	Land: low; Mar.: high
<Bryoz.>		50/5,000 (mainly Gymnolaemata): George 1979	Mar.: ALKAL.: β-phenylethylamine; indole; pyrrole; pyridine; isoquinoline; β-carboline. ISOPR.: Monoterp.: typical of terrestrial plants; few higher terpenes. POLYKET.: macrolides	Land: low; Mar.: high	Land: low; Mar.: high
Brachiopoda		0/100: May 1994	Mar.: β-alkoxy glycerol ethers and glycerol enol ethers.	low	low
<Arthropoda>	Chelicerata	> 10^6/40,000: Groom. 1992	Land: PEPT.: toxic pept. from Arach. — Mar.: ALKAL.: tetrapyrrole and guanidine luciferins. PROTEINS: adhesive cement proteins.		
	Uniramia	2-10·10^6/few: Speight 1999	Land: ALKAL.: mostly dietary from plants. ISOPR.: Diterp.	high	high
Annelida		Oligochaeta 3,500; Hirudinea 500/Polychaeta 13,000: Groom. 1992, May 1994	Mar.: ALKAL.: amines, pteridines; pyrrole. POLYKET.: coupled brominated phenolics.	Land: low; Mar.: high	Land and Mar.: low

Table 9.I. Contribution to natural product diversity by land-dwelling vs marine organisms

Phylum or division[a]	Class or order	Nr of species on land/sea	Biogenetic classes of natural products[b]	Bioactivity level	Mol. complexity
Nemert.		800, mostly marine: May 1994	**Mar.:** ALKAL.: pyridine.	low	low
Echiura		0/130: George 1979	Anthraquinones, chlorins		
Echin.		0/6,000: George 1979	ALKAL.: benzyl tetrahydroisoquinoline. ISOPR.: Steroids: polyhydroxylated and saponins	medium	medium
Hemich.		0/100: George 1979	**Mar.:** ALKAL.: brominated indole and pyrrole; pyrazino bis-steroidal. POLYKET.: bromophenols.	high	medium
Tunicata	**Ascid.**	0/1,250: George 1979	**Mar.:** ALKAL.: brominated β-carboline, guanidine disulfide, imidazole, (bis)indole, indolocarbazole, pept., phenazine, physostigmine, (long-side-chain) piperidine, pyridoacridine, pyrrole, pyrroloiminoquinone, (hydro)quinoline, (hydro)pyrrole, polypyrrole, (pyrrolo)quinoline, quinolizidine, and pyrazino bis-steroidal alkal.. ISOPR.: few acyclic mono-, sesqui-, (labdane) diterp., and meroterpenoids. POLYKET.: macrolides.	high	high
Vertebr.	Mamm.	overall 4,629	scents, mostly on land (www.nmnh.si.edu).	medium	medium
	Chondrich.	0/256	ALKAL.: steroidal.	high	medium

[a]Groups of organisms exclusive of either the sea or land are indicated in boldface; those present in both ecosystems are enclosed within angle brackets; production by macroorganisms is not distinguished from that of their microbial symbionts.
[b]Optional variants to the basic class are indicated within parentheses. Marine origin is always indicated in boldface; terrestrial origin is only indicated when the taxonomic group comprises both marine and land dwelling organisms.

Table 9.II. Molecular skeletal complexity $(S)^a$ and size $(H)^a$ metrics for marine (in boldface) vs land-produced secondary metabolites

Metabolite	Organism	S	S/H
Maitotoxin/polyether polyket. (C 7.2.FA/PO1) 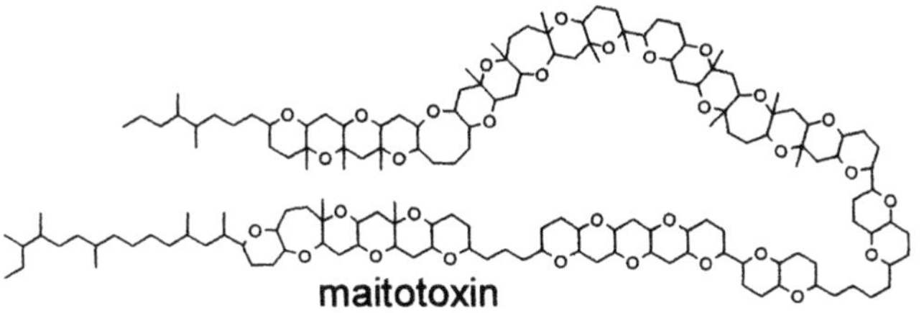	*Gambierdiscus toxicus* Adachi et Fukuio, 1979, Dinofl. from French Polynesia	296	0.6
maitotoxin			
Prymnesin-2/polyether polyket. (C 8.2.FA/PO) 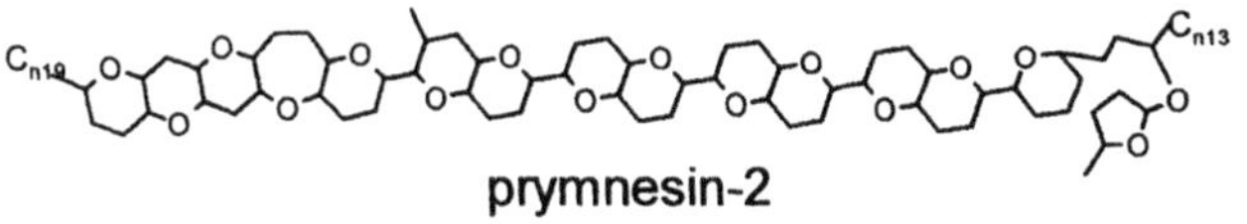	*Prymnesium parvum* Carter 1967, Prymnesiophyta, found in brackish waters	140	0.49
prymnesin-2			
Ciguatoxin/polyether polyket. (C 7.2FA/PO) 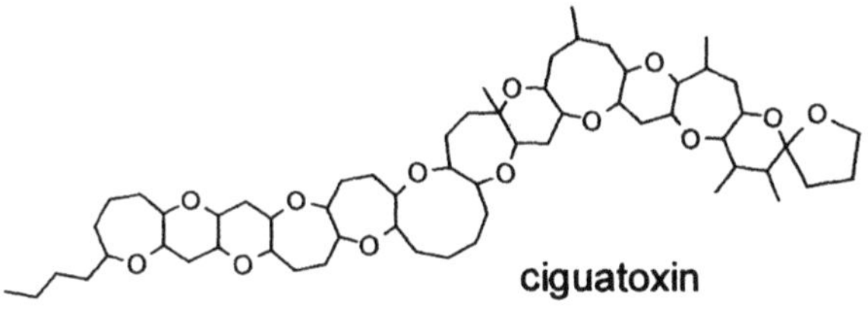	*Gymnothorax javanicus* (Blecker, 1859) (moray eel), Osteich., from French Polynesia	121	0.67
ciguatoxin			
Everninomicin/oligomeric carboh. deriv. (C 6.1.2.I/FA/PO/C) 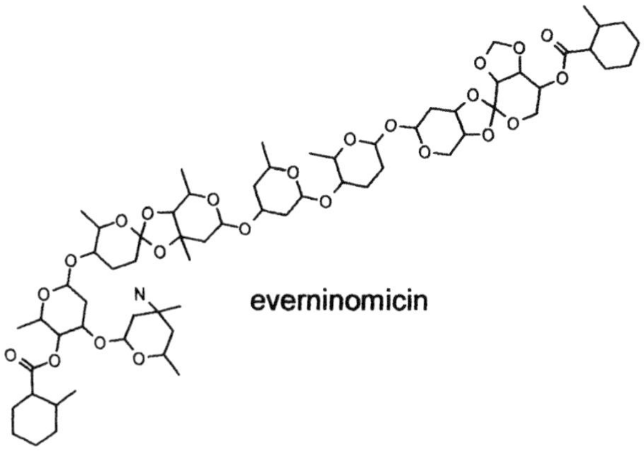	*Micromonospora carbonacea* var. *africana*, Bact., from Kenya	112	0.74
everninomicin			

Table 9.II. Molecular skeletal complexity $(S)^{a}$ and size $(H)^{a}$ metrics for marine (in boldface) vs land-produced secondary metabolites

Metabolite	Organism	S	S/H
Palytoxin/polyket. (C 7.2.FA/PO1) 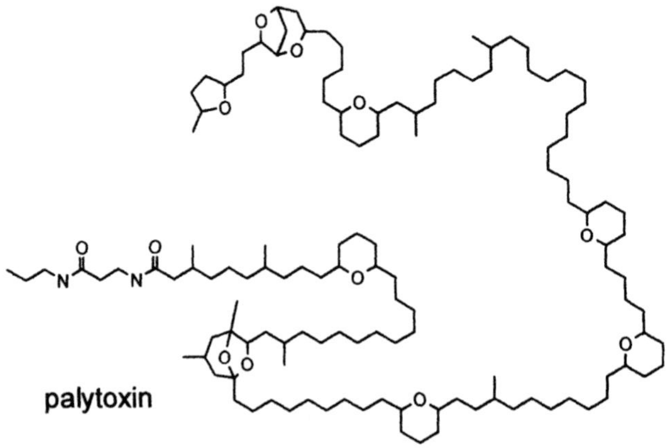	***Palythoa tuberculosa*** from Okin., ***P. toxica*** from Hawaii, and analogs from ***Ostreopsis siamensis*** Schmidt 1901, Dinofl., from Okin.	**108**	**0.28**
palytoxin			
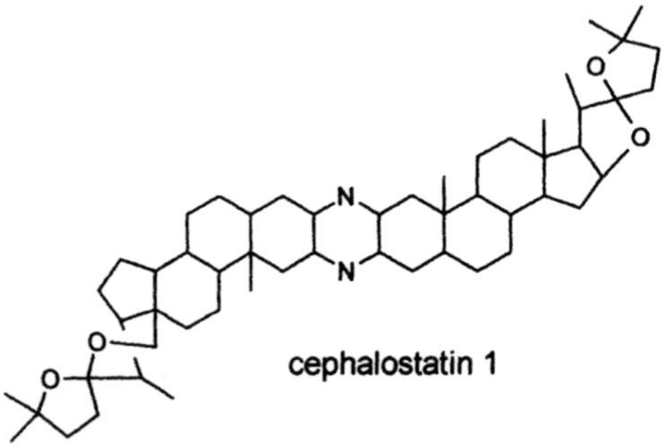 **Cephalostatin 1**/dimeric steroidal pyrazine alkal.	***Cephalodiscus gilchristi*** Ridewood, Hemich., from Indian Ocean off SE Africa	**102**	**0.65**
cephalostatin 1			
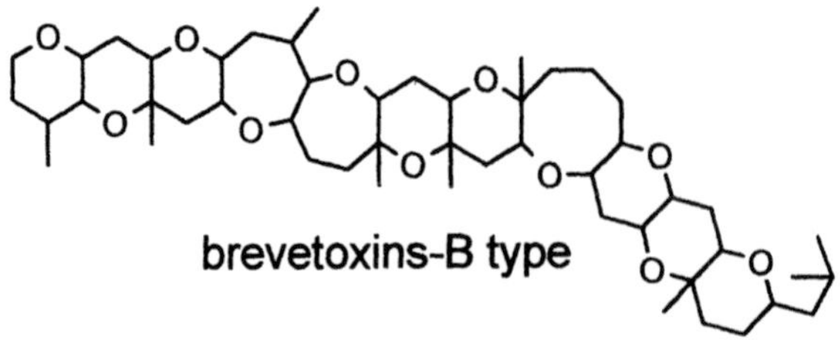 **Brevetoxin-B class**/polyether polyket. (C 8.2.FA/PO)	Gymnodiniales, Dinofl., from Florida and Moll. from NZ	101	0.67
brevetoxins-B type			

Table 9.II. Molecular skeletal complexity $(S)^a$ and size $(H)^a$ metrics for marine (in boldface) vs land-produced secondary metabolites

Metabolite	Organism	S	S/H
Kauluamine/manzamine alkal. dimer (see this Table at S=48 for monomer) 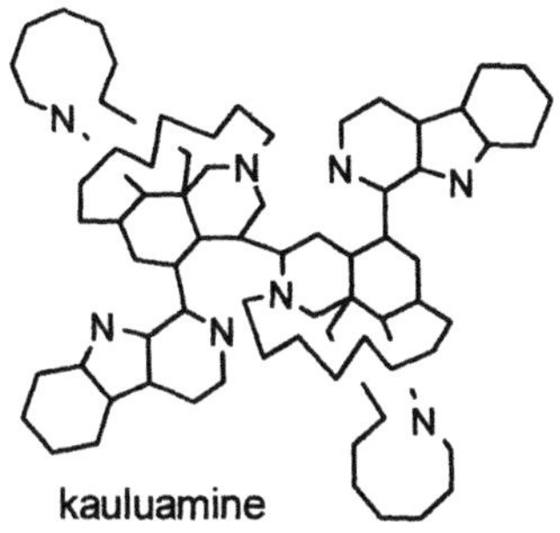	*Prianos* sp., Poecil., Porif., from Indonesia: Ohtani 1995	100	0.46

kauluamine

Vinblastine/dimeric form of vindoline-like alkal. 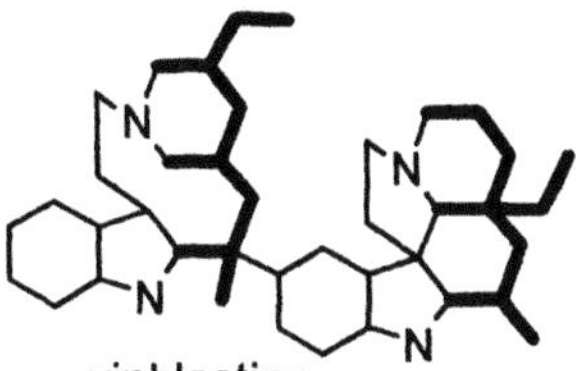	*Vinca rosea* L. (Madagascar periwinkle), Apocyanaceae, Ang.	68	0.57

vinblastine

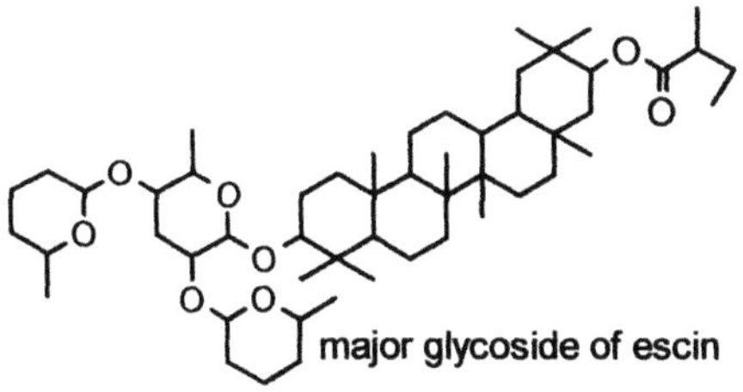 Escin/triterp. glycoside	*Aesculus hippocastanum* L. (horse chestnut), Hippocastanaceae, Ang., from Eurasia and temperate N America	66	0.48

major glycoside of escin

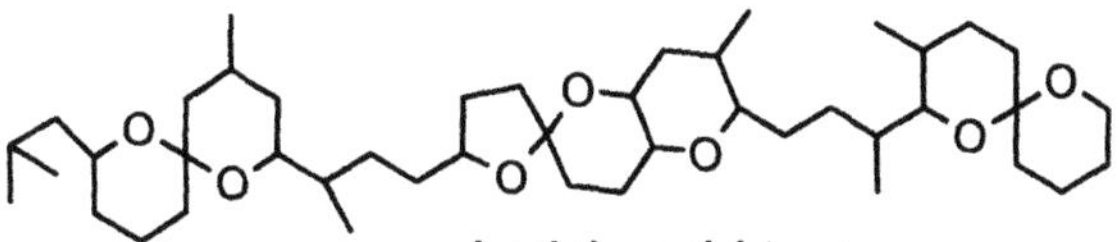 **Okadaic acid class**/polyether polyket. (C 8.2.FA/PO)	Prorocentrales (Dinofl.) and Demosp. hosts	61	0.47

okadaic-acid type

Table 9.II. Molecular skeletal complexity (S)[a] and size (H)[a] metrics for marine (in boldface) vs land-produced secondary metabolites

Metabolite	Organism	S	S/H
Manzamine/macrocyclic alkal. (C 7.2.A) 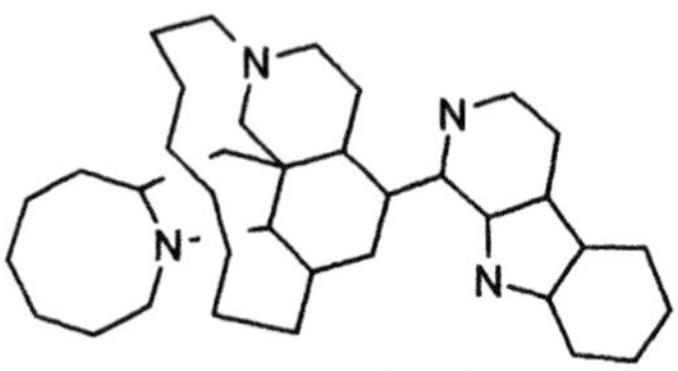	***Xestospongia*** sp. (Nepheliosp.), ***Haliclona*** sp. and ***Pellina*** sp. (Haplosclerida), Porif., from Okin., and *Pachypellina* sp. from Indonesia	48	0.44
Rapamycin/macrolide (C 6.2.FA/PO/C) 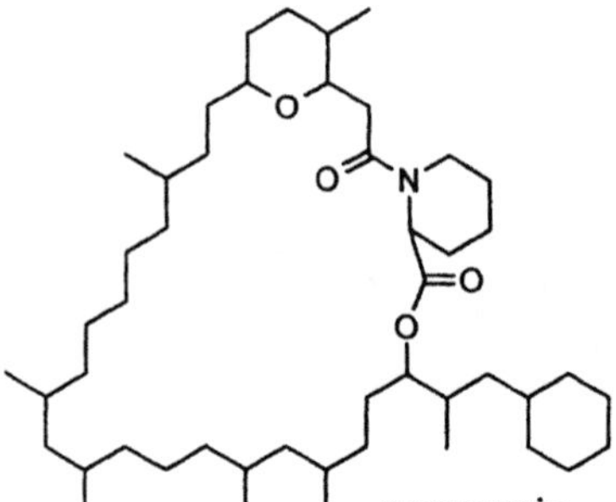	*Streptomyces hygroscopicus*, Actinom., Bact. from Easter Is.	43	0.29
Vindoline/secologanin indole alkal. (C 6.1.3.A) 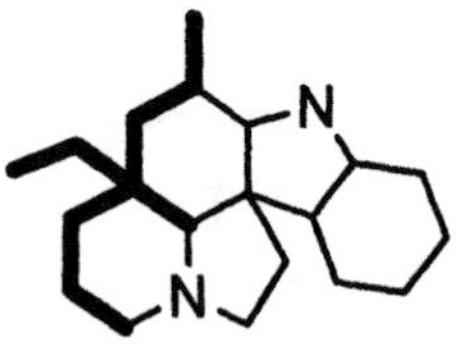	*Vinca rosea* L., Apocyanaceae, Ang. (Madagascar periwinkle)	36	0.6

manzamine A

rapamycin

vindoline

[a] S = complexity metric; H = bond count for molecular size (Whitlock 1998).

Table 9.III. Metabolites from deep-water marine organisms

Depth range	Species (specific area)	Metabolites[a]/bioactivity or technological use
Indo-Pacific		
80-120m	*Microscleroderma* sp., Lithistida, Porif. (Palau)	microsclerodermins (cyclic pept.), C 7.2.P3, also from New Caledonia/antifungal: Qureshi 2000

Table 9.III. Metabolites from deep-water marine organisms

Depth range	Species (specific area)	Metabolites[a]/bioactivity or technological use
200-350m	*Acanthogorgia* sp., Gorgonacea (Hawaii)	papakusterol (cyclopropyl bearing cholestane): Bonini 1983
	Agelas dendromorpha Lévi, Agelasida, Porif. (NC)	agelastatins (C_{11} alkal.), C7.2.A/potently cytotoxic on human tumor cells: D'Ambrosio 1994 (also from *Cymbastela* sp., Axinel., Porif. from shallow waters in W Australia: Hong 1998)
	Corallium sp., Gorgonacea, Cnid. (Hawaii)	coraxeniolides (xenicane diterp.): Schwartz 1981
	Dragmacidon sp., Axinel., Porif. (NC)	nortopsentin D (bis-indole alkal.), C 7.2.A/the permethylated derivative is strongly cytotoxic on human tumor cells: Mancini 1996
	Gerardia sp., Zoantharia, Cnid. (Hawaii)	pseudozoanthoxanthins (guanidine alkal.), C 8.2.A, also from shallow Mediterranean waters/DNA intercalation: Schwartz, 1979
	Microscleroderma sp., Lithistida, Porif. (NC)	microsclerodermins (cyclic pept.), C 7.2.P3, also from Palau/antifungal: Qureshi 2000
	Orina sp., Haplosclerida, Porif (NC)	gelliusine A and B (tris-indole alkal.)/strong affinity for the somatostatin and neuropept. Y receptors: Bifulco 1995A
	Phloeodictyon sp., Nepheliosp., Porif. (NC)	phloeodictyns (guanidine alkal.), C 7.2.A/cytotoxic on human tumor cells: Kourany-Lefoll 1994
	Placogorgia sp., Gorgonacea, Cnid. (Hawaii)	guaianolide yellow pigment (guaiane sesquiterp.): Li 1984
	Stylotella sp., Halichon., Porif. (NC)	stylotelline (isocyanosesquiterp.): Païs 1987
	Ritterella rete, Ascid. (NC)	hydroxylated furanosesquiterp.: Lenis 1998
	Zyzza massalis (Dendy), Poecil., Porif. (NC)	zyzzin (indole alkal.), C 7.2.A/spontaneously changing into a thermochromic pigment: Mancini 1994A
400-700m	*Cladocroce incurvata*, Haplosclerida, Porif. (NC)	cladocrocin and cladocrocic acid (acyclic polyket.): D'Auria 1993
	Corallistes fulvodesmus Lévi & Lévi, Lithistida, Porif. (NC)	corallistine (guanidine alk.), C 7.2.A: Debitus 1989
	Corallistes sp., Lithistida, Porif. (NC)	corallistin A-E (new free porphyrins)/cytotoxic on human tumor cells: D'Ambrosio 1993
	Corallistes undulatus, Lithistida, Porif. (NC)	new pteridine (alkal.).: Guerriero 1993

Table 9.III. Metabolites from deep-water marine organisms

Depth range	Species (specific area)	Metabolites[a]/bioactivity or technological use
	Deltocyathus magnificus Moseley, 1876, Scleractinia, Cnid. (NC)	new cholic-acid-like sterones: Guerriero 1996
	Erylus sp., Choristida, Porif. (NC)	erylosides C-D (lanostane-derived triterp. oligogalactosides): D'Auria 1992
	Gymnocrinus richeri, Crinoidea, Echin. (NC)	gymnochrome A-D (brominated phenanthroperylenequinone); also from terrestrial plants (not brominated)/antiviral agents (dengue): de Riccardis 1991
	Ircinia sp., Dictyocer., Porif. (NC)	polyprenyl hydroquinone alcohol sulfate and C_{31} furanoterpene/binding properties to the neuropept. Y receptor: Bifulco 1995B
	Mediaster murrayi, Aster., Echin. (Philippines)	new steroid glycosides/inhibition of cell division of fertilized sea urchin eggs: Kicha 1999
	Neosiphonia supertes Sollas, Lithistida, Porif. (NC)	superstolides and sphinxolides (macrolides), C 7.2.FA/PO2/cytotoxic on human tumor cells: D'Auria 1994; Carbonelli 1999 (sphinxolides also from shallow water Nudibr. from Hawaii: Guella 1989C)
	Pleroma menoui, Lithistida, Porif. (NC)	ethyl 6-bromo-3-indolcarboxylate and 3-hydroxyacetal-6-bromoindole (indole alkal.): Guella 1989B
	Poecillastra laminaris Van Soest & Stentoft, Choristida, Porif. (Philippines)	annasterol sulfate/inhibitor of glucanase: Makarieva 1995B
	Reidispongia coerulea Lévi & Lévi, Lithistida, Porif. (NC)	sphinxolide-like macrolides/cytotoxic on human tumor cells, C 7.2.FA/PO2: Carbonelli 1999
	Stelletta sp., Choristida, Porif. (NC)	new stigmastane sterols and sterones: Guerriero 1991
1,000-5,000m	*Alteromonas* sp., Bact. (Mariana Through)	*N*-acetylglucosamine-6-phosphate deacetylase (protein enzyme): Yamano 2000
	Streptomyces sp., Actinom., Bact. (Hawaii)	γ-indomycinone (pluramycin-like anthraquinone)/antibiotic: Schumacher 1995
	Styracaster caroli, Aster., Echin. (NC)	new polyhydroxy sterols and sterol sulfates: Iorizzi 1994
	Undeterm. Bact. (Hawaii)	caprolactin A and B (caprolactams)/cytotoxic on human tumor cells and antiviral: Davidson 1993

Table 9.III. Metabolites from deep-water marine organisms

Depth range	Species (specific area)	Metabolites[a]/bioactivity or technological use
	Caribbean	
80-120m	*Aplysina fistularis fulva*, Verongida, Porif.	aplysillin A(disulfate ester of 1,4-diphenyl 1,3-butadiene)/thrombin receptor antagonist: Gulavita 1995
	Batzella sp., Poecil., Porif.	discorhabdin P (pyrroloiminoquinone alkal.)/inhibitor of phosphatase activity of calcineurin; secobatzelline A (pyrroloquinoline alkal.)/inhibitor of peptidase activity of cysteine proteases, caspases: Gunasekera 1999
	Dercitus sp., Choristida, Porif.	dercitin (pyrido(4,3,2-*mn)*acridine alkal.)/disrupter of DNA and RNA synthesis: Ciufolini 1995
	Didiscus flavus, Halichon., Porif. (also local shallow waters)	curcudiol and curcuphenol (sesquiterp.)/cytotoxic on human tumor cells: Wright 1987
	Henricia downeyae, Aster., Echin.	new steroidal glycosides: Palagiano 1996
	Ircinia sp., Dictyocer., Porif.	sulfircin (sesterterpene sulfate with guanidinium counterion)/antifungal: Wright 1989
	Plakinastrella onkodes, Homoscl., Porif.	new peroxylactones/cytotoxic on human tumor cells: Horton 1994
	Spongosorites ruetzleri Van Soest & Stentoft, 1988, Halichon., Porif.	topsentin and related (bisindolyl)imidazole alkal.: Tsujii 1988
	Spongosorites sp., Halichon., Porif.	dragmacidin d (bis-indole alkal.)/cytotoxic on human tumor cells, antiviral, and antifungal: Wright 1992
150-300m	*Corticium* sp., Homoscl., Porif.	meridine (pyrido(4,3,2-*mn*)acridine alkal.)/antifungal (known from shallow-water *Amphicarpa meridiana*, Ascid., from S. Australia): McCarthy 1992
	Cribrochalina vasculum, Haplosclerida, Porif.	vasculyne (polyacetylene alcohol) (similar metabolites from the same nominal species from shallow-waters)/cytotoxic on human tumor cells: Dai 1996
	Discodermia polydiscus, Lithistida, Porif.	discodermindole (brominated (aminoimidazolinyl)indole alkal.)/cytotoxic on human tumor cells: Sun 1991
	Discodermia sp., Lithistida, Porif.	polydiscamide A (depsipept.)/cytotoxic on human tumor cells: Gulavita 1992
	Discodermia sp., Lithistida, Porif.	discobahamin A and B (pept.)/antifungal: Gunasekera 1994A

Table 9.III. Metabolites from deep-water marine organisms

Depth range	Species (specific area)	Metabolites[a]/bioactivity or technological use
	Dragmacidon sp., Axinel., Porif.	dragmacidin (bis-indole alkal.), C 8.2.A/cytotoxic on human tumor cells: Kohmoto 1988 (also from *Spongosorites* sp., Halichon., Porif. from deep-waters in S. Australia: Capon 1998)
	Scleritoderma cf. *paccardi*, Spirophorida, Porif.	24(R)-methyl-5α-cholest-7-enyl 3β-methoxymethyl ether (sterol methoxymethyl ether)/cytotoxic on human tumor cells: Gunasekera 1996B
	Spongia sp., Disctyoceratida, Porif.	isospongiadiol (diterp.)/cytotoxic on human tumor cells: Kohmoto 1987
	Verongula gigantea Hyatt, Verongida, Porif.	verongamine (bromotyrosine metabolite)/histamine H_3 receptor antagonist: Mierzwa 1994
330-400m	*Epipolasis reiswigi* Topsent, Choristida, Porif.	reiswigin A and B and isoreiswigin (new diterp.)/antiviral: Kashman 1991
	Poecillastra sollasi, Choristida, Porif.	sollasin a-f (sesquiterp. alkal.)/antifungal: Killday 1993
	Pseudosuberites hyalinus (Ridley and Dendy), Hadromerida, Porif.	new bromoindole alkal.: Rasmussen 1993 (a widespread class of metabolites in the sea)
	Strongylophora hartmani, Nepheliosp., Porif.	strongylin A (sesquiterp. hydroquinone)/cytotoxic on human tumor cells: Wright 1991

Mediterranean

Depth range	Species (specific area)	Metabolites[a]/bioactivity or technological use
80-120m	*Funiculina quadrangularis* pallas, 1776, Pennatulacea, Cnid.	funiculides (briarane diterp.): Guerriero 1995
	Scaphander lignarius, Cephalaspidea, Opisthobr., Moll.	lignarenone A and B (ω-phenyl conjugated trienones): Cimino 1989
200-300m	*Gryphus vitreous* (Born, 1778), Brachiopoda	new glycerol enol ethers: D'Ambrosio 1996C
	Madrepora oculata Linnaeus, 1758, Scleractinia, Cnid.	10-hydroxy substituted polyunsaturated C_{20} and C_{22} fatty acids: Mancini 1999B
500-700	*Hamacantha* sp., Poecil., Porif.	hamacanthin A and B (bis-indole alkal)/antifungal: Gunasekera 1994B
	Lophelia pertusa, Scleractinia, Cnid.	10-hydroxy substituted polyunsaturated C_{20} and C_{22} fatty acids: Mancini 1999B
1,000-4,000m	*Atolla wyvillei*, and other Schyphozoa, Cnid	protoporphyrin (free porphyrin): Bonnett 1979

North Pacific

Depth range	Species (specific area)	Metabolites[a]/bioactivity or technological use
80-120m	*Bacillus* sp., Bact.	halobacillin (iturin-class pept.)/cytotoxic on human tumor cells: Trischman 1994B

Table 9.III. Metabolites from deep-water marine organisms

Depth range	Species (specific area)	Metabolites[a]/bioactivity or technological use
	Linneus torquatus Coe, Nemert.	$\Delta^{5,22}$, $\Delta^{5,24(28)}$, and Δ^5 sterols: Ponomarenko 1995
	Polycitor sp., Ascid., Tunicata	varacin (aromatic polysulfide)/antimicrobial: Makarieva 1995A
	Pteraster tesselatus, Aster., Echin.	ophiuroid-class sulfated steroids: Levina 1998
	Zopfiella marina Furuya et Udagawa, Eumyc.	zofimarin (diterp. glycoside)/antifungal: Takeshi 1987
200-700m	*Aspergillus fumigatus*, Eumyc.	tryprostatin A and B, as well as spirotryprostatin A and B (prenylated diketopiperazine pept.), C 8.3.P/inhibitors of cell cycle progression of tsFT210 cells at the G2/M phase: Schkeryantz 1999
	Ophiopholis aculeata Ophiura sp., and *Stegophiura brachiactis*, Ophiuroidea, Echin.	new sulfated steroids: Fedorov 1994
1,000-4,000m	*Alteromonas haloplanktis*, Bact.	bisucaberin (siderophore)/human tumor cells are made susceptible to cytolysis: Kameyama 1987
	Bacillus sp., Bact.	3-amino-3-deoxy-D-glucose/antibiotic (produced also by terrestrial Bact.): Fusetani 1987
	Undeterm. Bact.	macrolactins (macrolides), C 7.4.P/I/PO/antiviral: Rychnovsky 1992

South Pacific

400-500m	*Pyrosoma* sp., Thaliacea, Tunicata	sterols: Ballantine 1977

Zealandic

80-200m	*Aciculites pulchra*, Lithistida, Porif.	pulchrasterol (C_{30} sterol, first example of double alkylation at C-26): Crist 1983
	Higginsia sp., Axinel., Porif.	new tricyclic diterp. and daucadiene sesquiterp.: Cassidy 1993
	Hippospongia sp., Dictyocer., Porif.	hippospongins A-F (furanosesterterpenes): Rochfort 1996
	Lamellomorpha strongylata, Choristida, Porif.	calyculins, calyculinamides, and swinholide: antitumor: Dumdei 1997
	Latrunculia brevis Ridley and Dendy, Hadromerida, Porif.	discorhabdins (pyrroloiminoquinones), C8.3.A/antimicrobial and cytotoxic on human tumor cells: Perry 1988
	Phorbas sp., Poecil., Porif.	phorbasin A, new skeleton diterp.: Vuong 2000
	Sigmosceptrella sp., Hadromerida, Porif.	norditerp. and norsesterterp.: Ovenden 1999

Table 9.III. Metabolites from deep-water marine organisms

Depth range	Species (specific area)	Metabolites[a]/bioactivity or technological use
	Spongia sp., Dictyocer., Porif.	cometins A-C (furanosesterterp.): Urban 1992
	Spongosorites sp. Halichon., Porif.	dragmacidin E (bis-indole alkal.)/inhibitor of serine-threonine protein phosphatases: Capon 1998
200-400m	*Arenochalina mirabilis* (Lendenfeld 1887), Poecil., Porif.	mirabilins A-F (ptilocaulin-like guanidine alkal.): Barrow 1996
	Luffariella geometrica Kirkpatrick, Dictyocer., Porif.	luffarins A-Z (sesterterp. and bis-norsesterterp.), C 7.10.A/PO: Butler 1992B
	Sarcotragus spinulosus, Dictyocer., Porif.	new chromenol sulfates: Stonik 1992
1000m	*Somniosus pacificus*, Chondrich., Vertebr., and other sharks	squalene and diacylglyceryl ethers: Bakes 1995

Antarctic and pre-Antarctic

Depth range	Species (specific area)	Metabolites[a]/bioactivity or technological use
80-120m	*Aplidium meridianum* Sluiter, 1906, Ascid., Tunicata	meridianins A-E (2-aminopyrimidine bearing indole alkal.)/cytotoxic on human tumor cells: Franco 1998
	Undeterm. Demosp. (likely in the genus *Latrunculia*)	epinardins A-C (discorhabdin-like pyrroloiminoquinone alkal.): D'Ambrosio 1996A
300-500m	*Leiopathes glaberrima*, Antipatharia, Cnid.	new hydroxydocosapentaenoic acid: Guerriero 1988
	Paragorgia arborea, Gorgonacea, Cnid.	arboxeniolide-1 (xenicane diterp.): D'Ambrosio 1984
	Psammopemma sp., Poecil., Porif.	psammopemmins A-C (indole alkal. bearing a 2-bromo-4-aminopyrimidine moiety), C 7.12.A/I: Butler 1992A
	Tedania charcoti Topsent, 1907, Poecil., Porif.	high cadmium and zinc level/antibacterial: Capon 1993
	Undeterm, Hexactin., Porif.	ergosta-4,24(28)-dien-9-one (present also in shallow water Demosp.): Guella 1988

[a] Underlining denotes metabolite(s) found also in shallow-water organisms.

Chapter 10. Life under extreme conditions

Temperatures low enough to prevent any form of life do not occur on Earth. The coldest areas, the Antarctic and Arctic, were considered in Chapters 7.12 and 7.11. High pressures in the oceans also do not prevent wide forms of life, rich of unusual secondary metabolites (Chapter 9).

Life is more susceptible to high temperature conditions. No multicellular plant or animal has ever been found to survive above 50 °C and no protist is known that can tolerate long-term exposure to above 60 °C. Conditions of high salinity are also very selective, allowing the growth of only extreme halophile prokaryotes, both Archaea and Bacteria (Kates 1993).

Hyperthermophilic archaeans and bacteria (Bhattacharya 1999) have colonized late upper volcanic areas (De Rosa 1977), non-volcanic oil reservoirs (Stetter 1993), and hydrothermal vents. The latter are special ecosystems found at mid-ocean ridges, where seawater penetrates into the fractures and becomes heated by the hot mantel material (van Dover 2000). The Galápagos hydrothermal vents were the first to be explored in 1977 with a submersible, revealing a richness of invertebrates and vertebrates, particularly mussels, tube worms, arthropods, and fishes. Life is sustained at hydrothermal vents by bacterial chemosynthesis in contrast with the photosynthetic activity of illuminated surface areas. The methanogenic archaeans are found not only in deep-sea hydrothermal vents (genera *Methanococcus* and *Methanopyrus*) but also in terrestrial fumaroles (genus *Methanothermus*) (Huber 1989).

Cold seeps are also areas of high hydrostatic pressure, rich in mollusks, particularly mytilids, which thrive on hydrogen sulfide from bacterial decomposition of sunken wood or whale bones (Distel 2000).

Studies into the metabolism of extremophiles have disclosed unusual adaptive metabolites. The extreme conditions have selected heat-stable enzymes and cell walls for the archaeans (Flamm 1994). Such enzymes are exploited for PCR procedures (Sigma 2000). The cell walls are stabilized by *sn*-2,3-*O*-diprenylated glycerols. In contrast, the eukaryotic cell walls incorporate the antipodal *sn*-1,2-*O*-diacyl glycerols, which do not confer thermal stability (Kakinuma 1988). Normally, these prenylated glycerols of the archaeans are fully saturated; a rare example of an unsaturated analogue, 2,3-di-*O*-dihydro-14,15-geranylgeranyl glycerol, was found in the cell walls of a *Thermococcus* species from deep-sea hydrothermal vents; it was postulated as an intermediate in the biosynthesis of saturated *sn*-2,3-*O*-diprenylated glycerols (Gonthier 2001).

Bacteria of the hydrothermal vents produce unique exopolysaccharides that have found use in food technology and wastewater treatment (Rougeaux 1996).

In spite of a rich life, no signaling or defensive secondary metabolite has ever been reported from organisms thriving on the hydrothermal vents. Lack of competition in these areas, where the hostile environment provides protection from invaders, has not stimulated the formation of defensive metabolites. Defensive heat-stable enzymes and cell walls were raised against the hostile environment.

Chart 10 Skeleton classes of metabolites from organisms thriving under extreme conditions (Smax=52, av=32; S/Hmax=0.2, av=0.15).

Isoprenyl glyceryl ethers: archaeol-like (archaeol C_{20}-C_{20}), comprising C_{20}-C_{25} and C_{25}-C_{25} homologues from extreme halophilic, thermoacidophilic, and methanogenic Archaea: Kates 1993; caldarchaeol-like (caldarchaeol monocyclic C_{40}-C_{40}), comprising bicyclic structures from thermoacidophilic and methanogenic Archaea: Kates 1993. (Archaea in boldface are not meant to be exclusively marine).

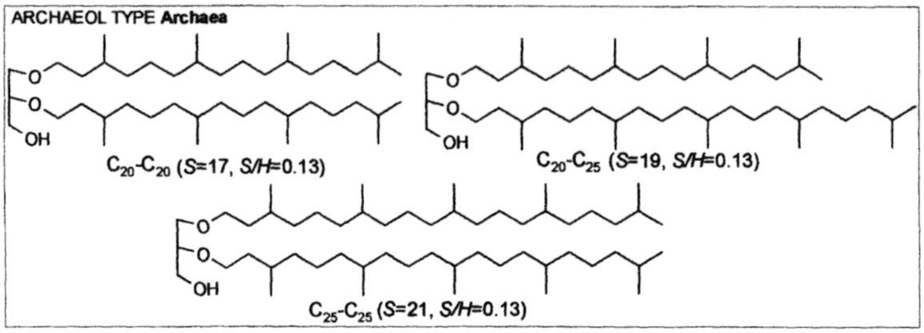

ARCHAEOL TYPE **Archaea**

C_{20}-C_{20} (S=17, S/H=0.13)

C_{20}-C_{25} (S=19, S/H=0.13)

C_{25}-C_{25} (S=21, S/H=0.13)

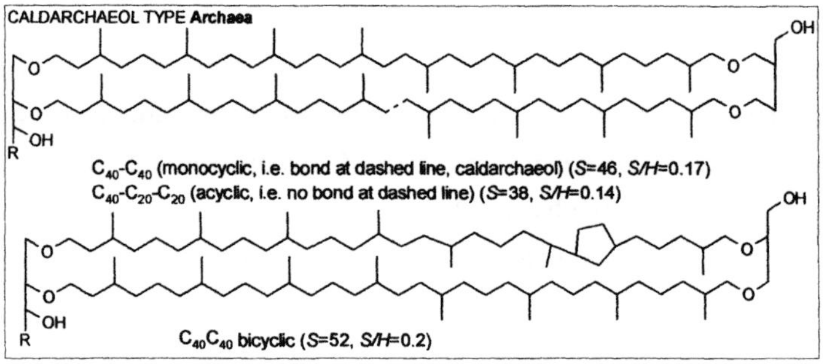

CALDARCHAEOL TYPE **Archaea**

C_{40}-C_{40} (monocyclic, i.e. bond at dashed line, caldarchaeol) (S=46, S/H=0.17)
C_{40}-C_{20}-C_{20} (acyclic, i.e. no bond at dashed line) (S=38, S/H=0.14)

C_{40}-C_{40} bicyclic (S=52, S/H=0.2)

Chapter 11. Graphic analysis of the skeletal diversity and complexity of natural products

The rationale underlying the representation of the variety of natural products through their skeletons has been outlined early in Part III. Then, the variety of natural products has been illustrated in charts

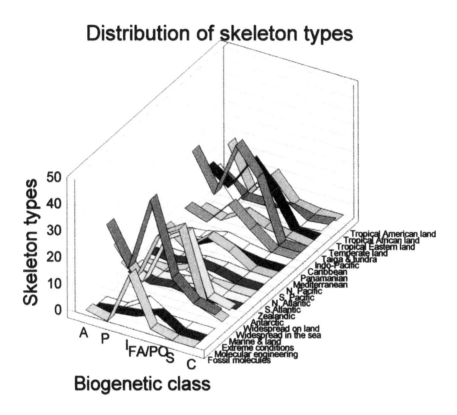

Figure 11.1. Distribution of molecular skeleton types, in the various biogenetic classes, per ecosystem
Z axis: skeletal variety (number of different skeleton types).
X axis: biogenetic class (A = alkaloids; P = amino acids, peptides, peptides, polypeptides and proteins; FA/PO = fatty acids and polyketides; S = shikimates; C = carbohydrates).
Y axis: ecosystem.

for the various ecosystems (Chapters 6-10), paying attention also to the taxonomy of the living organisms. These data are represented graphically in Fig. 11.1, allowing us a quick comparison of the various classes of metabolites and ecosystem types.

It is seen from Fig. 11.1 that, beyond any bias from the personal choice of the examples, the isoprenoids show the largest skeletal variety in all ecosystems. The absolute maximum is observed for the isoprenoids from the Indo-Pacific (dark green ribbon, sixth from the right), which represents a triumph of the natural product diversity of Indo-Pacific coral reefs. To its left, the pale violet ribbon for the Caribbeans shows a similar trend, although with lower values attributable to the more restricted area with less extensive and varied coral reefs.

Many terpenoids from land and the sea have the same skeleton. Although the skeletal distribution per taxon is not represented in Fig 11.1, a perusal of Charts 1-10 shows that the greatest skeletal similarities occur for the terpenoids from plants and cnidarians (Pietra 1995), which are comprised in the violet ribbon for "marine and land" (fourth from the left).

A high variety of isoprenoids is also observed for fossil molecules (Fig. 11.1, pink ribbon, first to the left). This suggests that the first cnidarians at Cambrian times have extensively exploited what the terpene chemistry allows (Pietra 1995). Although overshadowed by diagenetic transformations, and not comprising volatiles for obvious reasons, Chart 16.1 is suggestive of the structure of these early isoprenoids.

In measuring natural product diversity against biodiversity, it was decided early in Chapter 5 to take into account not only the variety of natural products but also their complexity. It was chosen to represent it through the structural complexity metric S and size metric H (Whitlock 1998). In the introduction to Part III the limitations and shortcomings in applying such metrics to the molecular skeletons were examined. Since the ratio S/H is largely independent from the size of the molecule, it may be called specific molecular (or skeletal) complexity. Values of S and S/H are shown for the molecular skeletons at the Charts in Chapters 6-10. The trend can be better appreciated from the graphic boxes in Fig. 11.2A-B for the peak values, and Fig. 11.2C-D for the average values. Fig. 11.2A shows that the Indo-Pacific biogeographic marine area is also peculiar for furnishing, with toxic polyethers from dinoflagellates, the most complex molecular skeletons (highest S values, dark green ribbon and Chart 7.2.FA/PO1). Low S/H values contrast these exceptionally high S values, reflecting the quasi-repetitiveness of structural elements of these polyethers (Chart 7.2.FA/PO1 and green ribbon in Fig. 11.2C). A different view is afforded by the average S values (Fig. 11.2B): the highest peaks are observed for the peptides, reflecting their homogeneously high molecular complexity.

High values of the specific skeletal complexity S/H are observed for isoprenoids from temperate land and alkaloids from both the eastern tropical land and the Indo-Pacific biogeographic marine area, or present both in the sea and on land (Fig. 11.2C). Comparing such a huge area as the temperate land with more restricted biomes is difficult. However, it emerges clearly that the average specific skeletal complexity (S/H)av (Fig. 11.2D) is high for the alkaloids from any ecosystem. Fossil molecules are the only exceptions, possibly because of the removal of nitrogen in diagenetic processes.

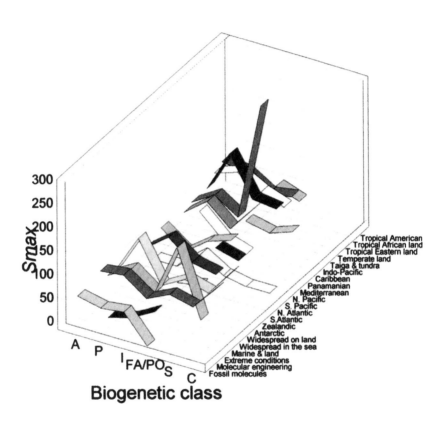

Highest skeletal complexity

Figure 11.2.A. Highest molecular skeletal complexity (Smax), in the various biogenetic classes, per ecosystem (meaning of axes as in Fig. 11.1)

Average skeletal complexity

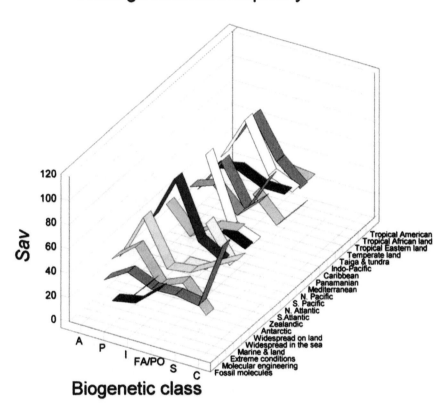

Figure 11.2B. Average molecular skeletal complexity (*Sav*), in the various biogenetic classes, per ecosystem (meaning of axes as in Fig. 11.1)

Highest specific skeletal complexity

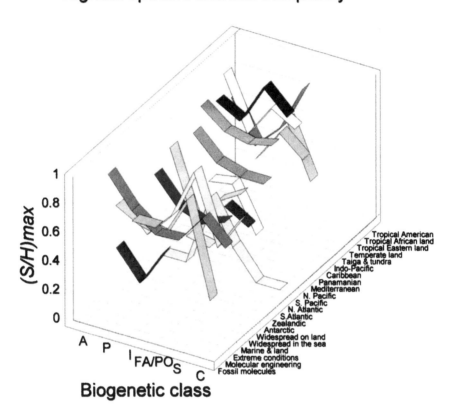

Figure 11.2C. Highest specific molecular skeletal complexity (*S/H*)max, in the various biogenetic classes, per ecosystem (meaning of axes as in Fig. 11.1)

Average specific skeletal complexity

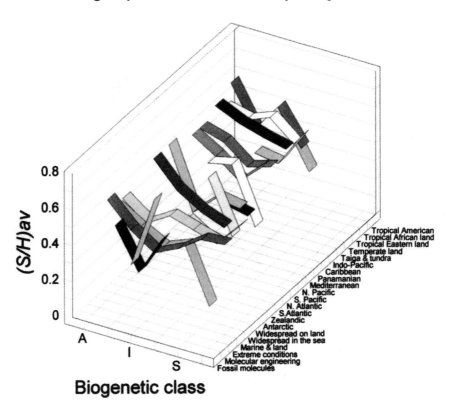

Figure 11.2D. Average specific molecular complexity (*S/H*)av, in the various biogenetic classes, per ecosystem (meaning of axes as in Fig. 11.1)

Part IV. Natural product diversity at functional level

And the fish that was in the river died;
and the river stank, and the Egyptians
could not drink of the water of the river;
and there was blood throughout all the
land of Egypt.
Exodus 7.21

Chapter 12. Signaling, defensive, and environmentally relevant metabolites

Physical means of information are found with all living organisms. Chemical signaling is typical of the lower organisms and, through the vomeronasal neurons, the animals (Leinders-Zufall 2000). The latter rely also on sound and colors as information systems. Sound is flatly repetitive, such as with birds, to avoid breeding between closely related species. With man (MacNeilage 2000), and perhaps also apes and dolphins (Nowak 2000), sound is syntactic. Man seems to have also inherited a rudimentary chemical information system, that is, sexual attraction exerted by pheromones released from the armpits (Stern 1998). However, the claim remains to be fully substantiated.

In any event, bioactive metabolites should be viewed in a wider perspective than merely the toxins and the defensive compounds raised in response to injuries or infections in organisms that lack an inducible immune system. Bioactive compounds have attracted much attention, first by the organic chemist and then by the ecological biochemist. The challenge continues apace. The organic chemist is faced with the ever increasing barrier to surmount in total synthesis, in a struggle to outperform nature by providing in substantial amounts rare metabolites and model compounds for supramolecular chemistry. The ecological biochemist is confronted with the ever increasing spatial and temporal scales to look at the events (Chapter 1.4).

These remain specialized areas of science, largely confined to de Solla Price's "invisible colleges" (de Solla Price 1963). On the other hand, the specialistic jargon of ecology makes things less intelligible. We need not be fully involved in such affairs. It suffices appreciating that semiochemicals (as the signaling natural products are called) are classified into pheromones (intraspecific agents) and allelochemicals (interspecific agents).

Interdisciplinary studies about signaling agents are of long date for terrestrial species (Harborne 1999A), and have yielded many practical applications, particularly for the control of pests and microbial infections. Marine signaling agents have been approached recently, but in a period of powerful analytical means, such as high performance liquid chromatography and soft mass and NMR spectrometry. Therefore, structures and functions of ecologically relevant marine metabolites have been deciphered much faster than in the past and on a much smaller amount of material (Paul 1992).

These powerful means have rapidly exhausted the most common sources of new metabolites, making the discovery of entirely new kinds of metabolites a rare event. The challenge has not lessened, however, since metabolites produced in trace amounts by rare species remain largely to be discovered. The function of the metabolites is also worth studying, not only along the organismic approach but, intersecting with the study of intracellular communication agents, also as to the fundamental intracellular events.

Graphs in Chapter 11 have shown that, at skeletal level, the diversity of natural products is highest for the Indo-Pacific area. Although a similar semiquantitative analysis for the actual metabolites is difficult to make, the trend is expected to be the same; because of the highly interacting forms of life, semiochemicals in coral reefs should prove an even more common system of communication than in the tropical rain forest.

12.1 Recruiting, alarming, and growth stimulating agents

Signaling agents (recruiting, alarming, and growth stimulating agents) are long known for insects and plants on land. Many arose from plant-insect coevolution, serving the intercommunication. Although semiochemicals released to the air are simple volatile molecules, finely tuned molecular mechanisms are involved in avoiding mis-communication.

Pheromones are classified in at least ten different categories. The most complex category involves enantiomers at a specific composition (Mori 2000), implying multiple receptors. Other complex networks of biological events involve terpenoids released by plants in response to grazing by herbivores. These metabolites can be used to attract parasitic wasps for pest control (Tumlinson 1993), or, like the strigolactones (Table 12.1), as useful germination stimulants of root parasites of plants in the families Scrophulariaceae and Orobanchaceae (Welzel 1999; Ryan 2001).

Marine lipophilic semiochemicals are well known, in particular the C_{11} hydrocarbon pheromones of brown algae (Oldham 1996, Chart 8.2.FA/PO). Hydrophilic semiochemicals, which are more difficult to identify, are also emerging. Those from tobacco plants are the first known multiple polypeptide hormones deriving in plants from a polyprotein precursor, like in animals and yeast (Pearce 2001).

Representative semiochemicals are illustrated in Table 12.1, especially alarming and growth stimulating agents of plants and animals, osmoprotecting compounds of bacteria that also prevent damages on dessication, and recruiting agents of insects.

Metabolites with a potential semiochemical role may also be found in organisms that do not need them anymore. This may be the case of coralloidolide A, a modified cembranoid diterpene isolated from a Mediterranean soft coral, *Alcyonium coralloides*. The structure of coralloidolide A recalls closely lophotoxin, a defensive agent of tropical gorgonians (Table 12.1). Mediterranean soft corals are not subjected to the strong predation of their counterparts in coral reefs, from which they probably derive. Therefore, coralloidolide A may be imagined as a vestigial metabolite from tropical alcyonacean precursors.

Table 12.1. Recruiting, alarming, and growth stimulating agents

Agent class	Name/Chemical class/Function/Structure		Source (Mori 2000)
Alarming	geosmin/polyket./warning about the presence of *Symploca muscorum*, Cyanobact.: MI; Slater 1983	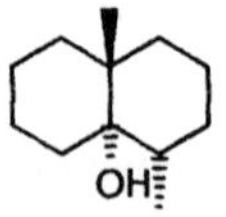 geosmin	beet (*Beta* spp., Ang) (a volatile metabolite that affords earthy flavor to waters and fish)

Table 12.1. Recruiting, alarming, and growth stimulating agents

Agent class	Name/Chemical class/Function/Structure	Source (Mori 2000)
Alarming	melatonin/indole alkal. / inducing reversion in the skin color of amphib. and Mamm., perhaps mediating photoperiodicity: MI 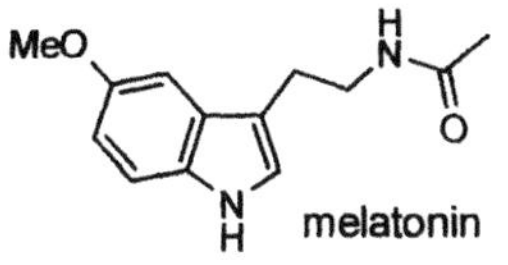 melatonin	widely present in animals
Alarming	(-)-phyllanthurinolactone / polyket. / chemotactic leaf-closing agent 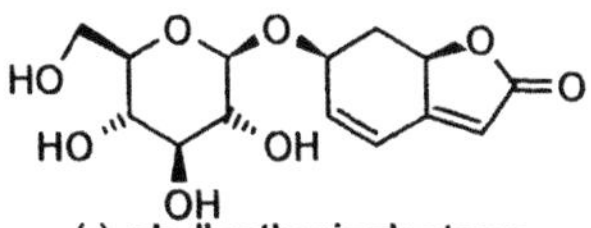 (-)-phyllanthurinolactone	*Phyllanthus urinaria,* Ang.
Growth stimulating	strigol and strigolactone/terpenoids / germination stimulants of crops in US, Africa, and Asia 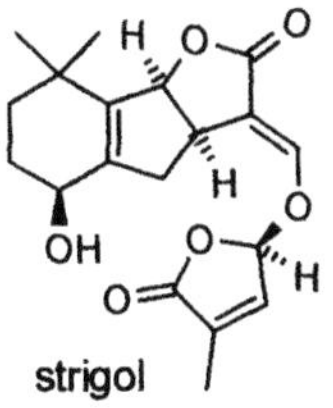 strigol	parasitic weeds, *Striga* and *Orobanche* spp., Ang.; structurally and behaviorally correlated to alectrol and (+)-orobanchol
Osmoprotecting	ectoine/pyrimidine alkal./protecting agent in Bact. against osmotic pressure changes and drying during storage 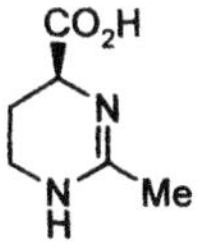 ectoine	*Streptomyces* spp. and other Bact.
Recruiting	citrus bug pheromone/acetogenin / aggregation pheromone 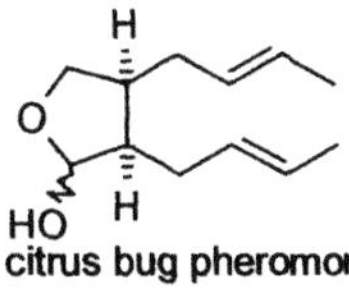 citrus bug pheromone	male spined citrus bug, *Biprolurus bibax*, Ins.
Recruiting	dimethyl oligosulfides/polyket./aggregation pheromones	bat-pollinated and sapromyophilous flowers: Harborne 1999A
Recruiting	(+)-*exo*-brevicomin/polyket./aggregating sex pheromone 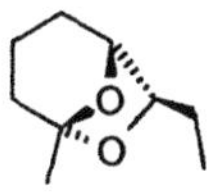 (+)-*exo*-brevi-comin	silkworm moth

Part IV. Natural product diversity at functional level

Table 12.1. Recruiting, alarming, and growth stimulating agents

Agent class	Name/Chemical class/Function/Structure	Source (Mori 2000)
Recruiting	(+)-disparlure/fatty acid deriv./sex aggregating pheromone 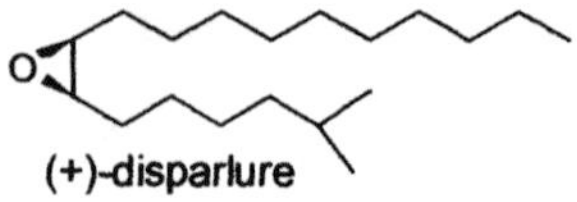 (+)-disparlure	gypsy moth, *Lymantria dispar*, Ins.
Recruiting	(+)-lineatin/polyket./sex aggregating pheromone 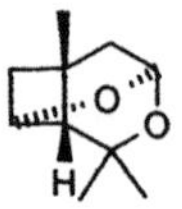 (+)-lineatin	pine beetle from western countries
Recruiting	3-methyl butanoic acid/acetogenin/sex aggregating pheromone	fetid-smelling fly-pollinated flowers: Harborne 1999A
Recruiting	glycinoeclepin/polyket./hatching stimulant for the soy bean cyst nematode agent 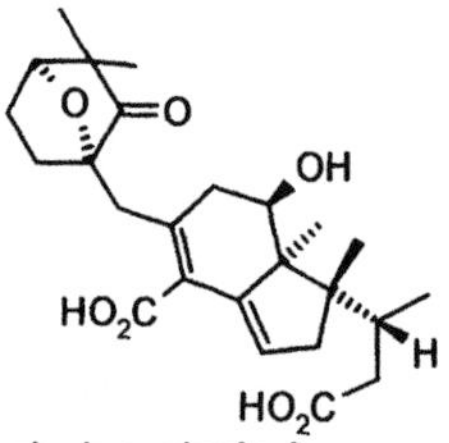 glycinoeclepin A	kidney bean, *Phaseolus* spp., Ang.
Recruiting	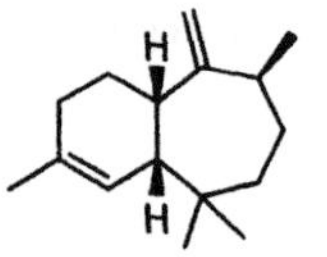 (S)-9-methylgermacrene-B and (1S,3S,7S)-3-methyl-α-himachalene/sesquiterp./male pheromones 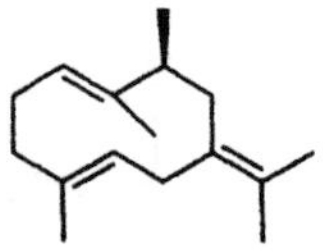 (S)-9-methylgerma- (1S,3S,7S)-3-methyl-crene-B α-himachalene	sandfly, *Phlebotomus* spp., Diptera, Ins. (vector of the leishmaniase disease)
Recruiting	monoterp., or 2-phenylethanol, methyl phenyl acetate, benzaldehyde, and indole and aromatic esters/pheromones	fly-pollinated flowers: Harborne 1999A
Recruiting	volicitin/amino acid ester/elicitor in seedlings of insect-attracting volatiles 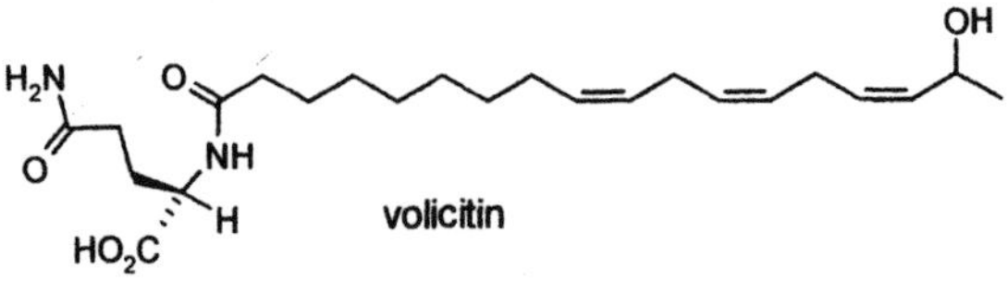 volicitin	oral secretions of beet armyworm caterpillars (*Spodoptera exigua*, Ins.): Alborn 1997

Table 12.1. Recruiting, alarming, and growth stimulating agents

Agent class	Name/Chemical class/Function/Structure	Source (Mori 2000)
Recruiting (excitant)	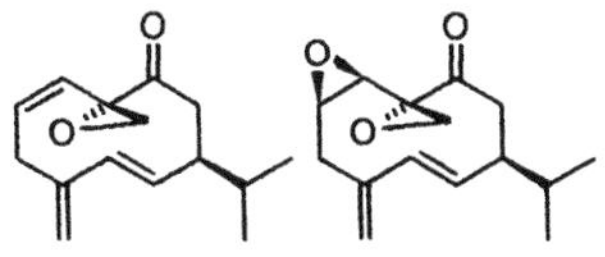 periplanone A and periplanone B / polyket. / female-produced sex pheromones	American cockroach, *Periplaneta americana* L., Ins.
	periplanone A periplanone B	
Recruiting	(-)-olean/polyket./sex aggregating pheromone 	female olive fly, *Bactrocera oleae,* Ins. / sex pheromone
	(-)-olean	
Vestigial	coralloidolide A/diterp./its recalls strongly bioactive lophotoxin from tropical *Lophogorgia* spp., Gorgonacea 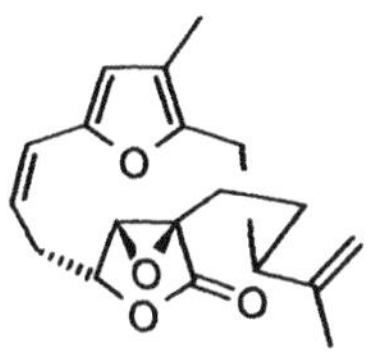 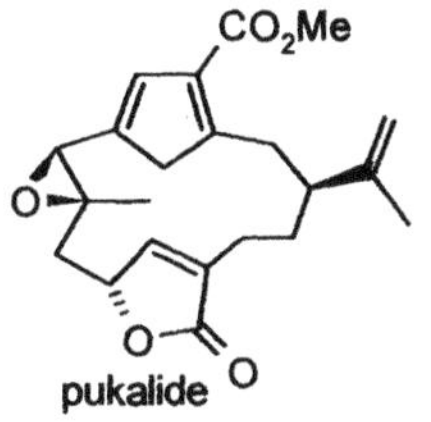	**marine:** *Alcyonium* (= *Parerythropodium*) *coralloides* (Pallas, 1766), Cnid.: D'Ambrosio 1987
	CO$_2$Me coralloidolide A pukalide	

12.2 Antifeedant and antimicrobial agents

Defensive agents occur widely in plants, except endemic species of isolated islands like Hawaii, reflecting the evolution of plants in the absence of herbivorous mammals. Why herbivorous insects did not raise defensive compounds in Hawaiian plants is not clear, however. Anyway, things have changed much in Hawaii: introduced plants sum up to an astonishing record of 43.8% of all local flora (Gábor 1997) and plenty of species of herbivorous mammals have been brought in.

Defensive agents are also well known from insects and chelicerates on land. The typically associative mode of marine life has also stimulated the biosynthesis of many defensive agents (Paul 1992).

A selection of marine and terrestrial defensive compounds is displayed in Table 12.2, comprising (a) plant metabolites against insects (pyridine and tetrahydroquinoline alkaloids, other than polyketides and fatty acids), mammal grazers (pyrrolizidine alkaloids, steroidal glycosides, and eremophilane and drimane sesquiterpenes, taxane diterpenes, and tetranortriterpenes), insects and mammal grazers (polyphenols and neolignans), and microbial pathogens (polyacetylenes), (b) dietary indolizidine alkaloids in tropical frogs against carnivorous predators, (c) plant pyrrolizidine and tropane alkaloids, cardenolides, and naphthalene, uptaken by insects and larvae against insectivorous birds, (d) benzoquinones in insects against arthropod predators, (e) polyazamacrolides in insects against other insects, (f) iridoid monoterpenes in insects against insectivorous animals, (g) triterpenes in marine gastropods against carnivorous predators, (h) bromophenols in seaweeds against herbivorous grazers.

Brominated compounds widely occur in marine organisms, particularly seaweeds and invertebrates. Many of them play a defensive role against parasites and predators, and are incorporated by opisthobranch mollusks to this scope from the diet. Bromine is uptaken by seaweeds and invertebrates from bromide dissolved at sparingly 1 mM concentration in seawater. The process is catalyzed by haloperoxidases, which have been characterized both as structure and function (Butler 1997).

The phytoalexins are a special class of defensive agents that has attracted much attention in view of agricultural applications. Perhaps the most advanced study concerns insecticidal and antimicrobial benzoxazins of the Gramineae, which have been elucidated to the gene level (Frey 1997). A survey of phytoalexins is presented in Chart 12.2.

Table 12.2. Antifeedant and antimicrobial agents (phytoalexins are grouped in C 12.2)

Activity/(potential) predators and pathogens	Chemical class/Name/Structure	Source (ref. to Harborne 1999A for land and Paul 1992 for marine, when not otherwise stated)
Antifeedant/carnivorous animals	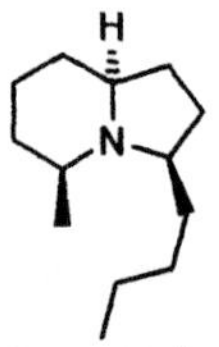 indolizidine alkal./ monomorine I monomorine I	fire ants, Ins.; dietary in Dendrobatidae frogs, Amphib.: Braekman 1998
Antifeedant/Ins.	 pyridine-pyrrolidine alkal./nicotine nicotine	induced by grazing in *Nicotiana* spp., Solanaceae, Ang.
Antifeedant/birds, spiders	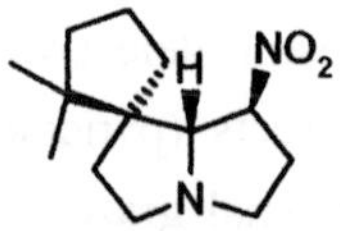 pyrrolizidine alkal./ (+)-nitropolyzonamine (+)-nitropoly- zonamine	danaid and ithomiine butterflies, chrysomelid beetles, Ins., and myriapods and fire ants, as dietary plant metabolites: Braekman 1998

Table 12.2. Antifeedant and antimicrobial agents (phytoalexins are grouped in C 12.2)

Activity/(potential) predators and pathogens	Chemical class/Name/Structure	Source (ref. to Harborne 1999A for land and Paul 1992 for marine, when not otherwise stated)
Antifeedant/herbivorous Mamm.	pyrrolizidine, pyrrolidine, and diterp. alkal. (Acnite class)/ senecionine *N*-oxide 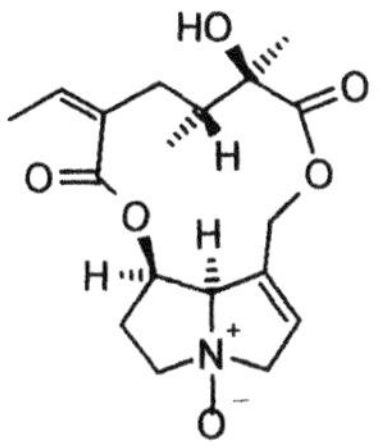 senecionine *N*-oxide	*Delphinium* spp., Ranunculaceae, and Compositae and Liliaceae, Ang.
Antifeedant/Ins. (ants)	polyazamacrolide alkal., 30-280 membered/ *Subcoccinella* 30-membered polyazamacrolide 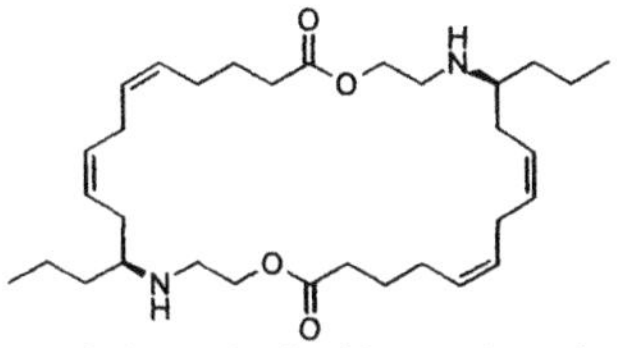 *Subcoccinella* 30-membered polyazamacrolide	ladybird beetle, *Epilachna borealis* (Fabricius), Ins. and squash beetles, Ins.: Schroeder 2000
Antifeedant/Ins. (ants), spiders, frogs	quinoline alkal.	Peruvian fire-stick, Ins.
Antifeedant/herbivorous Mamm.	steroidal glycosidic alkal./α-solamargine 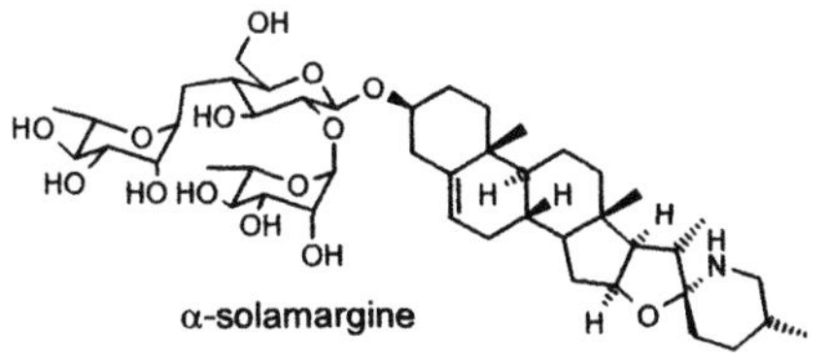 α-solamargine	*Delphinium* spp., Ranunculaceae, as well as Compositae and Liliaceae, Ang.
Antifeedant/Ins.	tetrahydroisoquino-line alkal./hydrastine 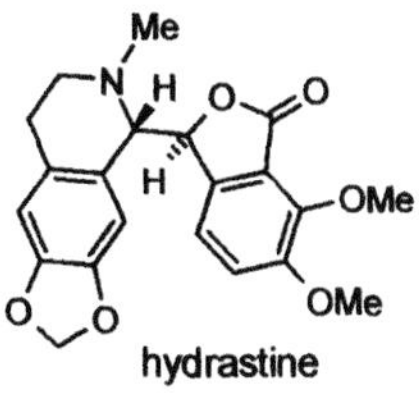 hydrastine	plants
Antifeedant/birds, monkeys	tropane alkal./C 6.1.A, 6.2.A2	dietary plant metabolites in the larvae, pupae, and adult ithomiine butterflies, Ins.

Table 12.2. Antifeedant and antimicrobial agents (phytoalexins are grouped in C 12.2)

Activity/(potential) predators and pathogens	Chemical class/Name/Structure	Source (ref. to Harborne 1999A for land and Paul 1992 for marine, when not otherwise stated)
Antifeedant/Ins., slugs, snails	isopr. (eremophilane and drimane sesquiterp.)/ C 8.3.I2)	Compositae, Ang., and Taxaceae, Gymn.
Antifeedant/Ins. and defensive/insectivorous animals	isopr. (tetranor triterp.)/azadirachtin 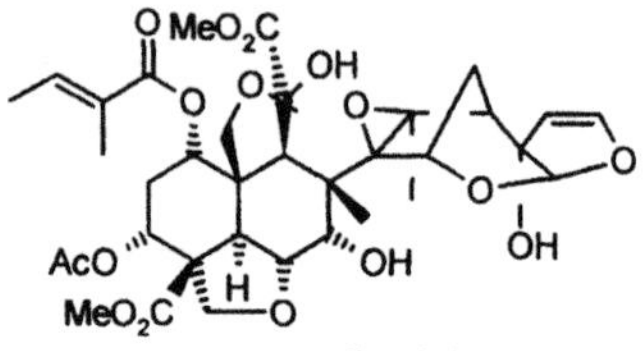 azadirachtin	neem tree, *Azadirachta indica* A. Juss., Meliaceae, Ang., and Ins. feeding on it: MI
Antifeedant and antimicrobial/Bact., sea urchin, abalone, fish	isopr. (prenylguaiane diterp.)	**marine:** seaweeds
Antifeedant/Ins.	isopr. (steroids)	phytoecdysteroids of Amaranthaceae, Chenopodiaceae, and Verbenaceae, Ang.
Antifeedant/Ins., slugs, snails	mixed biogenesis isopr. (saponins)	Compositae, Ang., and Taxaceae, Gymn.: MI
Antifeedant/birds	polyket. and fatty acids/palmitoleic, linoleic, and oleic acids	female social wasps, Ins.
Antifeedant/Ins. (ants), pathogens	polyket./naphthalene	Formosan termites, Ins.
Antifeedant/herbivorous Mamm., fish, Ins.	shikim. (polyphenols and neolignans)/ phloroglucinol meroditerp., and polyket.	Oleanaceae and Magnoliaceae, Ang.
Antifeedant/Ins.	shikim. (bisphenylpropanoids)/ (didesmethyl)rocaglamide 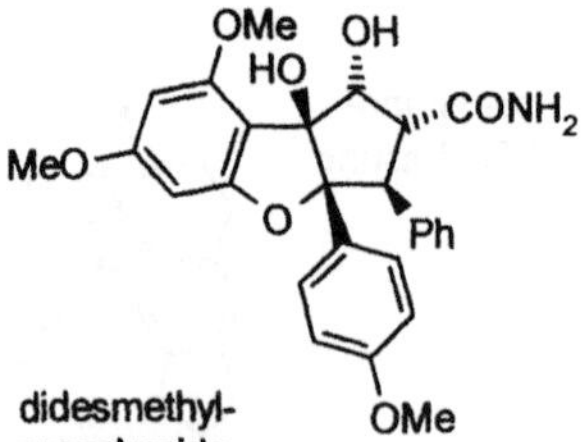 didesmethyl- rocaglamide bisphenylpropanoid	*Aglaja elliptica* and *Aglaja harmsiana*, Meliaceae, Ang., of potency rivaling azadirachtin
Antifeedant and antimicrobial/Bact., sea urchin, abalone, fish	shikim. or polyket. (bromophenols)	**marine:** seaweeds

Table 12.2. Antifeedant and antimicrobial agents (phytoalexins are grouped in C 12.2)

Activity/(potential) predators and pathogens	Chemical class/Name/Structure	Source (ref. to Harborne 1999A for land and Paul 1992 for marine, when not otherwise stated)
Antifeedant/birds	shikim./salicylaldehyde	salicyl alcohol of willow leaves as dietary metabolite in beetles, Ins.
Antifeedant/fish	various dietary metabolites: Paul 1992	**marine:** dietary metabolites from sponges, cnidarians, bryozoans, and ascidians in carnivorous Opisthobr., Moll.
Antifeedant/fish	various dietary metabolites: Paul 1992	**marine:** dietary metabolites from Cyanobact. and seaweeds in herbivorous Opisthobr., Moll.
Antimicrobial, insecticide/insectivorous animals	isopr. (iridoid monoterp.)/ iridomyrmecin 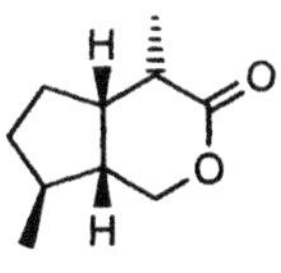 iridomyrmecin	Argentine ant, *Iridomyrmex humilis* Mayr., Ins.: MI
Antimicrobial/pathogens of Mamm.	polyket. (polyacetylenes)/thiarubrine A and B 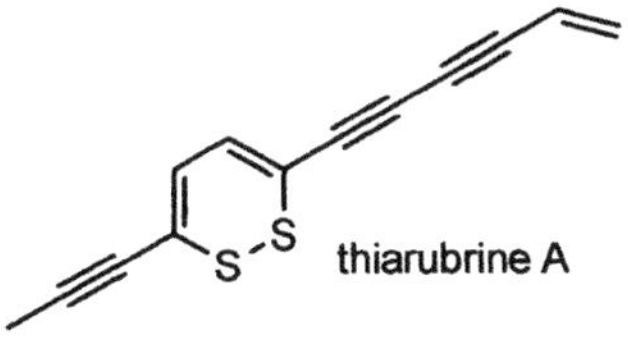 thiarubrine A	Asteraceae, Ang., from S and E Africa, whose leaves are eaten by the chimpanzees: MI
Defensive/arthropods and Vertebr.	polyket. (benzoquinones and hydroquinones)	African and eastern millepedes, Diplopoda: Deml 2000
Defensive/ants, birds, monkeys	isopr. (cardenolide steroids C 8.1.I)	danaid butterflies larvae, Ins., as plant dietary metabolites
Defensive/Ins.	polyket./naphthalene	*Magnolia* sp. flowers, Ang.

Chart 12.2 Phytoalexins (ref. to Harborne 1999B, unless otherwise stated) (Smax=39, av=14; S/Hmax=0.77, av=0.4).
Alkal.: benzoxazins: *Dianthus* phytoalexin (*Dianthus caryophyllus*, Caryophyllaceae phytoalexin) and open form, avenalumin I (Gramineae, Ang.), see also: Frey 1997; sulfurated indole: brassicanol A, camalexin, brassilexin (Cruciferae, Ang.).
Isopr.: MONOTERP.: isopropylcycloheptatrienones: 6-isopropyltropolone β-glucoside (*Cupressus sempervirens* L., Gymn.). SESQUITERP.: guaiane: lettucenin A (Compositae); eremophilane: (Solanaceae, like capsidiol); vetispirane: solavetivone (Solanaceae). DITERP.: prenylated isopropylcycloheptatrienones: 5-(3-hydroxy-3-methyl-*trans*-1-butenyl)-6-isopropyltropolone-β-glucoside (*Cupressus sempervirens* L., Gymn.; pimaranes: momilactone A from *Oryza sativa* L., Gramineae Ang.; cassanes): phytocassane A (Gramineae, Ang.)
Polyket.: cyclopenta-1,3-diones: (*Allium cepa* L., Liliaceae, Ang.); phenalenones and naphthalenes: (banana plant, *Musa acuminata*, Ang.); polyacetylenes: mycosinols E/Z (*Colostephus myconis*, Anthemidae, Ang.; wyerone of leguminosae; also in Solanaceae).
Shikim.: stylbenes: resveratrol (grapevine, Ang.); flavonoids: pinocembrin (*Pinus silvestris* L., Gymn.); Narcissus phytoalexins; luteolinidin (*Sorghum* spp. and sugar cane *Saccharum officinarum* L., Gramineae, Ang., also prenylated in Moraceae); isoflavonoids: pisatin and glyceollin II (pterocarpans typical of legumes, Ang.); phenylpropanoids: yurnelide (*Lilium maximowczii*, Ang.); lignans: hydroxymatairesinol (spruce, *Picea abies*, Gymn.); coumarins: scopoletin (sunflower, *Helianthus annuus* L., Compositae, Ang.; mostly represented in Umbelliferae, Ang.); benzofurans: vignafuran (tyical of legumes; also present in Moraceae); biphenyls: aucuparin, and eriobofuran, rare in Rosaceae, Ang.); benzoic acid like: salicylic acid (willow tree, *Salix* spp., Salicaceae, Ang.: Verberne 2000).

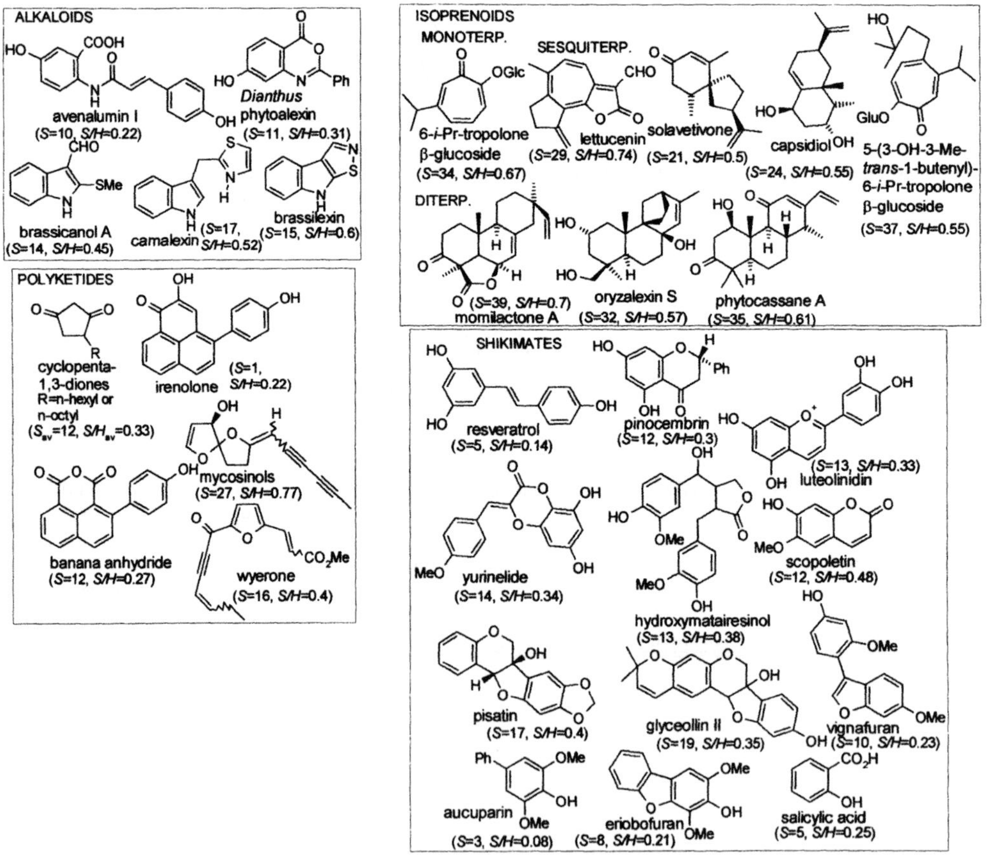

ALKALOIDS

avenalumin I (*S*=10, *S/H*=0.22)

Dianthus phytoalexin (*S*=11, *S/H*=0.31)

brassicanol A (*S*=14, *S/H*=0.45)

camalexin (*S*=17, *S/H*=0.52)

brassilexin (*S*=15, *S/H*=0.6)

POLYKETIDES

cyclopenta-1,3-diones R=n-hexyl or n-octyl (*S*av=12, *S/H*av=0.33)

irenolone (*S*=1, *S/H*=0.22)

mycosinols (*S*=27, *S/H*=0.77)

banana anhydride (*S*=12, *S/H*=0.27)

wyerone (*S*=16, *S/H*=0.4)

ISOPRENOIDS

MONOTERP.

SESQUITERP.

6-*i*-Pr-tropolone β-glucoside (*S*=34, *S/H*=0.67)

lettucenin (*S*=29, *S/H*=0.74)

solavetivone (*S*=21, *S/H*=0.5)

capsidiol (*S*=24, *S/H*=0.55)

5-(3-OH-3-Me-*trans*-1-butenyl)-6-*i*-Pr-tropolone β-glucoside (*S*=37, *S/H*=0.55)

DITERP.

momilactone A (*S*=39, *S/H*=0.7)

oryzalexin S (*S*=32, *S/H*=0.57)

phytocassane A (*S*=35, *S/H*=0.61)

SHIKIMATES

resveratrol (*S*=5, *S/H*=0.14)

pinocembrin (*S*=12, *S/H*=0.3)

luteolinidin (*S*=13, *S/H*=0.33)

yurinelide (*S*=14, *S/H*=0.34)

hydroxymatairesinol (*S*=13, *S/H*=0.38)

scopoletin (*S*=12, *S/H*=0.48)

pisatin (*S*=17, *S/H*=0.4)

glyceollin II (*S*=19, *S/H*=0.35)

vignafuran (*S*=10, *S/H*=0.23)

aucuparin (*S*=3, *S/H*=0.08)

eriobofuran (*S*=8, *S/H*=0.21)

salicylic acid (*S*=5, *S/H*=0.25)

12.3 Toxins and environmentally noxious metabolites

Most ichthyotoxins in the oceans are produced by a few species of dinoflagellates that bloom at certain periods, representing a deadly risk for humans. The blooming of toxic species, once limited to specific areas, has now extended to many coastal waters because of the spreading of the species through ballast water operations.

Blooming of dinoflagellates is a complex affair, contemplated in the "paradox of plankton". That is, at the equilibrium, resource competition models suggest that the number of coexisting species cannot exceed the number of limiting resources. In contrast, within nature, more species can coexist. A rationalization of these phenomena, possibly solving the paradox, may be found in species oscillations and chaos, without the need of advocating external causes (Huisman 1999).

Bacterial siderophores are another class of widespread toxins produced by bacteria. On land, mycobactins and exophilins are the best-known representatives. Those found in the sea, aquachelins, marinobactins, anguibactin, vulnibactin, and aerobactin (Table 12.3.I) are amphiphilic. This is a property that has received great attention for drug delivery problems and in relation to recognition processes (Testa 2000), devising efficient algorithms for computer calculation (Fisher 2000).

Toxin-producing species may serve as biological weapons. Best known to this regard is anthrax, a multicomponent toxin and rare disease deriving from a bacterium, *Bacillus anthracis*, whose spores are the most hardy and durable of all resting cells. The rapid multiplication of the bacterium makes protection from anthrax by antibiotics problematic; reportedly, the best one is ciprofloxacin (Table 15.III). A new therapy focuses on the mechanism of translocation of the toxin, which is also a way to contrast the resistance developed by pathogens to antibiotics (Sellman 2001). But what may happen if genetically modified *Bacillus anthracis* is used as a biological weapon?

Some natural products, or their degradation products, represent a hazard for mammals not because of general toxicity but for subtle, adverse properties, such as carcinogenicity and tumor promotion. They are best known from marine dinoflagellates (okadaic acid and structural analogues), filamentous fungi (trichothecenes and ochratoxins), and plants (pyrrolizidine alkaloids).

Endogenous toxins - which normally act against invader organisms during the immune response - are also known, like peroxynitrite, formed by macrophages from nitrogen monoxide (Chapter 12.5) and hypochlorite, formed by leucocytes (Herold 2001).

Environmentally relevant metabolites include hydrocarbons released from *Eucalyptus* forests. They undergo oxidation by aerial dioxygen to give carboxylic acids that form aerosols, influencing adversely the climate (Kavouras 1998).

Table 12.3.I. Toxins, ecotoxins that may reach man, and environmentally relevant metabolites

Name or class/Vectors	Chemical class/Structure	Producing species[a]
Aflatoxins/various crops, especially peanuts and corn	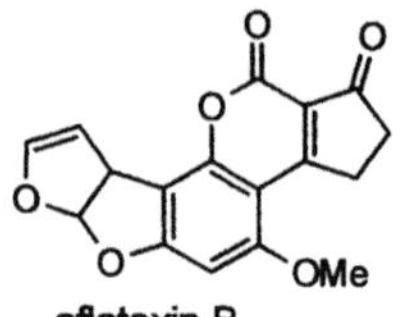 polyket., like aflatoxin B$_1$	*Aspergillus flavus* and *Aspergillus parasiticus*, Deuterom., Eumyc., growing on various crops worldwide: MI

aflatoxin B$_1$

| Amanitins (α- and β-amanitin) and phalloidin/mushroom | bicy-clic octa-pept. | *Amanita phalloides*, Basidiomyc.: MI |

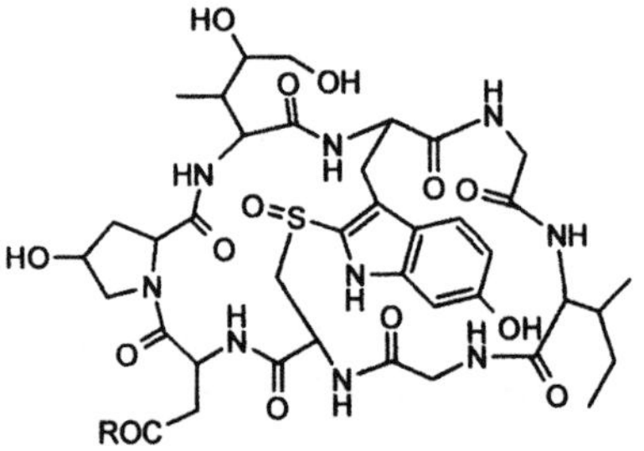

α-amanitin R = NH$_2$ all L-amino acids
β-amanitin R = OH   (*R*) sulfoxide

| Amnesic shellfish poison (domoic acid)/shellfish | 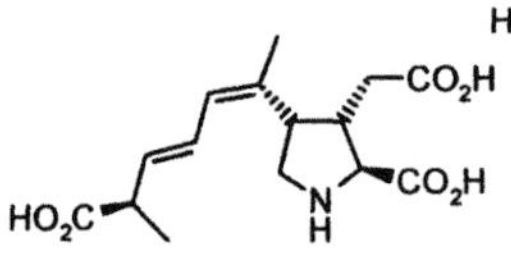 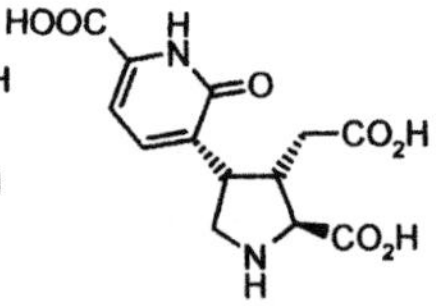 amino acid | **marine:** *Pseudo-nitzschia multiseries*, Diatom. of Atlantic Canada, *Pseudo-nitzschia australis* of Pacific US, and *Chondria armata* Kützing and other Japanese Rhodomelaceace, Rhodoph.; structurally related to, and agonized by, acromelic acids from *Clitocybe acromelalga*, Basidiomyc.: Baldwin 2000 |

domoic acid   acromelic acid A

| Anguibactin/fish | 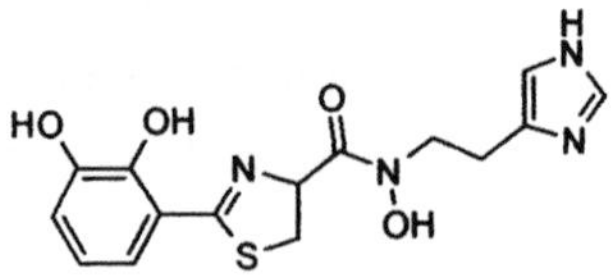 amphiphilic siderophore | **marine:** *Vibrio anguillarum* Bergeman 1909, Bact. |

anguibactin

| Anthrax/farm and wild animals | Capsular poly-D-glutamate and factors I-III (edema, protective, and lethal proteins) in conjunction | *Bacillus anthracis* Cohn 1872, Bact. |
| Apamin/honeybee | polypept. constituted of 18 amino acids and bearing 2 disulfide bridges | *Apis mellifera* (L.), Ins.: MI |

Table 12.3.I. Toxins, ecotoxins that may reach man, and environmentally relevant metabolites

Name or class/Vectors	Chemical class/Structure	Producing species[a]
Aquachelins, like aquachelin A/released to the medium	amphiphilic siderophores aquachelin A R =	**marine**: *Halomonas aquamarina* (ZoBell and Upham 1944) Dobson and Franzmann 1996, comb. nov., Bact.
Aristolochic acid/herb	polyket. aristolochic acid	*Aristolochia fangchi*, Aristolochiaceae, Ang.; it is transferred to butterflies from the diet: MI; acute kidney damage recently occurred in Belgium because of an erroneous substitution for the Chinese medicinal herb, *Stephanis tetranda*
Azaspiric acids/*Mytilus edulis*, Moll., Bivalvia in Killary Harbor cultures, Ireland	azaspiric acids R1, R2, R3, R4 various H, Me, OH azaspiro polyethers	very likely dinoflagellates: Forsyth 2001
Botulins/food	polypept.	*Clostridium botulinum*, Eumyc.: MI
Cholera toxins/water and food on land or copepods in the sea	polypept.	**marine**: *Vibrio cholerae* Pacini 1854, Bact., now worldwide diffused through ballast waters

Table 12.3.I. Toxins, ecotoxins that may reach man, and environmentally relevant metabolites

Name or class/Vectors	Chemical class/Structure	Producing species[a]
Ciguatera poisoning (ciguatoxins and analogues and maitotoxin)/fish	polyether polyket. 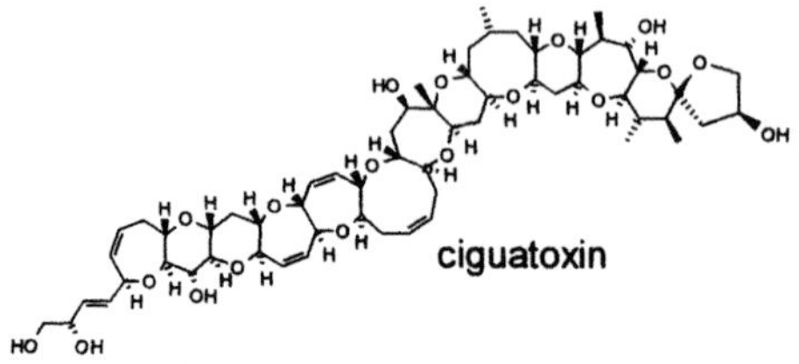 ciguatoxin	**marine:** isolated from the viscera of moray eel, *Lycodontis javanicus*, Osteichtyes from the Indo-Pacific
Ciguatera poisoning (ciguatoxins and analogues and maitotoxin)/fish	polyether polyket. 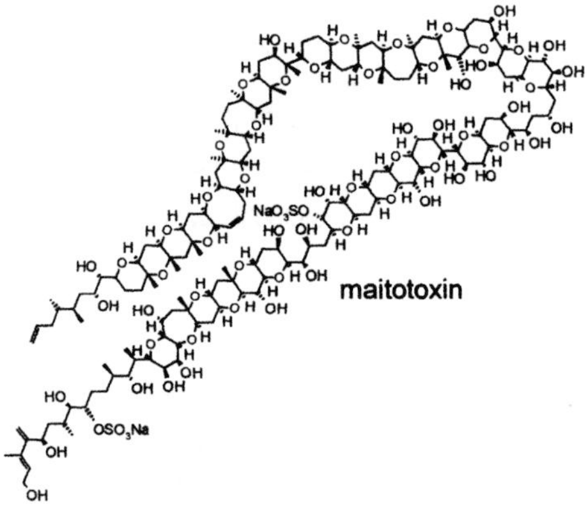 maitotoxin	**marine:** *Gambierdiscus toxicus* Adachi et Fukuio, 1979 from French Polynesia;
Cobratoxin and other snake toxins/snake	60-amino-acid polyept.	cobra, *Naja* spp., Serpentes, from tropical East: MI
Conotoxins (ω-conotoxin MVIIA)/Moll.; Kohno 1995, and its synthetic version SNX-111	polyept. 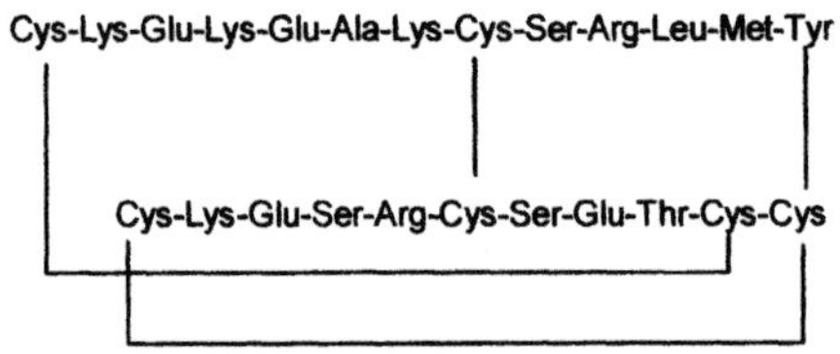 amino acid sequence and disulfide bridges of ω-conotoxin MVIIA	**marine:** *Conus magus* Linné, 1758, Neogastropoda, Moll.
Cyanotoxins (hepatotoxin)/released to the waters	microcystins/pept. (C 8.1.A/P)	*Microcystis, Nostoc, Nodularia,* and *Oscillatoria* spp., Dinofl. of worldwide distribution

Table 12.3.I. Toxins, ecotoxins that may reach man, and environmentally relevant metabolites

Name or class/Vectors	Chemical class/Structure	Producing species[a]
Cyanotoxins (neurotoxins)	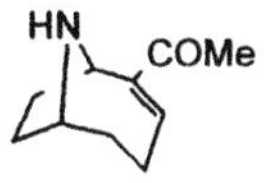 anatoxin-a/alkal.	*Anabaena flos-aquae*, Cyanobact.: Harada 1993
	HN COMe	
	anatoxin-a	
Diarrheic shellfish poisoning/shellfish	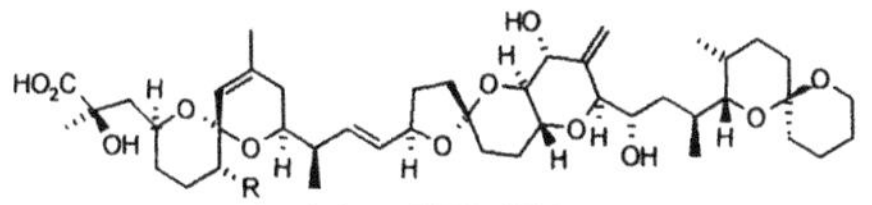 okadaic acid and dinophysis toxins/polyethers	**marine:** *Dinophysis* and *Prorocentrum* spp., Dinofl., now of worldwide distribution
	HO₂C ... OH ... R ... HO ... OH OH	
	okadaic acid R=OH 7-deoxy-okadaic acid R=H	
Ergot/mostly rye	indole alkal.	*Clavices purpurea*, parasitic Ascom.
Gastroenteritis toxins/shellfish and fish	polypept.	**marine:** *Vibrio parahaemolyticus* (Fujino *et al.* 1951), Bact.
Marinobactins, like marinobactin A	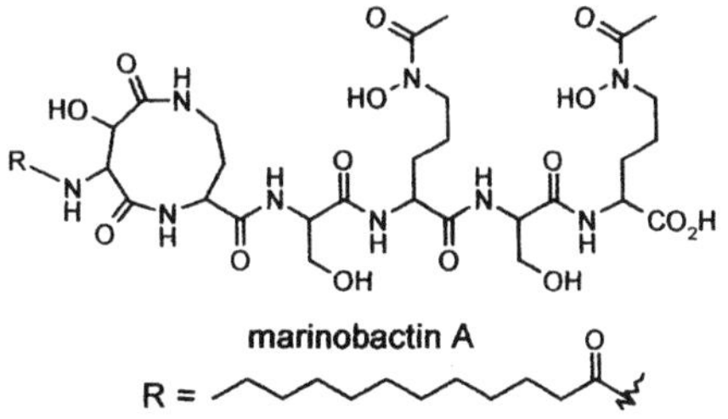 amphiphilic siderophores, released to the medium	**marine:** *Marinobacter* sp., Bact.
	marinobactin A R =	
Neurotoxic shellfish poisoning/shellfish	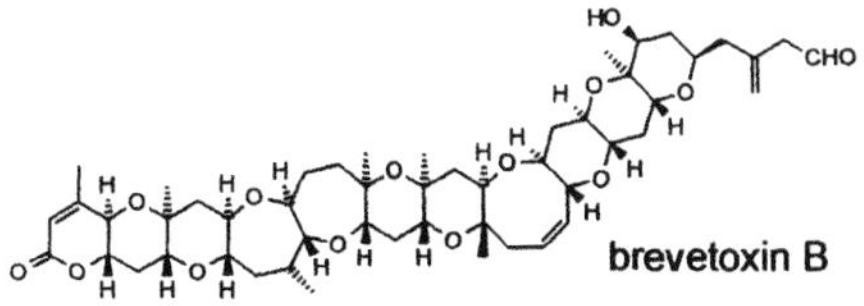 brevetoxin A and B/polyether polyket.	**marine:** *Gymnodinium breve* Davis [= *Ptychodiscus brevis* Davis], Dinofl. from the Gulf of Mexico, North Carolina
	brevetoxin B	
Nicotine/plant	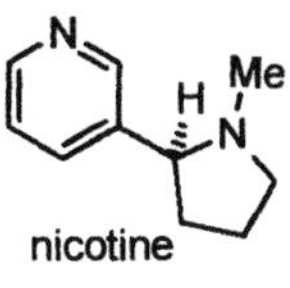 alkal.	*Nicotiana glauca* Graham and *Nicotiana rustica* L., Solanaceae, Ang.
	nicotine	

Table 12.3.I. Toxins, ecotoxins that may reach man, and environmentally relevant metabolites

Name or class/Vectors	Chemical class/Structure	Producing species[a]
Paralytic shellfish poisoning/fish and shellfish	alkal. 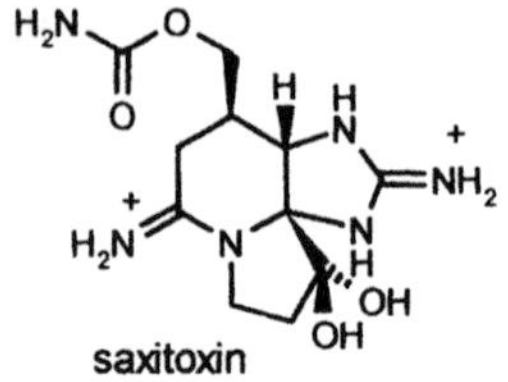 saxitoxin	**marine:** *Alexandrium, Gymnodinium,* and *Pyrodinium* spp., Dinofl. now of worldwide distribution; also *Aphanizomenon flos-aquae* (L.) Ralfs, freshwater Cyanobact.
Pinnatoxin A and B/shellfish	amino acid polyether 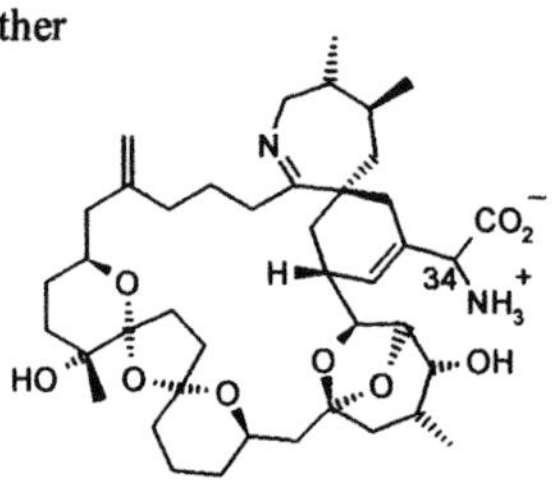 pinnatoxin B: 34*S* pinnatoxin C: 34*R*	**marine:** *Pinna muricata,* Bivalvia, Moll. from Okin.; analogues, containing a cysteine moiety, were found both in *P. muricata* and another bivalve mollusk from Okin., *Pteria penguin,* suggesting microbial origin (Takada 2001)
Plant toxins/plant and Ins. feeding on them	*Aconitum* diterp. alkal., like lycoctonine 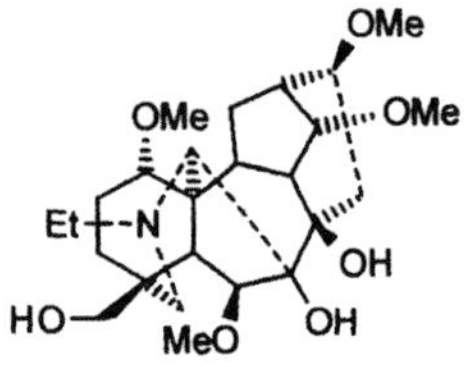 lycoctonine	*Delphinium* sp., Ranunculaceae, Ang. (tall larkspur in western US)
Plant toxins/plant and Ins. feeding on them	pyrrolidine alkal. (C 6.A1)	English bluebell, *Hyacinthoides nonscripta* (L.) Chouard ex Rothm., Ang.
Pufferfish poisoning/fish and crabs	tetrodotoxin/alkal. 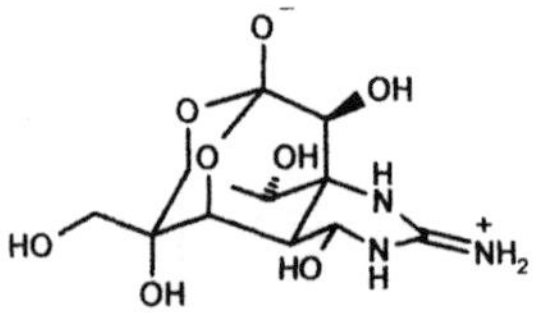 (-)-tetrodotoxin	**marine:** Bact. and Dinofl.: Kodama 1996; also puffers and newts and frogs (possibly from symbiotic Bact.)
Ricin/food	lectins/proteins	*Ricinus communis* L., Euphorbiaceae, Ang.

Table 12.3.I. Toxins, ecotoxins that may reach man, and environmentally relevant metabolites

Name or class/Vectors	Chemical class/Structure	Producing species[a]
Saponins/released to the waters	new and known triterp. saponins, responsible for the formation of foams on Rhine River in Switzerland: Wegner 2000	Aquatic *Ranunculus fluitans* L., Ranunculaceae, Ang.

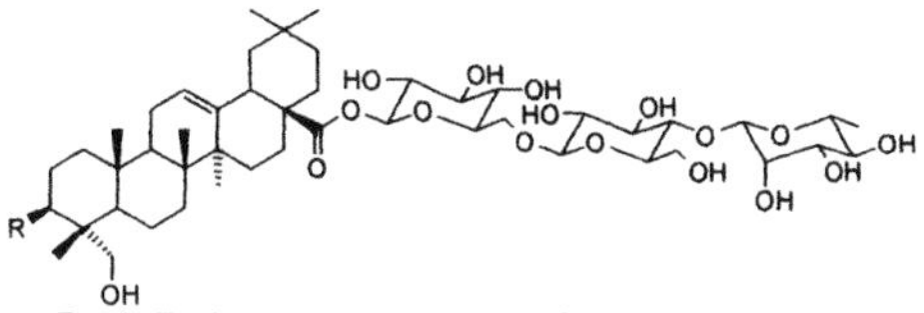

R = b-D-glucopyranosyl(1->3)a-L-arabinopyranosyl
new triterpene saponin from *R. fluitans*

Tarantula and other spider toxins/spider	polypept.	spiders, Araneae

[a]From land, unless marine origin is specified in boldface.

Table 12.3.II. Ecotoxins that cannot normally reach man

Name/Chemical class/Structure	Producing species[a]	Actual (or potential) target
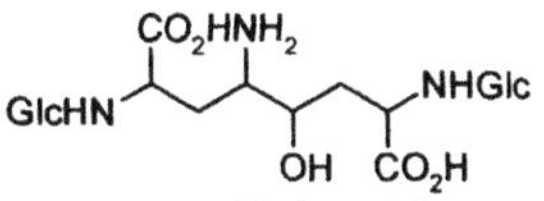 *Aschochyte* glucoside/amino acid glucoside amino acid glucosides	*Aschochyta caulina* (P. Karst) v.d Aa & v. Kest, Eumyc.	weeds
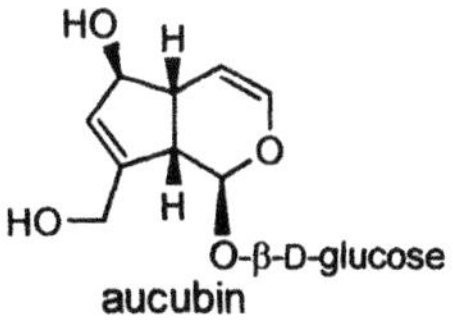 Aucubin/isopr. (irodoid monoterp.) aucubin	*Verbascum thapsus*, L., Scrophulariaceae, Ang.	barley seeds: anti-germina-tive
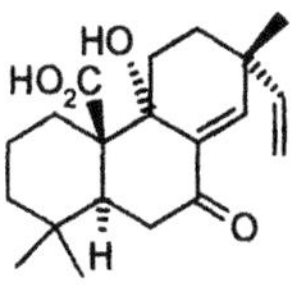 *Sphaeropsis* pimarene/isopr. (pimarane diterp.) *Sphaeropsis* pimarene	*Sphaeropsis sapinea* (Fr.) Dyko & Sutton, Eumyc.	cypress tree and various *Pinus* spp., Gymn.

Table 12.3.II. Ecotoxins that cannot normally reach man

Name/Chemical class/Structure	Producing species[a]	Actual (or potential) target
Anserinone A/polyket. 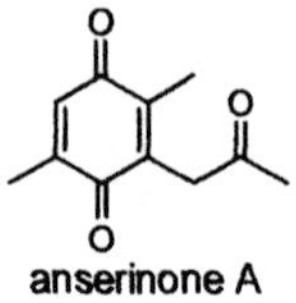 anserinone A	*Podospora anserina* (Ces.) Rehm, coprophilous fungus, Eumyc.	coprophilous fungi
Arugosin F/polyket. 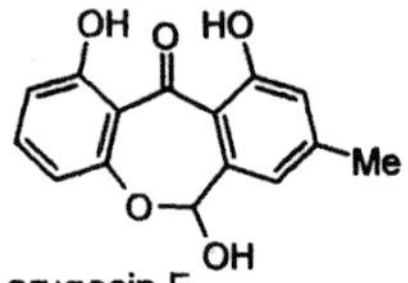 arugosin F	*Ascodesmis sphaerospora* Obrist, coprophilous fungus, Eumyc.	other coprophilous fungi
Aeollanthus acid/fatty acid deriv. 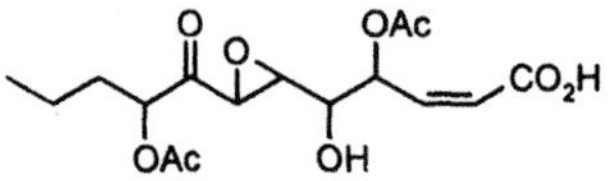 *Aeollanthus* toxin	*Aeollanthus parvifolius*, Labiatae, Ang.: Dellar 1996	phyto-pathogenic fungi, Eumyc.
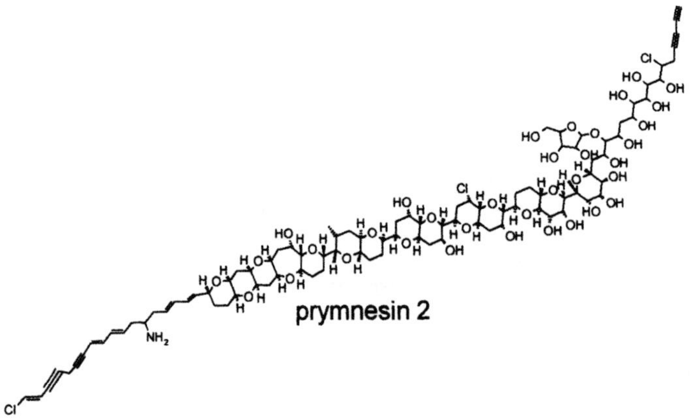 prymnesin 2 Prymnesin-1 and -2/polyether glycosides	*Prymnesium parvum* Carter 1937, **brackish waters** worldwide: Igarashi 1999	threatening fish farming
Putaminoxin B/polyket. 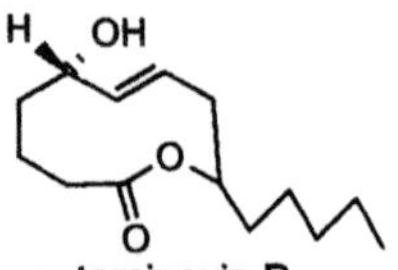 putaminoxin B	*Phoma putaminum* Spegazzini, Eumyc.: Evidente 1998	artichoke and other edible plants, Ang.

Table 12.3.II. Ecotoxins that cannot normally reach man

Name/Chemical class/Structure	Producing species[a]	Actual (or potential) target
Resorcinol/polyket. 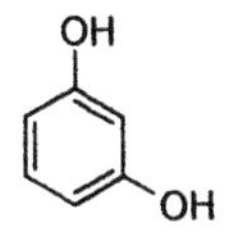 OH ... OH resorcinol	freshwater *Nuphar luteum* (L.) Sibth. & Sm., Nymphaeaceae, Ang.	Arthropoda, algae
Tellimagrandin II and other tannins/polyphenol, shikim. 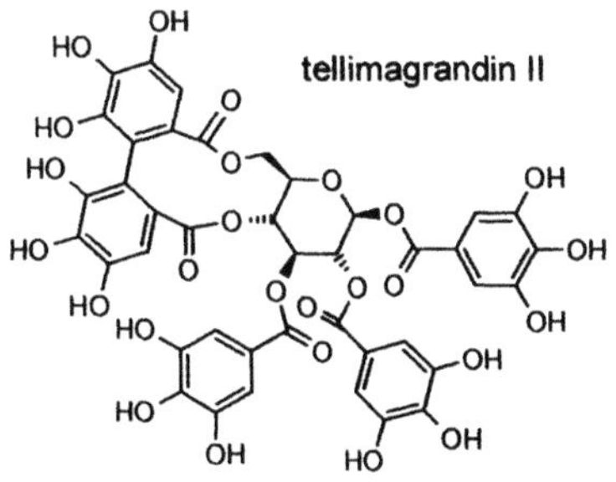 tellimagrandin II	Eurasian water milfoil, *Myriophyllum spicatum* L., Ang. (tannins decompose in the soil in a few weeks)	Chloroph.
Salicylaldehyde and 2-chlorophenoltannins/shikim.	cattail, *Typha domigensis* Typhaceae, freshwater Ang.,	water ferns

[a]Land ecotoxins, unless marine origin is indicated in boldface (from Harborne 1999A, unless otherwise indicated).

12.4 Messengers of biodiversity

That certain secondary metabolites from plants promote biodiversity has received experimental support. Thus, corn, *Zea mays*, responds to the release of an amino acid ester, volicitin (Table 12.1), by a caterpillar, *Spodoptera exigua*, with the emission, at the foraging period of the day, of volatile mixtures of indole, monoterpenoids, and sesquiterpenoids. These attract a caterpillar predator, the female parasitic wasp, *Cotesia marginiventris* (Alborn 1997).

Brown algae in the genera *Stypopodium* and *Sargassum*, in a kind of opportunistic association, provide a refugee for grazeable seaweeds, promoting biodiversity (Hay 1992, 1996). Distasteful metabolites released by the brown algae are responsible for these effects.

Other examples of messengers of biodiversity are shown in Table 12.4. However, observations as to the defensive role of metabolites are often circumstantial. Thus, tetrodotoxin, long thought to be a defensive compound in a tropical fish, *Fugu niphobles*, has recently been found to play a pheromonal role in the fish (Table 12.4). Similar conclusions have been drawn for prostaglandins in fish (Matsumura 1995).

Table 12.4. Messengers of biodiversity

Name/Function	Chemical class/Structure	Producing species[a]
Dinofl. toxic compounds/making a refugee for phytoplankton	presumably alkal., like saxitoxin 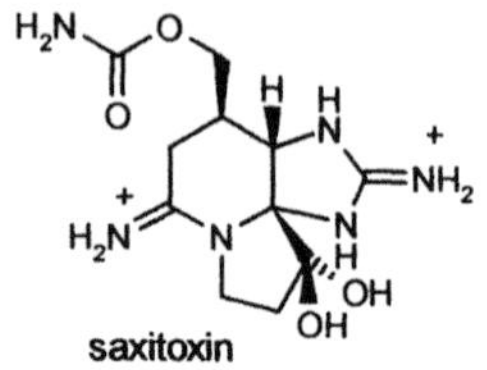	**marine:** toxic Dinofl.: Hay 1996

saxitoxin

| Isatin/making a refugee for shrimps | 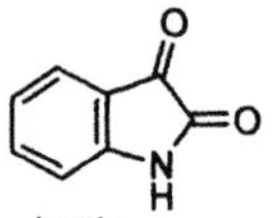 alkal. | **marine:** *Alteromonas* sp., Bact. |

isatin

| Pachydictyol A/making a refugee for amphipods | 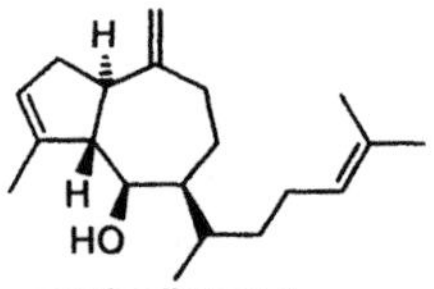 diterp. | *Dictyota bartayresii* in the Caribbean, and other Phaeoph.: Hay 1996 |

pachydictyol A

Sulfuric acid/making a refugee for grazeable *Macrocystis* spp.	inorganic/H_2SO_4	**marine:** *Desmarestia* spp. in Chilean kelp beds: Hay 1996
To be given/to make a refugee for grazeable seaweeds	to be identified	**marine:** *Stypopodium* and *Sargassum* spp., Phaeoph.: Hay 1992, 1996
Tetrodotoxin/pheromone and toxin, T12.3.II	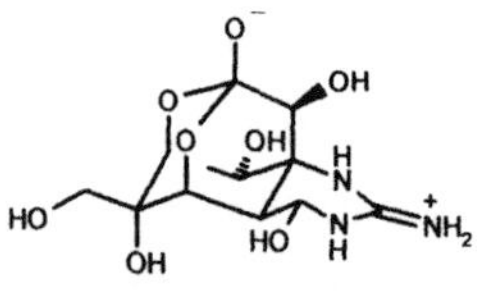 alkal.	**marine:** Bact. and Dinofl.: Kodama 1996; puffers; land newts and frogs (possibly from symbiotic Bact.)

(-)-tetrodotoxin

[a]From land, unless marine origin is indicated in boldface.

12.5 Mediators of signals

Certain proteins, peptides, steroids, and other small organic molecules, serve as cell messengers or mediators of signals. Hydrophilic mediators activate receptors at the surface of the target cell. Acetylcholine is a common example of such signaling molecules. It becomes bound on the exterior surface at the acetylcholine receptor of the nervous system, opening the channel.

Hydrophobic messengers, especially the steroids, can diffuse across the plasma membrane, activating intracellular receptors present in the cytosol or the nucleus. Cortisol (hydrocortisone) is a long known signaling molecule that enters the cell. This main glucocorticoid hormone, produced by the adrenal cortex, activates a protein receptor giving rise to both primary and delayed secondary responses.

The most common secondary messengers in signal transduction pathways, formed from the primary interaction, are calcium ions (Ca^{2+}). The receptor-hormone complex induces changes in the cell's metabolism, usually by affecting transcription or translation.

Cyclic AMP (adenosine 3',5'-cyclic monophosphate) is another secondary messenger that acts as an intracellular mediator for many different hormones, communicating the signal through the cyclic AMP-dependent protein kinase. This, in turn, phosphorylates other proteins at serine and threonine residues. Certain cell-surface receptors act by increasing the concentration of intracellular cyclic AMP. A long-duration sudden increase of intracellular cyclic AMP takes place with cholera toxins in intestinal epithelial cells. Other cell-surface receptors play the opposite role of decreasing the concentration of cyclic AMP.

Nitrogen monoxide, formed *in vivo* at nanomolar concentrations, acts as a messenger for the relaxation of blood vessels. At higher concentrations, produced by macrophages during the immune response, it may damage proteins and membranes by the way of peroxynitrite (Herold 2001).

Special messengers are involved in chemical genetics. Cyclosporin, rapamycin, and FK506 are examples from nature. Synthetic compounds with similar properties were derived from combinatorial libraries. All these agents become bound to proteins, altering their functions and serving to unravel their functions in the cell's trafficking. The efficiency of this methodology may be compared to gene knockout in genetics. Cells and organisms that are less prone to genetic manipulation can also be investigated (Haggarty 2000).

From these observations one may wonder whether the ecological role attributed to many natural products is but a shortcut of scientific research, while their fundamental role remains to be discovered. In any event, the ecological effects played by certain metabolites reflect the variety of receptors and interaction modes in nature. This explains why they may find pharmaceutical use.

Control of signaling pathways by small molecules in the cell is also at the basis of modern strategies of drug discovery. Much sought to this concern are small molecules that control IgE signaling pathways and are known to be involved in immune disorders.

That of mediators of signals is an area open to discovery. Endogenous cannabinoids are an example. Putatively involved in important functions with cannabinoid receptors on nerve cells, they remain elusive, however (Christie 2001).

Table 12.5. Small-molecule mediators of cell signals

Name/Function	Chemical class/Structure	Origin
Abscisic acid/plant growth hormone regulator that triggers abscission and closing of stomata	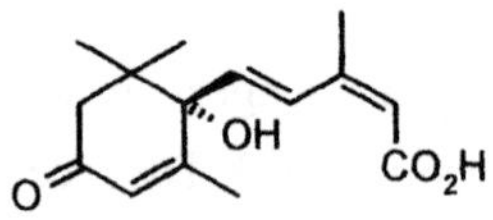 sesquiterp. abscisic acid	widespread occurrence in plant leaves: Schroeder 2001
Acetylcholine/on binding to the receptor, the channels are opened	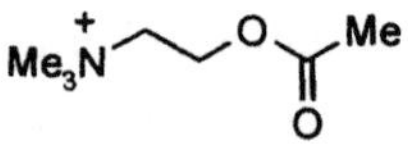 aminoalcohol acetylcholine	present in nervous tissues
Cannabinoids/on binding to the receptor, a CNS syndrome may be alleviated	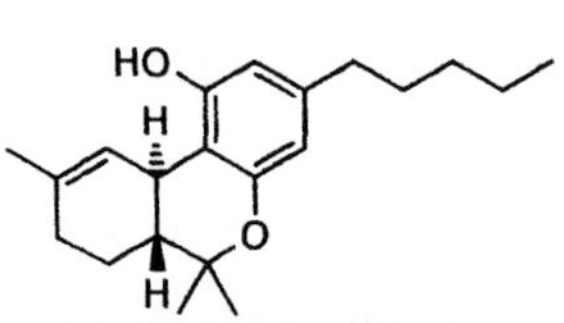 mero-terpenoid tetrahydrocannabinol	hemp (*Cannabis sativa* var. *indica* Auth., Moraceae, Ang.); cultivated
Cortisol/receptor activation causes primary and delayed secondary responses	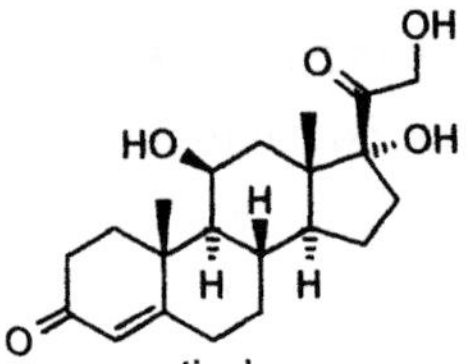 steroid cortisol	adrenal cortex of Mamm.
Cyclic AMP/intracellular mediator for many different hormones, communicating the signal by activating the cyclic AMP-dependent protein kinase	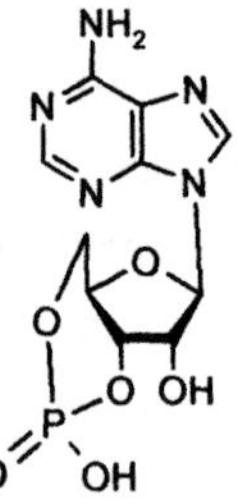 nucleoside phosphate cyclic AMP	bacteria, certain higher plants, and most animal cells

Table 12.5. Small-molecule mediators of cell signals

Name/Function	Chemical class/Structure	Origin
Cyclosporin A/binding to proteins serves to study cell trafficking and protein function: Haggarty 2000	cyclopept. 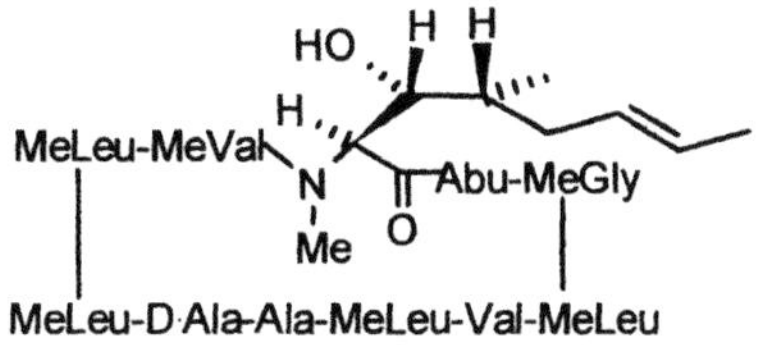 cyclosporin A	immunosuppressant agent from *Tolypocladium* Gams, Eumyc.
FK506 (tacrolimus)/ on binding to proteins it serves to unravel cell trafficking and protein function: Haggarty 2000	polyket. macrolide 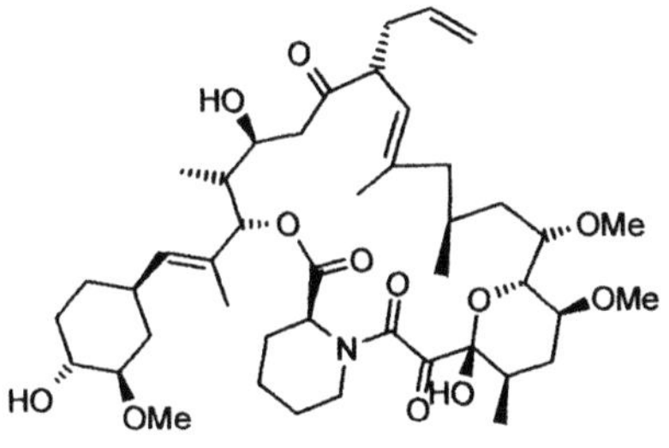 FK-506	immunosuppressant agent from *Streptomyces tsukubaensis*, Actinom., Bact.
Rapamycin/binding to proteins (used to unravel cell trafficking and protein function: Haggarty 2000)	polyket. macrolide 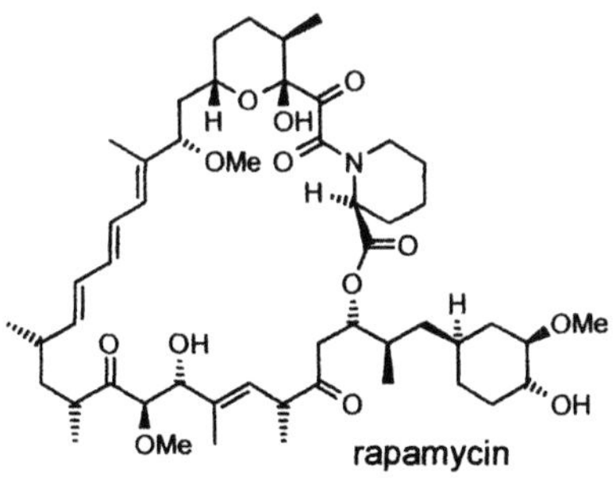 rapamycin	immunosuppressant from *Streptomyces hygroscopicus*, Actinom., Bact.

Chapter 13. Exploiting natural product diversity

In this chapter it is outlined how man has exploited the signaling, defensive, and environmentally relevant metabolites. That bioactive metabolites of phylogenetically distant organisms are detected by man's receptors finds rationalization in the origin of all organisms from a common ancestor and the limited possibilities of diversification of receptors and biosynthetic pathways. This is why the pharmaceutical industry is interested in chemical ecology (Caporale 1995).

Microbial life has contributed much to the drugs of the pharmacopeias. An astonishing one quarter of all biologically active secondary metabolites derives from filamentous fungi. This notwithstanding, the classification of the fungi is the most crude one could imagine, built around ecological observations. Molecular biology is a recent addition: attention begins to be paid to the phylogeny of ascomycetes from RNA data (Lumbsch 2000).

Filamentous fungi have been exploited so much since the second world war that the traditional niches are no more rewarding in the search for new drugs. The largest untapped resource is now believed to reside in endophytic fungi (Brady 2001; Tan 2001). Cryptic forms of marine fungi also remain to be explored, but things move sluggishly (Pietra 1997).

13.1 Food, food additives, and food processing from land and the oceans

Food from land and the oceans is largely made by proteins, fats, and carbohydrates. The composition is much the same when organisms belonging to a certain group are examined, without much geographical specialization, except the relative proportions of the components. This is particularly true for crops that, regrettably, have been reduced to single species, such as corn.

Certain crops require a special climate and soil. Plants native to the American tropical rain forest give coca (*Erythroxylum cocoa*), avocado (*Persea americana*), the American counterpart of the mango (*Spondias mombin*), guava (*Psidium guajava*), papaya (*Carica papaya*), the Brazilian nut (*Bertholletia excelsa*), and chewing-gum (latex from *Achras sapota*).

The Indo-Pacific rain forest is the mother ground of the banana (*Musa* spp.), mango (*Mangifera indica*), chewing gum (from *Dyera costulata*), and many spices (Table 13.1). With the advent of freezers, the spices have lost much importance as food preservatives, but their culinary role remains, albeit undermined by the fast food industry. Spices span a large variety of structures, from small isoprenoids to shikimates.

Kola (*Cola nitida* and *Cola acuminata*) is endemic of the African rain forest. It shows stimulating properties attributable to caffeine and theobromine present at high concentrations. Ethiopia is known for the largest variety of coffee seeds (of *Coffea* spp.) from an extremely varying micro-habitat, and spices.

The culture of edible seaweeds has a long tradition in the East. Particularly the red *Porphyra tenera* Kjelliman is the object of a large trade owing to a larger content of proteins, with a better balance of amino acids relevant to the human diet, than most other seaweeds. Certain marine mollusks, especially mytilids, are also widely cultured. With the decline of traditional fisheries, mariculture has much broadened, particularly in countries rich with protected bays, such as Japan.

The most renewed food adjuvants (health foods) are eicosapentaenoic acid (EPA), docosapentaenoic acid (DHA), astaxanthin, and *cis/trans*-β-carotene. EPA, an antithrombotic agent, is contained in appreciable amounts in most seaweeds, together with other antioxidants of the group mycosporine oxo acids. It can also be extracted from fish, or made by fermentation of certain marine bacteria. DHA - needed for the correct growth of the nervous system in mammals - is commercially made by the culture of a marine dinoflagellate, *Crypthecodinium cohnii*; thraustochitrids are a promising alternative source. Astaxanthin and *cis/trans*-β-carotene, being transmitted along the food chain, are less species-specific than the spices; they are best obtained by biotechnological processes or extraction from marine wastes. Flavoring agents for beverages, bread, drugs, and chewing gum span in structure from monoterpenes to steroid derivatives. Polysaccharides are used as food stabilizing agents and glycoproteins as sweetening agents.

Table 13.1. Food, food additives, and food processing

Area	Name/Chemical class/Structure	Source[a]
Health (antithrombogenic; DHA also CNS adjuvant: Mata de Urquiza 2001)	EPA (eicosapentaenoic acid), and DHA (docosahexaenoic acid)/polyunsaturated fatty acids EPA DHA	EPA: synthesis (Hoffmann-La Roche); **marine**: fermentation of *Alteromonas* f. *putrefaciens*, Bact. (Sagami). DHA: **marine**: culture of *Crypthecodinium conhii* (Dinofl.) (by Kawasaki Steel) or thraustochytrids (in development).
Health (food adjuvants)	astaxanthin/carotenoid astaxanthin (3S,3'S), (3R,3'R), (3R,3'S, meso) 1:2:1	Bact. and yeasts; also **marine**: Bact.: Yongmanitchai 1998
Health (food adjuvants)	1:1 9-*cis*-all-*trans* β-carotene/carotenoid β-carotene	**marine**: open air cultures of *Dunaliella* complex, Chloroph., mainly in Israel and Hawaii

Table 13.1. Food, food additives, and food processing

Area	Name/Chemical class/Structure	Source[a]
Health (food adjuvants)	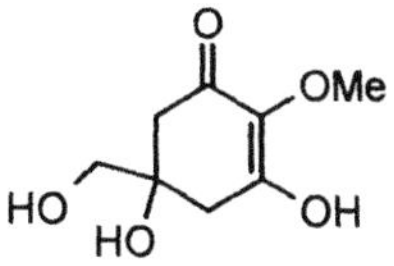 4-deoxygadusol/mycosporine derived antioxidant	**marine:** algae; in development at Topo Suisan Co for food processing and cosmetic applications

4-deoxygadusol

Health (vitamins)	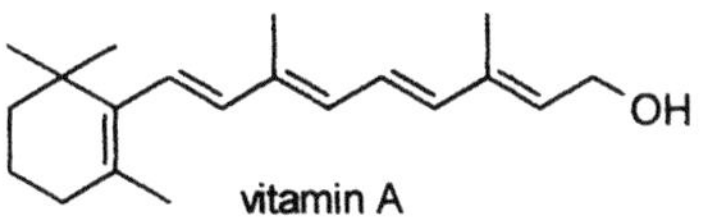 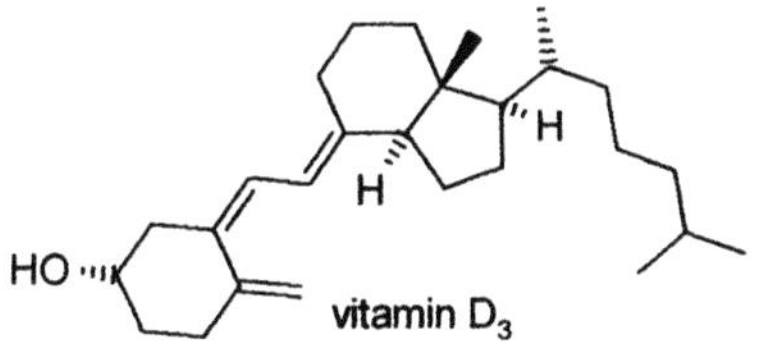 vitamins A and D$_3$, and ω3 fatty acids, EPA and DHA	**marine:** cod liver oil, mostly from cod from northwestern Atlantic

vitamin A

vitamin D$_3$

Flavoring agents (beverages)	 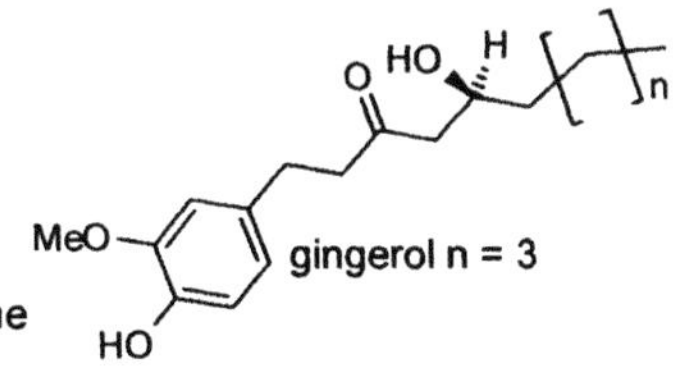 ginger: camphene/sesquiterp. and gingerol/shikim.	*Zingiber officinale* Roscoe, Zingiberaceae, Ang.

(-)-camphene   gingerol n = 3

Flavoring agents (bread)	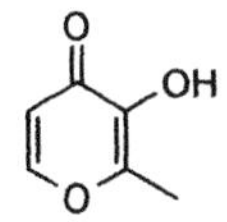 maltol/γ-pyrone	needles of *Abies alba* Mill. and bark of young *Larix decidua* Mill., Pinaceae, Gymn., from northern conifer forest

maltol

Flavoring agents (chewing gum)	Chicle	juice from *Manilkara zapotilla* (Jacq.) Gilly, Sapotaceae, Ang. from Central America and West Indies

Table 13.1. Food, food additives, and food processing

Area	Name/Chemical class/Structure	Source[a]
Flavoring agent (beverages and pharmacy)	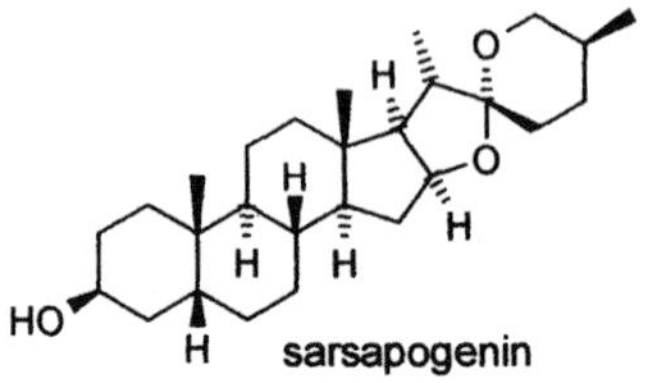 Mexican sarsaparilla/sarsaponin, whose aglycon is sarsapogenin	*Smilax aristolochiaefolia* Mill., Liliaceae, Ang. from Mexico
Food processing	ficin/proteolytic enzyme (C 6.1.3.A/P)	*Ficus* spp., Moraceae, Ang. from the East
Spices	cinnamon (white = canella): eugenol/shikim. 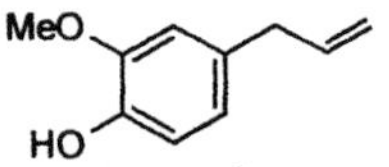	bark of *Canella alba* Murray [=*Canella winteriana* Gaertner], Canellaceae, Ang., from the Caribbean
Spices	cinnamon (Ceylon): cinnamaldeyde and tannins/shikim. 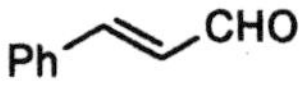	bark of *Cinnamomum* spp., Lauraceae, Ang. from Sri Lanka and Indonesia; widely cultivated
Spices	clove: eugenol and tannins/monoterp. and shikim.	flowers of clove, *Eugenia caryophyllata* Thunberg, Myrtaceae, Ang., native to Malaysia; cultivated in Indonesia and S America
Spices	curry: trigonellyne/alkal. (also **marine**); diosgenin/cardenolide C 8.1.I 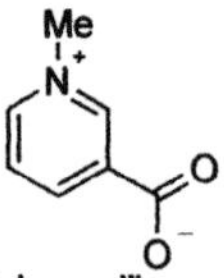	seeds of *Trigonella foenum graecum* L., Leguminosae, Ang. from Mediterranean regions and India

sarsapogenin

eugenol

cinnamaldehyde

trigonelline

Table 13.1. Food, food additives, and food processing

Area	Name/Chemical class/Structure	Source[a]
Spices	garlic: ajoene, allicin (garlic aroma), alliin (odorless (S)-2-propenyl-L-cysteine S-oxide): on crushing garlic, alliin is transformed by garlic alliinase into allicin through poorly known transient intermediates; on garlic maceration in oils, rather stable ajoene and other dithiins are formed from allicin; in onions the isomeric (S)-(E)-1-propenyl-L-cysteine S-oxide is transformed by onion alliinase into the lacrimatory factor, propanethial S-oxide/sulfur-bearing acetogenins and amino acids 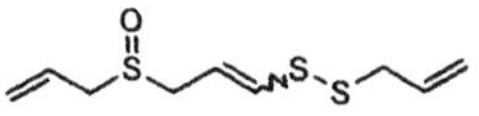 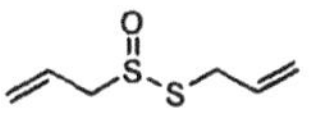 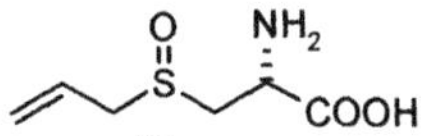	garlic, *Allium sativum* L., and Asiatic onion, *Allium cepa* L., Liliales, Ang;, cultivated: MI

ajoene (E/Z) 4:1   allicin

alliin $CH_3CH_2CH{=}SO$ propanethial S-oxide

Spices	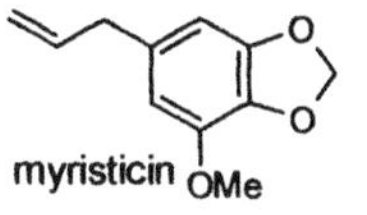 nutmeg: myristicin/shikim.	seeds of *Myristica fragrans* Houtt., Myristicaceae, Ang. from S Asia; cultivated in the tropics

myristicin OMe

Spices	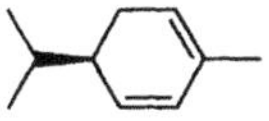 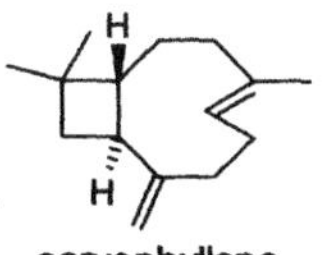 oil of pepper: α-phellandrene and caryophyllene/sesquiterp.	unripe black-pepper fruits, Ang.

(-)-α-phellandrene   caryophyllene

Spices	oil of Chinese cinnamon : cinnamaldehyde and eugenol/shikim.	oil from leaves of *Cinnamomum cassia* Nees, Lauraceae, Ang. from China

Table 13.1. Food, food additives, and food processing

Area	Name/Chemical class/Structure	Source[a]
Spices	pepper: piperine, clavicine, piperidine alkal., also bearing a cyclobutane ring, such as pipercyclobutanamide A: Fujiwara 2001 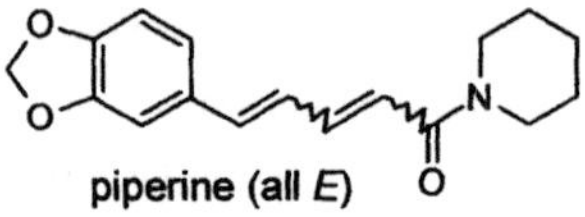 piperine (all *E*) chavicine (all *Z*) 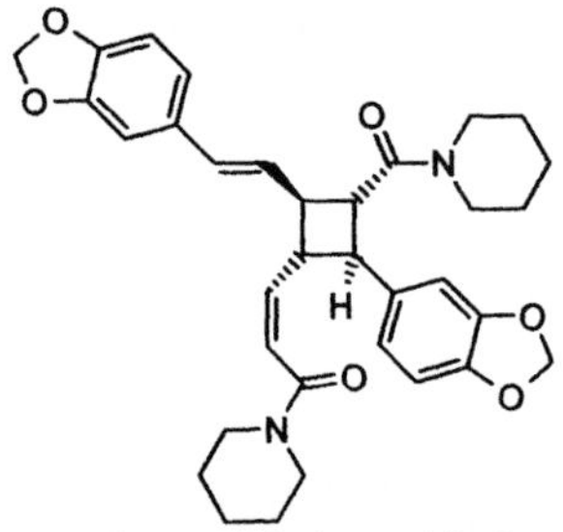 pipercyclobutanamide A	black pepper: *Piper* spp., including *Piper nigrum* L., Piperaceae, Ang. from India, Indonesia, and the Philippines
Stabilizing agents (beer, wine, ice cream, meat, bread)	carrageenans/sulfated polysacch.	Rhodoph.: *Chondrus crispus* Stackhouse in Brittany (Irish Moss) and Chile, and *Gigartina* spp. in NZ and Argentina
Sweetening agents	miraculin/ca. 44,000 MW glycoprotein	red berries of *Synsepalum dulcificum* (Schum.) Daniell, Sapotaceae, Ang. from W Africa.
Sweetening agents	thaumatin/ca. 22,000 MW protein, in different forms	fruits of *Thaumatococcus danielli* Benth., Marantaceae, Ang. from W Africa

[a]From land, unless marine origin is indicated in boldface.

13.2 Commercial natural drugs and folk medicines

Officially, about 25% of the most prescribed drugs derive directly or indirectly from natural products, which is probably an underestimate because of biopyracy (Carlson 1997).

Educated screening has superseded the random screening of living organisms in the search for new drugs (Houghton 1999). According to recent estimates, the chance is of one valuable substance in about 4,000 specimens (Cragg 1998). With the new targets provided by the sequence of the human genome (Lander 2001), the chance should be much higher. The 30,000-40,000 genes of the human genome are expected to raise the number of targets to 10,000 from the current 500 targets of the pharmacopeias (Drews 2000). Anyway, the natural product is rarely used directly as a drug. Most often it is just an advanced synthon, or a template, for the drug.

In a kind of educated screening, a few valuable drugs have been discovered through ethnobotanical practices. The active agent of arrow poison in Amazonia, tubocurarine, permits open-heart surgery. The active agent in witch ordeals in Nigeria, physostigmine, and its successor, pilocarpine, from South America, serve for the treatment of glaucoma. The difficulty in drug discovery through ethnobotanical practices lies in the secrecy on the procedures used by the sorcerer. Moreover, prevalent diseases that are found to occur in isolated populations, which consume raw, unpolluted, food in a clean atmosphere, may not coincide with the commonest diseases found in industrialized societies, where adulterated food is consumed, either in the form of fast food on the corner of the street, or in an environment polluted by vehicles. At any event, the disappearance of living species and indigenous knowledge at no pace, diminish the chances to get new remedies from tropical plants. Major problems can be foreseen, because more than 80% of the world's population - mostly in poor countries - relies directly on plants for primary health care (Cox 1997).

In the search for new drugs, most pharmaceutical companies have abandoned the natural product. Switzerland is a case in point: out of twenty start-up and spin-off pharmaceutical companies, seven make use of parallel and combinatorial synthesis, which is much less expensive than the search for bioactive natural products and is free from obligations toward the countries holding the rights for the germ plasm. Only one of the new Swiss companies relies on natural products, merely raw extracts for veterinary use (L'Epplatenier 2000).

The human genome sequence has also opened up new opportunities for the design from scratch of new drugs. There are already tangible signs to this regard with synthetic protease inhibitors of HIV, the agent of AIDS (Chapter 15). Even gene therapy is getting new opportunities (Abbott 2001). Perhaps, the major innovation in the post-modern drug area will be the genetic categorization of diseases (Henry 2001). Herceptin, an engineered monoclonal antibody by Genentech, is an example: an accompanying kit may reveal if the patient's breast cancer overexpresses the human epidermal growth factor receptor, a condition required for treatment by this drug. In any event, Genentech warns against possible devastating side effects, like ventricular dysfunction, congestive heart failure, and hypersensitive reactions.

Probably, however, it is the combined use of new discoveries in genomics, proteomics, and natural product chemistry that will provide the most benefit. An example is the control of the cholesterol level from the understanding of the receptors and genes involved. This requires small molecules that specifically interact with the receptors without side effects. Such agents are not comprised in the list of available drugs (Repa 2000) but may be found in the natural products, owing to their higher variety and complexity (Henkel 1999). Likewise, new agents are needed to alter the metabolism by interacting with RNA and modifying its conformation and ability to combine with proteins (Hermann 2000). Protein-small molecule interactions are also in the limelight (Chapter 12.5, Haggarty 2000).

In this postmodern approach to pharmaceuticals, chemistry is in a pole position for the construction of artificial chromosomes for gene therapy (Willard 2000) and the synthesis of regulatory agents devised through the computer.

Examples of natural products and derived products approved as drugs or drug components are found in Table 13.2.I. From left to right the column entries are the therapeutic area, the name, chemical class, and structure of the drug, the trade name and the producer or inventor, and, finally, the source. From top to bottom, Table 13.2.I shows agents active in the various diseases and symptomatic drugs; these are briefly illustrated in the following, making the therapeutic area visible by drawing a line underneath. The antibiotics are common fast-acting drugs. Antibacterial antibiotics are found in the peptides (penem alkaloids, bacitracin, and tuberactinomycins), polyketides (pseudomonic acid A) including macrolides (erythromycin), and aminoglycosides (neomycin B, nebramycin, and streptomycin). The antibacterial agents have lost much efficiency with the emergence of resistant strains. This aside, research into new antibiotics is lagging. According to a recent report, expired or expiring patents, western overproduction, and the intervention of emerging countries, have contributed to cut down the price of classical antibiotics, discouraging new enterprises. Major western producers have planned to shut down their plants or move to other therapeutic areas, especially food additives, which benefit from looser regulations (McCoy 2000). Less known is that the production of antibiotics has been moved by the major companies to emerging countries, like China, relying on loose regulations for environmental pollution (Chapter 16.2.3). Antifungal agents are rarer, being mostly found in the classes polyketides (amphotericin B) and terpenoids (pleuromutilin). Antiparasitic agents are found in the alkaloids, particularly of the tetrahydroisoquinoline (emetine) and terpenoid classes (quinine). The interferons (proteins) and carrageenans (sulfated polysaccharides) are typical natural antiviral agents. Agents active on blood, gout, and inflammation comprise phenethylisoquinoline alkaloids (colchicine), proteins (batraxobin and bromelain), polyketides (mevinolin), and shikimates (acetylsalicylic acid). Drugs for the chemotherapy of cancer are found in all biogenetic classes: alkaloids (pentostatin, vinblastine), peptides (bleomycins and actinomycin D), proteins (L-asparaginase, interferons, interleukin-2), terpenoids (paclitaxel and the meroterpenoid Δ^9-tetrahydrocannabinol), polyketides (daunomycin and doxorubicin), and shikimates (podophyllotoxin). Cardiovascular drugs are found in β-carboline alkaloids (deserpidine), indole alkaloids (reserpine), and cardenolide glycosides (digitalin and digoxin). Agents active on <u>cholinergic receptors</u> are based on tropane (atropine), indole (physostigmine), and imidazole alkaloids (pilocarpine). <u>CNS active agents</u> are found in purine (caffeine), benzylisoquinoline alkaloids (codeine), tropane (scopolamine and cocaine), and *Lycopodium* alkaloids (huperzine A), other than in the amino acids (levodopa). <u>Dermatological agents</u> are represented here by a terpenoid, retinoic acid. <u>Endocrine drugs</u> are mostly found in the isoprenoids (estrogens), polypeptides (insulin), and amino acids (levothyroxine). Drugs for the <u>metabolic disorders</u> are best represented by a γ-amino acid, levocarnitine. Chymopapain, a proteolytic enzyme, is a highly valued agent in <u>orthopedics</u>. Agents active on <u>smooth muscles</u> are found in isoquinoline (tubocurarine), phenethylamine (pseudoephedrine), purine (theophylline), and isoprenic alkaloids (ergot). The triterpene escin is the most widely known agent for the treatment of the degeneration of the <u>pheripheral vascular system</u>. <u>Symptomatic therapy</u> is carried out with the polycarboxylic agaricic acid.

In a biogeographic perspective, drugs from European and Asiatic plants are best represented by berberine (a gastric ailment and antipyretic agent from the angiosperm *Berberis vulgaris*, Ranunculales, and other trees) and aspirin. The latter is long made by total synthesis, while extraction from the willow tree (*Salix alba*, Salicales) is confined to historical interest. Middle East and northern Africa medicinal plants are best represented by *Silybum marianum*, Asterales, for activity against liver disorders, and *Urginea maritima*, Liliales, for scillaren A, a cardiotonic agent. From the East, L-ephedrine is a valuable bronchodilator and decongestant isolated from the gymnosperm *Ephedra sinica* from China. It served also as a template for many synthetic drugs that act on both smooth muscles and the cardiovascular system. The best examples from eastern mountain regions are conifers in the genus *Taxus* that give paclitaxel and related diterpenoids for the treatment of solid forms of human cancer.

Herbal products are used differently according to traditions and regulations in the different countries. Germany and France from one side - where herbal products are widely regarded as medicines - and The Netherlands from the other side - where a single herbal product is held as a medicine - represent the two extreme positions. Now the attention is focused on multicomponent crude extracts from different plants to achieve synergistic effects and mutual stabilizing effects. The list of herbal products provided in Table 13.2.II comprises agents active on pathogenic microorganisms (extracts from St Johns wart), CNS (Khat, Channa, and extracts from *Ginkgo biloba*), gastric ulcer (licorice), gastrointestinal disorders (Sangre de Grado), hypercholesterolemia (garlic), liver disorders (extracts from milk thistles), prostatitis (the saw palmetto), peripheral vascular disorders (bilberry and English Hawthorn), and worm infestation (seeds of the betel palm). Included are also extracts for symptomatic treatment (Siberian oil of fir and ginger) and panaceas (extracts from shiitake mushrooms).

Not everything is centered on drugs or gene therapy. Resurgence is also seen as a long abandoned idea in a difficult area of therapeutics, that is, viruses showing lytic properties against bacterial pathogens, called bacteriophages (Barrow 2000).

Table 13.2.I. Natural (or derived) products approved as drugs or drug components

Area	Name/chemical class/structure	Trade name/Co	Source[a]
Antibacterial	bacitracin A/pept. 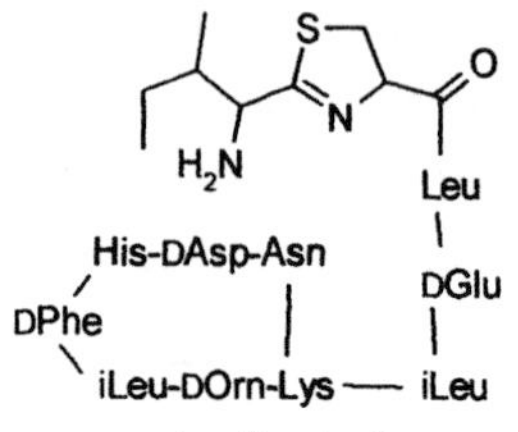	Bacitracin/ Merck	*Bacillus subtilis* (Ehrenberg 1835) Cohn 1872, Bact.

bacitracin A

Antibacterial	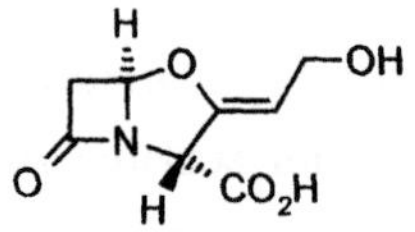 clavulanic acid and amoxillin trihydrate/penem alkal.	Augmentin/ SKB; Klavocin/Pliva	*Streptomyces clavuligerus* Higgens and Kastner 1971, Actinom., Bact.

clavulanic acid

Antibacterial	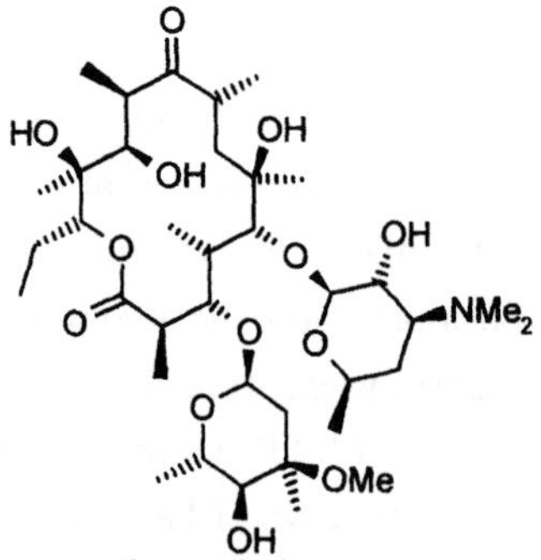 erythromycin A/polyket.	Ery-Tab/ Abbot; Erymax/Parke Davis	*Streptomyces erythreus* (Waksman 1923), Waksman and Henrici 1948, Actinom., Bact.

erythromycin A

Antibacterial	mupirocin (= pseudomonic acid A)/polyket. C 8.3.FA/PO 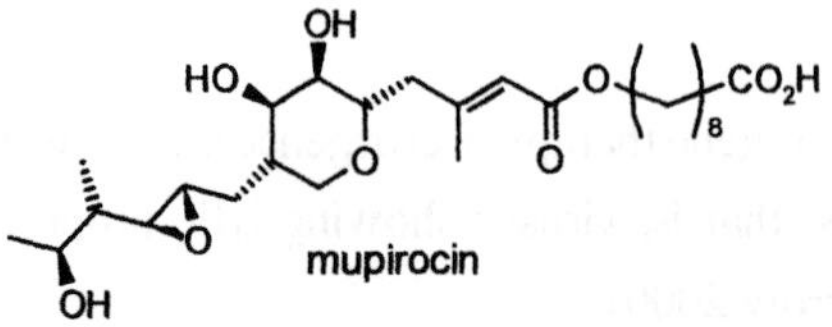	Bactroban/ SKB	*Pseudomonas fluorescens* Migula 1895, NCIB 10586, Bact.

mupirocin

Table 13.2.I. Natural (or derived) products approved as drugs or drug components

Area	Name/chemical class/structure	Trade name/Co	Source[a]
Antibacterial	neomycin A, B, C/aminoglycosides interfering with DNA: Hermann 2000 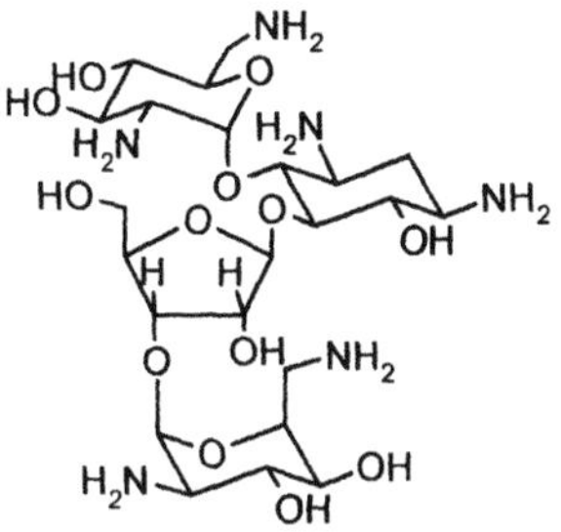	Mycifradin/ Upjohn	*Streptomyces fradiae* (Waksman and Curtis 1916) Waksman and Henrici 1948, Actinom., Bact.

neomycin B

Antibacterial	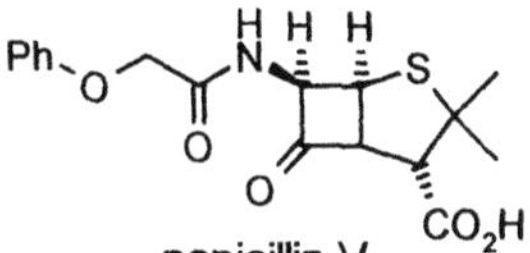 penicillin V/penem alkal.	Oracilline/ Rhône-Poulenc Rorer; V Cillin/Lilly	*Penicillium* spp., Deuterom., Eumyc.

penicillin V

Antibacterial	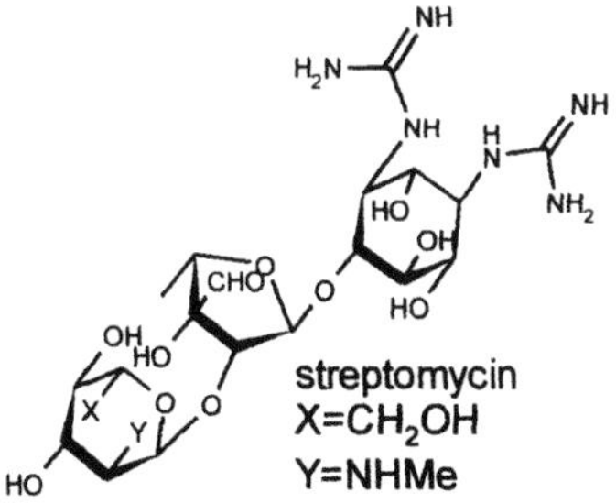 streptomycin/ alkal. glycoside	Pfizer, Merck & Co.	*Streptomyces griseus* (Krainsky) Waksman et Henrici, Actinom., Bact.

streptomycin
X=CH$_2$OH
Y=NHMe

Antibacterial	tobramycin (10% nebramycin)/aminoglycoside interfering with DNA: Hermann 2000 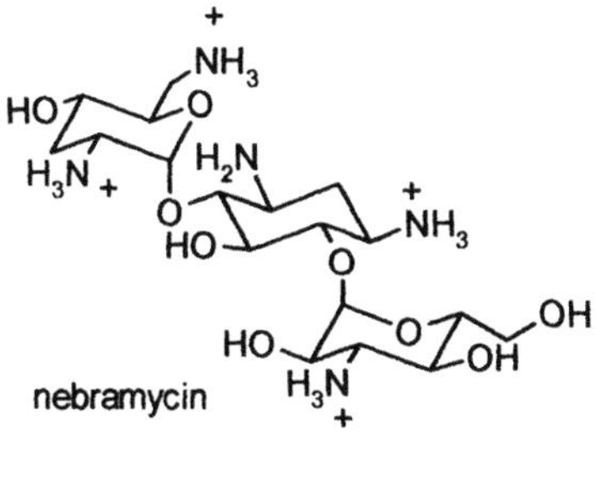	Gernebicin/ Lilly	*Streptomyces tenebrarius*, Actinom., Bact.

nebramycin

Table 13.2.I. Natural (or derived) products approved as drugs or drug components

Area	Name/chemical class/structure	Trade name/Co	Source[a]
Antibacterial	tuberactinomycin A, B, N, O/cyclopept., interacting with DNA: Hermann 2000	Tuberactin/ Toyo Jozo; Vionactan/ Ciba	*Streptomyces griseoverticillatus* (Shinobu and Shimada 1962) Witt and Stackebrandt 1991, var. *tuberacticus*, Actinom., Bact.

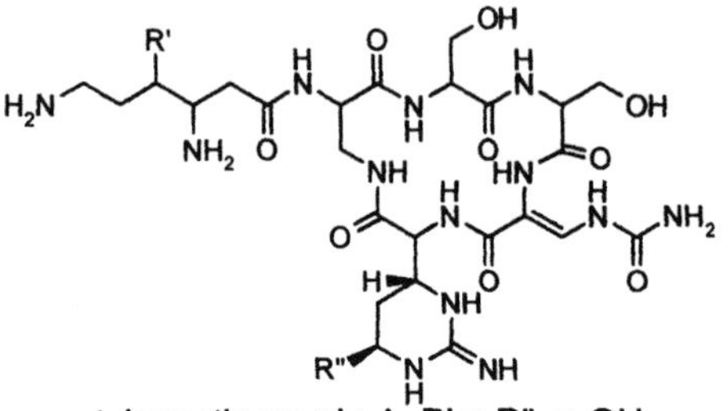

tuberactinomycin A   R' = R" = OH
tuberactinomycin B   R' = H, R" = OH
tuberactinomycin N   R' = OH, R" = H
tuberactinomycin O   R' = R" = H

Area	Name/chemical class/structure	Trade name/Co	Source[a]
Antifungal	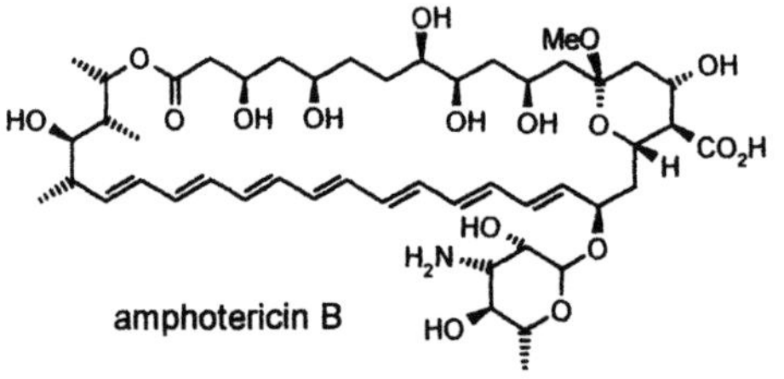 amphotericin B/ glycosidic polyene polyket.: MI	Fungizone/ Bristol-Myers Squibb; Amphocil/ Zeneca; nephrotoxic fungicidal agent which, on liposome formulation, acquires safety (AmBisome/ Glaxo)	*Streptomyces nodosus* Trejo 1961, from Orinoco soil, Actinom., Bact.

amphotericin B

Area	Name/chemical class/structure	Trade name/Co	Source[a]
Antifungal	pleuromutilin/labdane-derived diterp. 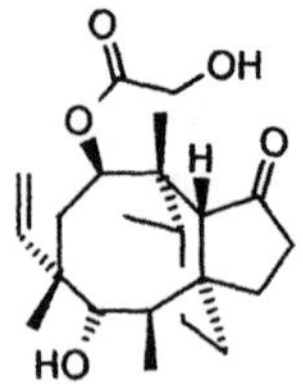	lead for many antibiotics: Erkel 1997	*Pleurotus mutilus* (Fr.) Sacc. and *Pleurotus passeckerianus* Pilat, Eumyc.: MI

pleuromutilin

Table 13.2.I. Natural (or derived) products approved as drugs or drug components

Area	Name/chemical class/structure	Trade name/Co	Source[a]
Antiparasitic	emetine tetrahydro-isoquinoline alkal. 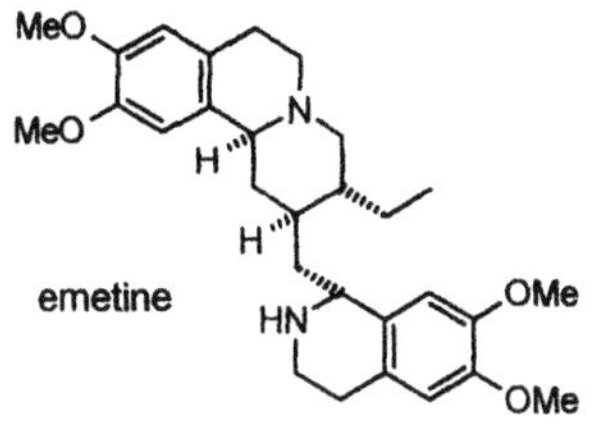	Hemometina/ Cusi	ipecac, *Uragoga ipecacuanha* (Brot.) Baill. [= *Cephaelis ipecacuahna* (Brot.) A. Rich.], Leguminosae, Ang. from tropical S America
Antiparasitic	quinine/terp. alkal. 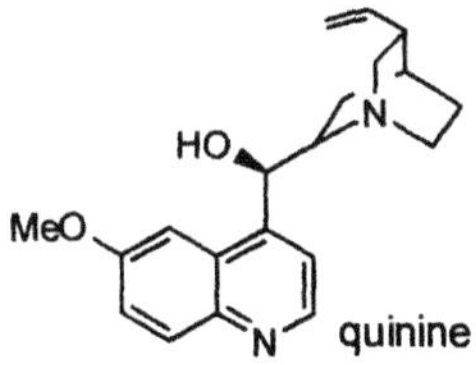	Biquinate/ Rhône-Poulenc Rorer; Dentojel/ Ayerst	*Cinchona* spp., Rubiaceae, Ang.
Antiviral	interferons/proteins	(γ) Immukin/ Boehringer Ing.	cloned
Antiviral	carrageenans/anti-HIV sulfated polysacch.	Carrageenan	red seaweeds: *Chondrus crispus* Stackhouse in Brittany (Irish Moss) and Chile, and *Gigartina* spp. in NZ and Argentina
Blood/gout inflam-mation	aspirin (2-(acetyloxy)benzoic acid = acetylsalicyclic acid) shikim. deriving from cinnamic acid 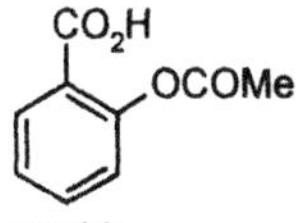	Aspro/ Nicholas; Arthrisin/ Sandoz	European willow tree, *Salix alba* L., Salicaceae, and *Spiraea* spp., Rosaceae, Ang.; now synthetic
Blood/gout/ inflam-mation	batraxobin/thrombin-like enzyme	Reptilase/ Knoll	*Botrox atrox* (L.) viper from tropical central and S America

Table 13.2.I. Natural (or derived) products approved as drugs or drug components

Area	Name/chemical class/structure	Trade name/Co	Source[a]
Blood/gout/ inflam- mation	bromelain/protein digesting enzymes	Ananase/Rorer (= Inflamen/ Hokuriku)	pineapple, *Ananas comosus* (L.) Merr., Bromeliaceae, Ang.
Blood/gout/ inflam- mation	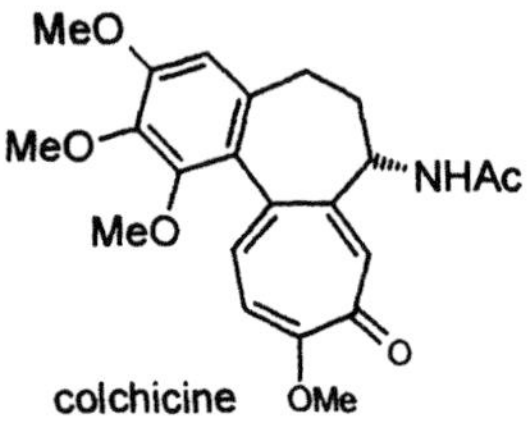 colchicine/phenethylisoquinoline alkal.	ColBenemid	autumn crocus, *Colchicum autumnale* L., Liliaceae, Ang.

colchicine

Area	Name/chemical class/structure	Trade name/Co	Source[a]
Blood/gout/ inflam- mation	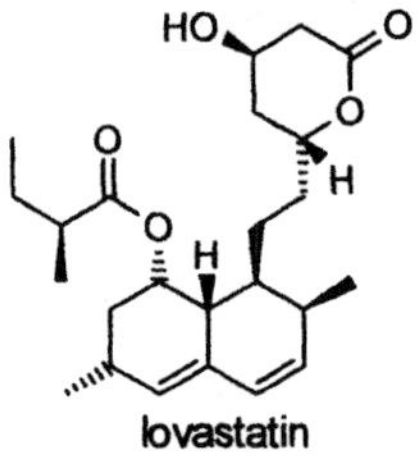 mevinolin (= lovastatin = methylcompactin)/polyket.	Mevacor/ Merck & Co; Lovalip/Riesel	*Aspergillus terreus* Thom. *and Monascus ruber* van Tieghem, Eumyc.

lovastatin =mevinolin

Area	Name/chemical class/structure	Trade name/Co	Source[a]
Anticancer	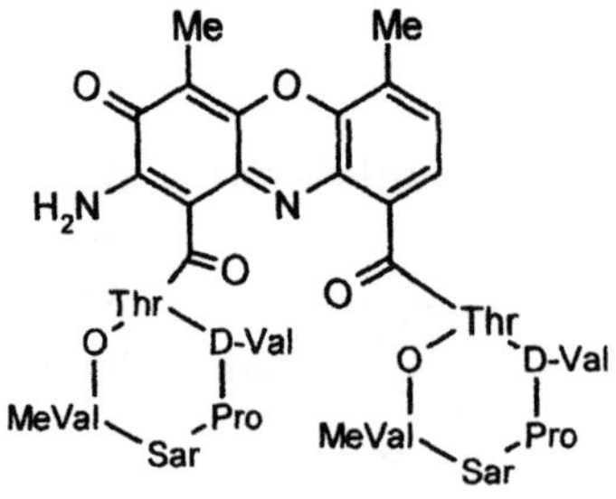 actinomycin D/phenoxazinone pept. alkal.	Cosmegen/ Merck & Co	*Streoptomyces* spp., Bact.

actinomycin D

Table 13.2.I. Natural (or derived) products approved as drugs or drug components

Area	Name/chemical class/structure	Trade name/Co	Source[a]
Anticancer	L-asparaginase/enzyme	Elspar/Merck & Co; Kidrolase/ Rhône-Poulenc	*Escherichia coli* (Migula 1895) Castellani and Chalmers 1919, Bact.
Anticancer	bleomycins (mainly A2)/ glycopept. 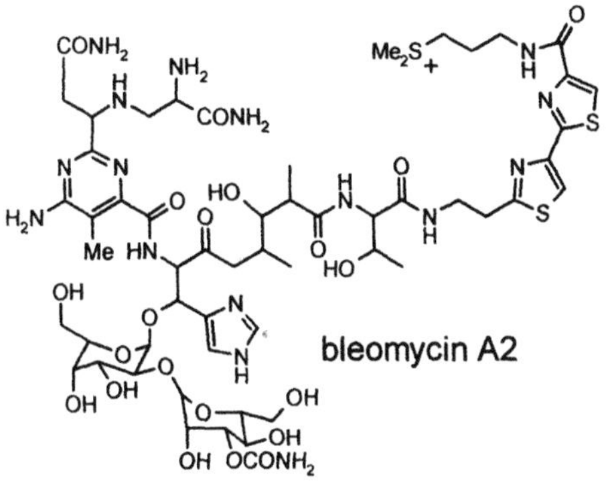	Blenoxane/ Bristol-Myers Squibb; Bleo/Nippon Kayaku	*Streptomyces verticillus*, Actinom., Bact.
Anticancer	daunomycin/polyket. anthracycline glycoside 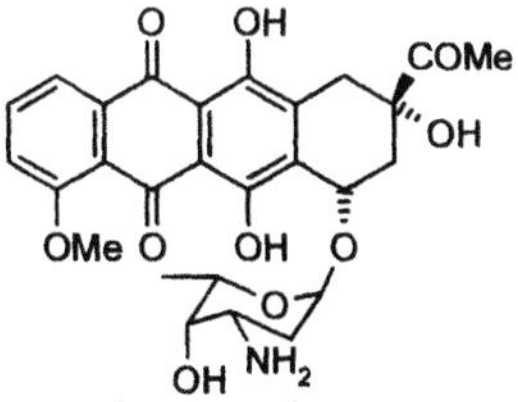	Daunoblastina/ Farmitalia; Cérubidine/ Rhône Poulenc Rorer	*Streptomycs peucetius*, Bact.
Anticancer	doxorubicin/anthracycline polyket. glycoside 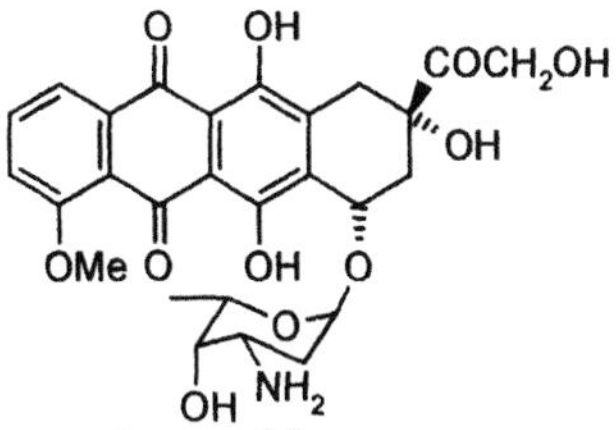	Adriamycin/ Farmitalia; Adriacin/ Kyowa	*Streptomyces peucetius* var. *caesius*, Actinom., Bact.

bleomycin A2

daunomycin

doxorubicin

Table 13.2.I. Natural (or derived) products approved as drugs or drug components

Area	Name/chemical class/structure	Trade name/Co	Source[a]
Anticancer	interferons/cytokine endogenous proteins	(γ) Immukin/ Boehringer Ing.	cloning
Anticancer	interleukin-2/T-cell cytokine protein (growth factor)	Leuferon-2/ Roche; Proleukin/ Chiron	microbial expression of human IL-2
Anticancer	mithramycin/polyket. glycoside 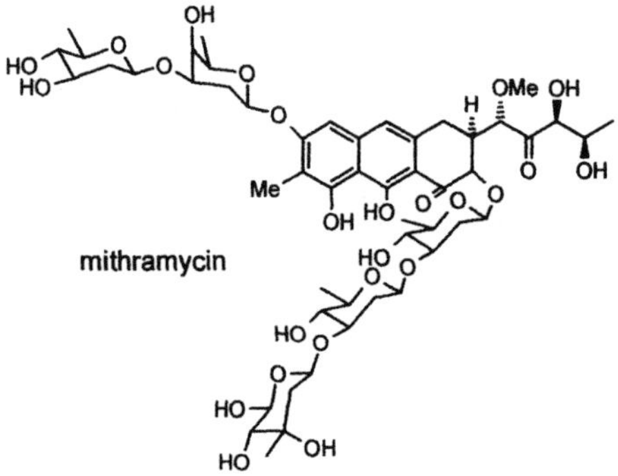 mithramycin	Mithracin/ Pfizer	*Streptomyces argillaceus* and *Streptomyces tanashiensis* Hata *et al.* 1952, Actinom., Bact.
Anticancer	neocarzinostatin/chromoprotein-class enediyne antitumor agent, composed of a protein and a labile chromophore	SMANCS (associated to polymer) in Japan	*Streptomyces carcinostaticus*, Actinom., Bact.
Anticancer (ovarian)	paclitaxel/diterp. 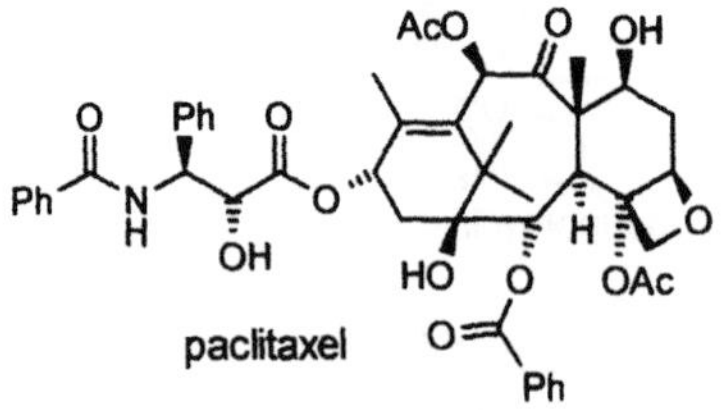 paclitaxel	Taxol/Bristol-Myers Squibb	Pacific yew, *Taxus brevifolia* Nutt., Taxaceae, Gymn.
Anticancer	pentostatin/imidazole- diazepine alkal. 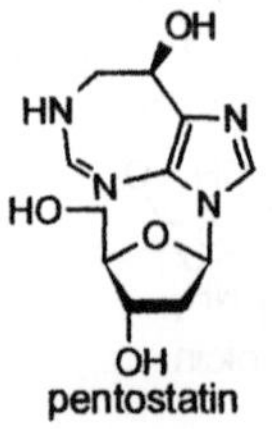 pentostatin	Nipent/Lederle	*Streptomyces antibioticus* (Waksman and Woodruff 1941) Waksman and Henrici 1948, Actinom., Bact.

Table 13.2.I. Natural (or derived) products approved as drugs or drug components

Area	Name/chemical class/structure	Trade name/Co	Source[a]
Anticancer	podophyllotoxin/mixed polyket. shikim. poisonous agent 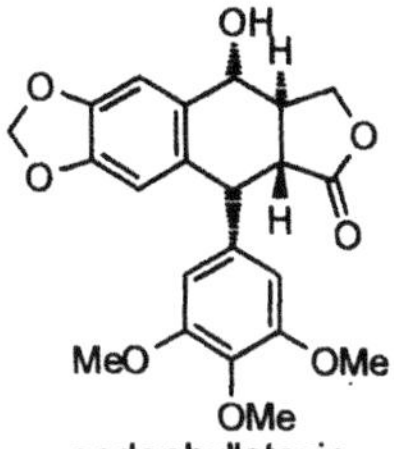 podophyllotoxin	Condyline/ Brocades; Condylox/ Oclassen	N American mayapple, *Podophyllum peltatum* L., Berberidaceae, Ang.; poisonous agent once used by the American Indians to commit suicide
Anticancer with many unwanted side effects (see Chapter 13.3)	Δ⁹-tetrahydrocannabinol/meroterpenoid 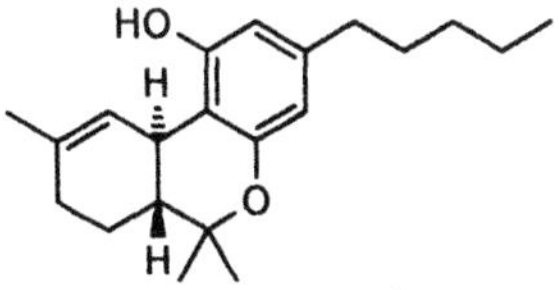 tetrahydrocannabinol	Marinol/ Roxane	hemp, *Cannabis sativa* var. *indica* Auth., Moraceae, Ang.; cultivated
Anticancer	vinblastine, vincristine/terp. alkal. 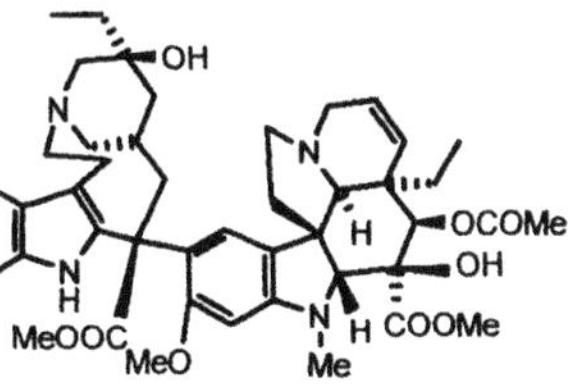 vinblastine	Velban and Oncovin/Lilly	periwinkle, *Vinca rosea* L. [= *Catharanthus roseus* (G. Don.)], Apocyanaceae, Ang.; trop., now cultivated
Cardio-vascular	deserpidine/β-carboline alkal. 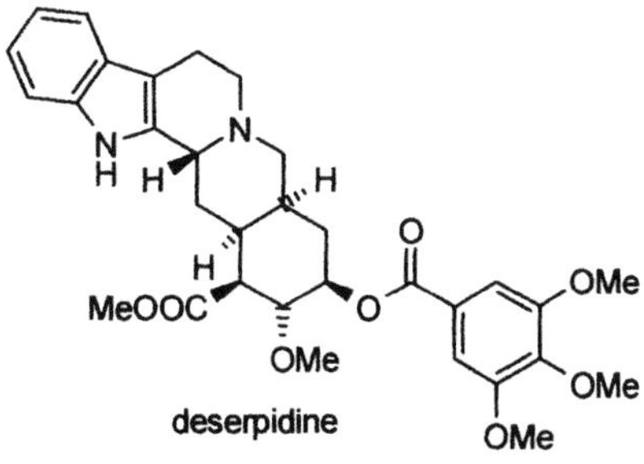 deserpidine	Harmonyl/ Abbott; Raunormine/ Penick	*Rauwolfia tetraphylla* L. or *Rauwolfia canescens* L., Apocynaceae, Ang.

Table 13.2.I. Natural (or derived) products approved as drugs or drug components

Area	Name/chemical class/structure	Trade name/Co	Source[a]
Cardio-vascular	digitalin/cardenolide glycoside 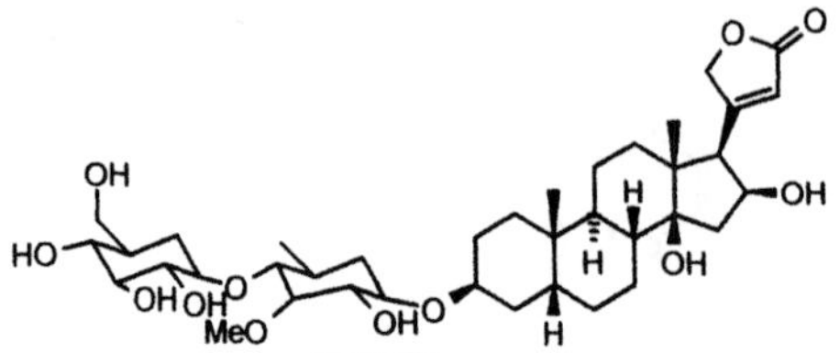 digitalin	drugstore	roots of *Digitalis purpurea* L., Scrophulariaceae, Ang., and seeds of *Adenium honghel* A. DC., Apocyanaceae, Ang.
Cardio-vascular	digoxin and digitoxin/cardenolide glycosides 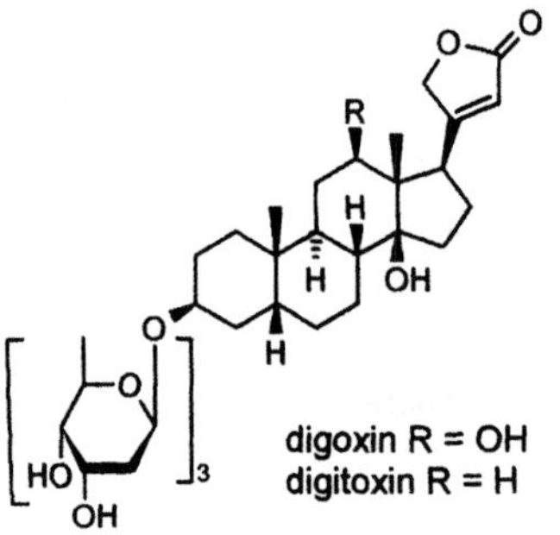 digoxin R = OH digitoxin R = H	Digacin/ Beiersdorf-Lilly; Lanoxin/Glaxo	foxglove, *Digitalis lanata* Ehrh. (digoxin) and *Digitalis purpurea* L. (digitoxin), Scrophulariaceae, Ang.
Cardio-vascular	reserpine, rescinnamine, ajmalicine/indole alkal. 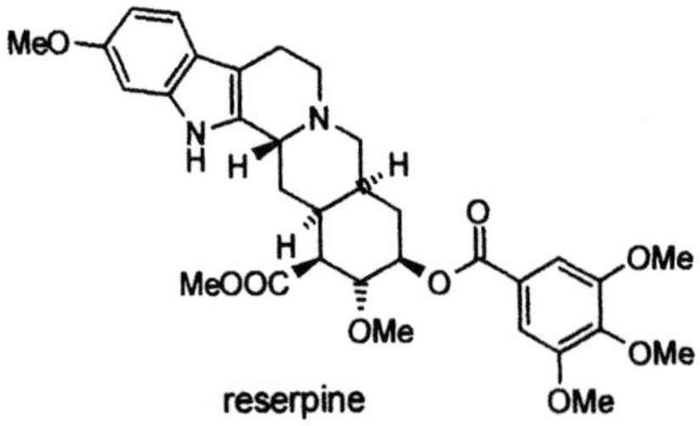 reserpine	Raused/ Bristol-Myers Squibb; Serfin/Parke-Davis); reag. grade: Sigma 2000 and Fluka	roots of Indian *Rauwolfia serpentina* L. Benth., Apocynaceae
Cholino-ceptor	(±)-atropine/tropane alkal. 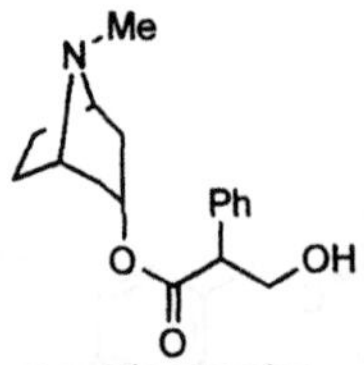 racemic atropine	Eumydrin/ Winthrop; Ekomine/ Hoechst	*Atropa belladonna* L. and *Datura stramonium* L., Solanaceae, Ang.

Table 13.2.I. Natural (or derived) products approved as drugs or drug components

Area	Name/chemical class/structure	Trade name/Co	Source[a]
Cholino-ceptor	physostigmine/indole alkal.	Cogmine/ Forest; Eserine/Alcon	*Physostigma venenosum* Balf., Leguminosae, Ang., native of Nigeria
Cholino-ceptor (anti-glaucoma)	pilocarpine/imidazole alkal.	Almocarpine/ Ayerst; Akarpine/Akorn	Brazilian *Pilocarpus jaborandi* Holmes, Rutaceae, Ang.
CNS	caffeine/purine alkal.; codeine/benzylisoquin. alkal.; aspirin/phenolic; butabarbital/oxadiazine	Fiorinal/ Sandoz	caffeine: *Coffea arabica* L., Rubiaceae; codeine: *Papaver sonniferum* L., Papaveraceae, Ang.; aspirin: *Salix alba* L., Ang.; butalbital (synthetic)
CNS	cocaine/tropane alkal.	restricted use in ophthal-mology	*Erythroxylum coca* Lamarck 1786, Erythroxylaceae, Ang. from hills of tropical SW America
CNS	huperzine A/*Lycopodium* alkal.	Chinese companies	*Huperzia serrata* (Thunb.) Trev. [= *Lycopodium serratum* Thunb.], Lycopodiophyta from China
CNS	levodopa/amino acid	Dopaston/ Sankyo; Laradopa/ Roche	Velvet bean, *Mucuna deeringiana* (Bort.) Merr., Leguminosae, Ang.

Table 13.2.I. Natural (or derived) products approved as drugs or drug components

Area	Name/chemical class/structure	Trade name/Co	Source[a]
CNS	scopolamine/tropane alkal. 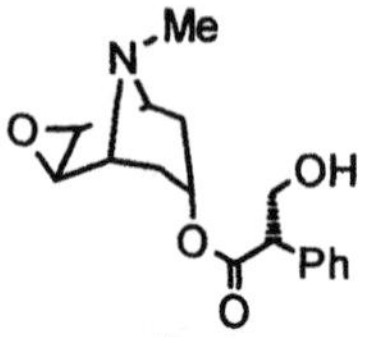 scopolamine	Scop/ Ciba-Geigy	henbane, *Hyosciamus niger* L., Solanaceae, Ang.
Dermatology	retinoic acid/degr. terpenoid 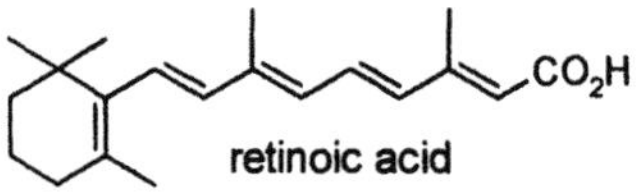 retinoic acid	Retin-A/ Ortho; Dermairol /Roche	vegetable oils (and many synthetic analogues)
Endocrine	estradiol, estradiol benzoate/steroids 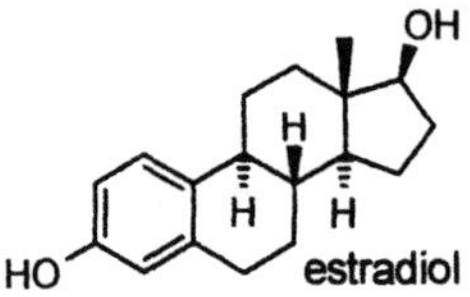 estradiol	Benzo- Gynestryl/ Roussel; analogues: Conestron/ Wyeth	follicular fluid of sow ovaries and urine of pregnant mares
Endocrine	insulin/polypept. hormone	Hypurin/CP Pharm (bovine)	pancreas of animals; now recombinant (Huminsulin/Lilly)
Endocrine	levothyroxine/amino acid 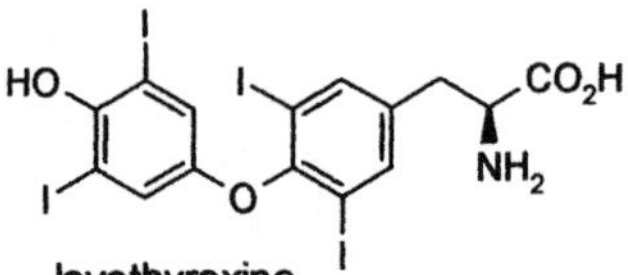 levothyroxine	Eltroxin/ Glaxo; Synthroid/ Sodium Flint	thyroid gland of farm animals; also synthetic
Metabolic disorders	levocarnitine/γ-amino acid betaine 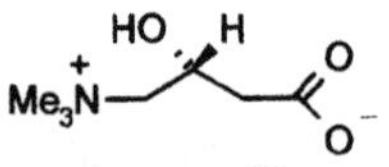 levocarnitine	Carnitor/ Sigma-Tau; Lefcar/Glaxo	meat extract
Orthopedics	chymopapain/proteolytic enzyme	Chymodiactin/ Smith; Discase/Boots	papaya, *Carica papaia* L., Caricaceae, Ang.; cultivated

Table 13.2.I. Natural (or derived) products approved as drugs or drug components

Area	Name/chemical class/structure	Trade name/Co	Source[a]
Smooth muscle	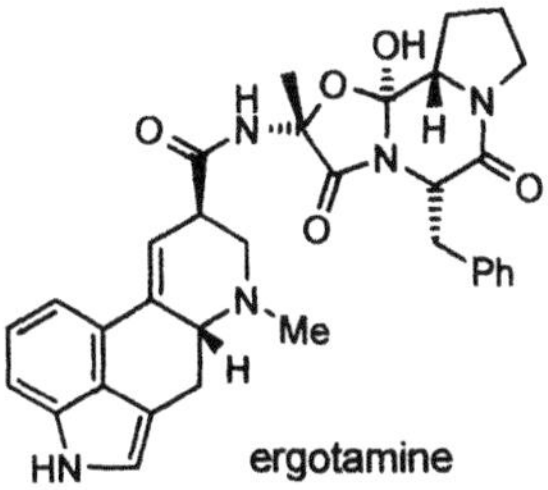 ergot (ergotamine, ergometrinine)/tryptophan-terp. alkal.	Ergostat/ Parke, Davis; Gynergen/ Sandoz	*Clavices purpurea* (Fries) Tul., Eumyc.

ergotamine

Smooth muscle	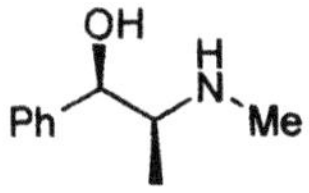 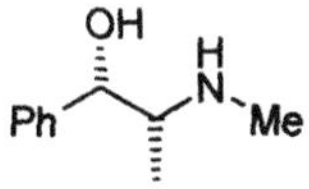 l- and d-pseudoephedrine/phenethylamine alkal.	Ephedral/ Sanedrin; Otrinol/ Ciba-Geigy;	*Ephedra* spp., Gnetaceae, Gymn. from China; l-(bronchodilator); d-(decongestant)

(-)-ephedrine   (+)-ephedrine

Smooth muscle	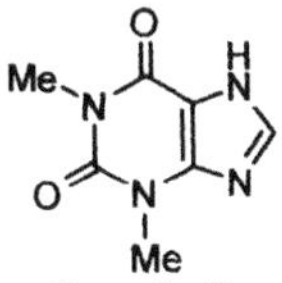 theophylline / purine alkal.	Oxyphyllin/ Astra	tea, *Camellia sinensis* (L..) Kuntze, Theaceae, Ang.; cultivated

theophylline

Smooth muscle	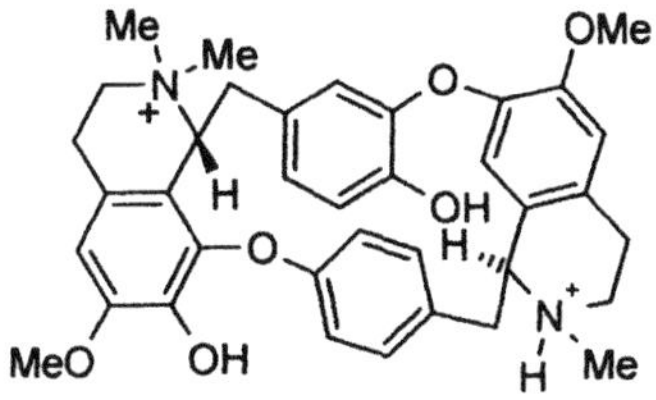 tubocurarine/isoquinoline alkal.	Tubarine/ Burroughs Wellcome; Tubalil/Endo	S American rain forest *Chondodendron tomentosum* R. and P., Menispermaceae, Ang.; now synthetic

(+)-tubocurarine

Sympto-matic (anhydrotic)	agaricic acid or raw agaric (C 6.3) 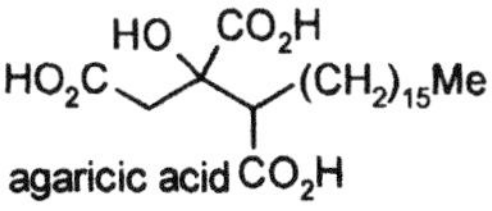	galenic/ polycarboxylic acid	*Fomes laricis* (Jacq.), Basidiomyc., Eumyc. from the northern conifer forest: MI

agaricic acid

Table 13.2.I. Natural (or derived) products approved as drugs or drug components

Area	Name/chemical class/structure	Trade name/Co	Source[a]
Vascular (peripheral)	escin/triterp. glycoside 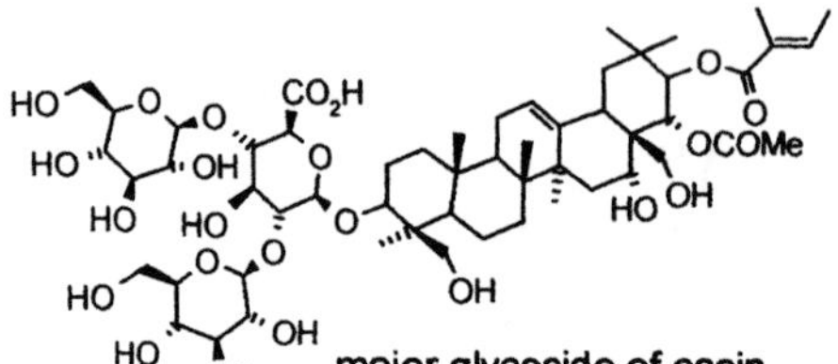	Flogensyl/ Parke-Davis	horse chestnut, *Aesculus hippocastanum* L., Hippocastanaceae, Ang. from Eurasia and temperate N America

major glycoside of escin

[a]From land, unless marine origin is indicated in boldface.

Table 13.2.II. Extracts and infusions or whole organisms in official or traditional use

Therapeutic area	Source	Preparation/Components	Use/Availability
Anthelmintic	*Areca catechu* L. (betel palm), Palmae from the East	arecoline/piperidine alkal. 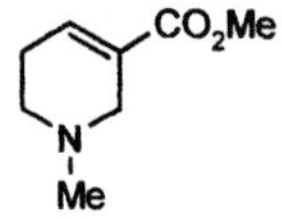	traditional chewing the seeds in the East as a vermifuge and myotic stuff
Antibacterial	*Hypericum uliginosum* H.B.K., Ang. (St Johns wart)	uliginosin A/phloroglucinol 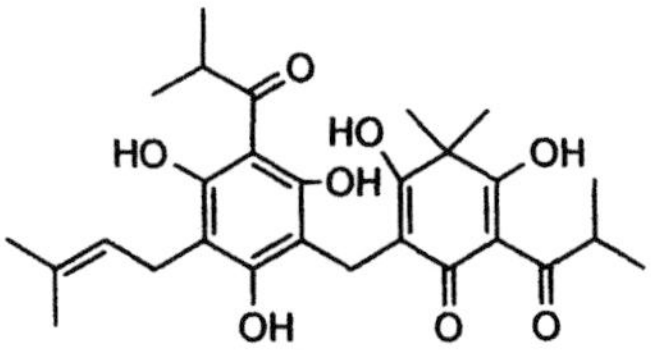	traditional use of the antibiotic plant extract

arecoline

uliginosin A

Table 13.2.II. Extracts and infusions or whole organisms in official or traditional use

Therapeutic area	Source	Preparation/Components	Use/Availability
Anticancer	shiitake, *Lentinus edodes*, (Berk.) Singer, Basidiomyc. (mushroom)	extract/polysacch. (3-D-glucan with 2β-1,6-glucopyranoside branching, such as 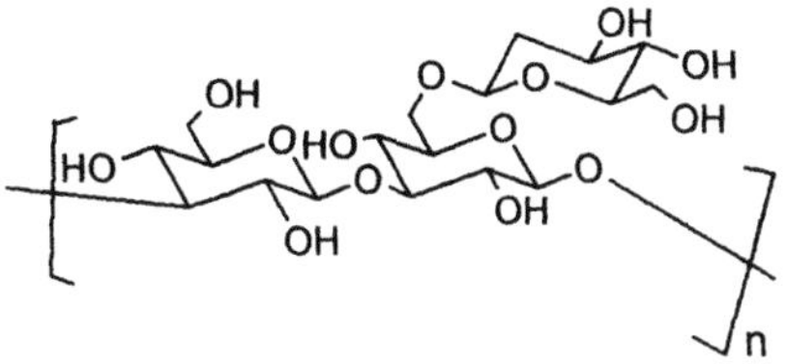 lentinan) lentinan (n =2,500-5,000)	against cancer and as a panacea against high blood pressure, high cholesterol level, and viruses/druggist in the East
CNS	*Catha edulis* Forsk., Celestraceae, Ang.	cathinone 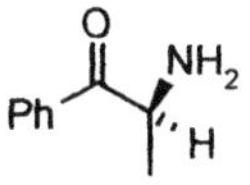 cathinone	traditional use (Khat) in E Africa as an amphetamine-like stimulant
CNS	*Sclettium expansum* L., Aizoaceae, Ang. from SW Africa	mesembrine/indole alkal. 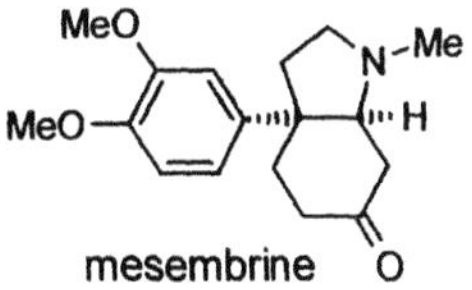 mesembrine	traditional use (Channa) in SW Africa as a CNS stimulant
CNS	*Gingko biloba* L., Ginkgoaceae, Gymn.	standardized extract/flavonoids and ginkgolide diterp. (C 6.2.I) 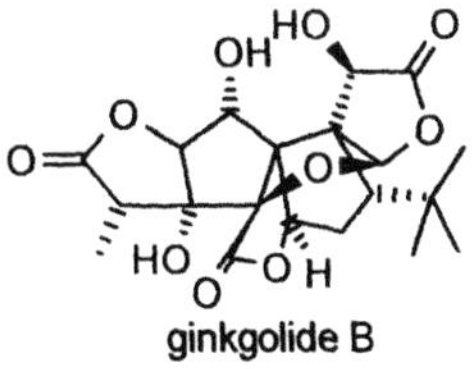 ginkgolide B	CNS and platelet aggregating factor antagonist/Lichwer Pharma/Schwabe
Gastric ulcer	*Glycyrrhyza glabra* var. *typica* or *glabra* (licorice), Leguminosae, Ang.	extract/glycyrrhizin 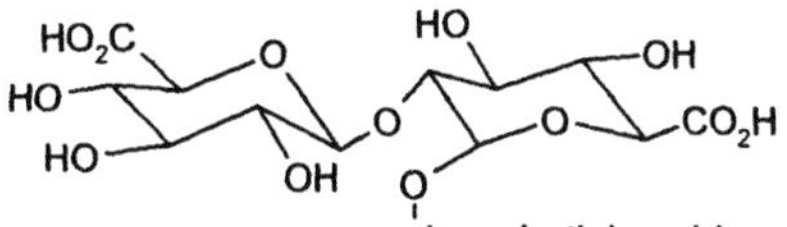 glycyrrhetinic acid glycyrrhizin	used for sweet taste and hormonal content, as well as against gastric ulcer/drugstore

Table 13.2.II. Extracts and infusions or whole organisms in official or traditional use

Therapeutic area	Source	Preparation/Components	Use/Availability
Gastro-intestinal	Sangre de Grado, *Croton lechleri* L., Euphorbiaceae Ang. from Ecuador and Peru	metabolites: α-calacorene, α-copaene, α-pinene, α-thujene, β-caryophyllene, β-elemene, β-pinene, betaine, borneol, calamenene, camphene, cuparophenol, D-limonene, dimethylcedrusine, dipentene, eugenol, euparophenol, γ-terpinene, γ-terpineol, lignin, linalool, methylthymol, myrcene, *p*-cymene, pectic acid, proanthocyanadins, tannins, taspine, terpinen-4-ol, and vanillin	traditional use of the plant extract; Shaman Pharm. brought anti-diarrheal Provir to clinical phase III, getting "fast track" FDA approval, but curative properties were later not proven and the substance was not approved; Provir is marketed via Internet; Shaman developed also Virend, a topical antiviral (herpes) drug
Hypercho-lesterolemia hypertension	garlic (*Allium sativum* L., Liliaceae, Ang.)	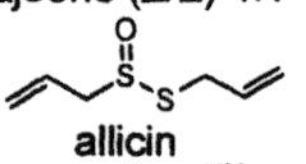 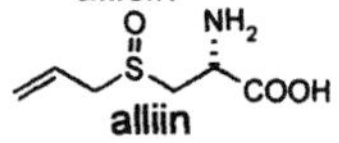 ajoene, allicin (garlic aroma), alliin (odorless (*S*)-2-propenyl-L-cysteine S-oxide): on crushing garlic, alliin is transformed by garlic alliinase into allicin through poorly known transient intermediates; on garlic maceration in oils, rather stable ajoene and other dithiins are formed from allicin 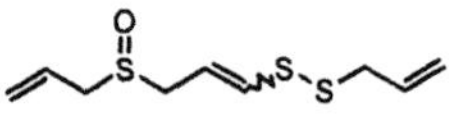	triglycerides, cholesterol, and LDL are kept at low levels, while HDL levels and fibrinolytic activity are increased, blood pressure is lowered, and platelet aggregation is inhibited

ajoene (*E/Z*) 4:1

allicin

alliin

Table 13.2.II. Extracts and infusions or whole organisms in official or traditional use

Therapeutic area	Source	Preparation/Components	Use/Availability
Liver disorders	milk thistle, *Silybum marianum* L. Gaertn. = *Carduus marianus L.*, Asteraceae, Ang.)	extract/flavolignans: silybin, silidianin, and silicristin  silybin	against inflammatory processes and liver intoxication/Apihepar from Asta and Silliver and Abbott
Prostatic benign hyperplasia	saw palmetto, *Serenoa repens* (Bartram) Small, endemic to southeastern US	extracts are rich in tannic acid, phytosterols, and fatty acids	for prostate health/drugstore
Symptomatic	*Abies alba* Mill. (*Abies sibirica* Ledeb.) Pinaceae, Gymn. from N conifer forest (Siberian oil of fir)	extract/pinene, bornyl acetate, limonene, camphene, β-phellandrene 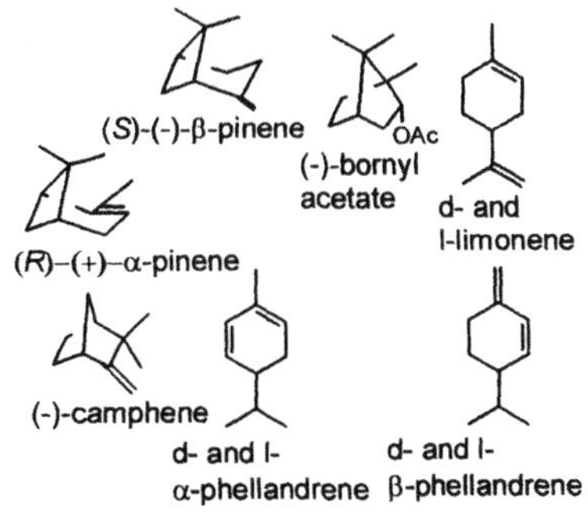 (S)-(-)-β-pinene, (-)-bornyl acetate, d- and l-limonene, (R)-(+)-α-pinene, (-)-camphene, d- and l-α-phellandrene, d- and l-β-phellandrene	expectorant/druggist
Symptomatic	ginger, *Zingiber officinale* Roscoe, Zingiberaceae, Ang.	extract/gingerol 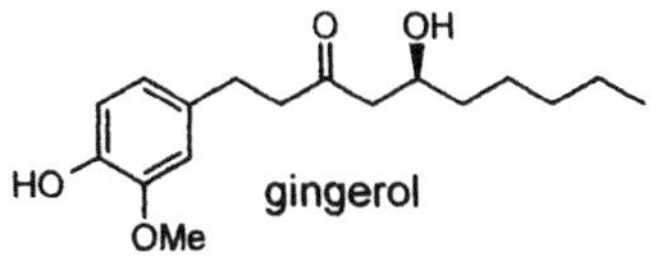 gingerol	against stomach upset caused by drugs/drugstore

Table 13.2.II. Extracts and infusions or whole organisms in official or traditional use

Therapeutic area	Source	Preparation/Components	Use/Availability
Vascular	bilberry, *Vaccinium myrtillus* L., Ericaceae, Ang.	extract/delphinin, myrtillin (aglycon: delphinidin)	in preventing capillary degeneration in the peripheral vascular system/drugstore
Vascular	Oneseed Howthorn or English Howthorn, *Crataegus monogyna* Jacq., Rosaceae, Ang.	extract/triterp. acids (mainly ursolic, oleanolic, and crataegolic acids)	vasodilator, cardiotonic/drugstore

13.3 Natural products, derivatives, and extracts in development as drugs

Examples of natural products, derivatives, and extracts in development as drugs are shown in Table 13.3, in the alphabetical order of therapeutic area.

Cancer chemotherapy has seen the largest number of projects, mostly fostered by the National Institute of Health. Promising antitumor metabolites of all biogenetic classes have been found in marine organisms: alkaloids (ecteinascidin 743 from ascidians), depsipeptides (aplidine from ascidians and dolastatin 10 from opisthobranch mollusks), polyketides, fatty acids and related compounds (bryostatin 1 from bryozoans, curacin A from cyanobacteria, and discodermolide, halichondrin B and agelasphin-9b from sponges), isoprenoids (eleutherobin and sarcodictyin A from anthozoans), and cartilage extracts from sharks, which are also in advanced trial in dermatology and ophthalmology.

Marine natural products are also on trial as CNS aids (SNX-111, a peptide from mollusks, which is a nonaddictive analgesic agent 100-1,000 fold more potent than morphine, now in clinical phase III) and against pathogenic fungi (15G256γ, a macrolide from filamentous fungi).

The problem with marine natural products is the scarce availability. A measure of the problem is given by bryostatin 1, a promising antitumor macrolide: recourse to expensive cultures of the productive strain of bryozoan on 1 m² panels on the open sea off the Californian coast was the only practical solution. Although the production of the active agent by the bryozoan is very low, the bioactivity is so high that a single panel furnishes enough bryostatin 1 for the full treatment of an adult. Unfortunately, the cultures suffered heavily from El Niño events. Many marine natural products have also failed during clinical trial for antitumor activity, like girolline, an alkaloid isolated from an axinellid sponge, *Cymbastela cantharella*, and didemnin B, a depsipeptide from a colonial ascidian, *Trididemnum cereum*.

Although failed as systemic drugs, several bioactive marine natural products have found use as topical ailments, laboratory tools, or models. The pseudopterosins, glycoside diterpenes of the gorgonian *Pseudopterogorgia elisabethae* from the Bahama Islands, have found topical use in dermatology. Manoalide, a sesterterpenoid isolated from the sponge *Luffariella variabilis* from Palau, is a useful model for new anti-inflammatory agents. The conotoxins, isolated from mollusks in the genus *Conus* from the Philippines, served as models for the synthesis of peptidic neuroprotective agents. These polypeptides form a β-sheet of three short antiparallel strands. Inside the folded structure are located four disulfide-bonded cysteine residues. An example is ω-conotoxin MVIIA, a 25-mer polypeptide with three disulfide bridges - and its synthetic version SNX-111 - which blocks N-type calcium channels (Table 15.I). The three-dimensional arrangement is characterized by the presence of a tyrosine group in the appropriate position for binding to N-type calcium channels (Kohno 1995).

Plant metabolites are on trial against several diseases: cancer (the alkaloids camptothecin and homoharringtonine, a simple acetogenin, 4-ipomeanol, the shikimate-derived NK-611, the diterpenoid prostratin, the meroterpene lapachol, and green tea, which is rich of flavonoids), bacterial infections (a flavolignan, 5'-methoxyhydnocarpin), cardiovascular dysfunctions (a polyacetylene, cunaniol), dermatological problems (ten-herb-extract), stomachic problems (columbin, a clerodane diterpene), wound healing (turmeric and Balsam Peru), and CNS dysfunctions (caffeine and nicotine). Δ^9-Tetrahydrocannabinol, extracted from hemp, *Cannabis sativa* var. *indica*, is also under examination in the CNS area. Interacting with cannabinoid receptors at nerve cells, it enters competition with the putative endogenous cannabinoids. It has many unwanted side effects, however, which has diverted the interest toward drugs that inhibit the transporters of these elusive endogenous cannabinoids (Christie 2001).

Herbal products are in large use in certain countries, particularly Germany and France (Chapter 13.2). However, they may interfere with prescribed drugs, if not with the metabolism. St John's wort, *Hypericum perforatum*, is a case in point because one of its metabolites, hypericin, affects neurotransmission in the brain, thus contrasting CNS drugs. Reportedly, hypericin also contrasts antiretroviral drugs, like digoxin, cyclosporin, and theophylline. On this basis, a more strict regulation of herbal products is demanded. On the good side, hypericin helps patients affected by cholestasis by breaking down bile acids (Piscitelli 2000; www.IBISmedical 2000).

Fungi are traditionally considered in the botanical area, although they are genetically closer to animals. An example of fungal polyketide of current great interest is zaragozic acid. Its derivatives are under evaluation for cardiovascular problems.

Metabolites in development from terrestrial animals include contortrostatin, an antitumor protein isolated from snakes, and SNX-482, a polypeptide active on the cardiovascular system, isolated from spiders.

Immunization has been the dream of the traditional Chinese medicine. In modern times, in the lack of time for active immunization, passive immunization with immunoglobulins has solved many urgent problems. Active immunization has prevented infectious mass diseases, such as diphtheria, hepatitis, influenza, measles, meningococcal, mumps, pneumococcal, polio, rabies, rubella, tetanus, typhus, varicella, and yellow fever, as well as several diseases of domesticated animals.

Two major diseases, malaria and AIDS, are still out of control: vaccines are not available, while the malaria parasite and the HIV virus, responsible for AIDS, have developed resistance to current drugs. Variability of the agent, lack of commercial interest, and perhaps also unconfessed political plans at population growth control, have been an obstacle to active immunization against malaria. The hope for an HIV vaccine is now from the engagement of Merck & Co (Conference 2001).

Table 13.3. Natural products, their derivatives, and raw extracts under development as drugs

Therapeutic area	Compound/Chemical class/Structure	Source[a]/function	Organization involved/Phase
Antibacterial	5'-methoxyhydnocarpin-D/flavolignan, shikim. 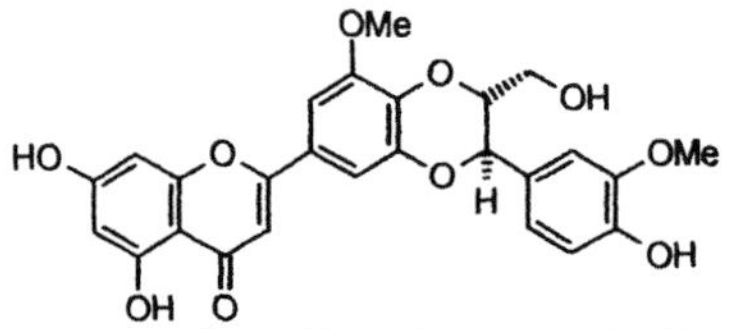 5'-methoxyhydnocarpin-D	*Berberis fremontii,* Berberidaceae Ang. of N American folk medicine/it acts by reversing multidrug resistance, as a multidrug pump inhibitor of the human pathogen, *Staphylococcus aureus,* thus acting synergistically with the plant antibiotics	Colorado State Univ.: Guz 2000/project
Anticancer	aplidine (= dehydrodidemnin B)/depsipept. 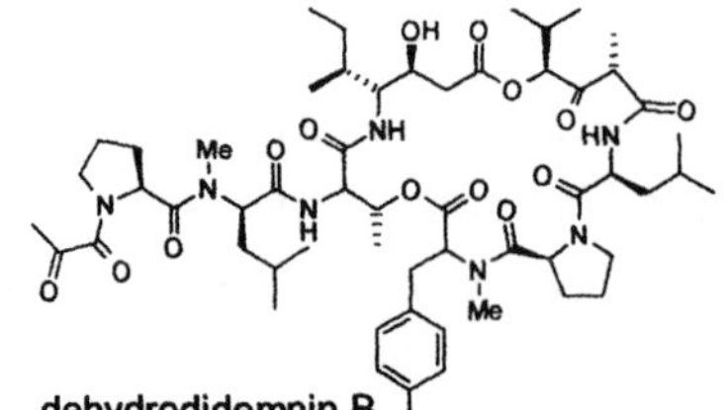 dehydrodidemnin B	**marine:** *Aplidium albicans* (Milne-Edwards, 1841), Mediterranean Ascid./prostate cancer (didemnin B was dismissed because of cardiotoxicity)	PharmaMar/ clinical phase: Rinehart 2000

Table 13.3. Natural products, their derivatives, and raw extracts under development as drugs

Therapeutic area	Compound/Chemical class/Structure	Source[a]/function	Organization involved/Phase
Anticancer	bryostatin 1/polyket. 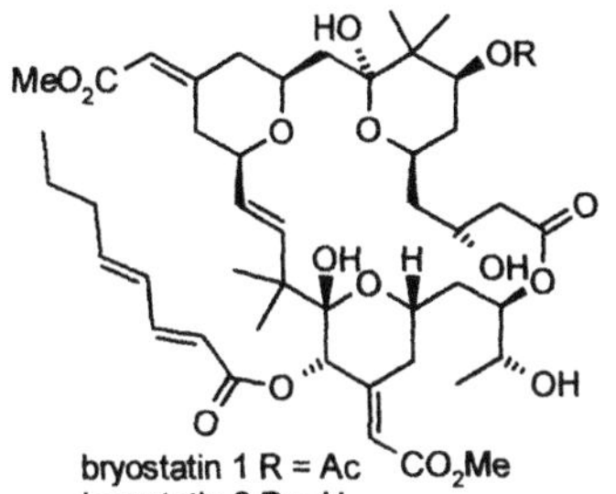 bryostatin 1 R = Ac CO₂Me bryostatin 2 R = H	**marine:** *Bugula neritina* (L.), Bryoz.	NCI and Bristol-Myers Squibb/phase II: Newman 2000
Anticancer	9-amino- and 9-nitrocamptothecin/indolizinoquinoline 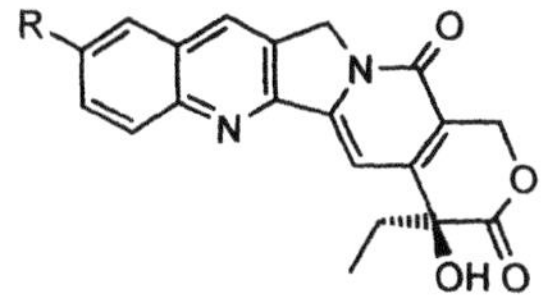 9-aminocamptothecin R = NH₂ 9-nitrocamptothecin R = NO₂ alkal.	chemical transformation of camptothecin from *Camptotheca acuminata* Decsne, Nyssaceae, Ang.	clinical trials: Newman 2000
Anticancer	contortrostatin/disintigrin protein	copperhead viper (*Agkistrodon contortix*, Serpentes, Reptilia)	Markland, F. ACS Boston Meeting, Sept. 1998/preclin.
Anticancer	curacin A/fatty acid deriv. 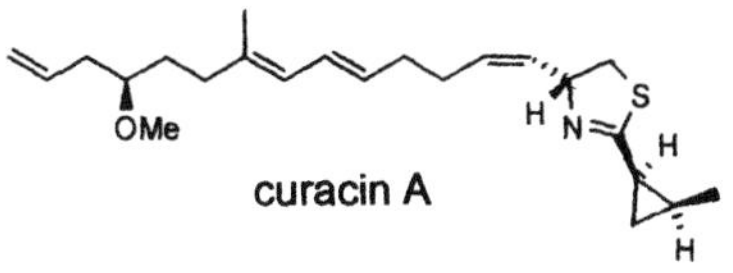 curacin A	**marine:** *Lyngbya majuscula* Gomont, Cyanobact.	preclinical

Table 13.3. Natural products, their derivatives, and raw extracts under development as drugs

Therapeutic area	Compound/Chemical class/Structure	Source[a]/function	Organization involved/Phase
Anticancer	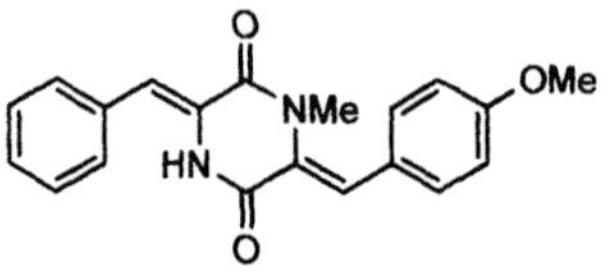 diketopiperazines/pept.	*Streptomyces* sp., Actinom., Bact.	Xenova/phase I: Chicarelli - Robinson 1997, p. 36
	exemplifying diketopiperazines		
Anticancer	(+)-discodermolide/polyket. 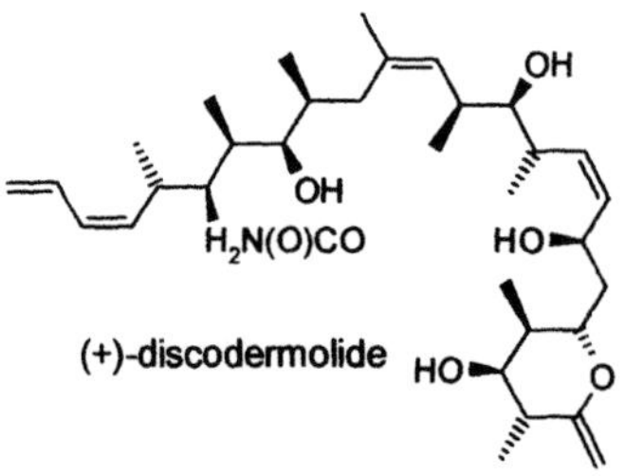	**marine:** *Discodermia dissoluta*, Lithistida, Porif./microtubule stabilization: Paterson 2001	licensed by Harbor Branch to Novartis Pharma AG/preclinical
	(+)-discodermolide		
Anticancer	dolastatin 10/pept. 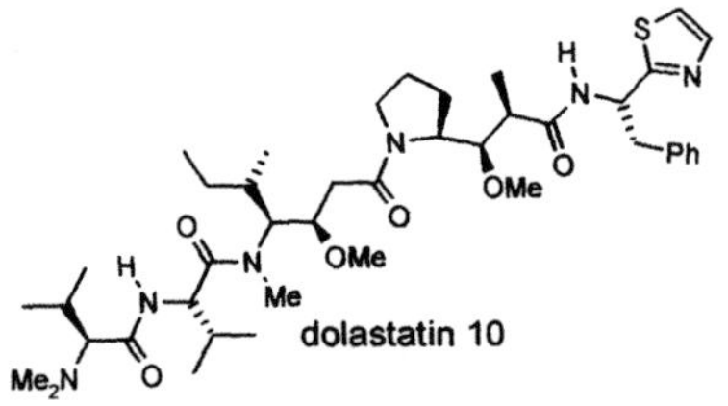	**marine:** *Dolabella auricularia* (Lightfoot, 1786), Anaspidea, Moll.	NCI/extended phase II
	dolastatin 10		
Anticancer	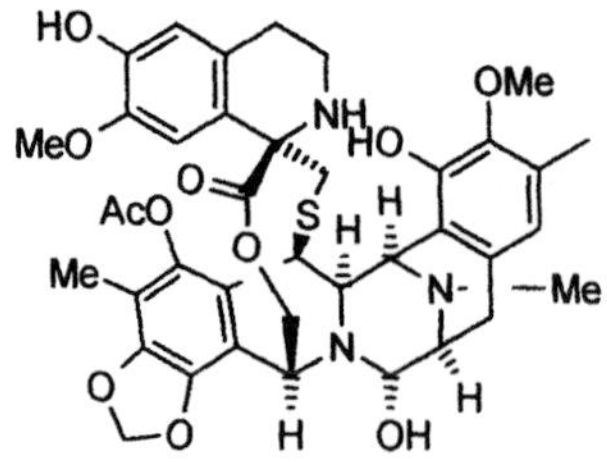 ecteinascidin 743/alkal.	**marine:** *Ecteinascidia turbinata* Herdman, 1880, Ascid. (rare metabolite: Rinehart 2000)	Pharma Mar/clinical phase II
	ecteinascidin 743		

Table 13.3. Natural products, their derivatives, and raw extracts under development as drugs

Therapeutic area	Compound/Chemical class/Structure	Source[a]/function	Organization involved/Phase
Anticancer	eleutherobin/diterp. glycoside 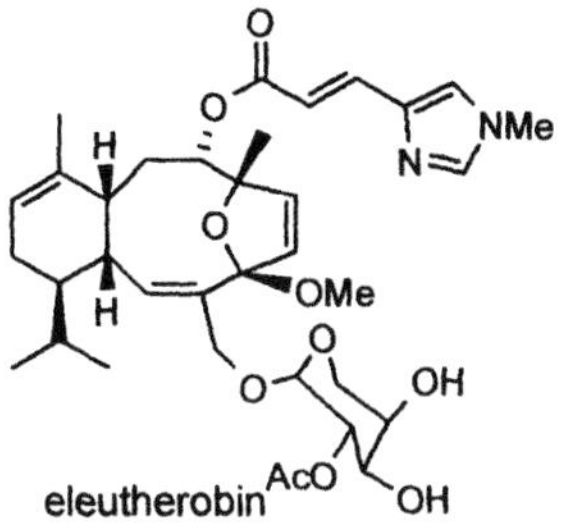 eleutherobin	**marine:** *Eleutherobia* cf. *albiflora*, Alcyonacea, Cnid.	licensed to Bristol-Myers Squibb
Anticancer	green tea infusion/flavonoids and other antioxidant polyphenols, notably epigallocatechin gallate	green tea (from incomplete fermentation of leaves of *Camellia sinensis* (L.) Kuntze, Theaceae, Ang.)	NCI /phase II for both epigallo-catechin gallate and tea extract
Anticancer	halichondrin B/polyket. 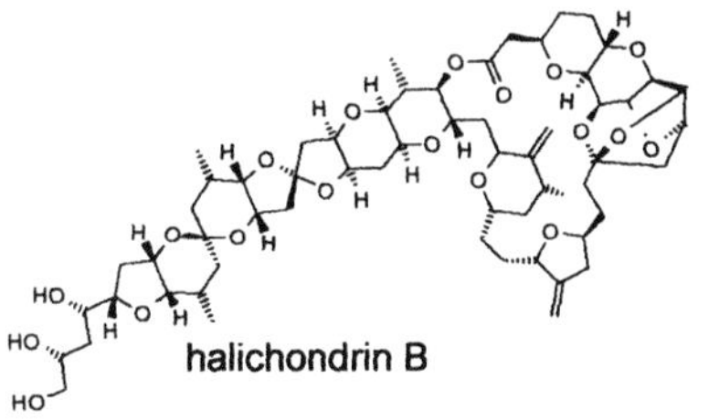 halichondrin B	**marine:** *Halichondria melanodocia* de Laubenfels, 1936, Halichon. and *Axinella* sp., Axinel., W Pacific, and *Lissodendoryx* sp., Poecil., NZ, Porif.	NCI/preclinical
Anticancer	homoharringtonine/alkal. 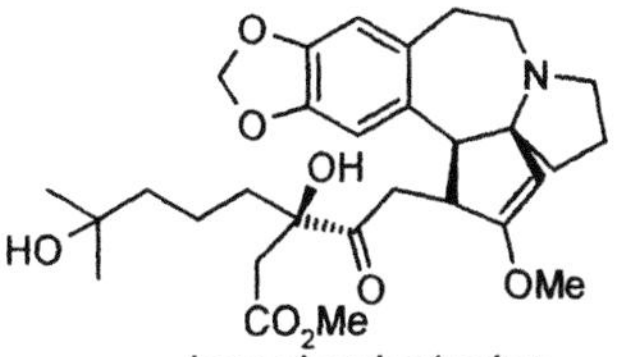 homoharringtonine	bark and roots of *Cephalotaxus harringtonii* var. *drupacea* (Sieb. & Zucc.) Koidz. 1930, Pinatae, Gymn., threatened in Vietnam/induction of complete hematologic remission in resistant myelogenous leukemia	Robin 2000

Table 13.3. Natural products, their derivatives, and raw extracts under development as drugs

Therapeutic area	Compound/Chemical class/Structure	Source[a]/function	Organization involved/Phase
Anticancer	4-ipomeanol/ acetogenin 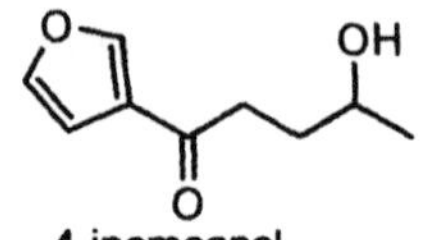 4-ipomeanol	Sweet potato (*Ipomoea batata* L., Convolvulaceae, Ang.)	NCI/preclinical
Anticancer	KRN7000/α-galactosylceramide, a methyl branched glycosphingolipid deriv. of agelasphin-9b 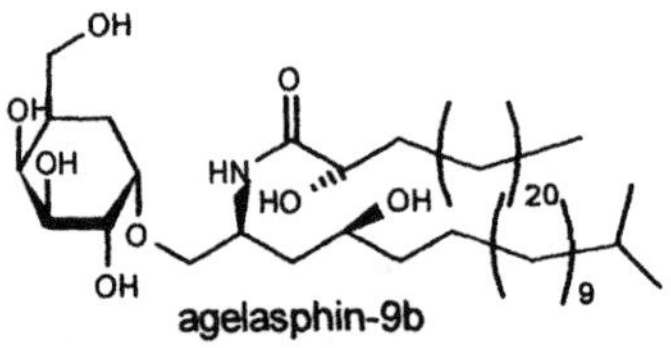 agelasphin-9b	**marine:** *Agelas mauritianus* Carter (1883), Agelasida, Porif. from Okin.	Stoter, G., Rotterdam; Kirin Co, Japan
Anticancer	lapachol and β-lapachone/hemiterpene-quinones 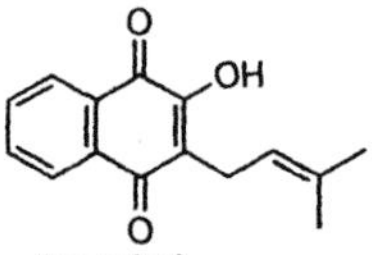 lapachol 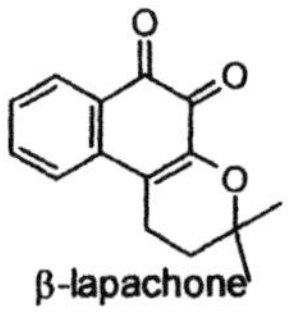 β-lapachone	Latin American lapacho tree, Pau d'Arco (*Tabebuia* spp., Bignoniaceae, Ang.), an Andean traditional tea	widely claimed to be curative, but assay methods may be at fault: Barrett 2000
Anticancer	Cartilage extract AE-941 (Neovastat in oncology, Psovascar in dermatology, and Neoretna in ophtalmology)	**marine:** cartilage extract from shark, Chondrich.	Æterna (Canada)/phase III
Anticancer	NK-611 (water soluble) 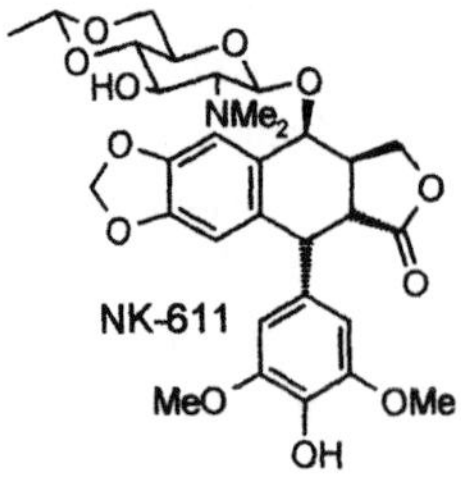 NK-611 podophyllotoxin	chemical transformation of epipodophyllotoxin (shikim. polyket.) from *Podophyllum peltatum* L., Ang. 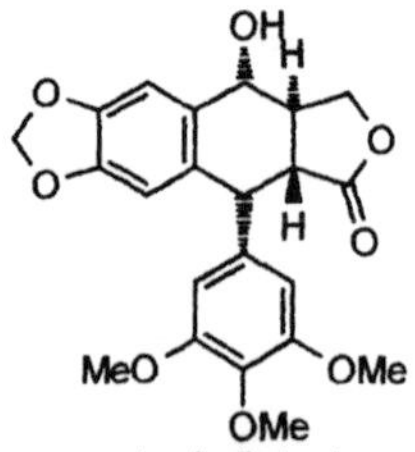	Nippon Kayaku/phase II: Newman 2000

Table 13.3. Natural products, their derivatives, and raw extracts under development as drugs

Therapeutic area	Compound/Chemical class/Structure	Source[a]/function	Organization involved/Phase
Anticancer	prostratin/phorbol class diterp. 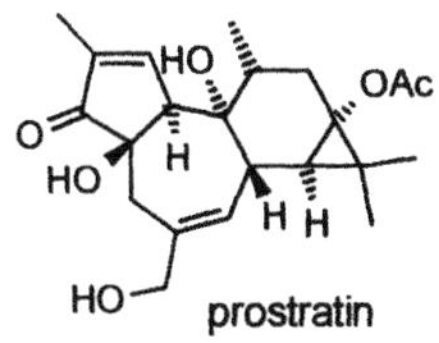	*Homalanthus nutans* Guill [= *Carumbium nutans* Muell. Arg.], Euphorbiaceae, Ang. from Samoa, where the leaves are used as a fish poison/activation of protein kinase C without tumor growth promotion: Gustafson 1992	waiting for clinical assays: Cox 2000
Anticancer	sarcodictyin A/diterp. 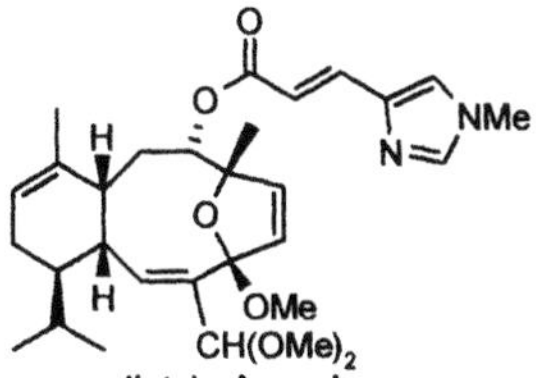	semisynthesis from sarcodictyin A, from **marine:** *Sarcodictyon roseum* (Philippi, 1842) [= *Rolandia rosea* (Philippi)], Stolonifera, as well as certain Pennatulacea and Alcyonacea, Cnid.	waiting for development: Nicolaou 1998
Anticancer	squalamine and derivs/steroidal alkal. 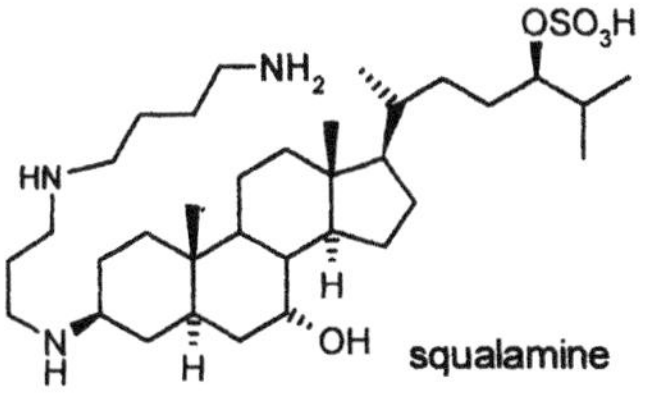	**marine:** *Squalus acanthias* L., Chondrich.	Magainin/ phase II (angiogenesis)
Antifungal	15G256γ/polyket. macrolide 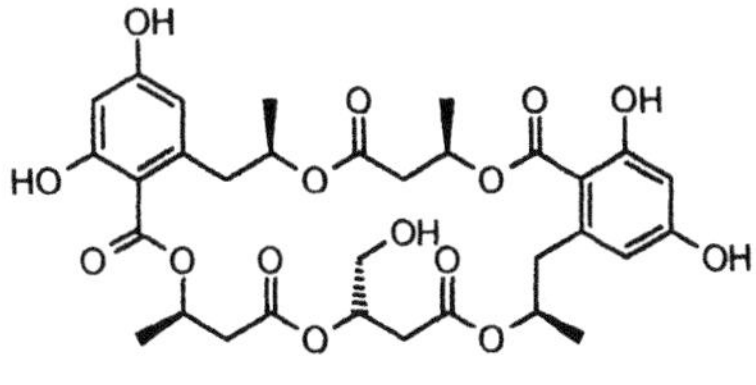	**marine:** *Hypoxylon oceanicum* S. Shatz, Eumyc./cell wall active antifungal agent	discovered by American Cyanamid: Albaugh 1998; licensed to Wyeth-Ayerst/ preclinical
Antiviral	interferons (proteins), mostly against hepatitis	recombinant	Merck

Table 13.3. Natural products, their derivatives, and raw extracts under development as drugs

Therapeutic area	Compound/Chemical class/Structure	Source[a]/function	Organization involved/Phase
Cardio-vascular	cunaniol (= ichthyothereol) (*trans-* and *cis*-2-(1-nonen-3,5,7-triynyl)-3-hydroxytetrahydropyran)/polyacetylene 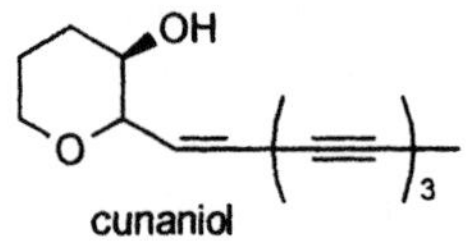	*Clibadium sylvestre* (Aubl.) Baill., Ang. from Central and South America and the West Indies, where the plant (Cunami) is used as a fish poison by natives: Gorinsky 1998	Zeneca and Glaxo; native countries have claimed intellectual property, however
Cardio-vascular	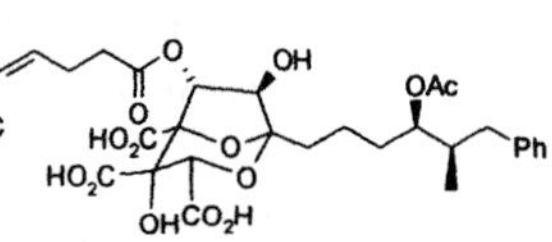 phomoidrides (CP-263,114) and zaragozic acid deriv.s/polyket. 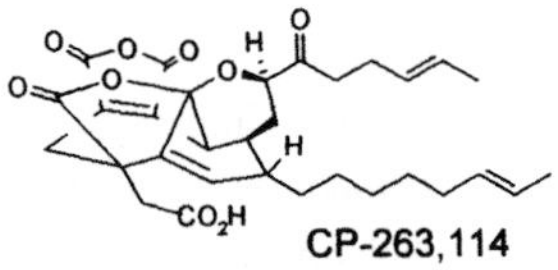	phomoid fungus from Texas	Pfizer: Hepworth 2000
Cardio-vascular	SNX-482/polypept. (Gly-Val-Asp-Lys-Ala-Gly-Cys-Arg-Tyr-Met-Phe-Gly-Gly-Cys-Ser-Val-Asn-Asp-Asp-Cys-Cys-Pro-Arg-Leu-Gly-Cys-His-Ser-Leu-Phe-Ser-Tyr-Cys-Ala-Trp-Asp-Leu-Thr-Phe-Ser-Asp)	venom of African tarantula, *Hysterocrates gigas* Pocock, 1897, Mygales, Araneae, Arach.	Neurex (Elan Pharm.)/preclinical; available from Peptides International
CNS	caffeine, theobromine/xanthine alkal. 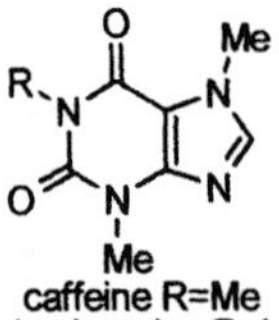	*Cola nitida* Schott and Endl., Sterculiaceae, Ang. from W Africa	traditional use in W Africa as a CNS stimulant (Kola)
CNS	nicotine/alkal. 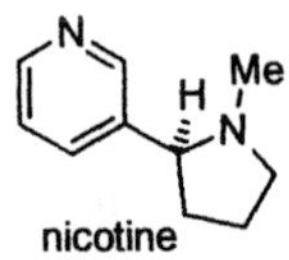	*Nicotiana* spp., Solanaceae, Ang.	University of South Florida/phase III

Table 13.3. Natural products, their derivatives, and raw extracts under development as drugs

Therapeutic area	Compound/Chemical class/Structure	Source[a]/function	Organization involved/Phase
CNS (analgesic)	SNX-111/pept.	**marine:** *Conus geographicus*, snail; now made by chemical synthesis	Neurex (Elan Pharm.)/phase III
Dermatology	AE-941Psovascar (shark cartilage extract)	**marine:** shark, Chondrich., cartilage	Æterna (Canada)/phase III
Dermatology (eczema)	non-steroidal mixture (Zemaphyte)	ten-herb extract, from Chinese traditional medicine	Phytopharm UK/phase III
Expectorant	Balsam Tolu/free cinnamic and benzoic acids	*Toluifera balsamum* L., Leguminosae, Ang. from mountains in tropical S America	traditional use
Hypotensive vasodilator	eledoisin/undecapept. Ala-Phe-Ile Gly-Leu-Met-NH$_2$ Asp-Lys-Ser-Pro-pGlu eledoisin	**marine:** salivary gland of *Ozaena* spp. [= *Eledone* spp.], Cephalopoda, Moll.	potential target for drug design/American Peptide Company
Ophthal-mology	AE-941Neoretna (shark cartilage extract)	**marine:** shark, Chondrich., cartilage	Æterna (Canada)/phase III
Stomachic	columbin and other clerodane terpenoids 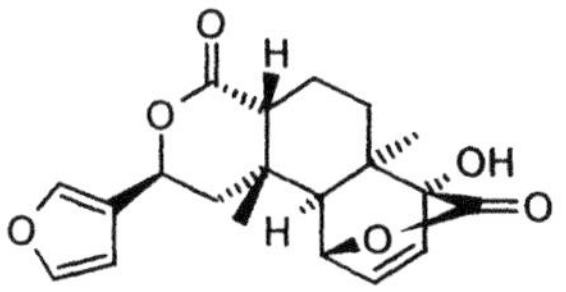 columbin	*Jatrorrhiza palmata* (DC.) Miers, Menispermaceae, Ang. from E Africa	traditional (Calumba) and modern use of plant extract; drugstore
Wound healing	Balsam Peru/cinnamic acid and benzoic acid esters	*Toluifera pereirae* (Klotzsch) Baill., Leguminosae, Ang. from coastal forests in San Salvador	traditional and hospital use of plant extract

Table 13.3. Natural products, their derivatives, and raw extracts under development as drugs

Therapeutic area	Compound/Chemical class/Structure	Source[a]/function	Organization involved/Phase
Wound healing	turmeric/p,p-dihydroxydicinnamoylmethane, p-hydroxycinnamoylferuloylmethane and related compounds	Indian *Curcuma longa* L., Ang.	US patenting was disclaimed because turmeric is a traditional ailment by the American Indians

[a]Marine organisms are indicated in boldface, even in obvious cases.
[b]Chemical modifications planned because of unsatisfactory results with the natural product.

13.4 Fragrances and cosmetics

The perfumes consist of raw fragrances dissolved in aqueous ethanol. Functional products are in a different group, comprising detergents, shampoos, shower gels, and oven cleaners. Functional products have an elaborated chemically-reactive matrix that does not allow using traditional fragrances; recourse is made to inexpensive, chemically inert, odoriferous materials.

The fragrance industry occupies a prominent position in the fine chemical industry. The technology involved can be compared with that for drugs, which explains the large means put in the research of perfumes. The Swiss fragrance industry uses dirigibles to place large traps on rain forest canopy trees, and the most advanced synthetic technologies in complex synthetic plans (Kaiser 2000).

The natural product diversity of the fragrances is modest, comprising low molecular weight terpenoids, acetogenins, mercaptans, and nitrogenous compounds, mostly oxime-*O*-alkyl ethers. It is the diversified composition of scents that complicates the matter, particularly in devising and building man-made scents.

Natural scents are exclusive products, used for the most highly-priced perfumes. The largest market of perfumes - particularly in the poor developing countries - is based on synthetic materials. Natural scents are mostly derived from flower plants, evolved to be recognized by pollinating insects. The orchids (flower plants in the monocotyledonous family Orchidaceae, which make about 10% of all flower plants) are highly rated for scents, particularly the woody scent afforded by a sesquiterpenoid, caryophyll-5-en-2α-ol.

A representative list of perfumes and fixatives is shown in Table 13.4, comprising alkaloids, terpenoids, acetogenins, and polyphenols from terrestrial plants. In the terpenoids the classical scents are ionone floral, white floral, rosy floral, woody scent, and oriental fragrances. Ionone floral scent is given by damascenone and β-ionone, isolated from freesias, particularly the African *Freesia* spp., Iridaceae. Rosy floral scent is given by citronellol, geraniol, and nerol, isolated from roses, in the family Rosaceae, from temperate and subtropical northern hemisphere. White floral consists of mixtures of terpenoids, acetogenins and alkaloids, obtained from jasmin oil, from *Jasminum* spp., Oleaceae, of temperate and warm areas. Spicy floral scent results from phenolic compounds, particularly vanillin and eugenol from *Eugenia* spp., Myrtaceae, and other plants. Phenolics, and phenyl propanoids, account also for oriental fragrances.

The classical scent of the sea is afforded by ambergris, which is also the best fixative for perfumes. Once secured from whales, ambergris can now be legally provided by chemical synthesis (Science whaling 2000). Other fixatives are obtained from land animals, such as castoreum, civet, and musk. The first one is a phenolic substance and the latter two macrocyclic ketones that have stimulated the early methodologies for the synthesis of macrocyclic compounds

Cosmetics are preparations (except soap and detergents) for the preservation and beauty of the body. In the rich countries domesticated animals also have their cosmetics, such as creams based on ω3-polyunsaturated fatty acids, for strengthening and repair of feet and noses of hounds and pointers, and tea tree oil (extracts from various plants) for very effective external hygiene.

The base for expensive cosmetics is agar, from red seaweeds, or stearyl alcohol, from animals. To this are added emulsifiers and stabilizing agents (sorbitan monolaurate, hydroxypropyl cellulose, egg oil, *Cydonia* gum, and carrageenans), emollients (cetyl alcohol, originally derived from whales), antioxidants (4-deoxygadusol, a cyclohexenone chemically derived from marine mycosporines), anti-inflammatory agents (α-bisabolol), and UV-protecting agents (astaxanthin and urocanic acid).

A doubtful place in cosmetics is occupied by Botulinum proteic toxins, which are claimed to remove, albeit temporarily, facial wrinkles.

Table 13.4. Components of commercial perfumes and cosmetics

Scent or common name	Components/class	Origin/properties	Commercial availability
Agar	Agar/polysacch.	**marine:** red seaweeds	base for cosmetics
Astaxanthin	astaxanthin/tetraterp. 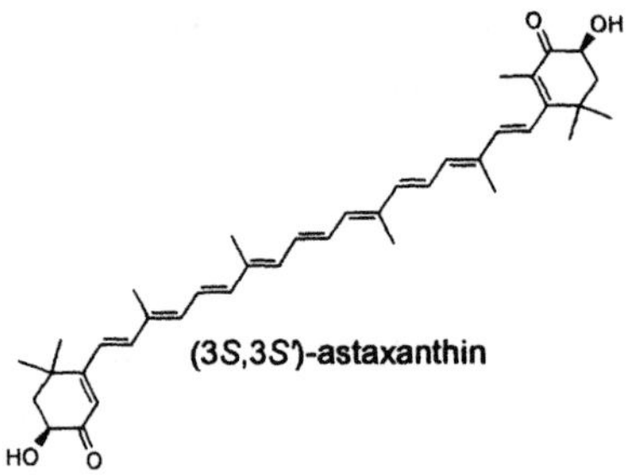	**marine:** wild salmon and shrimps, as a mixture of stereoisomers; the (3*S*,3'*S*) stereoisomer was isolated from *Agrobacterium aurantiacum*, Bact. and *Haematococcus pluvialis* Flotow, Chloroph. (Chumpolkulwong 1997); also of transgenic origin (C 14.1) and synthetic, from Hoffman-La Roche	other than as a food adjuvant, it is marketed as a mixture of stereoisomers for use in cosmetics (UV protecting cream, added of stabilizers, by Kosei Co., Japan)
α-Bisabolol	α-bisabolol/ sesquiterp. 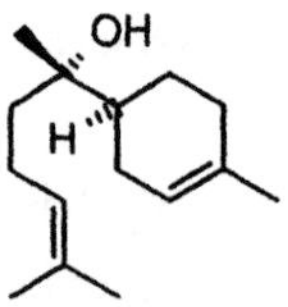	levorotatory sesquiterp. from *Matricaria chamomilla* L.; dextrorotatory from *Populus balsamifera* L., Ang.: MI	anti-inflammatory agent for cosmetics (Kamillosan by ASTA Pharma)
Botulinum toxins	proteins	*Clostridium botulinum* (van Ermengem 1896) Bergey *et al.* 1923, Bact./facial wrinkles are temporarily removed by relaxing facial muscles	Botox, prepared from botulinum toxins

(3*S*,3*S'*)-astaxanthin

(-)-α-bisabolol

Table 13.4. Components of commercial perfumes and cosmetics

Scent or common name	Components/class	Origin/properties	Commercial availability
Carrageenans	sulfated polysacch.	red seaweeds: *Chondrus crispus* Stackhouse in Brittany (Irish Moss) and Chile, and *Gigartina* spp. in NZ and Argentina	stabilizing agents for toothpaste; in air refreshing agents; dispersant for calcium-based pigments
Castoreum	*p*-methoxyphenol, *p*-ethylphenol, acetophenone, and 1,2-benzenediol/aromatics and degradation products	oily cream found in the sac of beavers *Castor* spp., Mamm./fixative in all perfumes, particularly leather, amber, and cypre	perfumes: Magie Noire (Lancôme), Emeraude (Coty), and Arpège (Lanvin)
Cetyl alcohol	C_{16} fatty acid	**marine:** first secured from the saponification of whale spermaceti/emollient	Ciba
Civet	civetone / polyket. 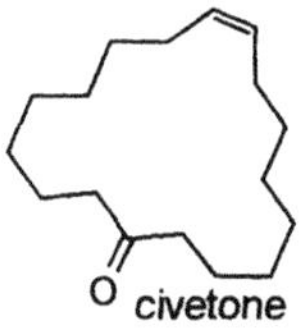	civet cat, *Civettictis* [= *Viverra*] *civetta* (Schreber, 1776), Mamm./fixative in expensive perfumery	total synthesis
Cydonia gum	gum containing emulsin, amigdalin, and cydonin/mucilages	seeds of *Cydonia oblonga* Mill., Rosaceae, Ang.	suspending and stabilizing agent for cosmetics
4-Deoxy-gadusol	4-deoxygadusol/derived from mycosporine-like amino acid 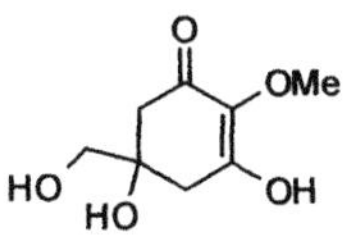	**marine:** algae/antioxidant for food processing and cosmetic applications	Topo Suisan Co
Egg oil	fatty glycerides, lecithin, and cholesterol	egg yolk/formulation of cosmetics	creams
Hydroxy-propyl cellulose	hydroxypropyl cellulose/polysacch.	by etherification of cellulose from plants	emulsifier for cosmetics

Table 13.4. Components of commercial perfumes and cosmetics

Scent or common name	Components/class	Origin/properties	Commercial availability
Ionone floral	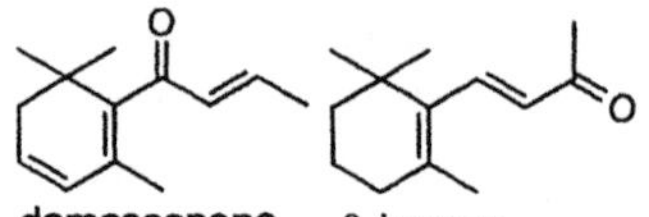 damascenone, β-ionone/degraded isopr.	*Freesia* spp., Iridacceae, Ang.	component of floral perfumes
	damascenone β-ionone		
Musk	muscone/polyket. 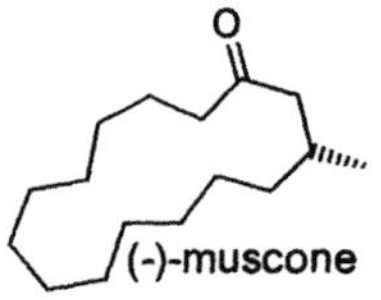 (-)-muscone	musk, from the preputial follicles of the musk deer, *Moschus moschiferus* Linnaeus, 1758, Mamm. from Himalaya	fixative in expensive perfumes
Oriental fragrance	patchouli alcohol and minor patchoulene, azulene, and eugenol/sesquiterp. and phenylpropanoids 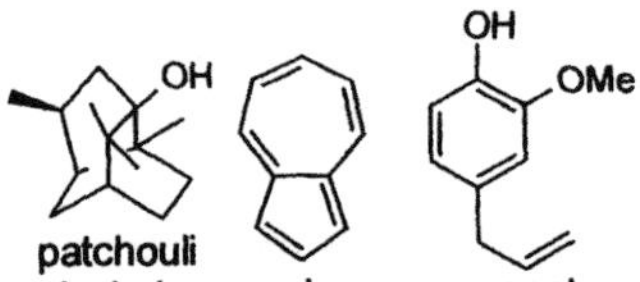 patchouli alcohol azulene eugenol	cultivated *Pogostemon cablin* (Blancho) Benth., Labiatae, Ang.	lasting oriental fragrance in expensive perfumes
Rosy floral	rose oil: free β-citronellol, geraniol, and nerol; geranium oil: geraniol esters/monoterp. 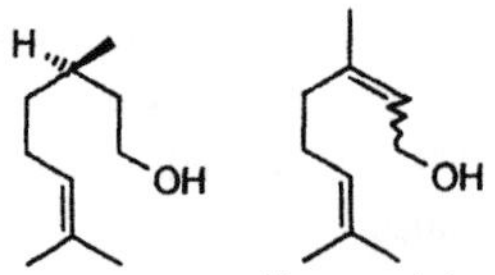 (+)-β-citronellols Z: geraniol E: nerol	rose (*Rosa gallica* L. and *Rosa damascena* Mill., Rosaceae, Ang.) and geranium oils (*Pelargonium odoratissimum* Ait., Geraniaceae, Ang.)	rosy floral scent in expensive perfumes, such as Anaïs Anaïs (Cacharel) and Eternity (Calvin Klein)

Table 13.4. Components of commercial perfumes and cosmetics

Scent or common name	Components/class	Origin/properties	Commercial availability
Scent of the sea	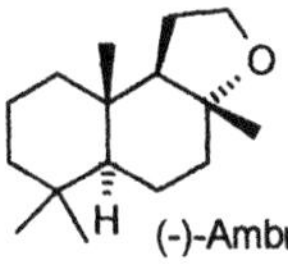 (+)-ambrein (the main component of ambergris)/triterp.; on air and sun it is degraded to (-)-Ambrox and other scents 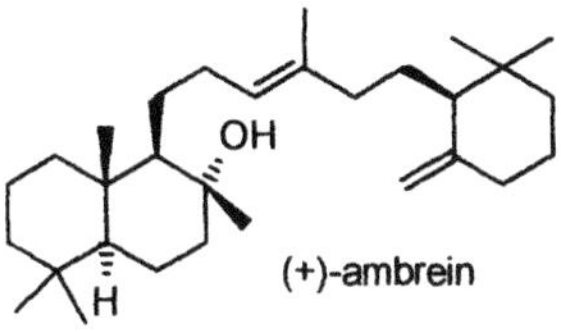	**marine:** ambergris is a concretion found in the intestine of the sperm whale, *Physeter macrocephalus* Linnaeus, 1758 [= *Physester catodon*]/scent of the sea, difficultly imitated with various compositions	Ambre (Firmenich), Adoxal (Givaudan), Algenone (Synarome), Arctander (IFF), Arpège (Lanvin), Cetonal (Givaudan), Oceano (Naarden), Signoricci (Nora Ricci 1965)
Sorbitan monolaurate	Sorbitan monolaurate, deriv. of sorbitol/carboh. ester and stabilizer in cosmetics	various grapes; also **marine:** algae/emulsifier, thickener, and stabilizer in cosmetics	Glycomul L (Lonza)
Spicy floral	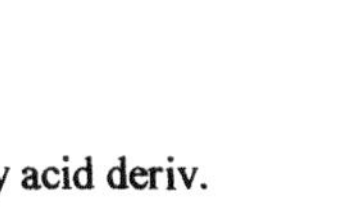 vanillin, eugenol/phenolics and phenylpropanoids 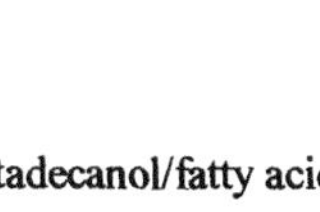	*Dianthus* spp., Ang.	spicy floral scent: Must (Cartier), Shalimar (Guerlain), Coco (Chanel), Tocade (Marcel Rochas), Vanilla Field (Coty), and Jean-Paul Gaultier (Jean-Paul Gaultier)
Stearyl alcohol	1-octadecanol/fatty acid deriv.	animal glycerides or by cotton seed reduction	base for vanishing cosmetic creams and emulsions
Urocanic acid	urocanic acid/alkal. 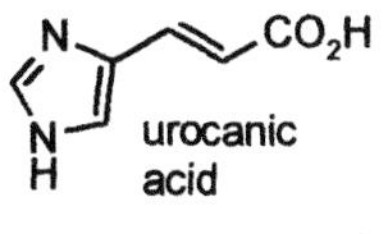	bacteria; also **marine**	UV-protecting cosmetic creams

Table 13.4. Components of commercial perfumes and cosmetics

Scent or common name	Components/class	Origin/properties	Commercial availability
White floral	linalool, nerolidol, (*E*)-2(3)-dihydrofarnesal, and minor amounts of indole, jasmone, and benzyl acetate 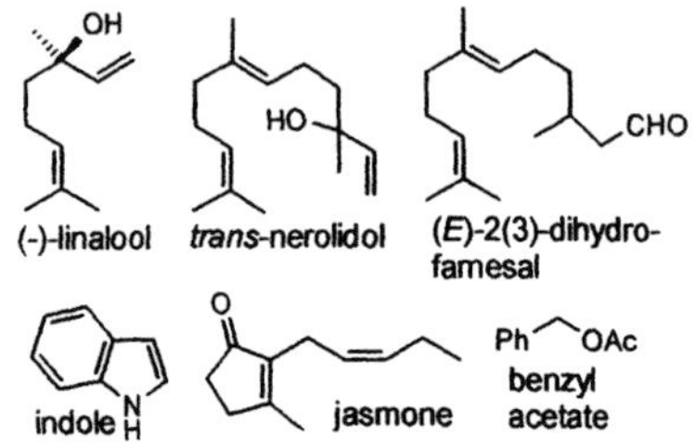	(*E*)-2(3)-dihydrofarnesal in *Citrus limon* (L.) Burm. f., and rain forest epiphytes, Ang.; linalool and nerolidol are commonly found in Ang.	Givaudan white floral perfumes: Kaiser 2000

(-)-linalool   *trans*-nerolidol   (*E*)-2(3)-dihydro-farnesal

indole   jasmone   benzyl acetate

| Woody scent | 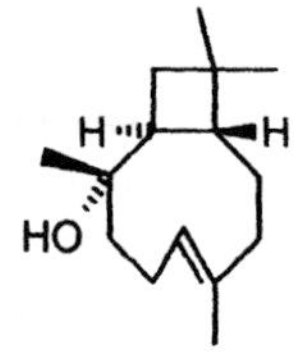 caryophyll-5-en-2α-ol/sesquiterp. | *Bollea coelestis* Reichenbach f., Orchidaceae, Ang., from Colombian rain forest: Kaiser 2000 | woody florals: Chanel No 19, Safari (Ralph Lauren) and White Linen (Estee Lauder) |

caryophyll-5-en-2α-ol

13.5 Technological compounds and laboratory tools

The technological use of natural products is as old as humanity, although mixtures of unknown composition were mostly used before the advent of chemistry. Mummification in ancient Egypt is one such example. In contrast, that of the Phoenician Tyrian Purple was an advanced technology based on a practically pure substance.

Technological compounds are listed in Table 13.5.I: agricultural and medical aids, pigments, solvents, and lacquers. The use of synthetic pesticides in agriculture is a major cause of biodiversity loss because to kill a few noxious species, useful species are exterminated, while the noxious species develop resistance. Control of pests for recruitment and mating through natural agents is a painstaking procedure, but one that has to be pursued if we want to preserve biodiversity. Prominent among the natural controlling agents in agriculture is azadirachtin, an antifungal tetranortriterpenoid from the neem tree. The European Patent Office has opposed patenting this natural product, on the allegation that raw extracts of the neem tree were already in use in India by farmers against phytopathogenic fungi. This appears to be an offence to chemistry, negating the value of finding and describing the active principle. Other insecticidal agents are found in the diterpenoids, phenolic amides, and alkaloidal glycosides. Selective herbicidal agents are much sought. A recent example is potassium 2,3,4-trihydroxy-2-methylbutanoate, a specific leaf-closing substance of *Leucaena leucocephalum*. It is expected to kill selectively this herb, which has infested crops in the East by releasing allelochemical substances (Ueda 2001).

That of the chitin is the oldest of the modern technologies of natural products. Chitin makes the cuticle of arthropods. It is available in large amounts for industrial purposes from the enormous crustacean wastes of marine fisheries. It is much sought as an environmentally safe nitrogen-containing polysaccharide, a commodity that serves as an absorbable surgical support. It is also useful in cleaning hairs and teeth, keeping food fresh, and in the manufacture of cloths and expensive paper. Chitosan, obtained by partial deacetylation of chitin, is useful in water treatment and for the removal of pollutants, besides also being a valuable food additive that prevents the body from absorbing too much fats. Chitosan production from chitin is a technology that dates to 1859 and is continuously improved and applied to derivatives, such as chitosan oligosaccharides made with fungal mutants (Chung 1997).

The area of natural adhesives has progressed enormously in recent years, spanning from medical applications, particularly in dentistry, to everyday articles, such as shoes. Byssus, from mytilids, has great strength and elasticity deriving from a polyphenolic proteic structure (Waite 1998). Proteins secreted by oyster pearls also form highly strong threads (Yamamoto 1995).

Technological compounds are obtained from plants of the tropical rain forest. The Indo-Pacific tropical rain forest is well known for traditional technologies of hunting and defense, like plant poisons used as fishing aids (from *Derris* spp.), or arrow poisons (from *Antiaris toxicaria*). It also provides tannins from the bark of *Uncaria* spp., resins for paint from *Agathis* spp., waxy mixtures from *Balanophora* spp., and fatty oils from palms, in particular *Orbignya martiana*.

Certain technological compounds find specialized use as laboratory tools for scientific research and routine analysis (Table 13.5.II). Unlike the drugs, they can be brought quickly to the market; many are already available from recent research on marine organisms. A revolution in the biological laboratory practice has been brought about by the green fluorescent protein. Isolated from the jellyfish *Aequorea victoria*, it is employed in the study of intracellular signaling and gene transcription, as a reporter for gene expression and protein localization (Tsien 1996). The amoebocyte lysate of the horseshoe crab, *Limulus polyphemus*, is used in assaying bacterial contamination in the clinic and pharmaceutical industry. Heat stable enzymes from archaeans have facilitated the heating and cooling cycles of PCR. Polyethers from marine dinoflagellates are key tools for the study of ion channels. Polypeptides isolated from the venomous mollusk *Conus magus* are also used to the latter scope: ω-conotoxin GVIA in identifying N-type calcium channels, and ω-conotoxin MVIA for the study of N-type calcium channels (Kohno 1995). Other areas of medical research that benefit from natural products include apoptosis, autonomic nervous system, blood, inflammation and gout, cancer, cell biology and biochemistry, CNS, enzyme inhibition, and molecular biology.

Understanding the mechanism of chemical reactions has shed rationality in organic chemistry. For dioxygen as a reactant, the reaction course depends on the spin conditions: a quantomechanical barrier protects the (singlet) molecules of life from reacting with (triplet) aerial dioxygen. As a unique case, such a barrier is broken under life-compatible conditions by dioxygen in the presence of the raspailynes, long-chain enol ethers of glycerol isolated from axinellid sponges in the genus *Raspailia* (Table 13.5.I) (Guella 1987). These enol ethers may serve as models to deepen our knowledge of the reactions of dioxygen with the molecules of life.

Table 13.5.I. Technological compounds from nature, semisynthesis, or patterned synthesis

Main use	Common/trade name	Chemical class/Structure	Source[a]
Agriculture (insecticide)	azadirachtin/a component of Align, Azatin, and Turplex (AgriDyne Technologies, Salt Lake City)	tetranortriterp. azadirachtin	neem tree, *Azadirachta indica* A. Juss., Meliaceae, Ang., and Ins. feeding on it: MI
Agriculture (pre-emergence grass herbicide)	cinmethylin/ Cinch (Du Pont)	monoterp. deriv. cinnmethylin	synthesis patterned on 1,4-cineole, a monoterp. component of *Eucalyptus* oil 1,4-cineole
Agriculture (insecticidal)	(2S)-1-cyano-2-hydroxybut-3-ene and phenylethyl-cyanide: Peterson 2000	fatty acid derivs.	*Crambe abyssinica* Hochst ex. R. E. Fries, Cruciferae, Ang.
Agriculture (insecticidal)	carbaryl/ Dicarbam (BASF)	phenolic amide carbaryl	synthesis patterned on physostigmine, an indole alkal. from Nigerian *Physostigma venenosum* Balf., Leguminosae, Ang. physostigmine

Table 13.5.I. Technological compounds from nature, semisynthesis, or patterned synthesis

Main use	Common/trade name	Chemical class/Structure	Source[a]
Agriculture (insecticidal)	echinacein (= neoherculin = α-sanshool)	fatty acid amide 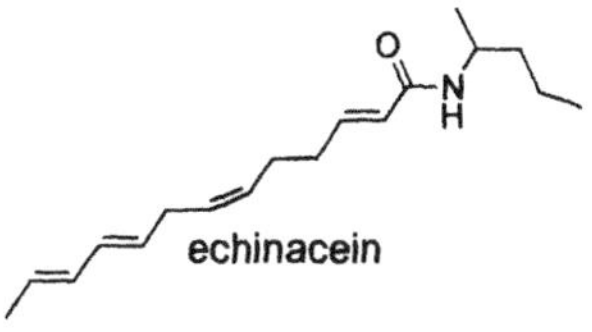 echinacein	*Echinacea* spp., Compositae; also Rutaceae, Ang.: Jacobson 1967
Agriculture (potential herbicide)	leaf-closing substances, such as potassium 2,3,4-trihydroxy-2-methyl-butanoate	hydroxyacid 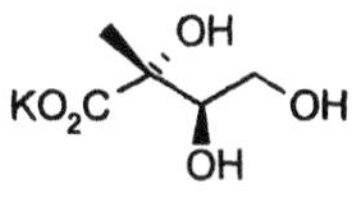 potassium 2,3,4-trihydroxy-2-methylbutanoate	*Leucaena leucocephalum*, Fabaceae, Ang., infesting cultures in East, by secreting allelochemicals that inhibit the growth of herbs: Ueda 2001
Agriculture (insecticidal)	ryanodine/ contained in Ryanex	diterp. 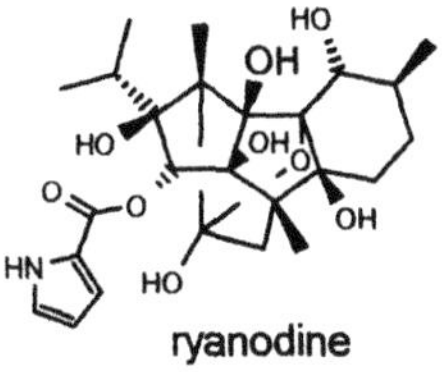 ryanodine	*Ryania speciosa* Vahl., Flacourtiaceae, Ang. from tropical America: MI
Agriculture (antibacteria land antifungal for crop protection)	streptomycin and oxytetracycline (Pfizer, Merck & Co)	alkal. glycoside, C 6.2.FA/PO/C 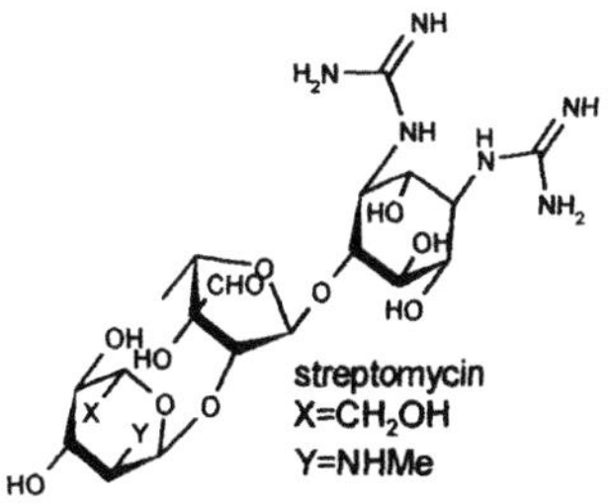 streptomycin X=CH$_2$OH Y=NHMe	*Streptomyces griseus* (Krainsky) Waksman et Henrici, Actinom., Bact.
Aquaculture	bioadhesives/ Research Corp. Techniques	proteins	**marine:** *Crassostrea virginica* Gmelin, Bivalvia, Moll.: Weiner 1991
Emulsion stabilizers	Gum Arabic	carboh. (sacch. and polysacch.)	*Acacia verek* Guill. and Perr. from Sudan, Africa

Table 13.5.I. Technological compounds from nature, semisynthesis, or patterned synthesis

Main use	Common/trade name	Chemical class/Structure	Source[a]
Food preservation	antifreeze: Ewart 2000	proteins	**marine:** Antarctic fish (also found in bacteria, fungi, plants and animals, especially insects)
Insect repellents	camphor/ monoterp.	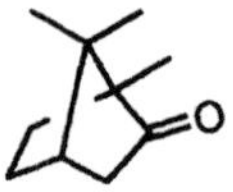 drugstore	*Cinnamomum camphora* T. Nees & Ebermeier, Lauraceae, Ang.

camphor

Medical (dentistry)	alginates/ Sorbosan, Maersk	polysacch.	**marine:** Phaeoph.
Medical (cryo-surgery)	antifreeze: Ewart 2000	proteins	**marine:** Antarctic fish (also found in bacteria, fungi, plants and animals, especially insects)
Medical (dentistry)	bioadhesives/ Genex Corp.	polyphenolic proteins	**marine:** *Mytilus edulis* Linnaeus, 1758, Bivalvia, Moll.: Maugh 1988.
Medical (neurology)	α-kainic acid	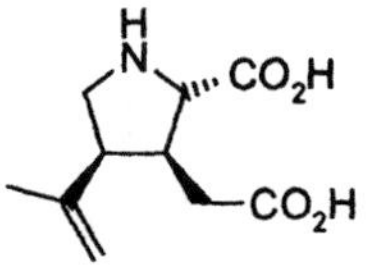 amino acid	**marine:** *Digenea simplex* Agardh, Rhodoph., of wide distribution but abundant enough for commercial purposes in Taiwan waters only

α-kainic acid

| Medical (surgery) | chitosan | 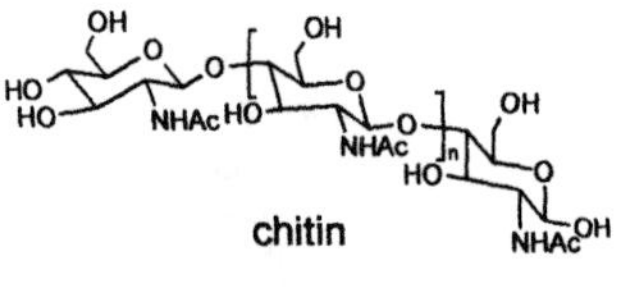 deacetylated chitin (aminopolysacch.) | chitin of crustaceans, also found in insects and fungi |

chitin

| Medical (surgery) | (+)-tubocurarine (the active agent of curare): Sigma 2000 | 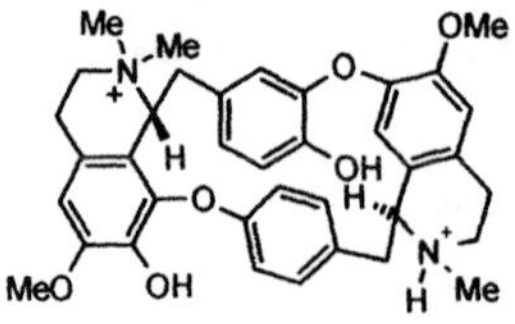 isoquinoline alkal. | bark of *Chondrodendron tomentosum* R. & P., Menispermaceae, Ang. from the American tropical rain forest: MI |

(+)-tubocurarine

Table 13.5.I. Technological compounds from nature, semisynthesis, or patterned synthesis

Main use	Common/trade name	Chemical class/Structure	Source[a]
Medical (ortho-pedics)	scleractinian skeletons (INOTEB)	calcareous or synthetic	**marine:** hermatypic scleractinian skeleton, Cnid.,: Christel 1993.
Medical (ortho-pedics)	Norian RSR (Norian Co)	synthetic polymer	**marine:** synthesis patterned on hermatypic scleractinian skeletons: Constanz 1995
Microscopy	Balsam Canada	oleoresin (abietic acid and volatile monoterp.)	*Abies balsamea* (L.) Mill., Pinaceae, conifers, Gymn. from the northern conifer forest: MI
Pigments (lacquers and varnishes)	dracorubin component of pigment/ Dragon's Blood	polyphenol deriv. 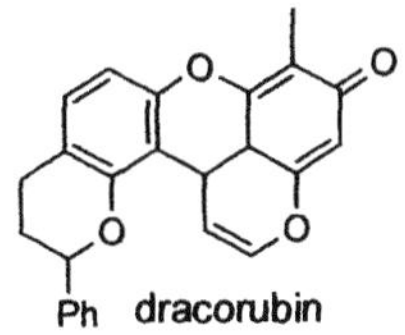	*Daemonorops propinquus* Becc., Palmae (Rattan palmae), Ang.: MI
Pigments (dying)	hematoxylin/ Hematoxylon	polyphenol 	*Haematoxylon campechianum* L., Leguminosea, Ang. from central America and the West Indies
Pigments	laccaic acid A, C, and D/Crimson lac dye	polyket. 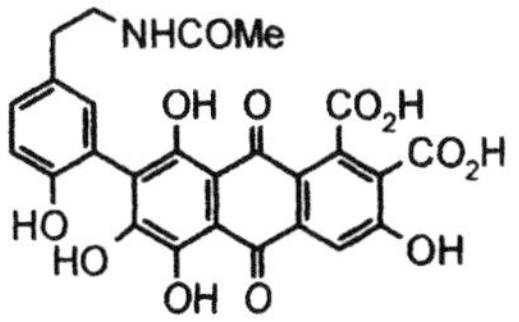	*Coccus laccae* [= *Laccifer lacca* Kerr], Ins., associated to certain trees in India: MI

dracorubin

hematoxylin

laccaic acid A

Table 13.5.I. Technological compounds from nature, semisynthesis, or patterned synthesis

Main use	Common/trade name	Chemical class/Structure	Source[a]
Reaction mechanisms	raspailyne A	long-chain ether of glycerol 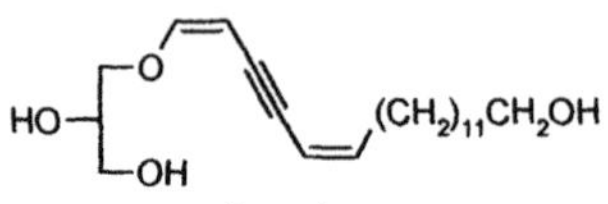 raspailyne A	*Raspailia pumila* (Bowerbank, 1866), Axinel., Porif., from Brittany: Guella 1987
Sensors	4(1*H*-indol-3-yl)-1*H*-imidazole-2,5-dione, deriv. of zyzzin by *S/O* exchange	indole alkal. 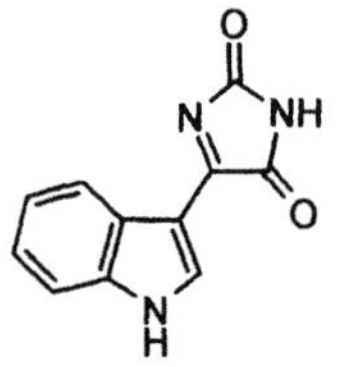 derivative of zyzzin by S/O exchange	*Zyzza massalis* (Dendy), Poecil., Porif. from NC: Mancini 1994A
Solvents / lacquers and varnishes	abietic acid	diterp. 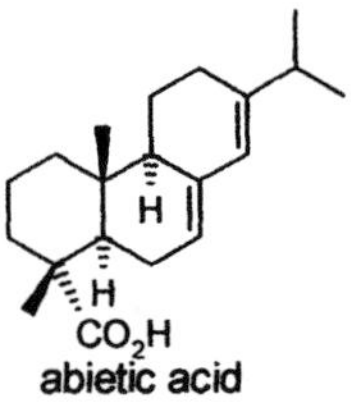 abietic acid	*Abies balsamea* L., Pinaceae, conifers, Gymn. from the northern conifer forest: MI
Solvents / lacquers and varnishes	Copaiba (Amazonian, African, and E Indian varieties)	oils, terpenoids	S American: *Copaifera* spp., Leguminosae. African: *Sindora supa* Merr., Leguminosae. E Indian: *Dipterocarpus* spp., Dipterocarpaceae, Ang.
Solvents / lacquers and varnishes	oil of turpentine	volatile terpenoids	*Pinus* spp. , Pinaceae, Gymn.
Textiles	cotton, linen, silk, wool	polysacch., proteins	cotton: *Gossypium* spp., Malvaceae, Ang.; linen: *Linum usitatissimum* L., Linaceae, Ang.; silk: *Bombyx mori* L.; wool: sheeps and other Mamm.

Table 13.5.I. Technological compounds from nature, semisynthesis, or patterned synthesis

Main use	Common/trade name	Chemical class/Structure	Source[a]
Tires, shoes	rubber	isopr. polymer (caoutchouc)	chiefly *Hevea brasiliensis* Muell. Arg., Euphorbiaceae, from the American tropical rain forest, and the Mexican shrub *Parthenium argentatum* Gray, Ang.
Traditional technologies (hunting)	(+)-tubocurarine, the active agent of curare/Sigma 2000	isoquinoline alkal. 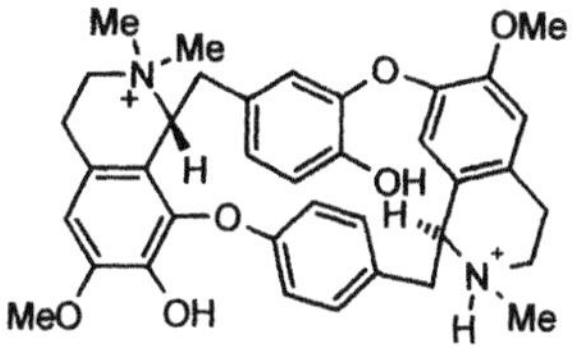 (+)-tubocurarine	bark of *Chondrodendron tomentosum* R. & P., Menispermaceae, Ang. from the American tropical rain forest: MI; traditionally used as an arrow poison
Traditional technologies (hunting)	strophantin (= strophantidin glycoside)	cardenolide 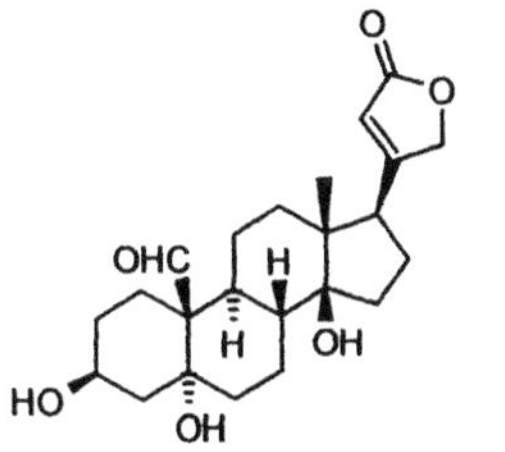 strophantidin	*Strophanthus* spp., Apocyanaceae, Ang. from eastern and central Africa; used as an arrow poison
Traditional technologies (hunting)	calotropin, and uscharidin	cardenolides 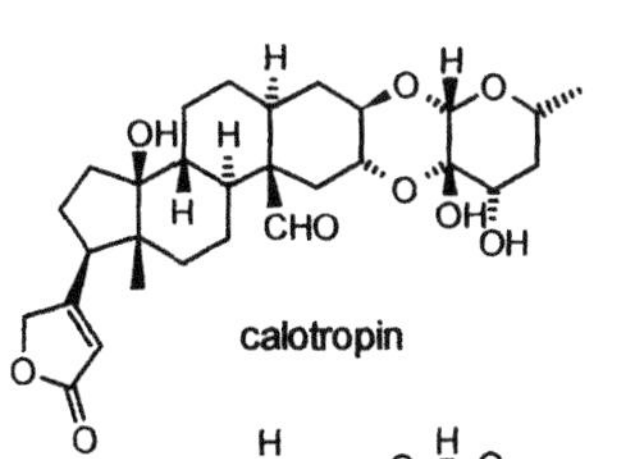 calotropin 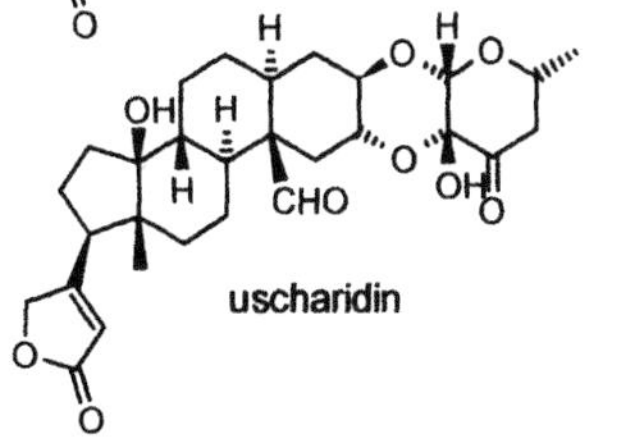 uscharidin	*Calotropis procera* Dryand., Asclepiadaceae, Ang. from tropical Africa; traditionally used as an arrow poison

Table 13.5.I. Technological compounds from nature, semisynthesis, or patterned synthesis

Main use	Common/trade name	Chemical class/Structure	Source[a]
Waste treatment	algae digesting enzymes (plastic digesting enzymes obtained from mutant)	proteins	*Amoeba* spp., Rhizopoda, Protozoa: Polne-Fuller 1988
Waste water treatment	chitosan/ drugstore	deacetylated chitin	chitin of crustaceans, also found in insects and fungi
Waxes (polishing and plasticizer)	Carnauba wax (= Brazil wax): Sigma 2000	carnaubic acid, cerotic acid, and melissyl cerotate	*Copernicia prunifera* (Muell.) H.E. Moore, Palmae, Ang.

[a]Marine organisms are indicated in boldface, even in obvious cases.

Table 13.5.II. Natural and semisynthetic laboratory tools[a]

Area	Name/Source	Chem class/Mode/Structure	Origin[b]
Analytical chemistry	Litmus (turnsole)	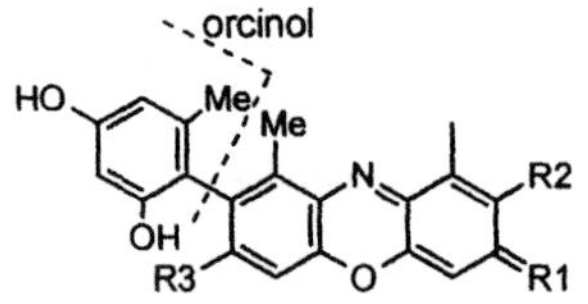 oligomer of lecanoric acid, orceins and erythrolein/used for litmus paper, microscopy, and beverage coloring	processing of lichens from the Scandinavian tundra, *Roccella tinctoria* DC. (1805), *Lecanora tartarea* (L.) Ach.

orceins R1 = O or NH;
R2 = H or orcinol; R3 = OH or NH$_2$

| Analytical chemistry | monensin/ Sigma 2000 | polyether/forming metal ion complexes 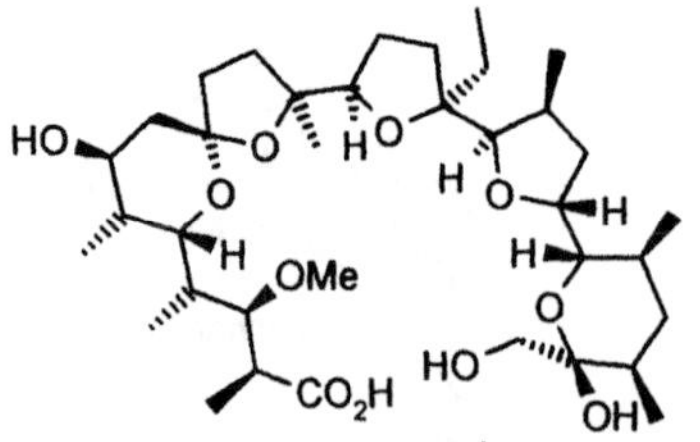 | *Streptomyces cinnamonensis* Okami 1952, Actinom., Bact. |

monensin

Table 13.5.II. Natural and semisynthetic laboratory tools[a]

Area	Name/Source	Chem class/Mode/Structure	Origin[b]
Apoptosis	catalases/ Sigma 2000	proteins/preventing apoptosis in rat ovarian follicles in culture	*Aspergillus niger* van Tieghem, 1867, Deuterom., Eumyc., or bovine or bison liver, Mamm.
Apoptosis	etoposide/ Sigma 2000	polyket. shikim. glucoside/involved in DNA breaking 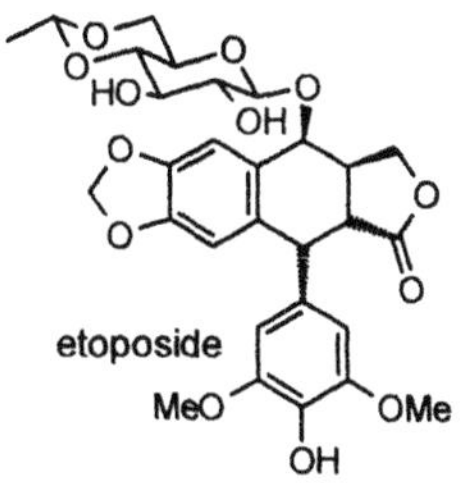	chemical modification of epipodophyllotoxin from N American *Podophyllum peltatum* L., Berberidaceae, Ang. 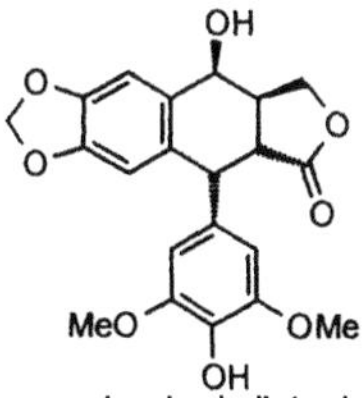

etoposide

epipodophyllotoxin

Area	Name/Source	Chem class/Mode/Structure	Origin[b]
Apoptosis	resveratrol/ Sigma 2000	shikim./platelet aggregation inhibitor 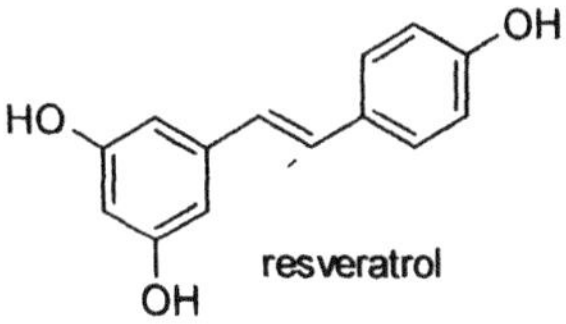	*Polygonum cuspidatum* Sieb. & Zucc., Polygonaceae, Ang.

resveratrol

Area	Name/Source	Chem class/Mode/Structure	Origin[b]
Autonomic nervous system	α-bungaro-toxins/Sigma 2000	polypept./high anticholinesterase activity	Formosan banded krait, *Bungarus multicintus*, Serpentes, Reptilia
Autonomic nervous system	Botulinum toxins/Sigma 2000	proteins/inducing the release of acetylcholine at neuromuscular junctions	*Clostridium botulinum* (van Ermengem 1896) Bergey *et al.* 1923, Bact.
Autonomic nervous system	dihydro-β-erythroidine	*Erythrina* alkal./competitive antagonist at nicotinic receptor/C 6.1.1.A	seeds of *Erythrina* spp., Leguminosae, Ang.: MI
Autonomic nervous system	erabutoxins A-B/Sigma 2000	polypept./post-synaptic action in blocking neuromuscular transmission	**marine:** *Laticauda semifasciata* (Reinwardt), Elapidae, Serpentes, Reptilia

Table 13.5.II. Natural and semisynthetic laboratory tools[a]

Area	Name/Source	Chem class/Mode/Structure	Origin[b]
Autonomic nervous system	fasciculin/ Sigma 2000	protein/potent cholinesterase inhibitor	Eastern Green Mamba, *Dendroaspis angusticeps*, Serpentes, Reptilia
Autonomic nervous system	methyl lycaconitine/ Sigma 2000	*Aconitum* alkal. (*N*-succinylanthranilic acid ester of lycoctonine)/binding to the α-bungarotoxin site	seeds of *Delphinium brownii* Rydb., Ranunculaceae, Ang.

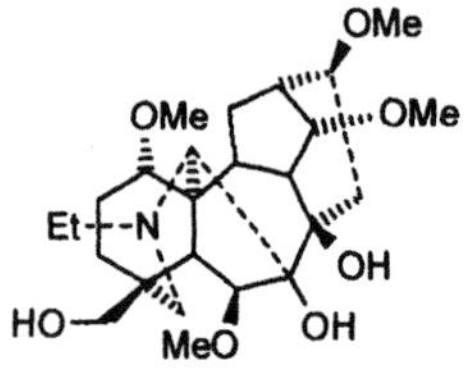

lycoctonine

| Autonomic nervous system | yohimbine/ Sigma 2000 | tryptophan-secologanin alkal./α$_2$-adrenergic antagonist | *Rauwolfia serpentina* L. Benth., Gentianales and *Corynanthe yohimbe* K. Schum., Rubiales, Ang. from S Asia |

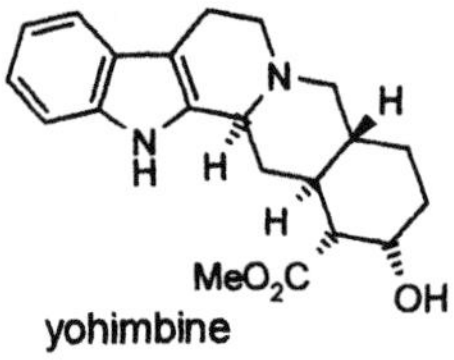

yohimbine

| Blood/gout/ inflamma- tion | botrocetin/ Sigma 2000 | polypept./fostering blood platelet aggregation | S American pit viper, *Bothrops jararaca* (Wied, 1824), Serpentes, Reptilia |
| Blood/gout/ inflamma- tion | digitonin/ Sigma 2000 | cardenolide glycoside/aiding the solubilization of membrane-bound proteins for cholesterol assay | *Digitalis purpurea* L., Scrophulariaceae, Ang. |

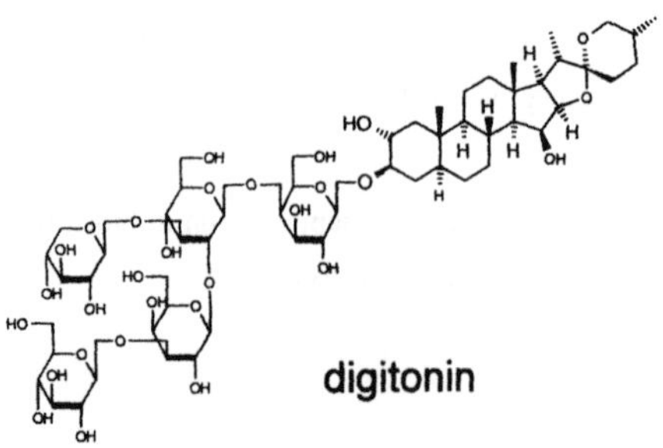

digitonin

Table 13.5.II. Natural and semisynthetic laboratory tools[a]

Area	Name/Source	Chem class/Mode/Structure	Origin[b]
Blood/gout/ inflamma- tion	epinephrine/ Sigma 2000	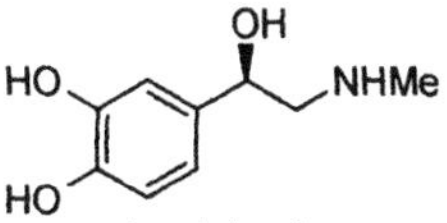 tyrosine alkal./platelet aggregation reagent	Mamm.; also **marine:** Cnid.: Pani 1994

epinephrine (= adrenaline)

Area	Name/Source	Chem class/Mode/Structure	Origin[b]
Blood/gout/ inflamma- tion	margatoxin/ Sigma 2000	pept./selective inhibitor of voltage dependent K⁺ channel in lymphocytes (eliciting immunoresponse in autoimmune and chronic inflammatory disorders)	scorpion *Centuroides margaritatus*, Arach.
Blood/gout/ inflamma- tion	nazumamide A/Sigma 2000	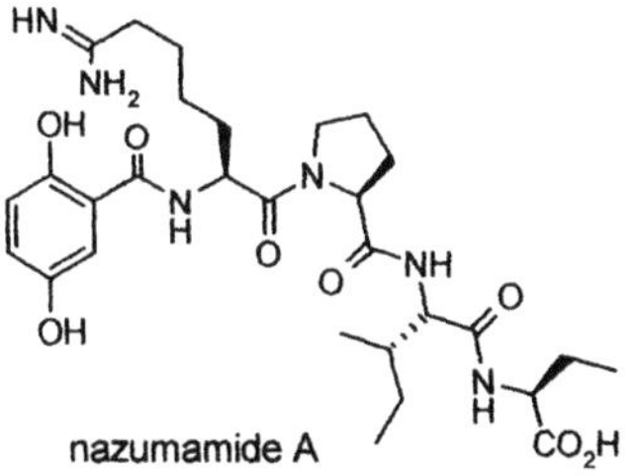 tetrapept./thrombin-like inhibition activity	**marine:** *Theonella swinhoei* Sollas 1888, Lithistida, Porif. from Japan

nazumamide A

Area	Name/Source	Chem class/Mode/Structure	Origin[b]
Blood/gout/ inflamma- tion	prosta- glandins, like (15S)-PGA$_2$/ Sigma 2000	fatty acid metabolite/involved in the control of blood pressure, renal blood flow, smooth muscle contraction, gastric acid secretion, and inflammation 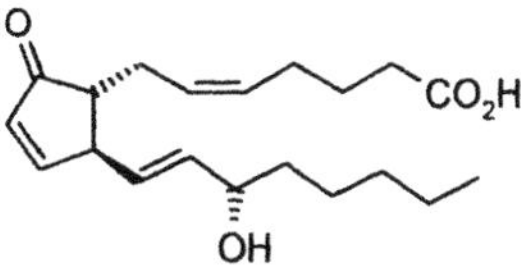	Mamm.; also **marine:** *Plexaura homomalla* (Esper), Gorgonacea, Cnid. from the Caribbean

(15S)-PGA$_2$

Area	Name/Source	Chem class/Mode/Structure	Origin[b]
Blood/gout/ inflamma- tion	*Oxyuranus* toxin/Sigma 2000	polypept./activator of prothrombin	E Asian and Australian *Oxyuranus* spp., Serpentes, Reptilia

Table 13.5.II. Natural and semisynthetic laboratory tools[a]

Area	Name/Source	Chem class/Mode/Structure	Origin[b]
Blood/gout/ inflamma- tion	thrombin-like enzyme/ Sigma 2000	protein/thrombin-like activity	Malayan pit viper, *Calloselasma rhodostoma* [= *Agkistrodon rhodostoma*], Serpentes, Reptilia
Bromination reactions	bromo- peroxidase Sigma 2000	enzyme/catalyst of electrophilic bromination	**marine:** *Corallina officinalis* (L.), Cryptonemiales, Rhodoph.
Cancer	adociasulfate 2	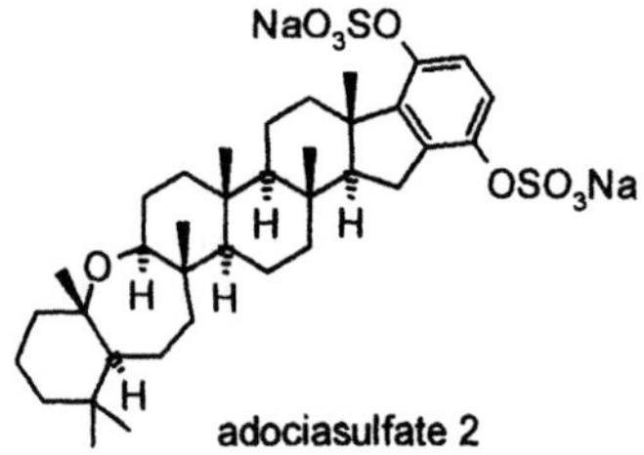 hexaprenoid quinone/inhibitor of kinesins	**marine:** *Haliclona* sp., Haplosclerida, Porif. from Palau, Micronesia: Sakowicz 1998

NaO$_3$SO

OSO$_3$Na

adociasulfate 2

| Cancer | aphidicolin/ Sigma 2000 | diterp./study of cell proliferation and differentiation 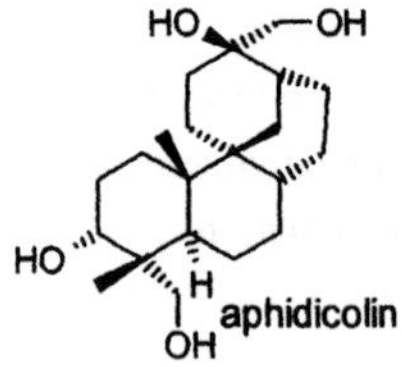 | *Cephalosporium aphidicola* Petch., Eumyc. |

HO—OH

HO

H
OH aphidicolin

Table 13.5.II. Natural and semisynthetic laboratory tools[a]

Area	Name/Source	Chem class/Mode/Structure	Origin[b]
Cancer	calphostin C/Sigma 2000	polyket. benzoquinone dimer/selective inhibitor of protein kinase C 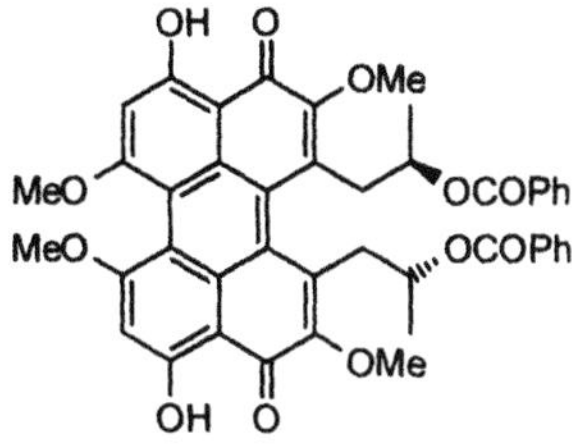	*Cladosporium cladosporioides* (Fresenius) de Vries, Eumyc.

calphostin C

Cancer	chelerythrine /Sigma 2000	protoberberine alkal./inhibitor of protein kinase C 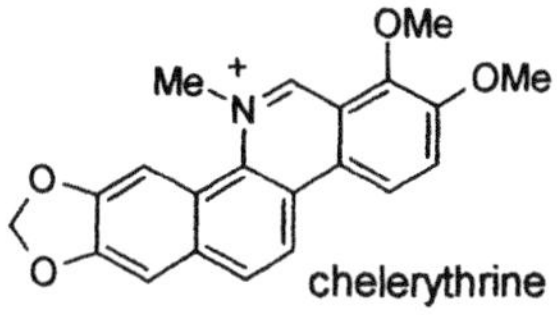	*Chelidonium majus* L. and *Macleaya* spp., Papaveraceae, Ang.: MI

chelerythrine

Cancer	colchicine/ Aldrich, and Fluka	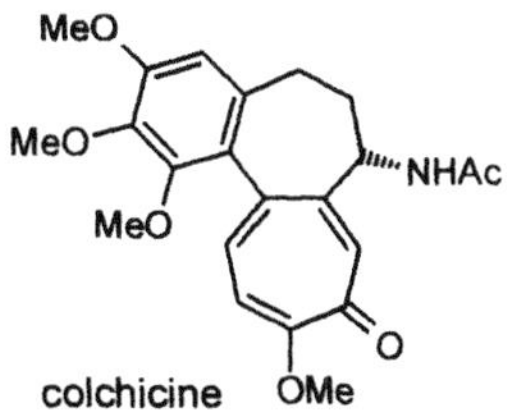 phenethylisoquinoline alkal./antimitotic	autumn crocus, *Colchicum autumnale* L., Liliaceae, Ang.

colchicine

Cancer	jaspamide	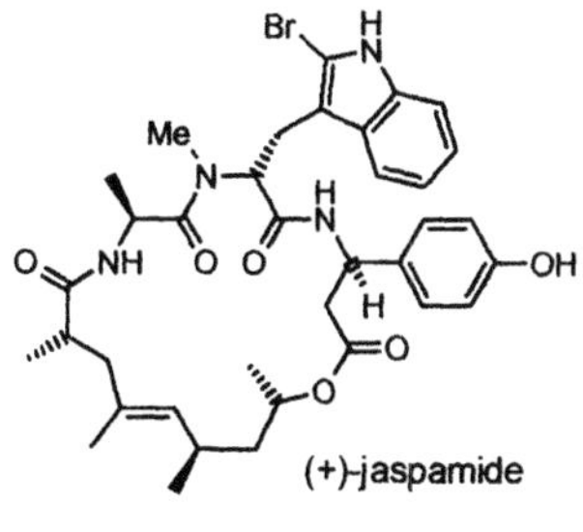 depsipept./affecting actin	**marine:** *Jaspis* spp., Choristida, and other sponges

(+)-jaspamide

Table 13.5.II. Natural and semisynthetic laboratory tools[a]

Area	Name/Source	Chem class/Mode/Structure	Origin[b]
Cancer	*Leiurus* venom/Sigma 2000	pept./in the development of drugs against glioma, the focus is on chlorotoxin for intracellular blocking of low conductance Cl⁻ channels	giant Israeli scorpion, *Leiurus quinquestriatum hebraeus*, Scorpiones, Arach.
Cancer	okadaic acid/Sigma 2000	polyket./tumor promoter that does not activate protein kinase C, while protein phosphatases-1 and -2A are inhibited; also contractile effects on smooth and heart muscles	**marine:** Demosp., Porif. and associated *Prorocentrum* spp., Prorocentrales, Dinofl.,

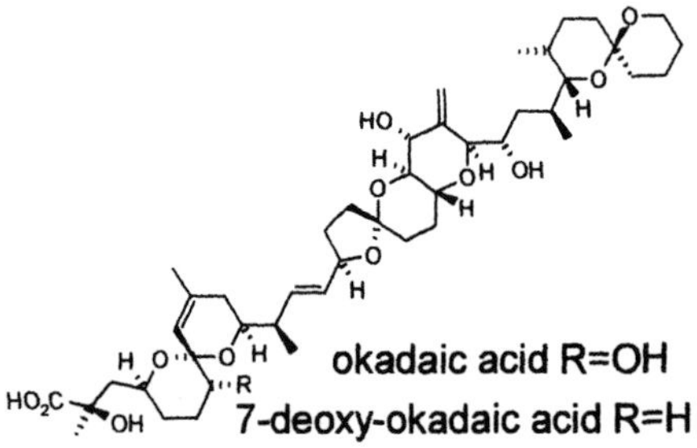

okadaic acid R=OH
7-deoxy-okadaic acid R=H

Area	Name/Source	Chem class/Mode/Structure	Origin[b]
Cancer	palytoxin/ Sigma 2000	polyket./phorbol ester-like tumor promoter and activator of the Na⁺ pump	**marine:** tropical *Palythoa* spp., Zoanthinaria, Cnid.

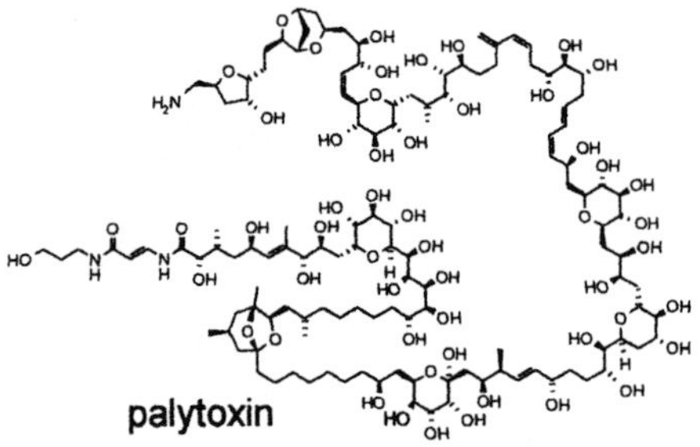

palytoxin

Table 13.5.II. Natural and semisynthetic laboratory tools[a]

Area	Name/Source	Chem class/Mode/Structure	Origin[b]
Cancer	staurosporine /Sigma 2000	indole alkal./inhibitor of protein kinases 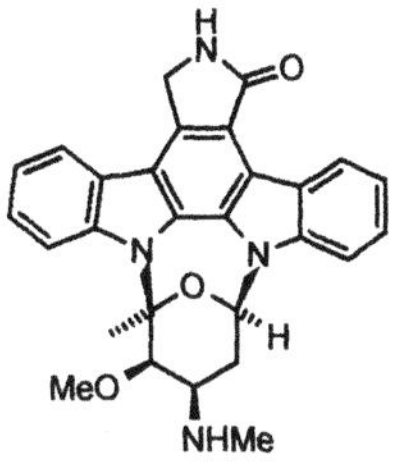	*Streptomyces staurosporeus*, Actinom., Bact.

staurosporine

Area	Name/Source	Chem class/Mode/Structure	Origin[b]
Cancer	swinholide A	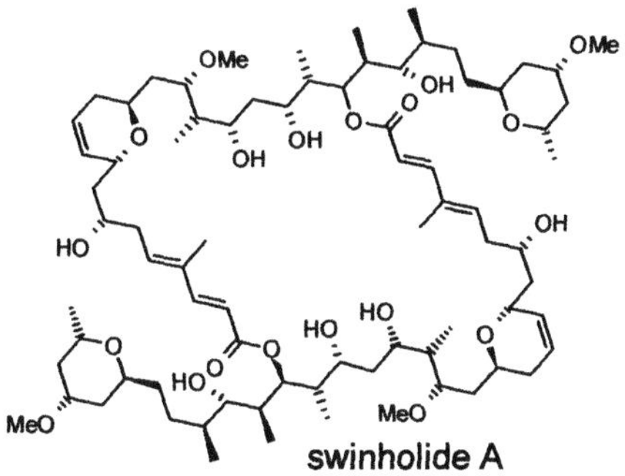 polyket. macrolide/affecting actin	**marine:** *Theonella swinhoei* Sollas 1888, Lithistida, Porif. from Okin.: Doi 1991

swinholide A

Area	Name/Source	Chem class/Mode/Structure	Origin[b]
Cell biology	ilimaquinone	merosesquiterp./inducing complete vesiculation of the Golgi apparatus 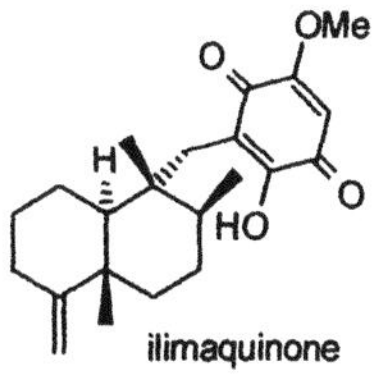	**marine:** *Hippospongia metachromia* Bergquist, 1967, Dictyocer., Porif. from Hawaii: Takizawa 1993

ilimaquinone

Area	Name/Source	Chem class/Mode/Structure	Origin[b]
Cellular biochemistry	efrapeptins/ Sigma 2000	modified pept./inhibitors of ATPase; useful in studies of oxidative phosphorylation	*Tolypocladium* spp., Eumyc.

Table 13.5.II. Natural and semisynthetic laboratory tools[a]

Area	Name/Source	Chem class/Mode/Structure	Origin[b]
Cellular biochemistry	forskolin/ Sigma 2000	labdane diterp./useful in the purification of adenyl cyclase 	*Coleus forskohlii* Briq., Labiatae, Ang.

forskolin

| Cellular biochemistry | glutathione/ Sigma 2000 | tripept./detoxifying agent in combination with glutathione *S*-transferase 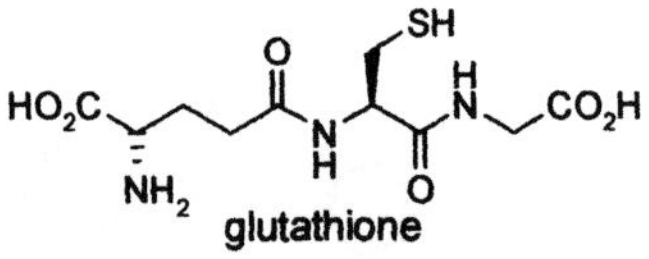 | Mamm. erythrocytes |

glutathione

| Cellular biochemistry | phospho-lipases/ Sigma 2000 | proteins/useful in removing phospholipids from cell membranes | from Bact., Ang., honey bee, venomous snakes, Mamm. pancreas |
| CNS | α-amino-3-hydroxy-5-methyl-isoxazole-4-propionic acid ((±)-AMPA HBr Sigma 2000) | ibotenic-acid analogue/excitatory amino acid which binds to kainate receptors | ibotenic acid from *Amanita pantherina* (DC.) Fr. and *Amanita muscaria* Fr., |

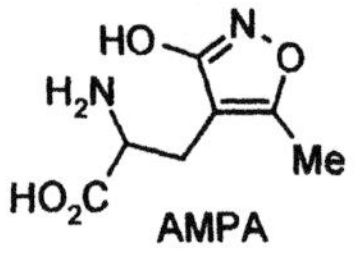

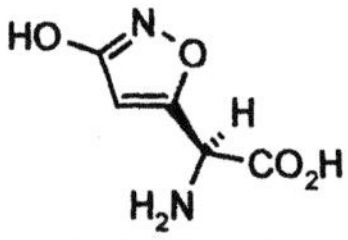

AMPA   ibotenic acid

Basidiomyc., Eumyc.: MI

Table 13.5.II. Natural and semisynthetic laboratory tools[a]

Area	Name/Source	Chem class/Mode/Structure	Origin[b]
CNS	(-)-bicucul-line methiodide/ Sigma 2000	tetrahydroisoquinoline alkal. derived from (+)-bicuculline/GABA$_A$ receptor antagonist	(+)-bicuculline: *Dicentra cucullaria* (L.) Bernh. and *Adlumia fungosa* (Ait) Greene, Fumariaceae, and

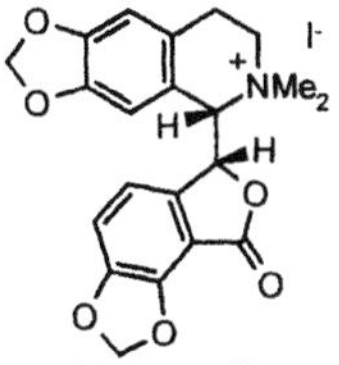

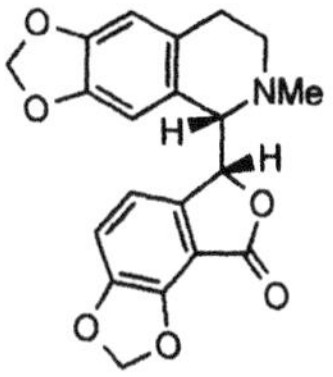

(-)-bicuculline
methiodide

(+)-bicuculline

Corydalis spp., Papaveraceae, Ang.: MI

| CNS | capsaicin/ Sigma 2000 | phenolic amide/activator of sensory neurons 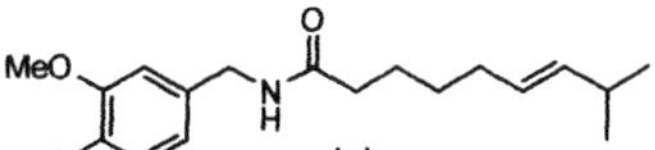 | Cayenne pepper (*Capsicum* spp., Solanaceae, Ang.: MI) |

capsaicin

| CNS | capsazepine/ Sigma 2000 | synthetic analogue of the alkal. capsaicin/capsaicin antagonist 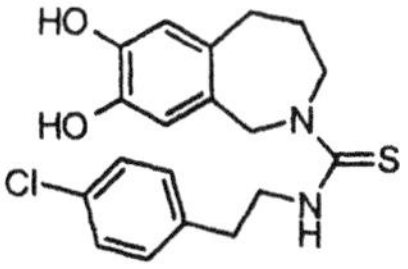 | synthesis patterned on capsaicin from *Capsicum* spp., Solanaceae, Ang.: MI |

capsazepine

Part IV. Natural product diversity at functional level

Table 13.5.II. Natural and semisynthetic laboratory tools[a]

Area	Name/Source	Chem class/Mode/Structure	Origin[b]
CNS	conotoxins/ Sigma 2000; Neurex (Elan Pharm.)	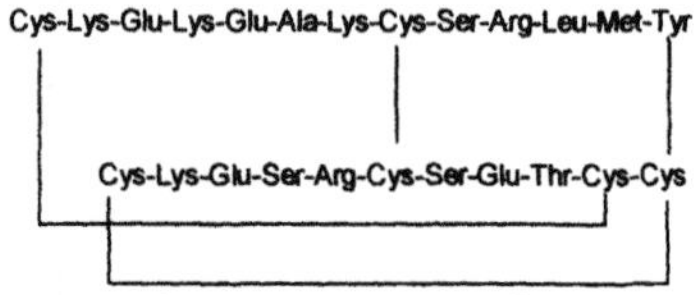 polypept., such as ω-conotoxin MVIIA/postsynaptic inhibition or blocking of nicotinic acetylcholine receptor, according to type: Kohno 1995, and its synthetic version SNX-111	**marine:** chemical modification of, or patterned synthesis on, peptides of *Conus magus* Linné, 1758, Neogastropoda, Moll.
		Cys-Lys-Glu-Lys-Glu-Ala-Lys-Cys-Ser-Arg-Leu-Met-Tyr Cys-Lys-Glu-Ser-Arg-Cys-Ser-Glu-Thr-Cys-Cys amino acid sequence and disulfide bridges of ω-conotoxin MVIIA	
CNS	epibatidine/ Sigma 2000	alkal./potent nicotinic agonist 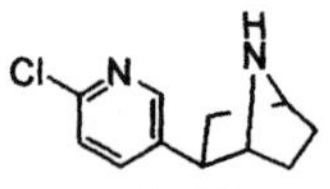 epibatidine	skin of Ecuadoran tree frog, *Epipedobates tricolor* (Boulenger, 1899), Amphib.: MI
CNS	ω-grammo-toxin/Zeneca	pept./inhibition of Ca^{2+} entry into nerve cell, by which atrial fibrillation is controlled	Chilean pink tarantula, *Grammostola spatulata* F.O.P. Cambridge 1897, Araneae, Arach.: Bode 2001
CNS	muscimol/ Sigma 2000	isoxazole alkal./GABA agonist 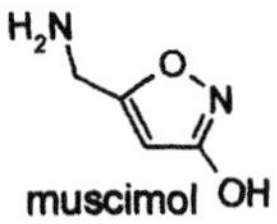 muscimol	*Amanita muscaria* Fr., Basidiomyc., Eumyc.
CNS	pardaxin/ Sigma 2000	pept. of structure resembling honey-bee melittin/inducer of neurotransmitter release	**marine:** Moses sole, *Pardachirus marmoratus* (Lacépède), Soleidae, Osteich. from the Red Sea
Drug discovery	cholera toxins/Sigma 2000	AB$_5$ hexameric proteins/triggering ADP-ribosylation or binding of ganglioside G_{M1} on cell's surface	*Vibrio cholerae* Pacini 1854, Bact.

Table 13.5.II. Natural and semisynthetic laboratory tools[a]

Area	Name/Source	Chem class/Mode/Structure	Origin[b]
Drug discovery	diphtheria toxins/Sigma 2000	proteins/ADP-ribosylation properties	*Corynebacterium diphtheriae* (Kruse 1886) Lehmann and Neumann 1896, Bact.
Drug discovery	luciferase and luciferin/ Sigma 2000	protein/high dynamic range receptor for binding studies (luciferase assay) in drug discovery	Bact., Ins. (Sigma 2000); also **marine:** *Renilla reniformis* (Pallas, 1766), Pennatulacea, Cnid.: Cormier 1994.
Enzyme activity evaluation	laminarin/ Sigma 2000	polysacch./substrate for evaluation of laminarase	**marine:** *Laminaria digitata* (Hudson) Lamouroux, Phaeoph.
Enzyme inhibition	castano-spermine/ Sigma 2000	indolizidine alkal./inhibitor of glycoside hydrolysis: MI 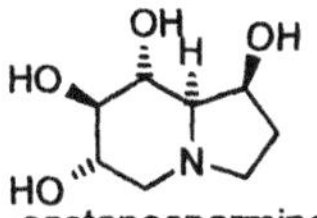	*Castanospermum australe* A. Cunn., Leguminosae, Ang.

castanospermine

Area	Name/Source	Chem class/Mode/Structure	Origin[b]
Enzyme inhibition	ouabain/ Sigma 2000	cardenolide/selective inhibitor of Na$^+$-K$^+$ ATPase and hybridoma reagent for cultures 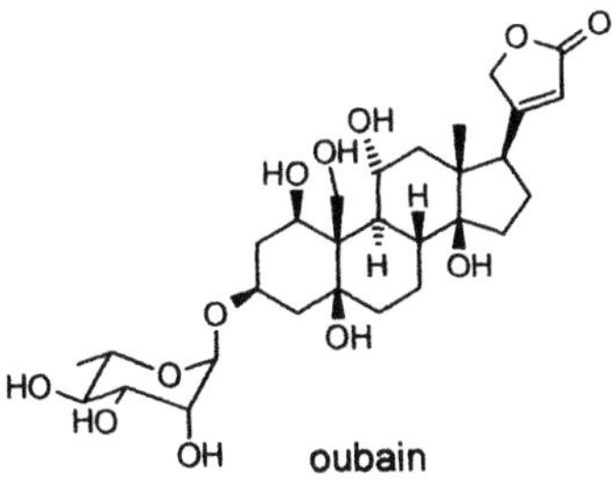	tropical *Strophanthus gratus* (Wallich & Hook. ex Benth.) Baill., and *Acokantera ouabaio* Cathel, Apocyanaceae, Ang.

oubain

Table 13.5.II. Natural and semisynthetic laboratory tools[a]

Area	Name/Source	Chem class/Mode/Structure	Origin[b]
Enzyme inhibition	thapsigargin/ Sigma 2000	sesquiterp./inhibitor of endoplasmic reticulum Ca^{2+}-ATPase, thereby allowing to manipulate intracellular Ca^{2+} levels 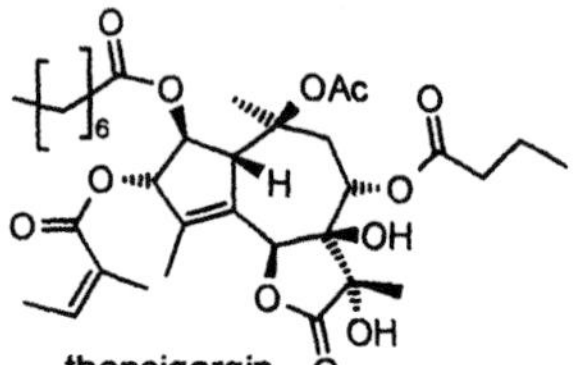	*Thapsia garganica* L., Apiaceae, Ang.

thapsigargin

Immune system	gliotoxin/ Sigma 2000	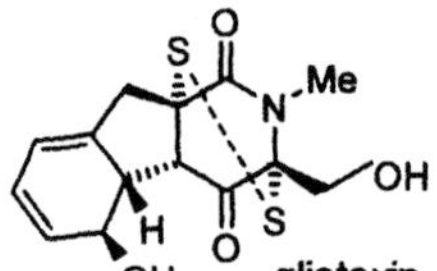 diketopiperazine / immunomodulator	*Gliocladium fimbriatum* J.C. Gilman et E.V. Abbott, Eumyc.

gliotoxin

Intracellular signaling	aequorin/ Sigma 2000	blue-emitting protein/in evaluation of Ca^{2+} level in serum and organelles; the chromophore is 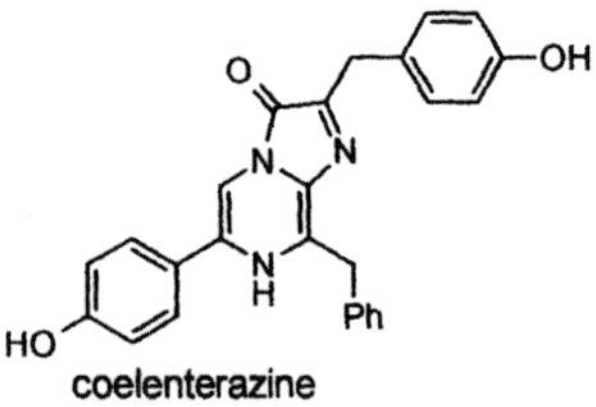 coelenterazine	**marine:** Pacific jellyfish, *Aequorea victoria* (Murbach & Shearer, 1902), Hydrozoa, Cnid.; cloned: Kendall 1998, and genetically engineered for enhanced thermal stability: Mochizuki 1998

coelenterazine

Intracellular signaling	N-(NBD-amino-hexanoyl)-sphingo-myelin/ Sigma 2000	semisynthetic sphingomyelins/fluorescent probes for plasma membrane transport	chemical modifications of Mamm.

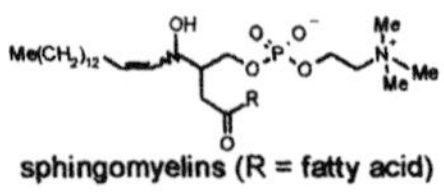

sphingomyelins (R = fatty acid)

sphingomyelins

Table 13.5.II. Natural and semisynthetic laboratory tools[a]

Area	Name/Source	Chem class/Mode/Structure	Origin[b]
Intracellular signaling	calcimycin/ Sigma 2000	benzoxazole alkal. polyether/acting as a Ca^{2+}-selective ionophore in isolated mitochondria: MI 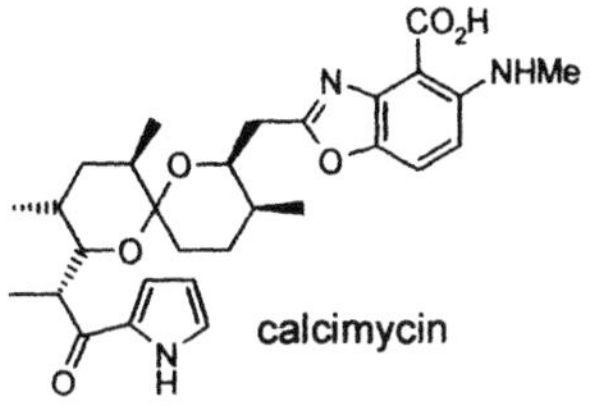	*Streptomyces chartreusensis*, Actinom., Bact.

calcimycin

Ion channels	brevetoxin A and B/Sigma 2000	lipid-soluble polyether polyket./activation of Na^+ channels 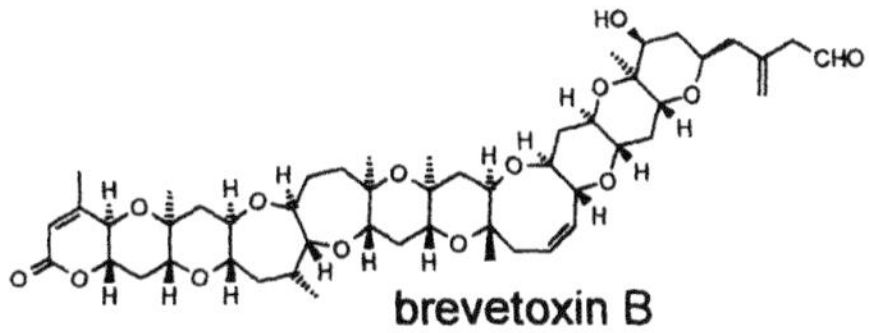	**marine:** *Gymnodinium breve* Davis [= *Ptychodiscus brevis* Davis], Dinofl.; also semisynthetic

brevetoxin B

Ion channels	conotoxins/ Sigma 2000; Neurex (Elan Pharm.)	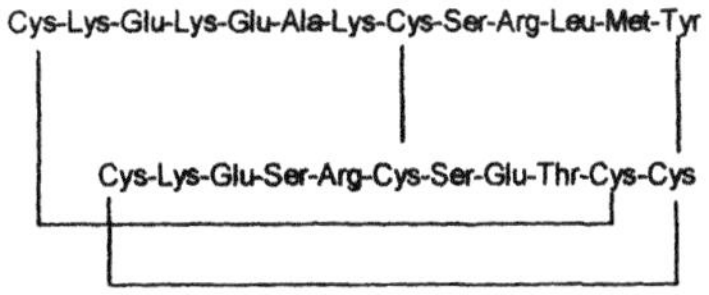 polypept., like ω-conotoxin MVIIA/blocking of Na^+ channels or voltage activated N-type Ca^{++} channels, according to type: Kohno 1995, and its synthetic version, SNX-111	**marine:** chemical modification of, or patterned synthesis on, pept. of *Conus magus* Linné, 1758, Neogastropoda, Moll.

amino acid sequence and disulfide bridges of ω-conotoxin MVIIA

Ion channels	maitotoxin/ Sigma 2000	polyether polyket./enhanced influx on extracellular Ca^{2+} 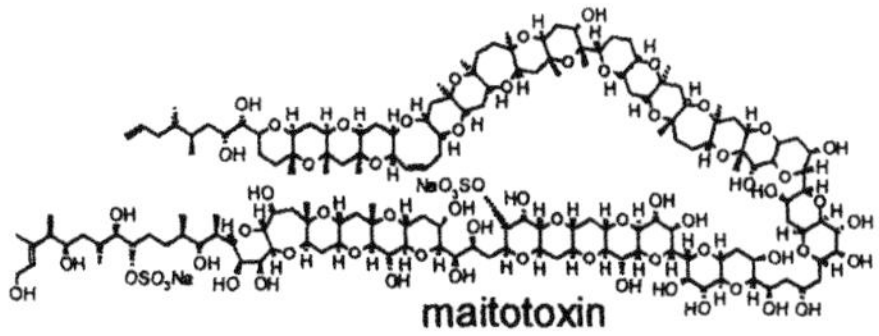	**marine:** *Gambierdiscus toxicus* Adachi et Fukuio, 1979, Dinofl.

maitotoxin

Table 13.5.II. Natural and semisynthetic laboratory tools[a]

Area	Name/Source	Chem class/Mode/Structure	Origin[b]
Ion channels	neomycin/ Sigma 2000	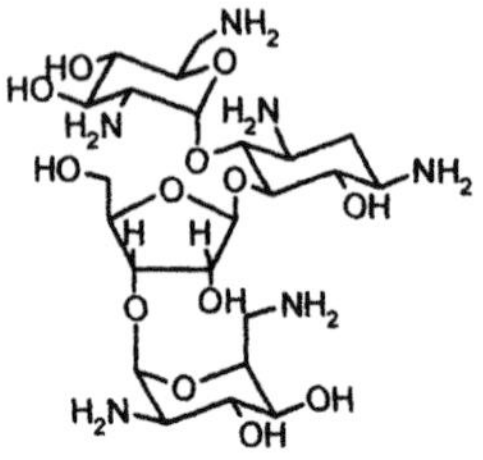 aminoglycoside/selective inhibition of Ca^{2+} channels	*Streptomyces fradiae,* Actinom., Bact.

neomycin B

| Ion channels | saxitoxin/ Sigma 2000 | water-soluble alkal./inhibition of Na^+ channels 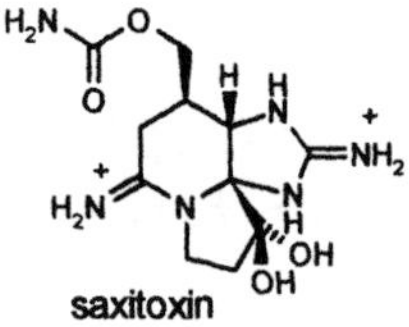 | **marine:** *Protogonyaulax* spp., Dinofl. |

saxitoxin

| Ion channels | tetrodotoxin/ Sigma 2000 | alkal./reversible blocking of Na^+ channels without affecting K^+-channel blockage; useful marker for Na^+ channels in excitable tissues; used in blocking Na^+ channels in multiple conducting systems 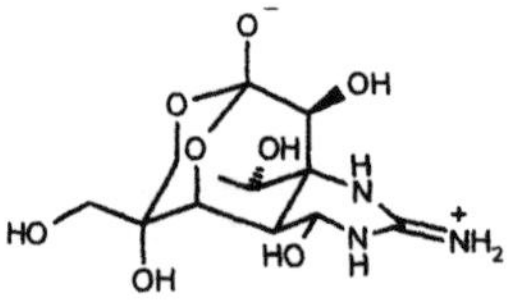 | newts, Amphib.; also **marine:** Bact., Dinofl., Moll., Crust., Nemert., puffers and parrotfish: Pietra 1995; Kodama 1996 |

(-)-tetrodotoxin

Table 13.5.II. Natural and semisynthetic laboratory tools[a]

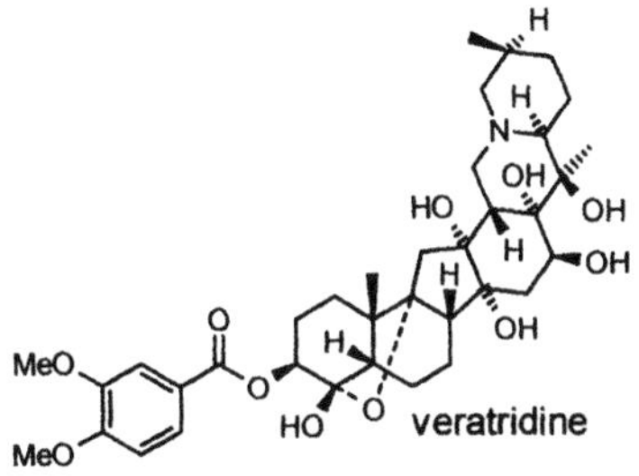

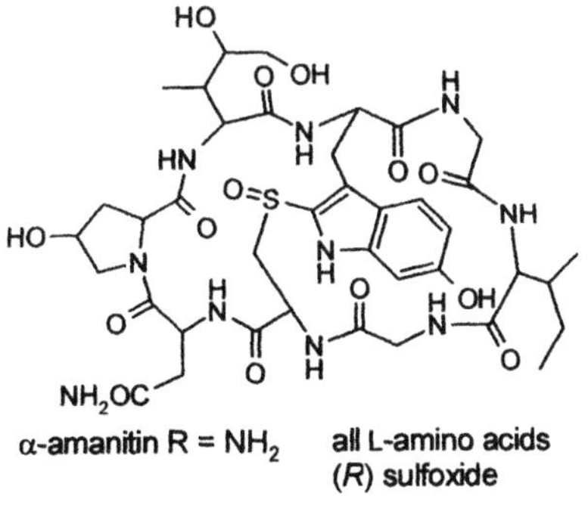

Area	Name/Source	Chem class/Mode/Structure	Origin[b]
Ion channels	veratridine/ Sigma 2000	*Veratrum* alkal./selective intracellular Ca[2+] enhancement	*Veratrum album* L. and *Schoenocaulon officinale* (Lindl.), Liliaceae, Ang.

veratridine

Metabolism	*Crotalus* venom/Sigma 2000	proteins/L-amino acid oxidase activity	tropical *Crotalus* spp., Serpentes, Reptilia
Molecular biology	α-amanitin/ Sigma 2000	cyclopept./selective inhibitor of eukaryotic RNA polymerases II and III	*Amanita phalloides* (Fr.) Seer., Basidiomyc., Eumyc.

α-amanitin R = NH$_2$ all L-amino acids
(R) sulfoxide

Molecular biology	(+)-brefeldin A/Sigma 2000	polyket./inhibitor of protein transfer from endoplasmic reticulum to the Golgi apparatus	*Eupenicillium brefeldianum*, Deuterom., Eumyc.

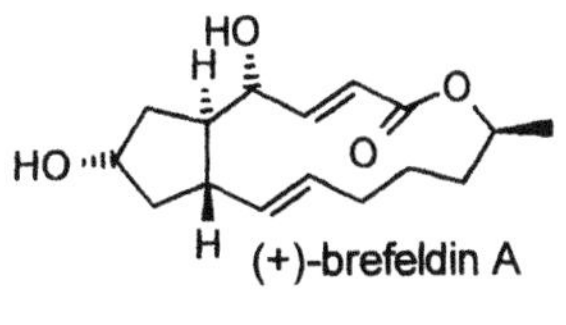

(+)-brefeldin A

Table 13.5.II. Natural and semisynthetic laboratory tools[a]

Area	Name/Source	Chem class/Mode/Structure	Origin[b]
Molecular biology	chloro-peroxidase/ Sigma 2000	protein/used for isotopic labeling of macromolecular compounds	*Caldariomyces fumago* Woron. (1927), filamentous Eumyc.
Molecular biology	ilimaquinone /Sigma 2000	merosesquiterp./triggering the reversible breakdown of Golgi membranes 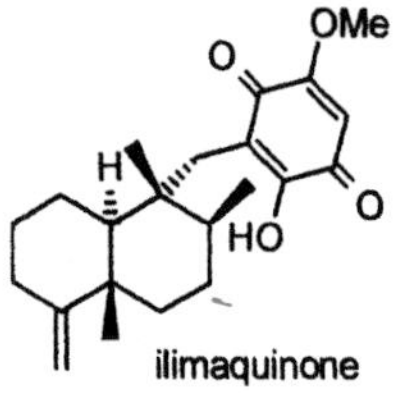	**marine:** tropical *Hippospongia* spp. and *Fasciospongia* spp., Dictyocer., Porif.

ilimaquinone

Area	Name/Source	Chem class/Mode/Structure	Origin[b]
Molecular biology	melittin/ Sigma 2000	protein/binding to calmodulin/reagent for protein analysis	honey bee (*Apis mellifera* (L.), Ins. venom
Molecular biology	phorbol esters/Sigma 2000	diterp./carcinogens, NO promoters; activators of protein kinase C 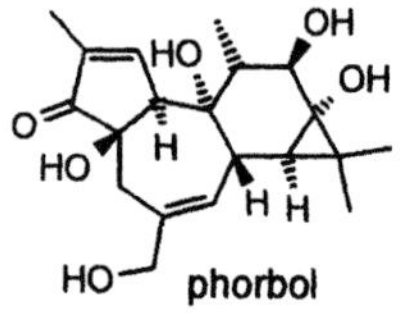	pantrop. and temperate Euphorbiaceae, Ang. (also a standard for phorbol-class receptor in cancer studies)

phorbol

Area	Name/Source	Chem class/Mode/Structure	Origin[b]
Molecular biology	proteases/ terrestrial only from Sigma 2000	proteins/used in cloning	*Streptomyces* spp., Actinom.; also **marine:** *Alteromonas* spp., Bact.: Domoto 1994, and shipworm Bact.: Griffin 1992
Molecular biology	*Pseudomonas* agarase/ Sigma 2000	agarose digesting enzyme protein/ used in the isolation of intact, high MW, DNA	**marine:** *Pseudomonas atlantica*, Bact.
Molecular biology	Taq DNA polymerase for PCR/Sigma 2000	proteins/used in PCR	**marine:** *Thermococcus littoralis*, Archaea

Table 13.5.II. Natural and semisynthetic laboratory tools[a]

Area	Name/Source	Chem class/Mode/Structure	Origin[b]
Signal transduction	quercetin/ Sigma 2000	shikim./inhibitor of mitochondrial ATPase and phosphodiesterase 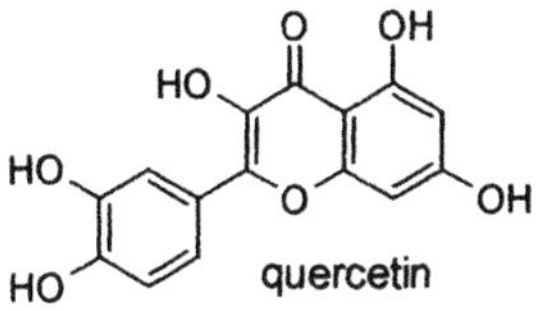	*Rhododendron cinnabarinum* HOOK. f., Ericaceae, Ang.

O OH
HO
HO O
HO O OH
HO quercetin

Area	Name/Source	Chem class/Mode/Structure	Origin[b]
Signal transduction	green fluorescent protein/BD Biosciences - PharMingen, BioWorld	green-emitting bioluminescent protein/protein localization	marine:: Pacific jellyfish (*Aequorea victoria* (Murbach & Shearer, 1902), Hydrozoa, Cnid.); cloned: Kendall 1998.
Smooth muscle	sarafotoxin S6c/Sigma 2000	pept./vasoactive, agonist of the ET_B endothelin receptor	*Atractaspis engaddensis*, Serpentes, Reptilia
Tissue culture media	carrageenans /Sigma 2000	sulfated polysacch./flexible gels, rigid gels, or no gels	**marine:** *Eucheuma cottonii* (under taxonomic revision: Doty 1988) or *Gigartina* spp., Gigartinales, Rhodoph.
Toxin assay (bacterial)	amoebocyte lysate/Sigma 2000	used in detection and quantitation of endotoxins in plasma: Cohen 1979	**marine:** horseshoe crab, *Limulus polyphemus* (L.), Chelicerata

[a]Many therapeutically useful drugs are also used as laboratory tools.
[b]Marine origin is indicated in boldface, even in obvious cases.

13.6 Drugs of abuse

Taking drugs to endure pains during the everyday work and the war, or to outperform in athletics, is recorded in the most ancient literature and paintings. Drugs like Ayahuasca, bufotenine, cocaine, ibogaine, mescaline, and psylocybin are bound to the history of people in America, like ibogaine is in Africa, and Khat, *Cannabis*, and morphine are in the East. With the commercialization of sport and the changing habits and laws of the society, drug abuse has become a major social problem.

Until recently, stimulants and narcotics were the most sought drugs in athletics. To circumvent the rules, drugs that do not affect the mood find now large use, like substances that increase the red blood cells count (blood doping), diuretics, peptide hormones, beta-blocker, and anabolic steroids.

A list of drugs of abuse is provided in Table 13.6. Numerically it is dominated by the alkaloids and products of synthesis patterned on alkaloids, like the amphetamines and Ecstasy. Meroterpenes have also an important place, like cannabinoids in the CNS cell trafficking (Christie 2001). Some of these metabolites, or their derivatives, find also medical use (Chapters 13.2 and 13.3).

Table 13.6. Drugs of abuse

Name	Origin	Effects/use	Components/chem. class
Amphet-amines	synthesis patterned on cathinone, from the leaves of *Catha edulis* Forsk., Crelestraceae, Ang., from highlands of Yemen 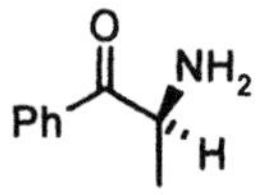	stimulants/doping in athletics	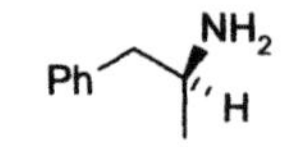 (+)-amphetamine/alkal. (+)-amphetamine cathinone
Ayahuasca	Ayahuasca, *Banisteriopsis caapi* (Spruce ex Griseb.) C. V. Morton); Chacruna (*Psychotria viridis* Ruiz & Pav.); Chaliponga (*Diplopterys cabrerana* (Cuatrec.) Gates [= *Banisteriopsis rusbyana*]), and other Ang. from the S American rain forest	hallucinogen inducing profound physical, mental, and spiritual effects/traditional drink in ceremonial settings in Amazonian tribes from at least 2,500 years	monoamine oxidase inhibitors (harmine and reduced analogues from *B. caapi*) and hallucinogenic *N,N*-dimethyltryptamine and the 5-methoxy analogue from *P. viridis* and *D. cabrerana*, resp./alkal. 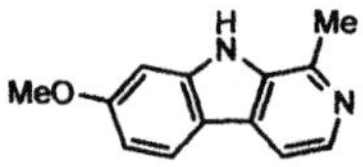 harmine 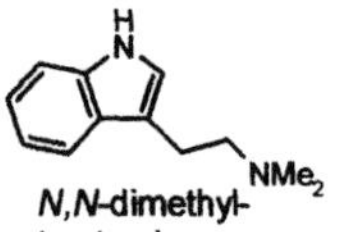 *N,N*-dimethyl-tryptamine
Bufotenine	toads, Amph., and seeds of *Piptadenia peregrina* Benth., Leguminosae, Ang.: MI	weak hallucinogenic agent, used by the indigenous of Trinidad at the times of the first Spanish explorations	bufotenine/alkal. 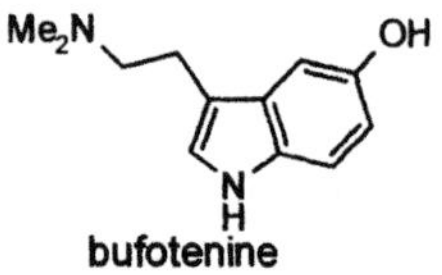 bufotenine
Caffeine	*Coffea arabica* L./Rubiaceae, Ang.	stimulant/doping in athletics	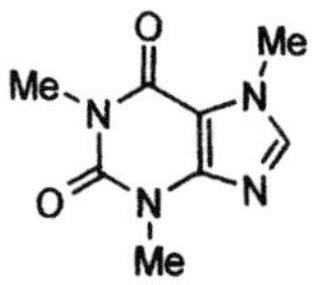 caffeine

Table 13.6. Drugs of abuse

Name	Origin	Effects/use	Components/chem. class
Cocaine	leaves of *Erythroxylum coca* Lam. from hills of Peru and other highlands of tropical S America	stimulant inducing forgiveness/traditional chewing the leaves against physical stress; ophthalmic anesthetic; doping in athletics as stimulant	cocaine/tropane alkal. 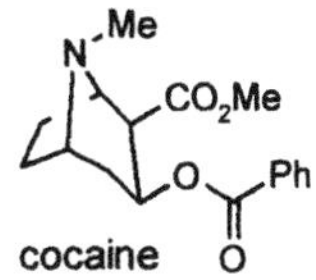
Ecstasy	synthesis patterned on cathinone from *Catha edulis* Forsk., Celestraceae, Ang. 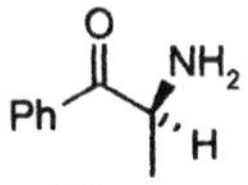	hallucinogen: MI	Ecstasy/ alkal. 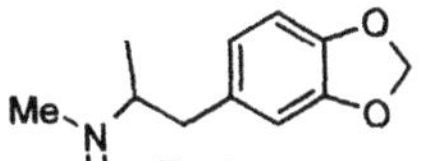
Ethanol	fermentation of carboh.	resulting in acute and chronic alcoholism	EtOH/lower alcohol
Ibogaine	roots and bark of *Tabernanthe iboga* Baill., Gentianales, Ang. from African rain forests	hallucinogen/plant extract are in traditional use by natives to retain mental alertness while constricted to motionless for a couple of days	ibogaine/ alkal. 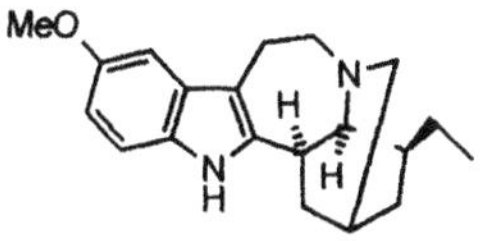
Khat	leaves of *Catha edulis* Forsk., Crelestraceae, Ang. from highlands of Yemen and E Africa	amphetamine-like stimulant/traditional chewing the leaves to endure hard life	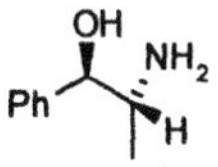 cathinone and minor components, such as norpseudoephedrin /alkal. 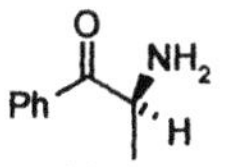
Lyser-gamide	*Ipomea tricolor* Cav. and *Rivea corymbosa* (L.), Convolvulaceae, Ang.	depressant	lysergamide/alkal. 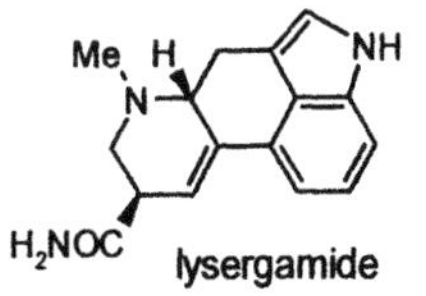

Table 13.6. Drugs of abuse

Name	Origin	Effects/use	Components/chem. class
Lysergic acid	from hydrolysis of ergot from *Clavices paspali* F. Stevens & J.G. Hall (1910), Eumyc.	depressant	lysergic acid/alkal 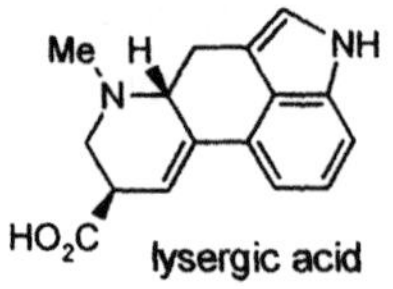 lysergic acid
Lysergic acid diethyl-amide (LSD)	*Clavices paspali* F. Stevens & J.G. Hall (1910), Eumyc., grown over hydroxyethylamide	hallucinogen	lysergic acid diethylamide/alkal. 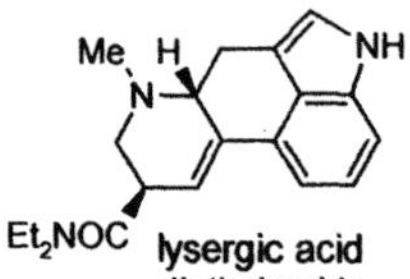 lysergic acid diethylamide
Mescaline	peyote: mescal buttons of *Lophophora williamsii* (Lemaire) Coult. [= *Anhalonium lewinii* Henn.], Cactaceae, Ang. from northern arid highlands of Mexico	hallucinogen/traditional use in Mexico since remote times	mescaline/alkal. 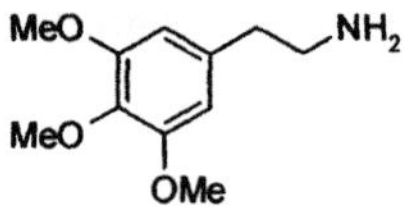 mescaline
Morphine	opium from *Papaver somniferum* L. or *Papaver album* Mill., Papaveraceae, Ang., from Middle East and other regions; biosynthesized in Mamm.: Herbert 2000	analgesic/traditional use in beverages since early historical times; it is now an analgesic for last aid	morphine/alkal. 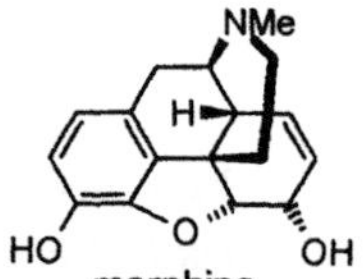 morphine
Nicotine	leaves of *Nicotiana tabacum* and *Nicotiana rustica*	sedative leading to addiction/tobacco smoking	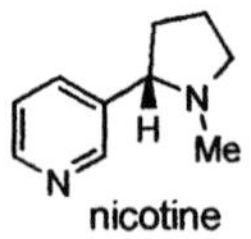 nicotine alkal. nicotine

Table 13.6. Drugs of abuse

Name	Origin	Effects/use	Components/chem. class
Psilocybin and, in trace amounts, psilocin	*Psilocybe mexicana* Heim., Agaricaceae, Ang.	hallucinogen/traditional use in Mexico	psilocybin/alkal. 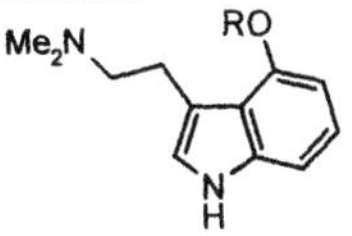
Tetra-hydro-cannabinol	hemp, *Cannabis sativa* var. *indica* Auth., Moraceae, Ang.; cultivated	hallucinogen/hashish for traditional beverage and seeds for smoking; used in the Middle East since remote times; doping in athletics; research on therapeutic use as antiemetic and anticancer approved in UK	tetrahydrocannabinol Δ^9-THC/meroterp. 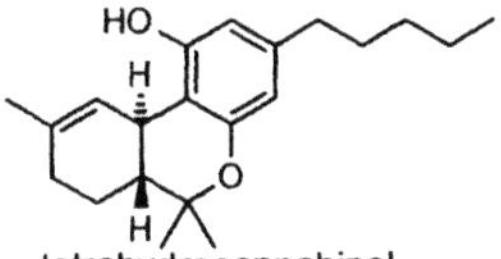

psilocybin R = PO(OH)$_2$
psilocin R = H

tetrahydrocannabinol

Part V. Biotechnology and chemical synthesis of natural products

And Noah began to be an husbandman,
and he planted a vineyard. And he drank
of the wine, and was drunken
Genesis 9.20-21

Chapter 14. The role of biotechnology

The raison d'être of natural product diversity, and the exploitation of natural products by man, were delineated in Chapters 12 and 13. Man went further, in an endeavor to improve on nature by making available rare bioactive compounds and synthetic analogues with a better performance (Liu 1999).

The history of the technology, embedded in the philosophy of science, is too vast and complex to be treated as a whole. This is mirrored in the early education of people: in Europe the child is taught about the revolution caused by the invention of the wheel, while in Oceania the equivalent is chosen in the invention of the double-hulled sailing canoe. Biotechnology is an old branch of technology, but one that has changed much with the times. At the beginning it was a purely empirical activity, though absolutely not naïve, such as with wine making by biblical Noah, soil preparations in folk medicine, and indigo dyeing. In the latter, a thermophilic bacterium in the genus *Clostridium* was unconsciously put to use (Nikki Paden 1998). Modern times, which have seen the genetic modification of microorganisms by such blind practices as UV irradiation or chemical methylation, are also over. In our postmodern society, even wine making demands much skill in molecular biology to compete in flourishing economies (Beazley 2001). The transfer of genetic material to, and over-expression in, other viable organisms has become a common practice. Postmodern biotechnology solicits also much philosophical interest by reopening the question of determinism caused by genes against free will (Rose 2000), if the problem is correctly posed at all.

In principle, biotechnology is any molecular practice that makes use of living systems. In practice, however, the concept is restricted to tasks involving the analysis, planned modification, and control of biological agents. Any company in the pharmaceutical, agricultural, and environmental sectors must comply with this trend. In the therapeutic area a strict link exists with genomics (such as in developing a drug that specifically controls gene expression) and proteomics (in dealing with receptors, enzymes, and other proteins that are not directly associated with genes, or that may have undergone post-translational modifications). Biotechnology makes use of all tricks of chemistry, such as in conjugating a drug with a biomacromolecule to bring the drug specifically to the target. To this end, monoclonal antibodies are used to block hormones. Diagnostics is another successful line, like in the imaging of pathological cells through isotope labeling.

In the agricultural sector, genetically modified crops find many obstacles. Recent regulations pose everything in doubt. Major problems were raised about StarLink corn, banned for planting purposes on the allegation that Cry9C (one of its heat-stable and hard-to-digest proteins) may have allergenic properties. The bovine spongiform encephalopathy also makes things much blurring. Resistance by the major producers of soy beans, who see their market endangered by genetically modified alternatives, poses further problems. Should biotechnology fail in its purposes, both the pharmaceutical and the agricultural industry will probably face a financial crisis, which might be quite serious for countries where these activities are at the top of postmodern technology.

In any event, biotechnology has already brought about a revolution in the pharmaceutical, agricultural, and environmental sectors. In the treatment of ailments there is no more room for small research-oriented pharmaceutical companies. International treaties also make difficult pursuing ethnobotanical practices, while the poorest nations can now circumvent the patent rights for certain drugs. This may undermine the pharmaceutical research. A better route would have been to cut down the price of key drugs for the poorest nations.

Science will scarcely feel this revolution in the short term because of the secrecy in industrial projects and corporate-funded projects at the university. But man's privacy and trust are seriously at risk following the elucidation of the sequence of the human genome. Defective persons may face discrimination by employment and health insurance companies. This may pose problems especially to the US citizens because of the lack of a national health care system. Protests against globalization are a sign of the soaring preoccupation by the most exposed, or conscious, society.

14.1 Biotechnology and natural products

The two fundamental branches of postmodern biotechnology are genomics and proteomics (Liu 1999). Genomics is concerned with the structure of DNA. Major projects have already been completed, like the first published sequence of a bacterial genome, *Haemophilus influenzae* (Fleishmann 1995). Then came the sequence of a bacterial plant pathogen, *Hylella fastidiosa* (Simpson 2000). In these days, the first draft of the human genome has been completed (Lander 2001). This project has gathered such powerful groups as the US National Human Genome Research Institute, Britain's Wellcome Trust, their consortia with pharmaceutical companies, and the Japanese Genomic Science Center.

Plant genome projects, notably the rice genome project, have also attracted large companies, like Monsanto, Novartis, and DuPont. *Arabidopsis thaliana*, a brassicacean plant that has become a model in plant genetics, was the first flowering plant to have the genome sequence completed (*Arabidopsis* Initiative 2000).

Unculturable microorganisms have recently been investigated for the possibility of isolating genes and cloning and over-expressing them in viable microorganisms, to new natural products (Pei 1991). Unculturable microorganisms should not be confused with those that resist culture because of inadequate culture media or conditions. Truly unculturable microorganisms are probably to be found in the strict symbionts and parasites that have renounced to part of their genome, relying on that of the host to supplement their limited metabolic capabilities. So far, no documented success in this area has appeared, however.

Genomics is followed by functional genomics. Now that the human genome has been sequenced, a harder problem is understanding the function and regulation of genes. The circuitry of the system, i.e. the cell signaling and regulation, needs to be unraveled. The DNA chip technology may help. The *Arabidopsis thaliana* genome project gives a measure of the gap of knowledge: although the whole sequence of the plant genome is known, only about 1,000 of the 25,900 genes have been assigned a function. According to optimistic forecasts, the function of all the genes of this plant will be assigned in the next decade (Somerville 2000). The important question of gene regulation will remain open anyway.

Proteomics embraces problems of structure and function of the proteins, in particular those lacking any direct link with DNA, or having undergone post-translational modifications (Liu 1999). Since the enzymes of the secondary metabolism are encoded in fundamentally the same way as those for the primary metabolism, both genomics and proteomics are of concern to the natural product chemist. Understanding how natural products get their imprint from DNA, and the possibility of directing their biosynthesis, is an aspect of functional genomics that attracts the pharmaceutical industry and could redirect our approach to natural products. When everything about the genome of the organisms of interest is clarified, natural product chemistry will be quite different from what we know. Elucidation of the enzymes of plant secondary metabolism is already a subject of study, in particular for plant pigments (Nakayama 2000). It is far from being a simple task, however, since variations in single genes may not affect the phenotype. Any future biotechnological approach to drug discovery may require the simultaneous variation in several critical genes (Bailey 1999). Things are even more complex with animals, where even the structure of the pigments is often obscure (Kopp 2000).

Representative examples of the biotechnological approach to natural products are shown in Chart 14.1 and are discussed below, in the same sequential order of classes of compounds of the charts in Part III.

14.1a Alkaloids

Gene transfer from the angiosperm *Catharanthus roseus*, and over-expression in the bacterium *Escherichia coli*, yielded the synthase for strictosidine, a known alkaloid of the tryptophan-secologanin class (Scott 1992). A similar strategy has clarified the biosynthesis of hydrogenobyrinic acid, an advanced precursor of vitamin B_{12} (Scott 1994).

14.1b Peptides and proteins

The production of somatostatin, human insulin, α-interferon, the growth hormone, and erythropoietin are landmarks of the recent past of genetic engineering.

The green fluorescent protein is a specialized protein that has found many applications. Isolated from jellyfish, it has also been obtained by inserting the jellyfish gene in the genome of the tobacco plant (Boevink 1999). However, the process is still far from the efficiency needed by industry.

Other important applications are illustrated in the following. Optical reporter genes have been exploited for photoprotein imaging in drug discovery (Contag 1999). *In vitro* evolution has been attempted by the technique of ribosome display (Hoffmüller 1998). Denileukin diftitox is a recombinant fusion protein approved by FDA in 1999. The receptor binding domain has been replaced by the IL-2 polypeptide hormone in a modified diphtheria toxin, commercialized under the trade name Ontak by Seragen (Ligand).

Small peptides have also been obtained. Thus, the construction of a cosmid library in *Escherichia coli*, and cloning and heterogously expressing individual biosynthetic pathways, led to the discovery

Chart 14.1 Natural and unnatural products from genetic engineering (P: S=20; S/H=0.53. I: Smax=34, av=24; S/H max=0.33, av=0.31. FA/PO: S max=61, av=41; S/H max=0.59, av=0.53) (unnatural, or previously unidentified, compounds are shown in small caps; marine origin is indicated in boldface)

Alkal.: indole: strictosidine (*Catharanthus roseus* Ang. → *Escherichia coli* Bact.: Scott 1992); corrins: hydrogenobyrinic acid (*Pseudomonas denitrificans* Bact. → *E. coli*: Scott 1994).

Pept.: PANTOCIN B (*Erwinia herbicola* 318 Bact. → *E. coli*: Sutton 2001), difficult to isolate from natural sources.

Isopr.: Sesquiterp.: pentalenene and, by mutation, Δ⁶-protoilludene and germacrene A (*Streptomyces* sp. → *E. coli*: Seemann 1999); trichodiene, (Z)-α- and β-bisabolene, α-cuparenene, β-farnesene, and, by mutation, ISOCHAMIGRENE (*Fusarium sporotrichoioides*, Eumyc. → *E. coli*: Cane 1996). Triterp.: β-amirin and lupeol together (chimeric proteins *Panax ginseng* ↔ *Arabidopsis thaliana*: Kushiro 1999). Tetraterp.: astaxanthin and ASTAXANTHIN β-D-DIGLUCOSIDE (engineering of *Escherichia coli* with **marine** *Agrobacterium aurantiacum*: Yokoyama 1998). Rearr. isopr.: CITREOANTHRASTEROIDS and CITREOHYBRIDONES (protoplast fusion of different *Penicillium citreo-viride* strains: Nakada 2000).

Fatty acids and polyket.: eicosapentaenoic acid: (**marine** *Shewanella putrefaciens* → plants) and docosahexaenoic acid: (**marine** *Vibrio marinus* → plants): Facciotti 1998; α-tocopherol: (γ-TMT from *Synechocystis* sp. Cyanobact. → *Arabidopsis thaliana*: Shintani 1998); 6-deoxyerythronolide analogues: KOSO15-22A and KOSO15-22B (whole module transformation by plasmid pKOSO15-22 → *Streptomyces coelicolor*: Liu 1997, and similarly for ERYTHROMYCIN D: Staunton 1997; Wu 2000); high-performance production of 6-deoxyerythronolide has been achieved by fermentation of a metabolically engineered strain of *Escherichia coli*: Pfeifer 2001; methymycins, calicheamicins, and pikromycins: METHYMYCIN-CALICHEAMICIN-CLASS and PIKROMYCIN-CALICHEAMICIN-CLASS: *Micromonospora echnospora CalH* → *Streptomyces venezuelae* mutant: Zhao 1999).

Shikim.: shikimic and quinic acids (genetically engineered *Escherichia coli*: Draths 1999).

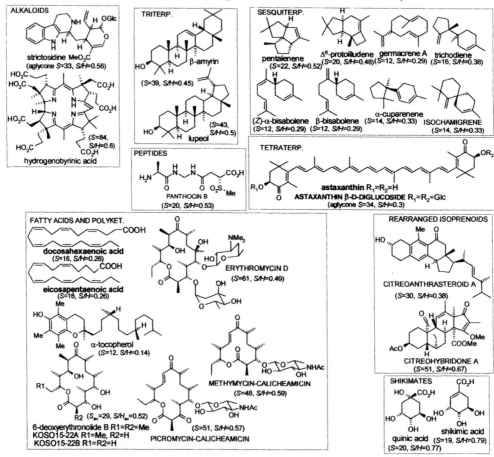

of pantocin B, an antibacterial peptide of the plant-protecting bacterium *Erwinia herbicola* (Sutton 2001). This small peptide had defied all previous attempts at isolation from bacterial cultures. Moreover, the technique of hairy roots induced by the Ri plasmid of *Agrobacterium rhizogenes* allows the production of simple opines, which derive from the condensation of amino acids with carbohydrates (Saito 1992).

14.1.c Isoprenoids

Pentalenene synthase from *Streptomyces* sp. was cloned and overexpressed in *Escherichia coli* getting pentalenene, a known sesquiterpene; site-directed mutants yielded other known sesquiterpenes, Δ^6-protoilludene and germacrene A (Seemann 1999).

Chimeric enzymes constructed from two different angiosperms (*Panax ginseng*, Araliaceae, and *Arabidopsis thaliana*, Brassicaceae) yielded mixtures of triterpenoids, β-amirin and lupeol, at a composition depending on the particular chimera; methyl scrambling was observed for lupeol only (Kushiro 1999). A few differences in the amino acids at the active site are responsible for these effects. This implies that the high variety of plant triterpenoids owes more to chimeric enzymes than product-specific triterpene synthases. It was proposed that these chimeric enzymes act as multifunctional triterpene synthases (Kushiro 1999).

Astaxanthin, the lobster red pigment used for the pigmentation of fish and shellfish in aquaculture, is produced in Japan by fermentation of a genetically engineered terrestrial bacterium. This gave also unnatural astaxanthin glucosides (Yokoyama 1998). In these experiments, the genes for astaxanthin production were derived from *Agrobacterium aurantiacum*, a bacterium isolated from the sea at Okinawa (Yokoyama 1998). In laboratory experiments, astaxanthin has also been obtained by engineering the tobacco's carotenoid biosynthetic pathway (Mann 2000).

cis/trans β-Carotene - which is a more valuable food additive that the all-*trans* analogue from plants or industrial synthesis - can be cheaply obtained by open-air fermentation of marine chlorophycean algae of the *Dunaliella* complex. Plants have been mainly set up in Israel, Hawaii, and eastern Australia (Bongiorni 1995). With respect to astaxanthin, *cis/trans* β-carotene is a less powerful, though cheaper, antioxidant.

Genetically engineered rice containing β-carotene is also long available from the US company Monsanto. Recently, novel carotenoids have been obtained by breeding bacteria through a DNA-shuffling procedure (Schmidt-Dannert 2000), or combining into plasmids genes from various bacteria (Albrecht 2000).

Protoplast fusion has also been successfully used to produce novel mold isoprenoids, such as citreoanthrasteroids and citreohybridones (Nakada 2000). Long chain aldehydes, valuable as a bio-flavor, have also been secured along similar procedures from thalli cultures of the green seaweed *Ulva pertusa*, regenerated from protoplasts (Fujimura 1990).

14.1d Fatty acids and polyketides

The biosynthetic capacity of seeds of the Californian plant *Umbellularia californica* for capric acid (10:0) and lauric acid (12:0) has been transferred to another angiosperm, *Arabidopsis thaliana*, where it performed efficiently, giving these medium-size fatty acids at the expense of long-chain fatty acid production (Voelker 1992). This procedure has been extended to valuable food additives, such as eicosapentaenoic acid (EPA) and docosapentaenoic acid (DHA), by gene transfer to plants from marine bacteria, *Shewanella putrefaciens* for the first, and *Vibrio marinus* for the latter (Facciotti 1998). This procedure is an alternative to getting DHA by extraction from fish liver oil or the culture of marine microorganisms: the dinoflagellate *Crypthecodinium conhii*, thraustochytrids, or the bacterium *Alteromonas* f. *putrefaciens* (Bongiorni 1995).

Owing to good antioxidant properties, α-tocopherol is also a much sought target for enhanced production. This was achieved by enhancing the level of the α-tocopherol biosynthetic precursor, γ-tocopherol, and over-expressing γ-tocopherol methyltransferase in seeds of the plant *Arabidopis thaliana* (Shintani 1998).

Following pioneering genetics-based work in UK on the biosynthesis of polyketides (Hopwood 1997), much progress has been made in understanding polyketide synthases (PKS) (Staunton 2001). These enzymes have been classified into types I and II. The latter is present in plants and bacteria for the production of aromatic polyketides, where the catalytic sites are borne on a separate enzyme subunit. Type I in fungi is characterized by catalytic sites borne as domains along the length of the multifunctional enzymes.

Modular type-I PKS, present in actinomycetes for the synthesis of macrolides, has attracted great interest for the possibility of shuffling around whole modules in a combinatorial manner, getting chimeric systems that may yield unnatural macrolides of new skeleton. The idea was correct in the functional domain substitution of a modular PKS of *Streptomyces coelicolor*. The new natural product diversity that has resulted from these studies is quite limited, however. At the time of writing, it is represented by only a few analogues of 6-deoxyerythronolide B, KOSO15-22a and KOSO15-22b (Liu 1997). This may be an intrinsic limitation, because modular type-I systems are highly evolved (Hopwood 1997), which could not be circumvented by mixing genes from different species that bear non-modular systems (Kulowski 1999). This contrasts with nature's versatility, in shuffling strategies that lead to a variety of mixed-biogenesis metabolites. Any limitation in scaling up the fermentation of engineered actinomycetes, as required for industrial processes, has been recently overcome for the erythromycins by a metabolically engineered strain of *Escherichia coli* (Pfeifer 2001).

Another area where heterologous gene combination was successful concerns antitumor agents of macrolide-enediyne structure. Thus, the *calH* gene of *Micromonospora echinospora*, which directs the formation of the amino group in 4-amino-4,6-dideoxyglucose of calicheamicin, needed for interaction with DNA, was amplified by PCR and cloned. The resulting plasmid was introduced into a mutant strain of *Streptomyces venezuelae* and the genetically engineered actinomycete was cultured. New macrolides of methymycin-calicheamicin class were obtained that combine the capacity of the genus *Streptomyces* for erythromycins with that of the genus *Micromonospora* for amination of the carbohydrate at C-4. It is hoped to adapt these processes to combinatorial biosynthesis (Zhao 1999).

These studies have also shed light on a protein-based self protection of *Micromonospora echinospora* from its own cytotoxic metabolites (Whitwam 2000). A gene cluster for an enediyne antibiotic, C-1027, was also cloned (Liu, W. 2000).

Experiments of site-specific mutagenesis were also carried out, leading, after enzymatic glycosylation, to unnatural erythromycin D and analogues (Khosla 1997).

These studies have much contributed to deepen our understanding of the biosynthetic mechanisms for polyketide production, particularly with the fundamental discovery of a novel RppA chalcone-synthase-related enzyme in bacteria (Funa 1999).

14.1e Shikimates

Shikimic acid and quinic acid are useful synthons for the preparation of drugs (Chapter 15). Both are plant products: shikimic acid from *Illicium* spp. and quinic acid from *Cinchona* spp. Yields are quite low, however. Moreover, quinic acid is stored in the bark, the removal of which kills the plant.

These limitations have been overcome by engineering a bacterium, *Escherichia coli*, with the plant genomic portion of shikimate-synthesizing SP1.1/pKD12.112 (Draths 1999). This technology has also been experimentally employed for the overproduction of a phytoalexin, salicyclic acid (Chart 12.1), in the tobacco plant. The technique should be applicable also to crops of commercial interest (Verberne 2000).

14.1f Inhibition of biosynthetic pathways

So far we have examined the biotechnological approach to natural products by either the activation of new biosynthetic pathways or the improvement of pathways for rare metabolites in nature. Another biotechnological strategy - a long-term trick in classical biosynthetic studies - consists in suppressing or limiting the routes toward certain metabolites. This is the case of caffeine and theophylline, as an alternative to drug-free beverages with better perspectives than those currently obtained from solvent or supercritical carbon dioxide extraction; preliminary results have appeared of cloning the gene that encodes caffeine synthase, involved in the final steps of caffeine production from purine nucleotides (Kato 2000).

Animal cells are highly differentiated and much less amenable to culture than plant cells, in particular when the focus is on the secondary metabolism. Even setting up transgenic animals is problematic. An example is the rhesus monkey that does not glow in spite of carrying the gene for the green fluorescent protein (Chan 2001). Whole-organism cultures are mostly limited to colonial invertebrates, such as bryozoans for bryostatin 1 (Chapter 13.3). The multi-chromosome system of invertebrates precludes also using most techniques of molecular engineering. The approach to receptors is easier, as far as it can be judged from the gene structure for tyrosine kinase receptors of the sponge *Geodia cydonium* (Müller 1997).

The biotechnological approach to nature is not limited to the secondary metabolism. Often what is sought is modifying the organism in its behavior, homeostatic properties, and interaction with other organisms. An example is a recombinant bacterium of agricultural interest, *Escherichia coli*, which has been engineered with a chitinolytic gene from a marine bacterium, *Alteromonas* sp. This transgenic *E. coli* can grow side to side to phytopathogenic fungi, destroying their chitin cell wall

(Hirayae 1996). Other examples concern crops genetically engineered to get resistance to herbicides and pests. The tobacco plant (Beetham 1999), and edible plants for genetically modified food (Zhu 1999), are common such targets. Worth mention are also current endeavors at engineering the stomatal response to abscisic acid (Table 12.5), aimed at controlling the carbon dioxide intake and water loss from plants (Schroeder, 2001). The biotechnology of forest products (Bruce 1998) is also focused more on sustaining timber production than harnessing secondary metabolites.

Unfortunately, these transgenic practices are often done to the disregard of the profound changes that secondary metabolic pathways may undergo, interfering with coevolving species, and thus possibly threatening biodiversity. A notable exception is a recent technique developed to insert a DNA/RNA hybrid that contains a copy of the gene to be modified and the desired mutation (chimeraplast) into the organism to modify, especially a plant. The plant recognizes this insert as a mistake and starts repairing the DNA on the basis of the chimeraplast; if everything works, a mutation is induced (Zhu 1999). This genetic modification, being closer to nature's strategies than the blind techniques discussed above, should help removing obstacles to transgenic plants (Beetham 1999).

Gene therapy is also in the biotechnological perspectives. Though much promising, it faces difficulties to inserting genes into human cells and getting them expressed. Nonetheless, and in spite of hostility by FDA caused by recent tragic events, there are at this moment a thousand of patients undergoing gene therapy.

Skeptic views have been expressed about the possibility of correlating everything to genes. It has been warned that "the correlation between gene and protein expression is low" and that "many aspects of protein biology are not encoded at genetic level". The analysis of what happens at protein level following drug delivery is believed to be a complementary approach (Ashton 1999), lending importance to proteomics. These remarks may be extrapolated to the biotechnological approach to the secondary metabolism. A poor correlation between genes and proteins means a poor correlation between genes and secondary metabolites. Worse, identifying genes that encode proteins may be an affordable enterprise, but unraveling the multienzyme system for secondary metabolites is a much harder goal to achieve, at least for eukaryotic organisms, for years to come.

14.2 Biocatalysis

Enzyme catalysis of reactions (biocatalysis) is a branch of biotechnology (Hauer 1999; Crameri 1999). The superiority of biocatalytic methods of synthesis, particularly if carried out in a continuum (Orsat 1999), is often manifestly clear, only limited by the cost of replacing the old chemical plants (Pachlatko 1999; Schmid 2001). Illustrative examples of biocatalytic plants are illustrated in Chart 14.2.

Vitamins made by biocatalysis include cyanocobalamin (Rhône-Poulenc and Merck & Co), vitamin A (Hoffmann-La Roche: Orsat 1999), ascorbic acid from sorbitol via, directly, ketogulonic acid (BASF), riboflavin (Hoffmann-La-Roche, BASF, and Rhône-Poulenc/Archer-Daniels), and niacinamide from synthetic materials (Lonza). The list includes a vitamin-like nutrient, L-carnitine from butyrobetaine (Lonza).

In the drug area, β-D-glucuronides, which are important drug vectors, are made by biocatalysis at Novartis (Pfaar 1999). Biotransformation of the synthetic drug CGP 62706, an inhibitor of the EGF-receptor tyrosine kinase, is carried out at Novartis (Kittelman 1999). In the hands of Merck, natural immunosuppressant agents, like FK-506 and ascomycin, gave scarcely active derivatives which proved useful in shedding light on the mechanism of immunosuppression (Chen 1999).

Precursors of technological compounds are also obtained by biocatalysis: acrylamide (for

Chart 14.2 Examples of compounds derived from, or used for, biocatalysis.
Vitamins: cyanocobalamin = vitamin B$_{12}$ (Rhône-Poulenc and Merck & Co); vitamin A (Hoffmann-La Roche); ascorbic acid (BASF), riboflavin (Hoffmann-La-Roche, BASF, and Rhône-Poulenc/Archer-Daniels), and niacinamide (Lonza).
Vitamin-like nutrients: L-carnitine (Lonza).
Drugs and drug-adjuvants: β-D-glucuronides (Novartis: Pfaar 1999). Biotransformation of FK-506 (Merck, Rahway, NJ: Chen 1999) and synthetic CGP 62706 (Novartis).
Precursors for technological compounds: 1,3-propanediol (DuPont); cyclic amino acids (Lonza); acrylamide (Nitto); (S)-1-methoxypropan-2-amine (CELGRO).

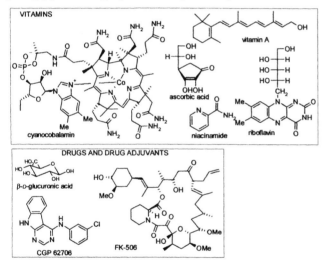

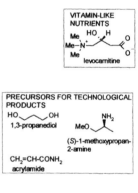

polyacrylamides and as a copolymer, Nitto), 1,3-propanediol (for new polyesters, DuPont), cyclic amino acids (for the pharmaceutical and agricultural industry, Lonza) (Petersen 1999), and (S)-1-methoxypropan-2-amine (for herbicides, CELGRO) (Matcham 1999).

Chapter 15. The role of chemical synthesis

Metabolic transformations are characterized by high speed and yield, as well as high regio-, diastereo- and enantio-specificity. Errors in the stereochemistry of the molecules that serve to construct the genetic material are smaller than for the planetary motions. With secondary metabolites, however, enantiomerically impure compounds are also encountered, typically with monoterpenes and alkaloids from terrestrial plants; even antipodal pathways in the same organism have been found, albeit as rare events (Guella 1998).

These processes are normally enzyme-catalyzed. Purely chemical processes are seldom encountered with carbon compounds in nature. The few exceptions include the very act of phenol coupling (by which racemic compounds are obtained), cyclization reactions of polyprenyl compounds (which benefit from the preferred conformation of the reaction partners, suitable for the cyclization, Wendt 2000), and Diels-Alder cycloadditions. The latter have been advocated for the biosynthesis of celastroidine A (= volubilide) from a lupane triterpene and an abietane diterpene in two different plants, *Hippocratea celastroides* Kunth from Mexico (Jiménez-Estrada 2000) and *Hippocratea volubilis* Linnaeus (Alvarenga 2000).

With man, the synthesis of natural products took a planned avenue. No major revolution in strategy was observed with respect to the early twentieth century, but it was greatly elaborated on the concepts of retrosynthetic analysis (Corey 1961; Ireland 1969; Corey 1989), biomimetic synthesis (Mulzer 1998), and cascade reactions (Neuschütz 1998), explicitly embodied for the first time in Robinson's tropinone synthesis (Robinson 1917). Many new reactions were also invented that allowed implementing the new strategies (Barton 1994). Natural products of any complexity have been synthesized, only differing from the metabolites in nature for the isotopic composition, which depends on the fractionation along the particular biosynthetic route. Total synthesis has not become easier, however, since ever more complex targets have to be attacked, at increasing efficiency, to stay at the forefront of the research.

Since the last world war, US, Japan, and UK, in the decreasing order of the resources invested, have performed outstandingly in the total synthesis of natural products. Examples of completed, or formal, total synthesis are shown in Table 15.I, where, unlike a recent account (Nicolaou 2000A), no historical perspective, nor details on either the retrosynthetic analysis or the execution of the synthesis, are given. The basis of selection of the examples was also different. Following the intents of this book, particular attention was paid to the source organism (last column of Table 15.I), its taxonomy, rarity, difficult accessibility, growth in protected areas as an endangered species, or low productivity, which determine the rarity of the metabolite, marked by an asterisk.

The target (1st column in Table 15.I) is also described by the structural complexity metric S and size metric H (4th and 5th columns), as defined in Chapters 5 and 11 (Whitlock 1998). A more complex molecule does not necessarily require a more complex synthesis, although it is often so, particularly when the specific complexity S/H is high and the starting materials are simple molecules.

Easily available advanced synthons, such as the carbohydrates, amino acids, hydroxyacids, and terpenoids, make the synthetic task easier than the complexity metrics of the target suggests; this is especially true for the glycosides, if the carbohydrate portion can be introduced intactly. It must also be borne in mind that the S metric is counted in a linearly additive fashion, neglecting interactions between the functional groups (Whitlock 1998); such interactions are not treated adequately by any method so far proposed to calculate the molecular complexity. Moreover, no attention was paid here to the graphic analysis of the synthesis plan based on the molecular complexity of the intermediates; these aspects have recently been reviewed (Bertz 1993; Whitlock 1998; Chanon 1998).

An arbitrary classification of the strategy of the synthesis and its scope appear in the second column of Table 15.I, bearing in mind that success in the total synthesis of natural products is always the result of a clever design, involving basic principles of chemistry. The strategy aimed at the most rapid solution of the problem (MRS) is focused on the target, seeking it by whichever synthetic method, in a race against time; less importance is given to the number of steps and overall yield than the timing of the project. The synthesis plan is less frequently centered on the application of new reactions (ANR) or new strategies (ANS), which relaxes the execution, since such plans can be rarely the same in different laboratories. Industrial application of total synthesis is usually prevented by the large number of reaction steps, some of a poor yield. These problems are addressed in a few projects of industrial total synthesis (ITS), where the number of steps cannot exceed fifteen, each of a good yield.

The availability of cheap advanced synthons that carry the required chirality is an advantage, particularly in projects aimed at industrial total synthesis. Natural products are often used as synthons, ideally from a renewable source, such as microbial fermentations. In a few cases, biotechnology has become an alternative source. The total syntheses of the antitumor agent esperamicin A and the immunosuppressant FK-506 are examples. In both cases, the synthon was quinic acid (Barco 1997), cheaply obtained by biotechnology (Chapter 14.1.e) rather than from the environmentally noxious extraction from the bark of *Cinchona* spp. Used to build up combinatorial libraries, quinic acid has gained further importance in organic synthesis (Phoon 1999).

Fatty acids have also great potential as synthons for the preparation of secondary metabolites. Renewable sources include rapeseed (*Brassica rapa*), sunflower (*Helianthus annuus*), soy bean (*Glycine max*), peanut (*Arachis hypogaea*), and linseed (*Linum usitatissimum*). These plants can be genetically engineered to furnish fatty acids in high-purity (Bierman 2000).

Biotechnological techniques are also frequently used to make key reaction steps highly regio- and stereoselective.

The synthesis of nucleic acids and polypeptides by the conventional flask method is time consuming. Automation, such as in Merrifield's peptide synthesis, was the trick that brought these processes to the industrial standard. Carbohydrate units have many more positions of attachment than amino acids. This is why oligosaccharides, in spite of high medical interest, have long resisted automated synthesis. It was the use of protecting groups that finally allowed us streamlining their synthesis (Plante 2001).

New concepts in the strategy of the synthesis of drugs rarely appear, such as from the observation that microorganisms often get resistance from enzymes that inactivate the drug through phosphorylation. To avoid the problem, the aminoglycoside antibiotic kanamycin A was modified in a way that it was re-obtained whenever it was modified by the microorganism resistance enzymes (Haddad 1999).

Total synthesis - particularly along a biomimetic approach - may also serve the fundamental role of clarifying natural processes at molecular level (Hinterding 1998), although cautious notes have been raised to this concern (Lauffenburger 1998).

Two other areas of chemical synthesis are of our concern. One is the chemical transformation of natural products in view of products with improved performance (Table 15.II). The semisynthesis of drugs or technologically useful compounds from easily available natural products is industrially attractive. If the source is renewable, the whole process may become environmentally safe. This is the case of the synthesis of paclitaxel from 10-deacetylbaccatin II, which is found in the needles of *Taxus baccata* (Colin 1988), and homoharringtonine from cephalotaxine of *Cephalotaxus* spp. (Robin 2000). Removal of bark for the extraction of paclitaxel, or roots for homoharringtonine, kills endangered gymnosperms in the genera *Taxus* and *Cephalotaxus*. The only problem that remains unsolved is the disposal of organic solvents used in the synthesis; this is why organic reactions in largely aqueous mixtures become increasingly attractive (Engberts 2001). Cell cultures of *Taxus* spp. offer an alternative route to paclitaxel, triggered by methyl jasmonate (Yukimune 1996). This, however, does not remove the problem of disposal of huge amounts of wastes.

Semisynthesis has also found much scope for antifungal agents. These rare drugs are much needed because weakening of the immune system by HIV has paved the way to pathogenic fungi. The status of antifungal agents can be briefly summarized. Amphotericin B, produced by the actinomycete *Streptomyces nodosus*, is the classical fungicidal agent; on binding to ergosterol (the typical sterol of fungi), it alters the permeability of the cell, resulting in cell death. Amphotericin B is highly nephrotoxic, however, a problem relieved only in part by formulation into liposomes. Synthetic azoles are safer, although their mode of action (impairing the biosynthesis of ergosterol) attains only a fungistatic level. New natural agents, echinocandins and pneumocandins, have emerged in the candin class of peptide inhibitors of fungal cell wall biosynthesis. Similar examples are caspofungin acetate, built on pneumocandin B(o), and LY303366 (V-echinocandin), built on echinocandin B; both were recently approved by FDA. (Table 15.II). The sordarins are another group of newly emerging natural antifungal agents, in particular a semisynthetic derivative, GM237354 (Table 15.II). Its mechanism of action is centered on the specific inhibition of fungal protein synthesis by stabilizing an addition complex between EF2 protein and ribosome. GM237354 has a broader spectrum of antifungal activity than sordarin, although both are inactive against *Aspergillus fumigatus* (Fostel 2000).

Patterned synthesis is also an active line (Table 15.III), best illustrated by ephedrine, which served as a model for many synthetic CNS-active drugs (Table 15.III). Patterned synthesis is aimed at retaining the properties of the model, such as the mode of interaction with receptors, with the minimum of side effects, in particular toxicity or resistance developed by pathogens. In the antiviral agents, an old example is Ara-A, adenine arabinoside, from synthesis patterned on nucleosides isolated from various demosponges in the early days of marine natural products (Table 15.III). The area has been greatly expanded with anti-HIV drugs, such as azidothymidine (AZT) and the recently FDA-approved dideoxycytidine, dideoxyinosine, and 2',3'-didehydro-3'-deoxythymidine (Table 15.III). The sugar portion of the nucleoside has also been imitated by cyclopentene (abacavir succinate) or thiofuran rings (lamivudine) (Table 15.III). Even non-nucleoside inhibitors have been synthesized. An example is indinavir, obtained from synthesis patterned on an actinomycete peptide, pepstatin (Table 15.III). Synergistic mixtures of these nucleoside analogues are also commercialized, like Combivir (lamivudine and azidothymidine) and Trizivir (abacavir succinate, azidomethydine, and lamivudine).

Peptidomimetics is an active branch of patterned synthesis. A notable family of tumor-avid peptides was derived from synthesis patterned on neurotensin, a peptide constituted of 13-amino acids. These synthetic peptides bind to the NT receptors expressed by many forms of cancer, which may serve to map the tumor, delivering radionuclides *in situ* (Waibel 2000). Molecules that, like the peptides, are constituted of repetitive blocks, are much simpler to synthesize than those composed of non-repetitive units. Given the powerful bioactivity, and improved stability on systemic administration, synthetic peptides may become rewarding if, as it has been estimated, one in two men, and one in three women, will get cancer in their lifetime.

Biomacromolecules and biopolymers have also been put to use as templates for the synthesis of new materials, or models for the study of molecular mechanisms in nature. Examples are nucleic acid synthetic alternatives (Eschenmoser 1999), β- and γ-peptide synthetic analogues of natural α-peptides (Seebach 2001), and analogues of poly-(R)-3-hydroxybutanoic acid. The latter can be obtained from chemical synthesis (Seebach 2001) or the culture of engineered *Pseudomonas* strains (Kessler 2001).

At the borderline between semisynthesis and patterned synthesis, versatile total syntheses may provide access to diversified bioactive structures. Classically, analogues of the active drug are sought that avoid the resistance developed by pathogens and show reduced side effects. In lucky cases, chemical simplification has afforded the wonder drug. This is the case of sodium valproate (2-propylvaleric acid sodium salt), which was synthesized in the fall of the XIX century and nearly a hundred years later became a leading drug in the treatment of epilepsy.

Chemical synthesis may be limited to a portion of the natural product imagined to determine the binding properties. An example is the right half diol of halichondrin, which is as strongly cytotoxic as natural halichondrin, while more amenable to total synthesis (T 15.II, Stamos 1997). On a similar vein, phthalascidin was obtained as a simplified analogue of ecteinascidin 743, without much loss in cytotoxicity (Martinez 2000). These, however, are exceptional cases. Usually, no part of the molecule is merely a burden, all parts serving important functions, such as aiding to cross the membranes by sheltering polar groups.

In any event, bringing a product from nature or synthesis to the pharmacy is a complex matter that poses heavy problems to companies, health authorities, and ultimately patients. An example is TROVAN (Table 15.III), a wide-spectrum antibacterial agent from synthesis patterned on natural antibacterial quinolones. According to press news, Pfizer conducted trials in Nigeria with this drug before getting approval in the US. This promoted an inquiry. TROVAN, associated with serious liver damage, has recently been approved as a last resort for patients with cerebrospinal meningitis.

Diversification of molecules in view of medical applications is also carried out by combinatorial synthesis, either in solution or solid phase. With the advent of high throughput procedures for screening, this has attracted the pharmaceutical industry, particularly for the simplification of synthetic procedures with polymer supported reagents. Starting from simple molecules, success has been lacking for years, however. The elaboration of complex natural products is more promising, such as with curacin A (from marine cyanobacteria: Wipf 2000B, Table 15.III), the sarcodictyins (from anthozoans; Nicolaou 2000A, Table 15.III), β-lapachone (from terrestrial plants; Nicolaou 2000C, Table 13.3), and psammaplin A, a tyrosine metabolite from several marine demosponges (Nicolaou 2001). A recent variant in this area is dynamic combinatorial chemistry, which makes use of mixtures of constituents in dynamic equilibrium, relying on the capacity of self-assembling processes (Lehn 2001).

Heavy spreading of AIDS in Africa, and menaces to rich countries, has stimulated pharmaceutical companies to go beyond the natural product. Transition state analogues, such as Saquinavir, and molecules from rational drug design, such as Amprenavir (Table 15.III), are already on the market, either individually or as synergistic mixtures. In making these drugs, the key site of the protein to inhibit was taken as a model. This is premonitory of a future when the limits of the secondary metabolite will be overcome, based on a detailed knowledge of the molecular structure of the target. In the drug discovery area, real-time PCR instruments are already on the market for the rapid identification of genome sequences.

Table 15.I. Total synthesis of environmentally, medically, and technologically relevant natural products (* = rarity of the natural product; ANR = application of new reactions; ANS = application of new strategies; ITS = industrial synthesis; MRS = most rapid solution; S and S/H = molecular and specific molecular complexity, respectively)

Target/class	Strategy/ synthetic precursors	Target structure	S	S/H	Origin/function
*adocia-sulfate 1/ hexaprenoid	MRS/ geraniol, 2-hydroxy-5-methoxy-benzaldehyde: Bogenstätter 1999	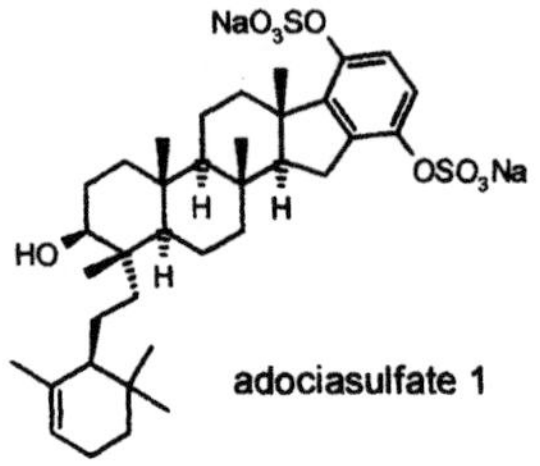	53	0.5	**marine:** *Haliclona* sp., Haplosclerida Porif./inhibitor of kinesin motor proteins
alterobactin A/ pept.	MRS/β-keto esters, methyl cinnamate, and *N*-Boc-glycine trichloro-ethyl : Deng 1995	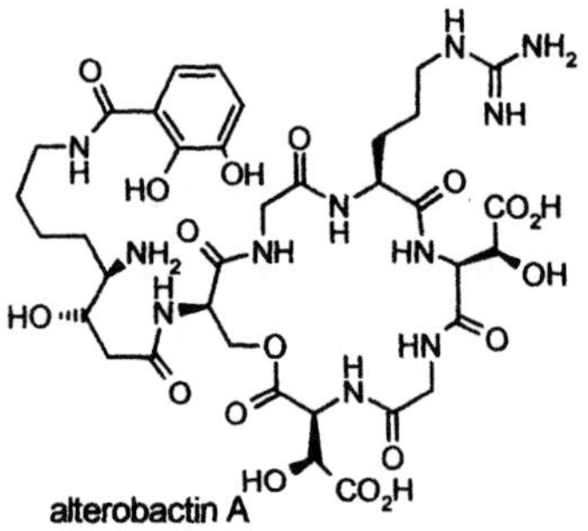	65	0.5	**marine:** *Alteromonas luteoviolacea*, Bact./siderophore
ambrein/ triterp.	MRS/bicyclic diol from enzyme reduction: Tanimoto 1997	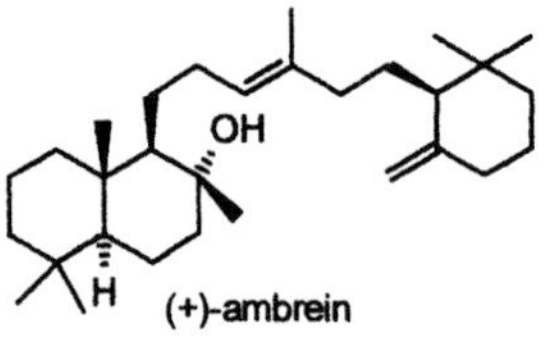	27	0.3	**marine:** sperm whale, *Physeter macrocephalus* Linnaeus, 1758 [= *P. catodon* L.], Odontoceta, /fixing agent in perfumery
Ambrox/ degr. of triterp. ambrein	ANR/homo-farnesol via nonenzymatic cyclization: Ischihara 1999	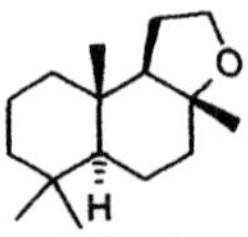	17	0.4	**marine:** (+)-ambrein: sperm whale, *Physeter macrocephalus* Linnaeus, 1758,/fragrance

Target structures:

adociasulfate 1

alterobactin A

(+)-ambrein

Ambrox®

Table 15.I. Total synthesis of environmentally, medically, and technologically relevant natural products (* = rarity of the natural product; ANR = application of new reactions; ANS = application of new strategies; ITS = industrial synthesis; MRS = most rapid solution; S and S/H = molecular and specific molecular complexity, respectively)

Target/class	Strategy/ synthetic precursors	Target structure	S	S/H	Origin/function
amphotericin/ polyene macrolide	MRS/(+)- xylose: Nicolaou 2000A	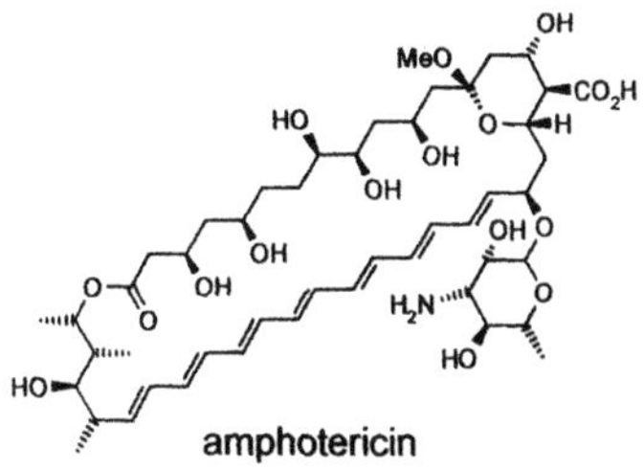 amphotericin	82	0.6	*Streptomyces nodosus* Trejo 1961, Actinom., Bact./fungicidal, albeit toxic
*agelastatin A/alkal.	ITS (14 steps, 7% yield)/ cyclopentadiene: Stien 1999	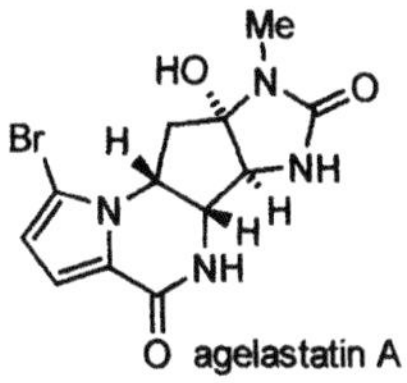 agelastatin A	32	0.9	**marine:** *Agelas dendromorpha* Lévi, Agelasida, Porif./cytotoxic on tumor cells
aphidicolin/ isopr.	MRS: Nicolaou 2000A	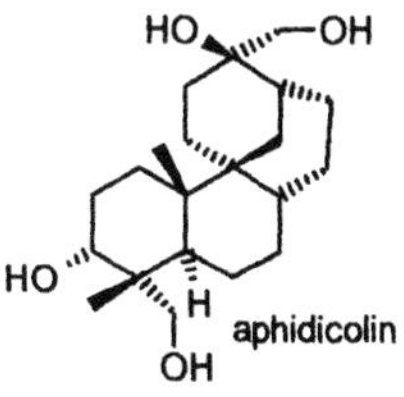 aphidicolin	36	0.6	*Cephalosporium aphidicola* Petch., Eumyc./tool for cancer study: MI
aspidophytine/ indole alkal.	ANR and MRS/3- methoxy- cyclopent-2- en-1-one: Nicolaou 2000A	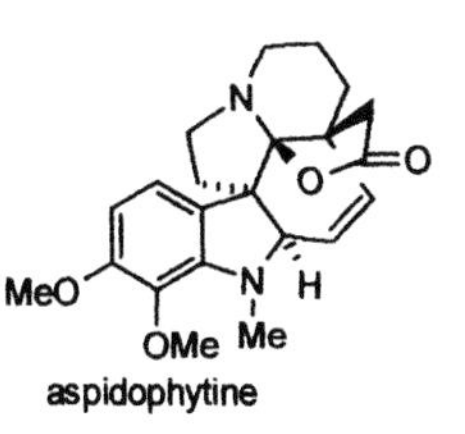 aspidophytine	38	0.6	*Haplophyton cimicidum* A. DC., Ang., a Mexican shrub, "La hierbe de la cucaracha", used since the Aztec as an anticokroach/insecticidal

Table 15.I. Total synthesis of environmentally, medically, and technologically relevant natural products (* = rarity of the natural product; ANR = application of new reactions; ANS = application of new strategies; ITS = industrial synthesis; MRS = most rapid solution; *S* and *S/H* = molecular and specific molecular complexity, respectively)

Target/class	Strategy/ synthetic precursors	Target structure	S	S/H	Origin/function
avermectin B₁ₐ/ glycosidic macrolide	MRS: White 1995	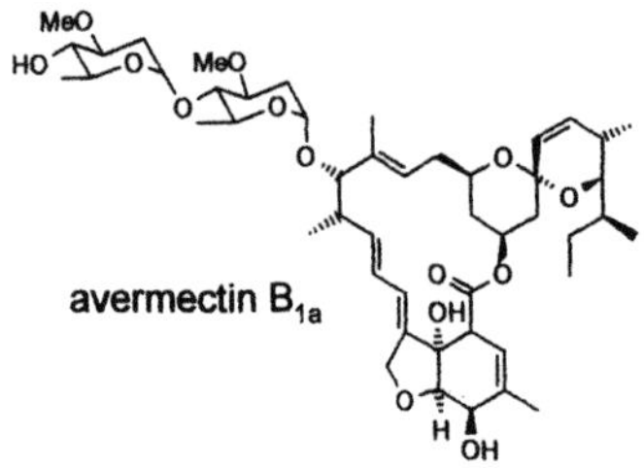 avermectin B₁ₐ	94	0.7	*Streptomyces avermitilis*, Actinom., Bact./insecticidal
batracho- toxinin A/steroid	MRS: Nicolaou 2000A; Daly 2000	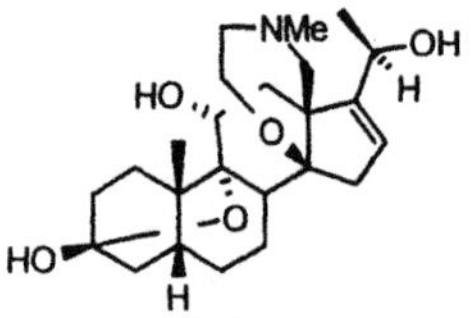 batrachotoxinin A	44	0.6	S American *Phyllobates* spp., Dendrobatidae frogs/lab tool for the study of Na⁺ channels
bisorbicil- linol/bisor- bicillinoids	MRS: Nicolaou 2000B	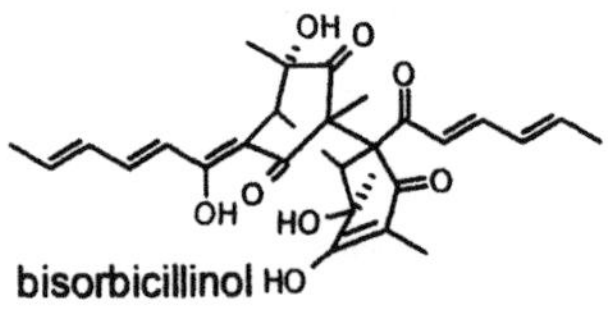 bisorbicillinol HO	52	0.7	*Trichoderma* sp. Eumyc. /antioxidant
brevetoxin A and B/polyether polyket.	ANR and MRS: Nicolaou 2000A	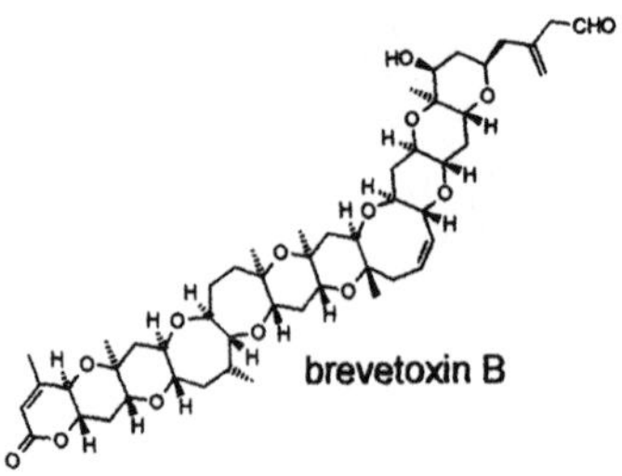 brevetoxin B	112	0.7	**marine:** *Gymnodinium breve* Davis [= *Ptychodiscus brevis* Davis], Dinofl./ichthyotoxins and lab tools for the study of ion channels

Table 15.I. Total synthesis of environmentally, medically, and technologically relevant natural products (* = rarity of the natural product; ANR = application of new reactions; ANS = application of new strategies; ITS = industrial synthesis; MRS = most rapid solution; S and S/H = molecular and specific molecular complexity, respectively)

Target/class	Strategy/ synthetic precursors	Target structure	S	S/H	Origin/function
*bryostatin 7/polyket. macrolide	MRS: Kageyama 1990	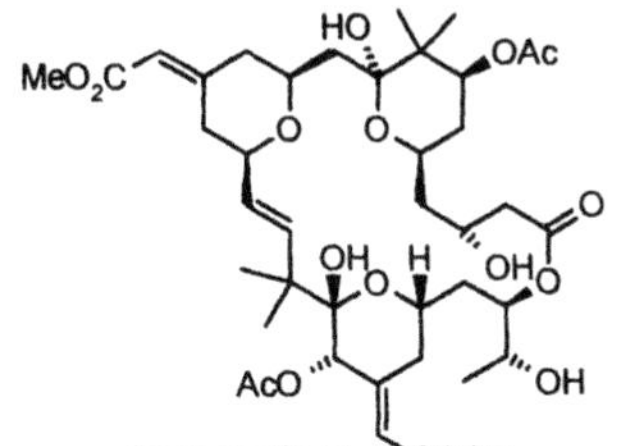 bryostatin 7	71	0.6	**marine:** *Bugula neritina* (L.), Bryoz./antitumor
*bryostatin 2 and 1/macrolide polyket.	MRS/simple materials and aldol technology: Evans 1999A	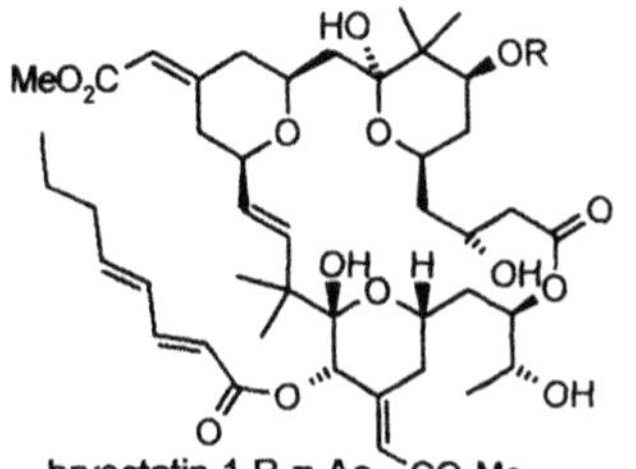 bryostatin 1 R = Ac bryostatin 2 R = H	1: 75; 2: 72	0.5	**marine:** *Bugula neritina* (L.), Bryoz./antitumor
calichea- micin γ/ acetogenin glycoside	MRS/simple materials, followed by glycosidation: Nicolaou 2000A	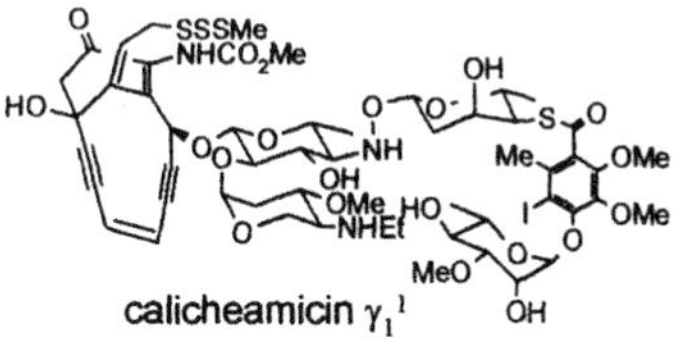 calicheamicin γ_1^1	107	0.6	*Micromonospora echinospora*, Bact./antitumor

Table 15.I. Total synthesis of environmentally, medically, and technologically relevant natural products (* = rarity of the natural product; ANR = application of new reactions; ANS = application of new strategies; ITS = industrial synthesis; MRS = most rapid solution; *S* and *S/H* = molecular and specific molecular complexity, respectively)

Target/class	Strategy/ synthetic precursors	Target structure	*S*	*S/H*	Origin/function
*calyculins A and C/isoxazole alkal.	MRS/simple materials along convergent routes: Ogawa 1998, Smith 1998	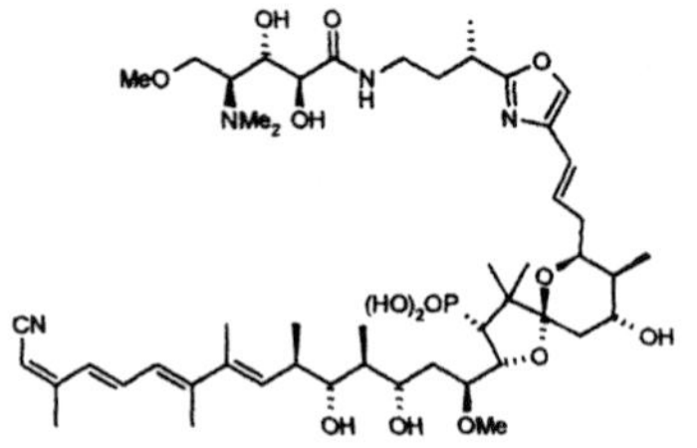 (-)-calyculin A	75	0.5	**marine:** *Discodermia calyx,* Lithistida, Porif./lab tool as an okadaic acid-like tumor promoter
*cephalo- statin 1/pyrazine alkal. steroid dimer	MRS/ preformed steroids: LaCour 1998	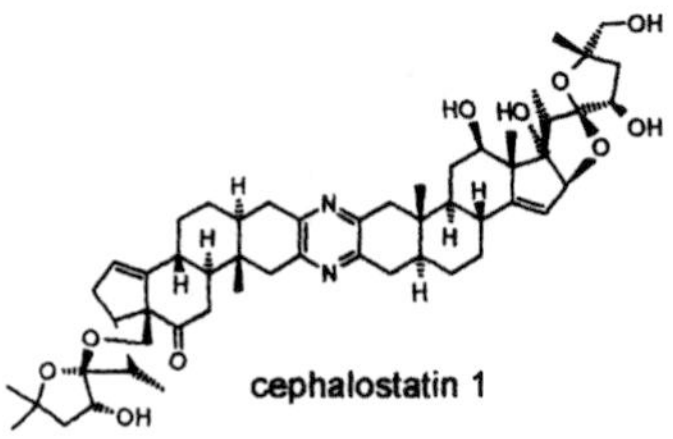 cephalostatin 1	108	0.7	**marine:** *Cephalodiscus gilchristi* Ridewwod, Hemich./antitumor
chloram- phenicol/ shikim.	ITS: MI	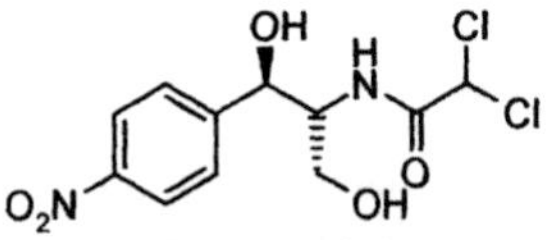 chloramphenicol	15	0.4	*Streptomyces venezuelae* from a soil sample near Caracas, Venezuela (1947), Actinom., Bact.; also **marine:** moon snail, *Lunatia heros,* Moll.: MI/antibacterial
compactin (6-demethyl mevinolin)/p olyket.	MRS/(*R*)-3- acetyl- cyclohex-2- en-1-ol : Hagiwara 1995	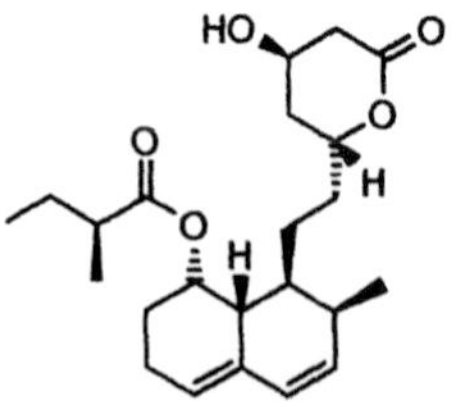 compactin	39	0.6	*Penicillium citrinum* Thom (1910), Eumyc./anti- cholesterolemic

Table 15.I. Total synthesis of environmentally, medically, and technologically relevant natural products (* = rarity of the natural product; ANR = application of new reactions; ANS = application of new strategies; ITS = industrial synthesis; MRS = most rapid solution; S and S/H = molecular and specific molecular complexity, respectively)

Target/class	Strategy/ synthetic precursors	Target structure	S	S/H	Origin/function
α-conotoxin SI/polypept.	MRS: Munson 1993	ω-conotoxin MVIIA, a 25-mer specific antagonist of N-type Ca^{++} channels with three disulfide bridges; for the three dimensional structure in solution, see: Kohno 1995 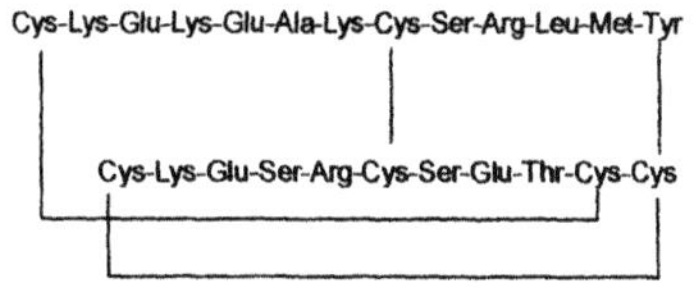 Cys-Lys-Glu-Lys-Glu-Ala-Lys-Cys-Ser-Arg-Leu-Met-Tyr Cys-Lys-Glu-Ser-Arg-Cys-Ser-Glu-Thr-Cys-Cys amino acid sequence and disulfide bridges of ω-conotoxin MVIIA	208	0.5	**marine:** *Conus magus* Linné, 1758, Neogastropoda, Moll./Ca^{++} channel blocker cardioprotective and neuroprotective
cyano- cobalamin (= vitamin B_{12})/corrin	ANR and ANS/ elaborated pyrroles: Eschenmoser 1977	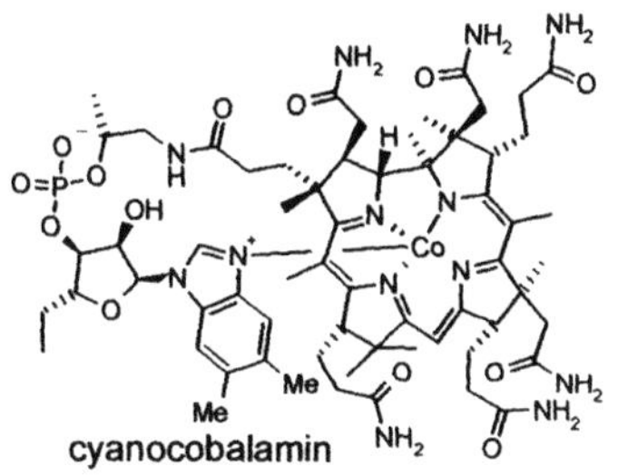 cyanocobalamin	109	0.5	Bact.: MI/dioxygen carrier
*didemnin A and B/lipopet.; analogues: Pfizenmayer 1999	MRS/2- hydroxy- isovaleric acid, (*R*)-*allo*- isoleucine, (*S*)–leu-OH, and (*S*)-Tyr-OH: Hamada 1989	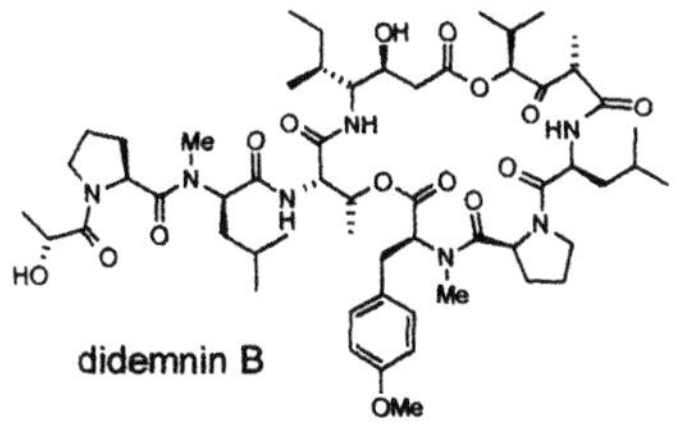 didemnin B	78	0.4	**marine:** *Trididemnum cereum* (Giard, 1872), Didemnidae, Ascid./cytotoxic on tumor cells, though toxic

Table 15.I. Total synthesis of environmentally, medically, and technologically relevant natural products (* = rarity of the natural product; ANR = application of new reactions; ANS = application of new strategies; ITS = industrial synthesis; MRS = most rapid solution; S and S/H = molecular and specific molecular complexity, respectively)

Target/class	Strategy/ synthetic precursors	Target structure	S	S/H	Origin/function
*(+)-disco-dermolide natural enantiomer/ polyket.	ITS/c.a. methyl (S)-3-hydroxy -2-methyl-propionate to a common intermediate for the molecule three parts: Smith 2000; aldol technology: Paterson 2001	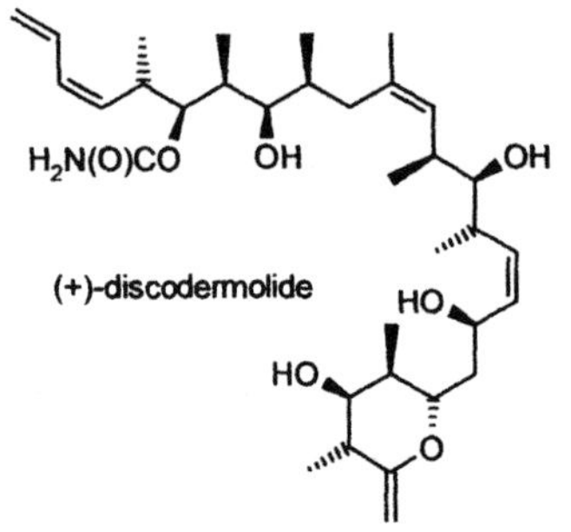	41	0.4	**marine:** *Discodermia dissoluta* Lithistida, Porif./a conformationally flexible antitumor agent that forms a complex with tubulin: Monteagudo 2001
*disco-rhabdin C/pyrrolo-imino-quinone alkal.	ANR/ hypervalent iodine induced spiro-annulation: Kita 1992		35	0.7	**marine:** Porif./cytotoxic and antibacterial
*dolastatin 10/pept.	MRS: Miyazaki 1995	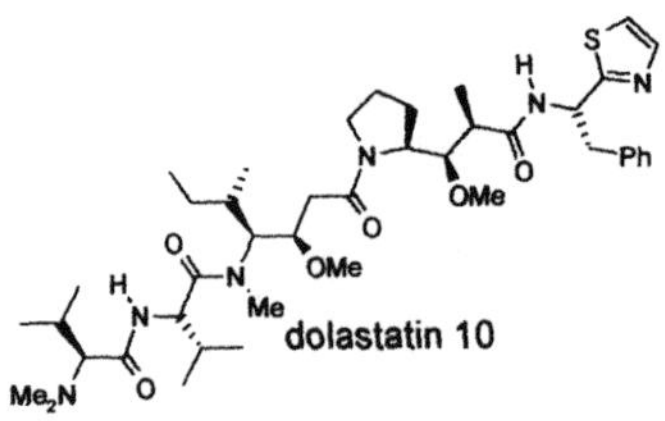	36	0.3	**marine:** *Dolabella auricularia* (Lihtfoot, 1786), Opistobr., Moll./antitumor

Table 15.I. Total synthesis of environmentally, medically, and technologically relevant natural products (* = rarity of the natural product; ANR = application of new reactions; ANS = application of new strategies; ITS = industrial synthesis; MRS = most rapid solution; S and S/H = molecular and specific molecular complexity, respectively)

Target/class	Strategy/ synthetic precursors	Target structure	S	S/H	Origin/function
*dysidiolide/ sesterterp.	ANR and MRS: Corey 1997B; Magnuson 1998; Takahashi 2000	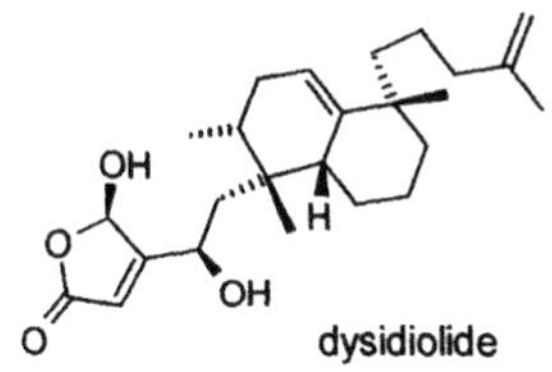 dysidiolide	36	0.5	**marine:** *Dysidea etheria* de Laubenfels, Dysideidae, Porif./inhibitor of dephosphorylation of *p*-nitrophenol phosphate by cdc25A protein phosphatase; antimitotic
*dysiherba-ine/amino acid	MRS: Masaki 2000	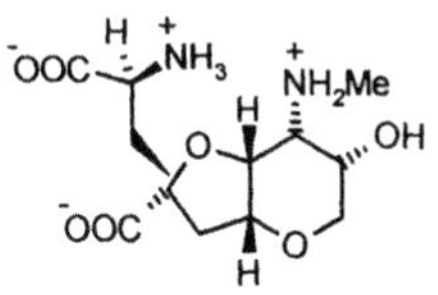 dysiherbaine	32	0.7	**marine:** *Dysidea herbacea* (Keller), Dictyoc., Porif. from Micronesia/neuroexci tatory
*ectein-ascidin 743/tetra-hydroiso-quinoline alkal.	ITS/3,4-(methylene-dioxy)phenyl methoxy-methyl ether: Martinez 2000	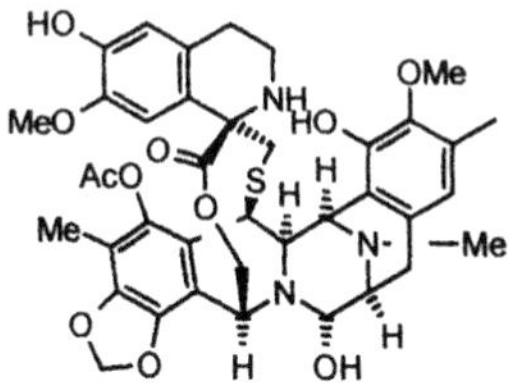 ecteinascidin	52	0.5	**marine:** *Ecteinascidia turbinata* Herdman, 1880, Ascid./antitumor

Table 15.I. Total synthesis of environmentally, medically, and technologically relevant natural products (* = rarity of the natural product; ANR = application of new reactions; ANS = application of new strategies; ITS = industrial synthesis; MRS = most rapid solution; S and S/H = molecular and specific molecular complexity, respectively)

Target/class	Strategy/ synthetic precursors	Target structure	S	S/H	Origin/function
*eleuthe-robin/diterp. glycoside	MRS/ (+)-carvone: Nicolaou 2000A	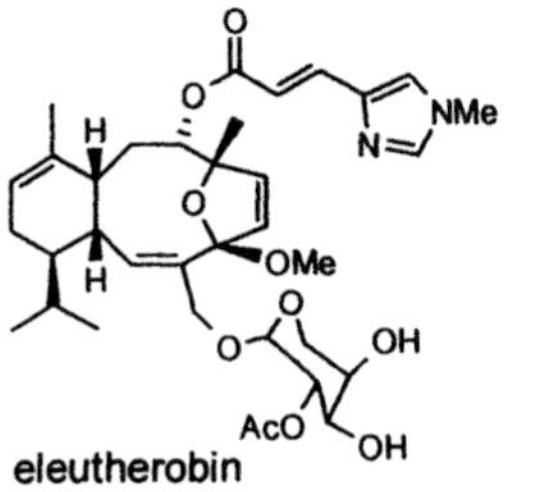 eleutherobin	55	0.5	**marine:** *Eleutherobia* spp., Alcyonacea from W Australia and *Erythropodium caribaeorum* Durchassaing & Michelotti, Gorgonacea, Cnid. from the Caribbean : Cinel 2000/paclitaxel-like activity on tubulin
epothilone A/polyket. macrolide	ANS solid-phase olefin metathesis: Nicolaou 2000A; epothilones B and D: White 2001	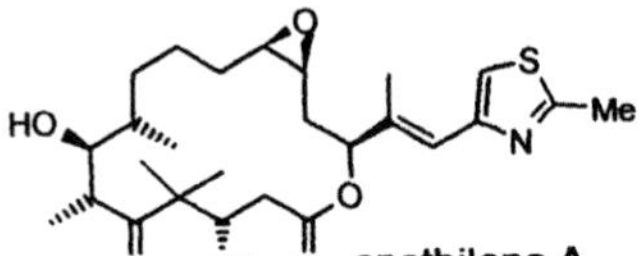 epothilone A	32	0.4	*Sorangium cellulosum,* Myxobact./paclitaxel-like activity on tubulin: Altmann 2000
erythrono-lide B/polyket. macrolide	ANR/2,3,6-trimethyl-phenol along doubly-activated 2-pyridine-thiol ester macro-cyclization: Nicolaou 2000A	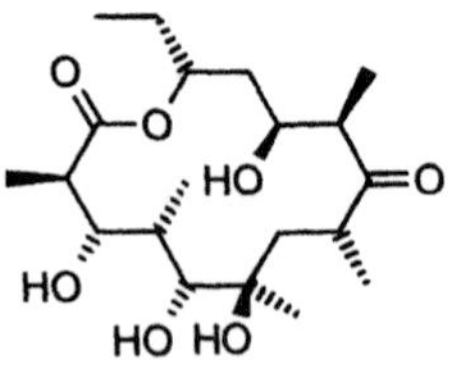 erythronolide	33	0.5	aglycon of erythromycin B of *Streptomyces* spp., Actinom., Bact./antibacterial

Table 15.I. Total synthesis of environmentally, medically, and technologically relevant natural products (* = rarity of the natural product; ANR = application of new reactions; ANS = application of new strategies; ITS = industrial synthesis; MRS = most rapid solution; *S* and *S/H* = molecular and specific molecular complexity, respectively)

Target/class	Strategy/ synthetic precursors	Target structure	S	S/H	Origin/function
everninomicin 13,384-1/ orthosomycin glycoside antibiotic	ANR/tin acetal-based induced 1,1'-disaccharide bond formation: Nicolaou 2000A	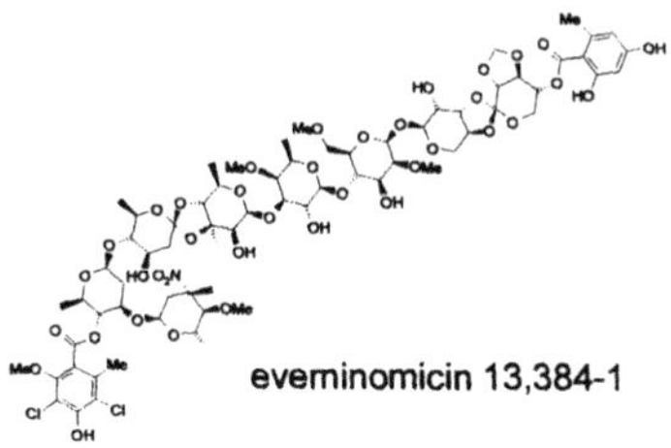 everninomicin 13,384-1	154	0.7	*Micromonospora carbonacea* var. *africana*, Bact. from banks of Nyio River, Kenya/antibacterial
FK-506 (Tacrolimus) /polyket. macrolide macrolactam	MRS/highly convergent: Ireland 1996	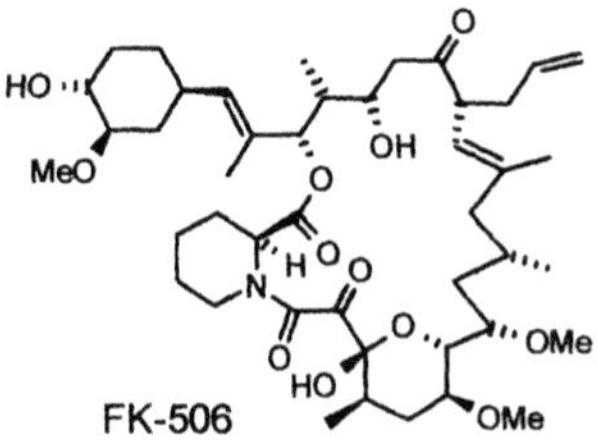 FK-506	68	0.5	*Streptomyces tsukubaensis*, Actinom., Bact. from near Tokyo, Japan,/immunosuppressant
*ginkgolide B/diterp.	MRS enantio-selective: Corey 1988; racemate: Crimmins 1999	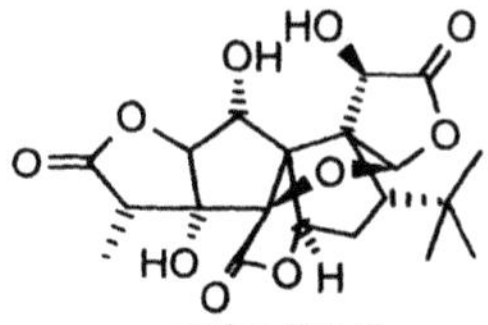 ginkgolide B	62	1	*Ginkgo biloba* L., Ginkgoaceae, Gymn./for the treatment of asthma and severe sepsis
*halichondrin B/polyether macrolide	MRS: Stamos 1997	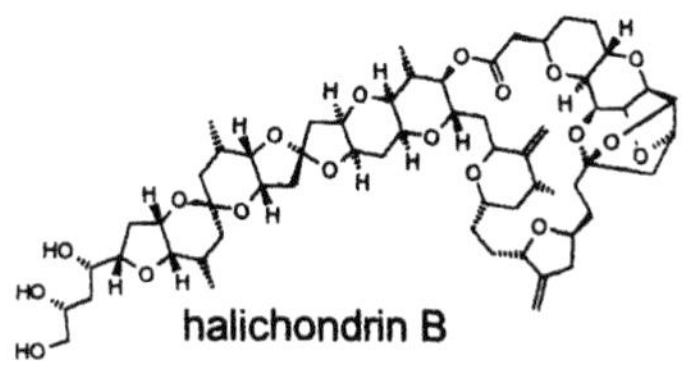 halichondrin B	143	0.8	**marine:** Demosp./antitumor

Table 15.I. Total synthesis of environmentally, medically, and technologically relevant natural products (* = rarity of the natural product; ANR = application of new reactions; ANS = application of new strategies; ITS = industrial synthesis; MRS = most rapid solution; S and S/H = molecular and specific molecular complexity, respectively)

Target/class	Strategy/ synthetic precursors	Target structure	S	S/H	Origin/function
halomon/ monoterp.	MRS/but-2-yne-1,4-diol via [3,3]-sigma-tropic route to α-chlorovinyl group Schlama 1998	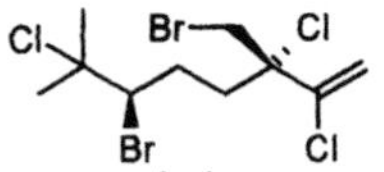 halomon	11	0.4	**marine:** *Portieria hornemannii* Lyngbye, Gigartinales, Rhodoph./cytotoxic on tumor cells
*hennox-azole A/oxazole alkal.	MRS: Wipf 1996, Williams 1999	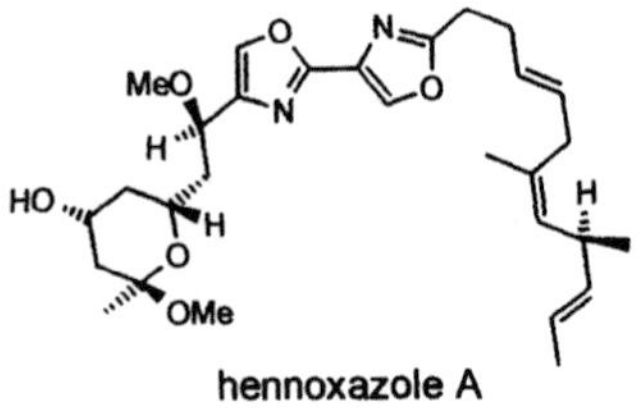 hennoxazole A	26	0.3	**marine:** *Polyfibrospongia* sp. Dictyocer., Porif./antiviral and peripheral analgesic
*homo-harringto-nine/alkal.	ITS/from abundant renewable cephalo-taxine: Robin 2000	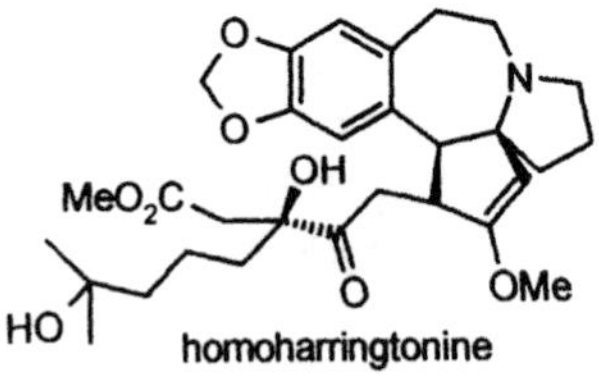 homoharringtonine 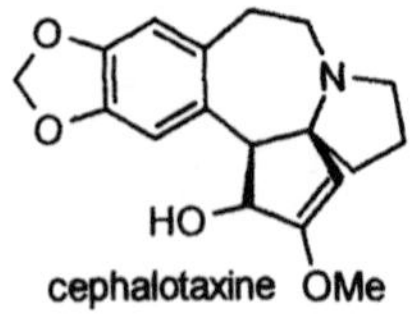 cephalotaxine OMe	29	0.6	barks and roots of *Cephalotaxus harringtonii* var. *drupacea* (Sieb. & Zucc.) Koidz. 1930) Pinatae, Gymn., endangered Chinese tree/antitumor

Table 15.I. Total synthesis of environmentally, medically, and technologically relevant natural products (* = rarity of the natural product; ANR = application of new reactions; ANS = application of new strategies; ITS = industrial synthesis; MRS = most rapid solution; *S* and *S/H* = molecular and specific molecular complexity, respectively)

Target/class	Strategy/ synthetic precursors	Target structure	*S*	*S/H*	Origin/function
kainic acid/amino acid	total synthesis has circumvented the problem of shortage from natural sources: Xia 2001	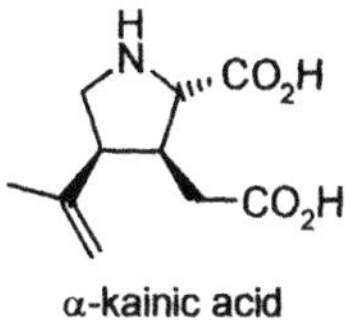 α-kainic acid			**marine:** *Digenea simplex* Agardh, Rhodoph. of wide distribution, though concentrated enough for commercial extraction in Taiwan waters only
*lamellarin O/pyrrole alkal.	ANR/versatile 1,2,4,5-tetrazine cycloaddition with alkynes: Boger 1999A	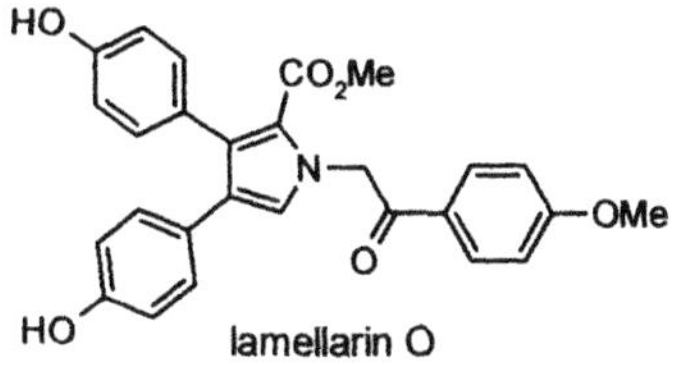 lamellarin O	11	0.2	*Dendrilla cactos,* Dendroc., Porif. from Bass Strait, S Australia/cytotoxic on multidrug resistant tumor cells
*leucascan-drolide A/polyket. macrolide	MRS: Hornberger 2001	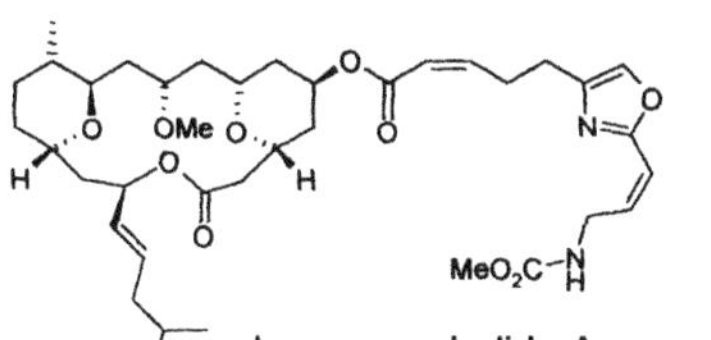 leucascandrolide A	42	0.4	*Leucascandra caveolata* Borojevic and Klautau, Calc., Porif./antifungal and cytotoxic on tumor cells: D'Ambrosio 1996B
luzopeptin B/depsipept.	MRS: Boger 1999B	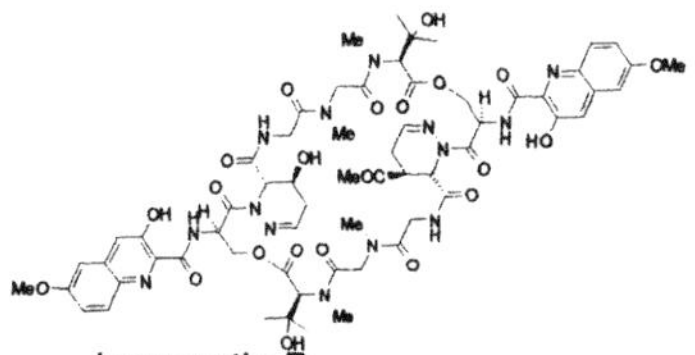 luzopeptin B	83	0.4	*Actinomadura luzonensis,* Actinom., Bact./antitumor and inhibitor of HIV reverse transcriptase

Table 15.I. Total synthesis of environmentally, medically, and technologically relevant natural products (* = rarity of the natural product; ANR = application of new reactions; ANS = application of new strategies; ITS = industrial synthesis; MRS = most rapid solution; *S* and *S/H* = molecular and specific molecular complexity, respectively)

Target/class	Strategy/ synthetic precursors	Target structure	*S*	*S/H*	Origin/function
*manzamine A and B, ircinal A/ macrocyclic alkal.	MRS: Nicolaou 2000A; Martin 1999. Biogenetic intra-molecular Diels-Alder cycloaddition Baldwin 1999	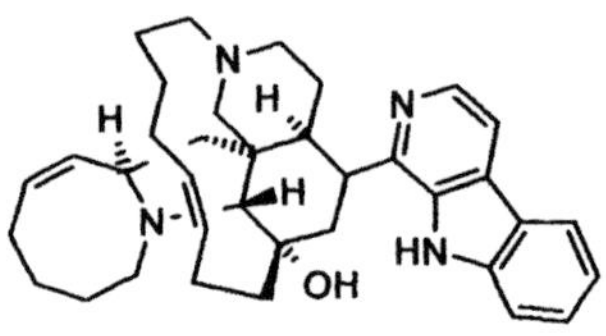 manzamine A	43	0.4	**marine:** Demosp., Porif./cytotoxic on tumor cells
monensin/ ionophore polyket. polyether	ANS/early example of acyclic stereocontrol: Fukuyama 1979, and chelation-controlled aldol technology: Collum 1980	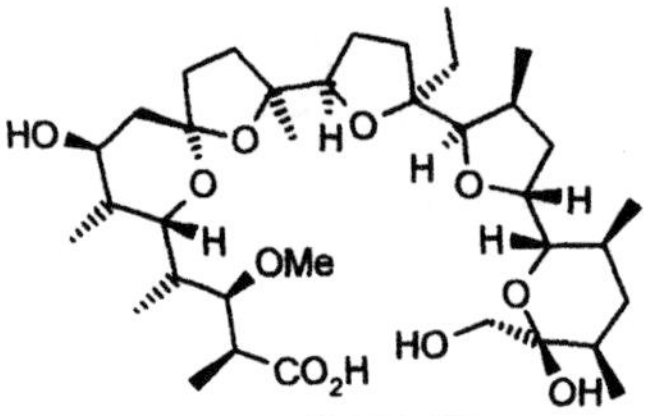 monensin	65	0.6	*Streptomyces cinnamonensis* Okami, 1952, Actinom., Bact./antibacterial
*motuporin/ cyclopept.	MRS/along Ugi four-component condensation: Bauer 1999	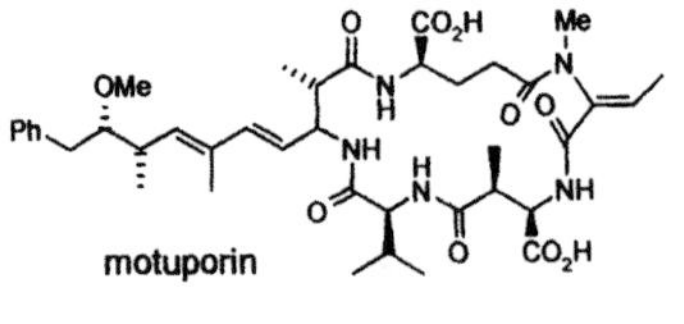 motuporin	54	0.4	**marine:** *Theonella swinhoei* Sollas 1888, Lithistida, Porif./inhibitor of phosphatase 1
*mycalolide A/alkal. macrolide	MRS/chiral silane technology: Panek 2000	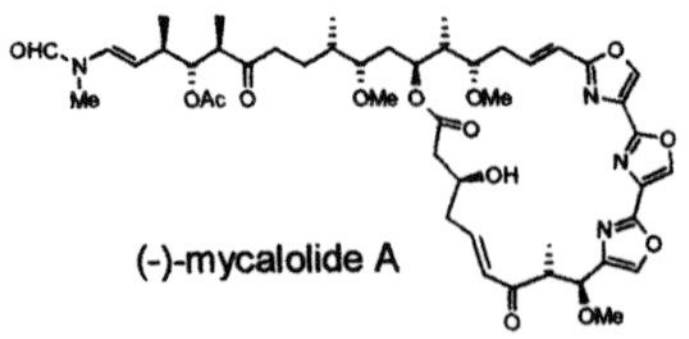 (-)-mycalolide A	55	0.4	**marine:** *Mycale* sp. Poecil., Porif./antifungal on pathogenic strains and cytotoxic on tumor cells

Table 15.I. Total synthesis of environmentally, medically, and technologically relevant natural products (* = rarity of the natural product; ANR = application of new reactions; ANS = application of new strategies; ITS = industrial synthesis; MRS = most rapid solution; *S* and *S/H* = molecular and specific molecular complexity, respectively)

Target/class	Strategy/ synthetic precursors	Target structure	*S*	*S/H*	Origin/function
okadaic acid and 7-deoxy-okadaic acid/polyket. polyether	MRS: Ley 1998 (okadaic); Dounay 2001 (7-deoxy)	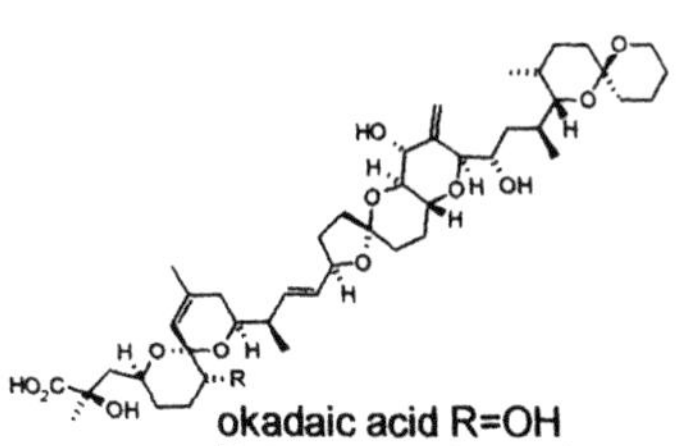 okadaic acid R=OH 7-deoxy-okadaic acid R=H	82	0.6	**marine:** *Prorocentrum* spp. Dinofl./reference-standard tumor promoter; PP-1 and PP-2A phosphatase inhibitor
oligo-saccharides	automated synthesis: Plante 2001	4 to 9 sugar units			Mamm.
paclitaxel/di terp.	ANS: various approaches summarized in Nicolaou 2000A	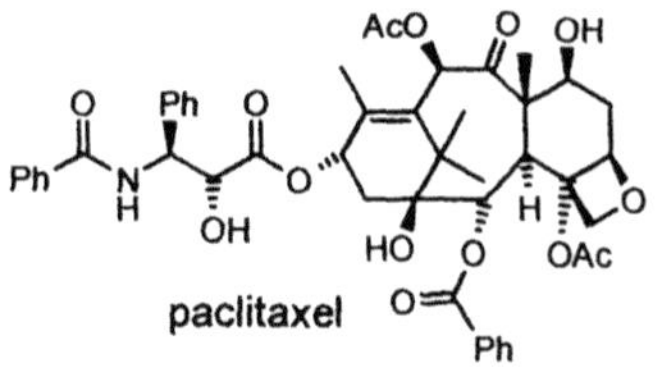 paclitaxel	57	0.5	bark of Pacific yew, *Taxus brevifolia* Nutt., Taxaceae, Gymn./inducing non-functional polymerization of tubulin; antitumor
*palau'-amine/alkal.	ANR/intra-molecular azomethine imine cycloaddition: Overman 1997	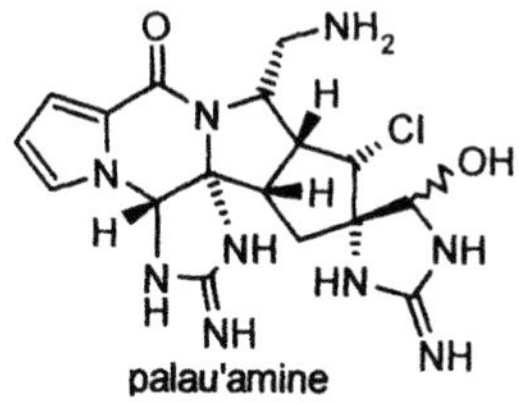 palau'amine	54	0.8	**marine:** *Stylotella aurantium* Kelly-Borges and Bergquist, Halichon., Porif./immunosuppressant, antibacterial, and toxic to tumor cells

Table 15.I. Total synthesis of environmentally, medically, and technologically relevant natural products (* = rarity of the natural product; ANR = application of new reactions; ANS = application of new strategies; ITS = industrial synthesis; MRS = most rapid solution; S and S/H = molecular and specific molecular complexity, respectively)

Target/class	Strategy/ synthetic precursors	Target structure	S	S/H	Origin/function
*palytoxin/ polyket. polyether	MRS: Kishi 1989	palytoxin	231	0.5	**marine:** *Palythoa* spp., Zoanthinaria, Cnid./activator of the Na^+ pump
*pateamine A/alkal. bis-macrolide	ANR/ β-lactam based macrocyclization: Romo 1998	pateamine A	34	0.4	**marine:** *Mycale* sp., Poecil., Porif./antiviral and cytotoxic on tumor cells
phomo-idrides (CP molecules)/ acetogenins	ANS/ anhydride cascade reaction: Nicolaou 2000A	CP-263,114	57	0.6	sterile *Phoma* sp., Eumyc./inhibitor of both squalene synthase and RAS farnesyl transferase
poly-cavernoside A/polyket. macrolide glycoside	MRS: Paquette 2000	(-)-polycavernoside A	75	0.6	**marine:** *Polycavernosa tsudai* [= *Gracilaria edulis*], Gigartinales, Rhodoph./toxin

Table 15.I. Total synthesis of environmentally, medically, and technologically relevant natural products (* = rarity of the natural product; ANR = application of new reactions; ANS = application of new strategies; ITS = industrial synthesis; MRS = most rapid solution; S and S/H = molecular and specific molecular complexity, respectively)

Target/class	Strategy/ synthetic precursors	Target structure	S	S/H	Origin/function
prosta-glandin E₂/fatty acid derv.	MRS: Nicolaou 2000A	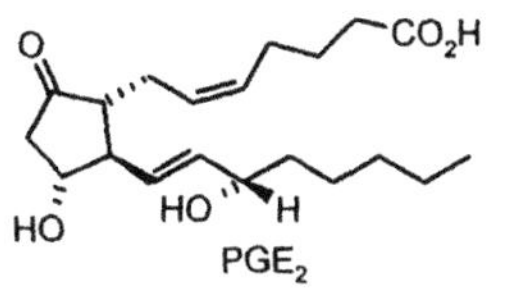 PGE₂	23	0.4	prostate fluid of animals; also **marine:** *Gracilaria verrucosa* (Hudson) Papenfuss, 1950, Rhodoph./ abortifacient
promo-thiocin A/thiopept. antibiotic	MRS: Bagley 2000	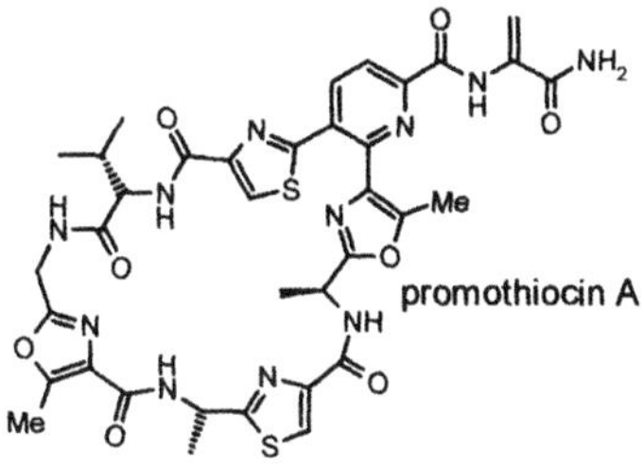 promothiocin A	43	0.4	*Streptomyces* sp. SF2741, Actinom., Bact./antibacterial
*pseudo-pterosin A/glycoside diterp.	ITS (*S*)-(-)-limonene : Corey 1998	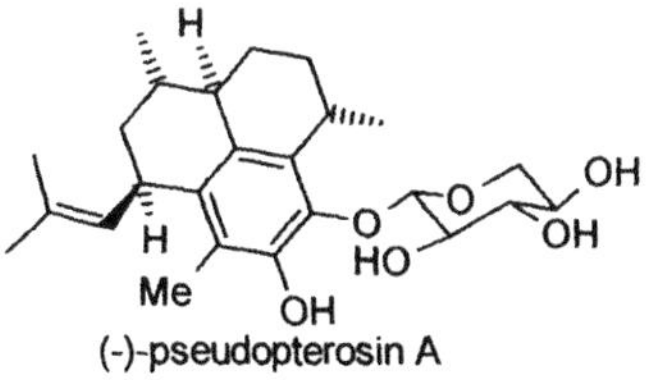 (-)-pseudopterosin A	36	0.5	**marine:** *Pseudopterogorgia elisabethae* Bayer, Gorgonacea, Cnid. /PLA₂ inhibitor with anti-inflammatory activity 50 times greater than indomethacin

Table 15.I. Total synthesis of environmentally, medically, and technologically relevant natural products (* = rarity of the natural product; ANR = application of new reactions; ANS = application of new strategies; ITS = industrial synthesis; MRS = most rapid solution; S and S/H = molecular and specific molecular complexity, respectively)

Target/class	Strategy/ synthetic precursors	Target structure	S	S/H	Origin/function
*ptilo-mycalin A and cram-bescidins/ guanidine alkal.	ITS/tethered Biginelli condensation: Coffey 2000	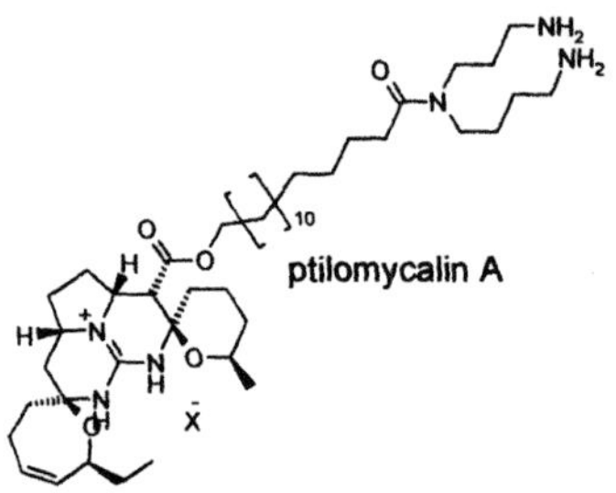 ptilomycalin A	49	0.3	**marine:** Poecil., Porif./antifungal, toxic to tumor cells, and HIV antiviral
dercitin/ pyrido-(4,3,2-*mn*)-acridine alkal.	ANR and ANS/new pyridine forming reaction and nitrene cyclization: Ciufolini 1995	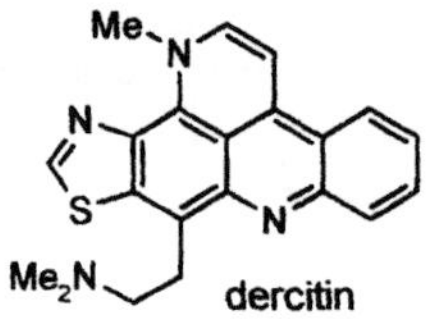 dercitin	27	0.5	**marine:** *Dercitus* sp., Choristida, Porif. from deep sea in the Bahamas/disrupter of DNA and RNA synthesis; only modestly active as antitumor *in vivo*
rapamycin/ polyket. alkal. macrolide	MRS/acyclic stereo-selection; macro-cyclization by Stille coupling: Nicolaou 2000A	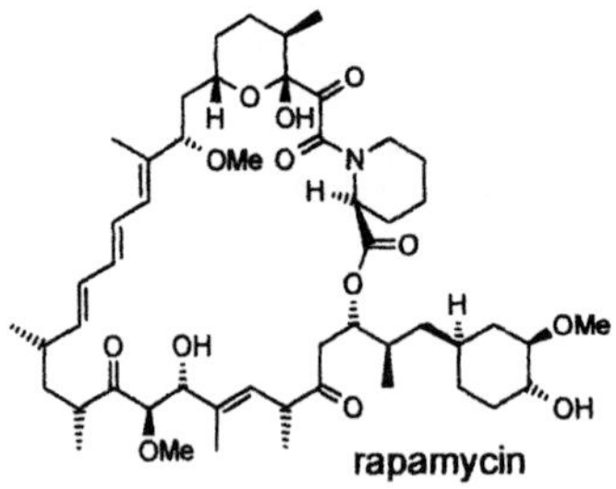 rapamycin	80	0.5	*Streptomyces hygroscopicus*, Actinom., Bact. from Rapa Nui soil in Easter Is./immunosuppressant

Table 15.I. Total synthesis of environmentally, medically, and technologically relevant natural products (* = rarity of the natural product; ANR = application of new reactions; ANS = application of new strategies; ITS = industrial synthesis; MRS = most rapid solution; *S* and *S/H* = molecular and specific molecular complexity, respectively)

Target/class	Strategy/ synthetic precursors	Target structure	*S*	*S/H*	Origin/function
sanglifehrin A/depsipept.	MRS/Stille coupling, following previous strategies: Duan 2001	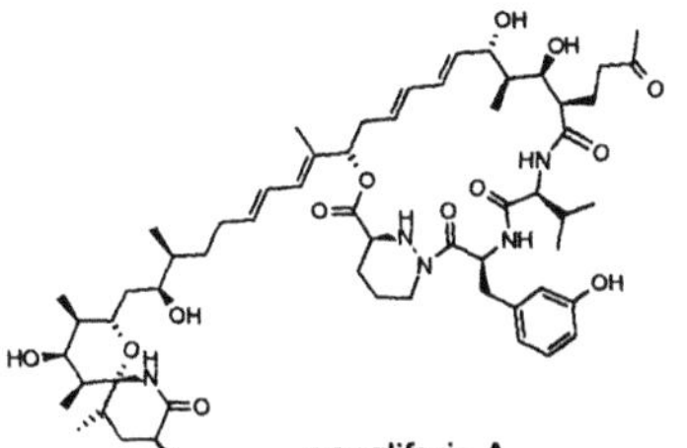 sangliferin A	84	0.5	*Streptomyces* sp., Actinom., Bact. from soil in Malawi, SE Africa/lab tool for the study of immunosuppression mechanisms, which differs from that of cyclosporin A and FK506
*sarco- dictyin A/diterp.	ANS/ prototype combinatorial libraries from solid phase technology with complex natural substrates: Nicolaou 2000A	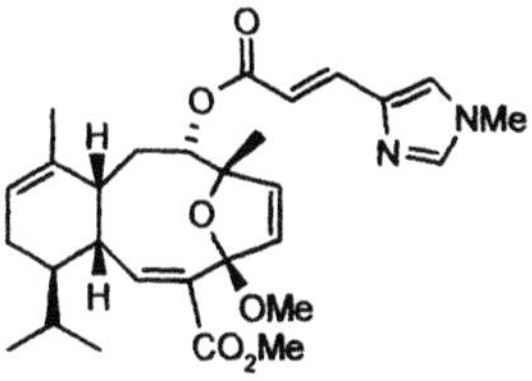 sarcodictyin A	42	0.5	**marine:** *Sarcodictyon roseum,* (Philippi, 1842) [= *Rolandia rosea* (Philippi)], Stolonifera, and certain pennatulaceans and alcyonaceans, Cnid./paclitaxel-like activity on tubulin: Ciomei 1997
scalarene- dial/ scalarane sesterterp.	ITS/polyene poly- cyclization: Corey 1997A	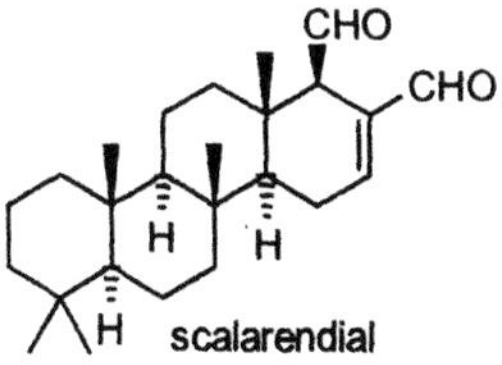 scalarendial 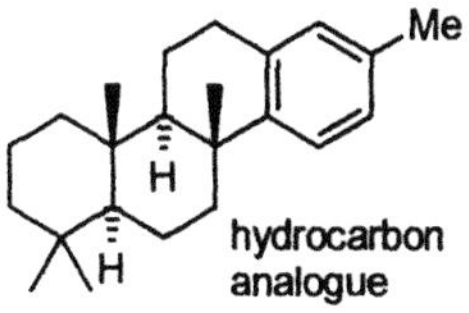 hydrocarbon analogue	scal. : 38; hyd. : 20	scal. : 0.54 ; hyd. :0.3	**marine:** Dictyocer., Porif./scalarendial is an antifeedant agent and the hydrocarbon analogue is a marker for sedimentary petroleum

Table 15.I. Total synthesis of environmentally, medically, and technologically relevant natural products (* = rarity of the natural product; ANR = application of new reactions; ANS = application of new strategies; ITS = industrial synthesis; MRS = most rapid solution; S and S/H = molecular and specific molecular complexity, respectively)

Target/class	Strategy/ synthetic precursors	Target structure	S	S/H	Origin/function
*spongi- statins 1-2/polyket. macrolide polyethers	MRS: Evans 1999B; Smith 2001	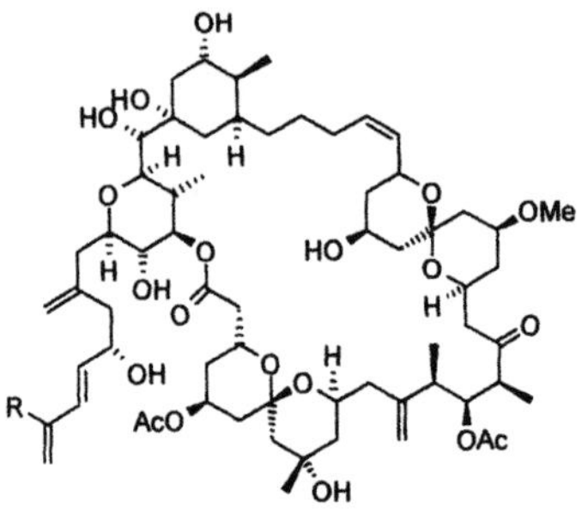 spongistatin 1 (= altohyrtin A) R=Cl spongistatin 2 (= altohyrtin C) R=H	104	0.5	**marine:** *Hyrtios* sp., Dictyocer., Porif./antitumor
squalamine/ steroidal alkal.	MRS: Zhang 1998	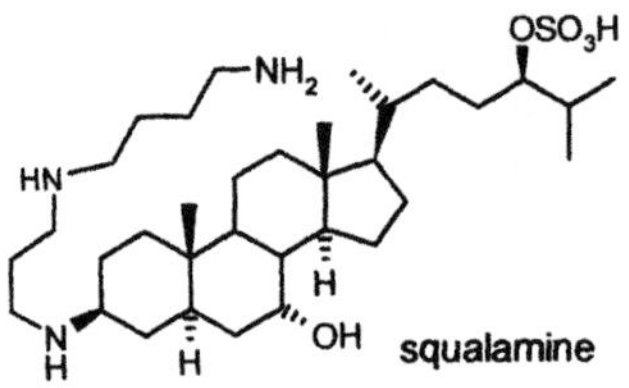 squalamine	50	0.5	**marine:** *Squalus acanthias* L., Chondrich., sharks/angiogenesis inhibitor and antibacterial
squalestatin S1 (= zaragozic acid A)/ acetogenin	MRS: Nicolaou 2000A	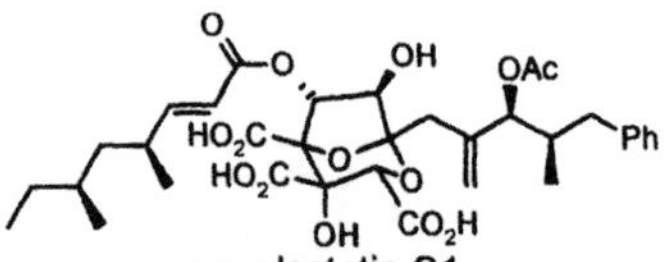 squalestatin S1	53	0.5	*Phoma* sp., Eumyc/inhibitor of squalene synthase and hence anticholesterolemic
strychnine	ANS (cobalt mediated [2 + 2 + 2] cycloaddition: Eichberg 2001	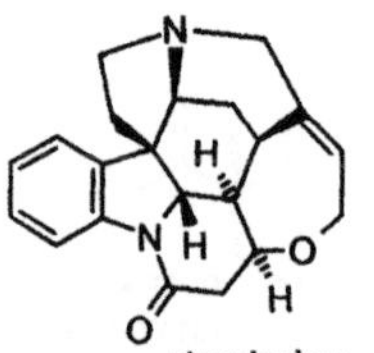 strychnine	44	0.7	Southeast Asian *Strychnos* spp., Loganiaceae, Ang./rodenticide, animal stimulant, and ligand for glycine receptor studies

Table 15.I. Total synthesis of environmentally, medically, and technologically relevant natural products (* = rarity of the natural product; ANR = application of new reactions; ANS = application of new strategies; ITS = industrial synthesis; MRS = most rapid solution; *S* and *S/H* = molecular and specific molecular complexity, respectively)

Target/class	Strategy/ synthetic precursors	Target structure	*S*	*S/H*	Origin/function
sucrose/ disacch.	ITS (80% yield; extensible to rare sugars: Oscarson 2000	sucrose	37	0.8	sugar cane, *Saccharum officinarum* L., Gramineae and sugar beet, *Beta vulgaris* L., Chenopodiaceae, Ang./sweetening agent
teurilene/ triterp.	ANS/ asymmetric synthesis of achiral molecules: Hoye 1987	teurilene	37	0.4	**marine:** *Laurencia* spp., Rhodoph./cytotoxic
thiocoralline /2-fold symmetric octa- depsipept.	MRS: Boger 2000	thiocoralline	82	0.6	*Micromonospora* sp. Bact. /cytotoxic on tumor cells
spiro- tryprostatin B and tryprostatin B/pept. (diketo- piperazines)	MRS: Sebahar 2000; von Nussbaum 2000; Schkeryantz 1999	spirotryprostatin-B / tryprostatin-B	34	0.5	**marine:** *Aspergillus fumigatus* Fres, Eumyc. from deep-sea at the mouth of Oi river, Japan/inhibitors of the cell cycle progression of tsFT210 cells at the G2/M phase

Table 15.I. Total synthesis of environmentally, medically, and technologically relevant natural products (* = rarity of the natural product; ANR = application of new reactions; ANS = application of new strategies; ITS = industrial synthesis; MRS = most rapid solution; S and S/H = molecular and specific molecular complexity, respectively)

Target/class	Strategy/ synthetic precursors	Target structure	S	S/H	Origin/function
vancomycin/ glycopept. antibiotic	ANR and MRS:/ triazene- driven biaryl ether synthesis: Nicolaou 2000A	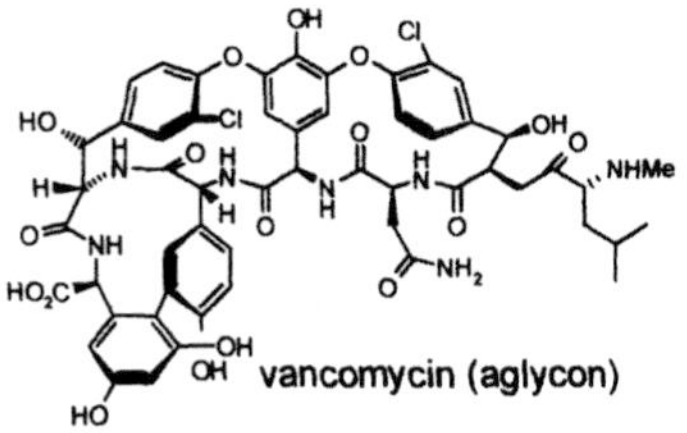 vancomycin (aglycon)	68	0.4	*Streptomyces orientalis* [= *Amycolatopsis orientalis* subsp. *orientalis* (Pittenger and Brigham 1956) Lechevalier *et al.* 1986], Actinom., Bact./antibacterial
*volvatellin/ sesquiterp.	MRS: biogenetically patterned carbonyl-ene route: Mancini 2000	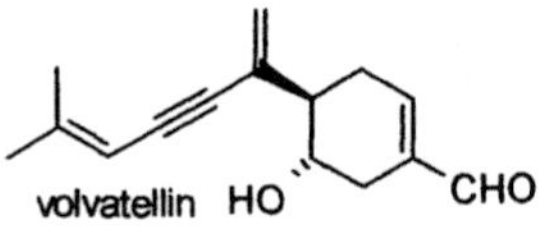 volvatellin HO CHO	22	0.5	**marine:** *Volvatella* sp., Opisthobr., Moll./antifeedant candidate
*xesto- spongins A and C, and 9'-aragu- spongin B/oxaquino- lizidine alkal.	ANS: biogenetically patterned route from 3-alkyldi- hydropyridine dimer: Baldwin 1998	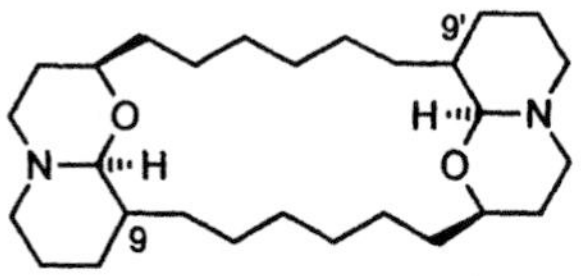 (-)-xestospongin A (9R,9'R) (-)-xestospongin C (9S,9'R) (-)-araguspongin B (9S,9'S)	32	0.4	**marine:** Nepheliosp. and Haplosclerida, Porif./somatostatin inhibitors

Table 15.II. Semisynthesis of drugs or technologically relevant agents from natural products

Area	Common or scientific name/Chemical class/Structure	Trade name/Company/phase	Precursor/Chemical class
Anti-bacterial agents	cefprozil/cephem alkal. deriv. 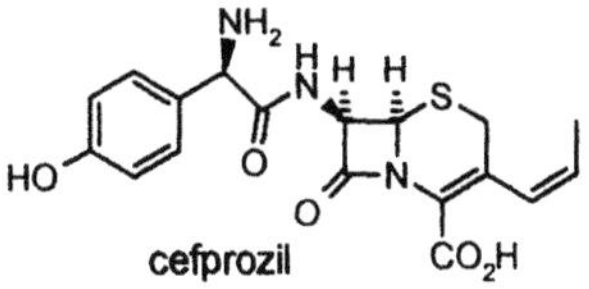 NH₂ ... cefprozil CO₂H	Cefzil/ Bristol-Myers Squibb	cephalosporins/cephem pept. alkal.
Anti-bacterial agents	cefuroxime 1-acetoxyl ester/cephem alkal. deriv. 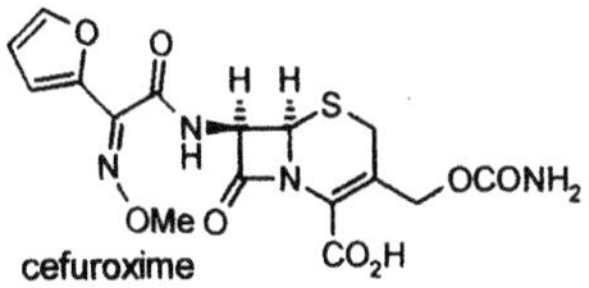 cefuroxime CO₂H OCONH₂	Ceftin/ Allenand-Hanburys; Zinnat/Glaxo	cephalosporins/cephem pept. alkal.
Anti-bacterial agents	clarithromycin/polyket. 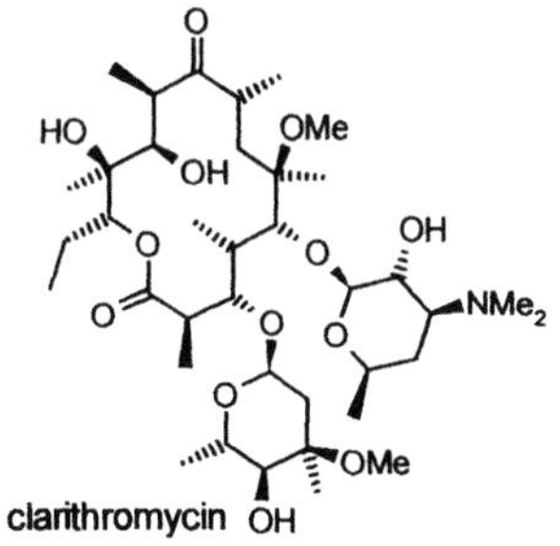 clarithromycin	Biaxin (= Klacid)/ Abbott/approved	erythromycin A (C 6.1.3.FA/PO) from *Streptomyces* spp., Actinom., Bact./glycosidic polyket. 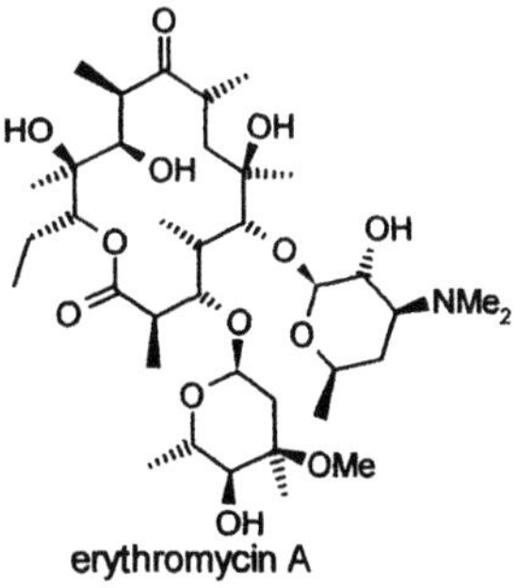 erythromycin A

Table 15.II. Semisynthesis of drugs or technologically relevant agents from natural products

Area	Common or scientific name/Chemical class/Structure	Trade name/Company/phase	Precursor/Chemical class
Anti-bacterial agents	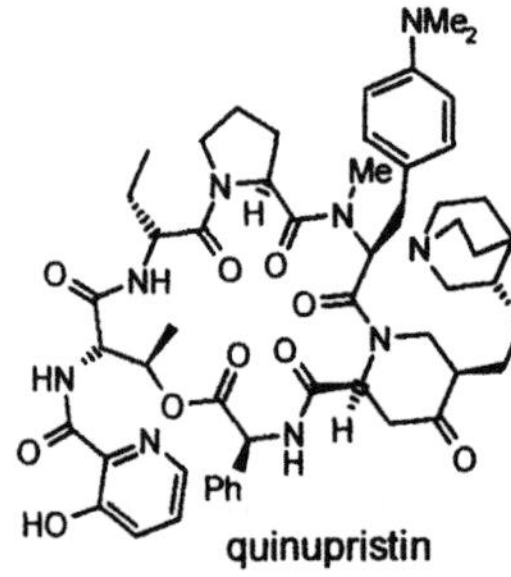 dalfopristin and quinupristin/streptogramin-like injectable 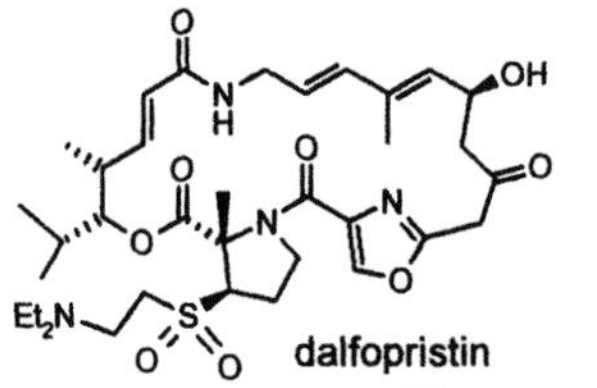	Synercid (mixture of the two antibiotics)/ Rhône-Poulenc Rorer (approved in 1999 by FDA against bloodstream infections by vancomycin-resistant *Enterococcus faecium* and in UK against nosocomial pneumonia)	streptogramins from *Streptomyces* sp., Actinom., Bact. C 6.2.P/pept.
Anti-bacterial agent that requires less frequent adminis-tration than vanco-mycin	V-glycopept./glycopept. deriv. 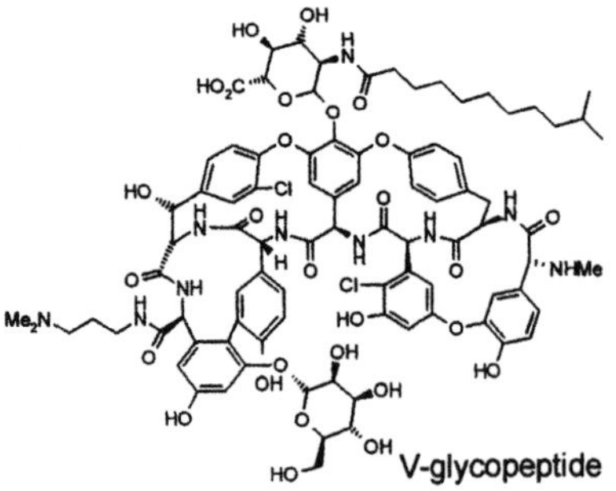	in development/ Versicor, phase II	vancomycin/glycopept. 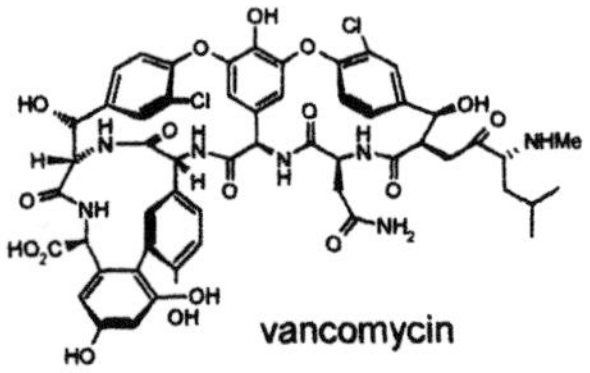

Table 15.II. Semisynthesis of drugs or technologically relevant agents from natural products

Area	Common or scientific name/Chemical class/Structure	Trade name/Company/phase	Precursor/Chemical class
Antifungal agents (specific inhibitor, such as sordarin, of fungal protein synthesis)	GM237354/ glycoside of sordarin diterp. aglycon	Glaxo-Wellcome/ preclinical for candidasis: Fostel 2000	sordarin from *Sordaria araneosa* Cain, Eumyc./diterp. glycoside: Coval 1995
Antifungal agents	LY303366 (= V-echinocandin)/low-toxicity lipopept. based on enzymatically deacylated echinocandin B (inhibitor of (1,3)-β-D-glucan synthesis, needed for cell wall biosynthesis)	in development/ Eli Lilly & Co	echinocandin from *Aspergillus nidulans* var. *echinulatus*, Eumyc./lipopet.: Debono 1995
Anticancer drugs	CGP41251/alkal. deriv.	in development/ clinical trials by Novartis: Newman 2000	staurosporine from *Streptomyces staurosporeus*, Actinom., Bact./alkal.

Table 15.II. Semisynthesis of drugs or technologically relevant agents from natural products

Area	Common or scientific name/Chemical class/Structure	Trade name/Company/phase	Precursor/Chemical class
Anticancer drugs	docetaxel/diterp.: Colin 1988 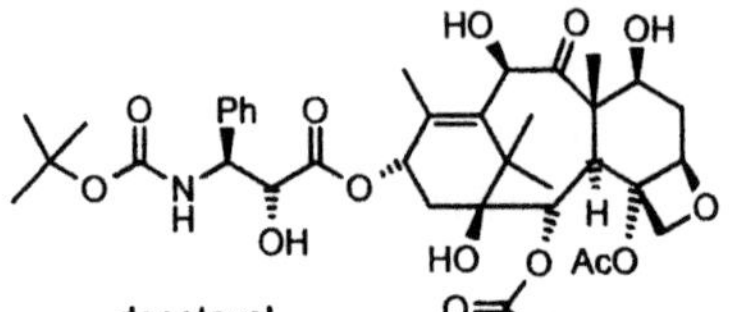	Taxotere/ Rhône Poulenc	10-deacetylbaccatin II from needles of the European yew tree, *Taxus baccata* L., Gymn./diterp. 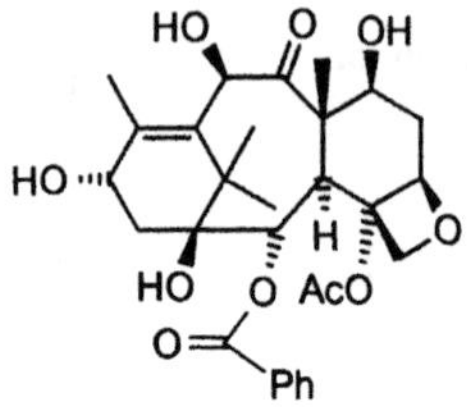
Anticancer drugs	etoposide/shikim. polyket. glucoside 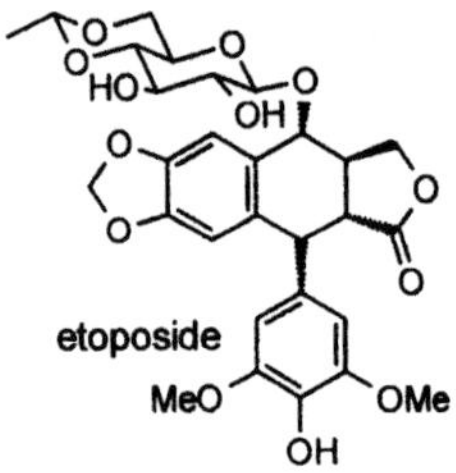	VePesid/ Bristol Myers Squibb (1970 US patent to Sandoz)	epipodophyllotoxin from N American *Podophyllum peltatum* L., Berberidaceae, Ang. 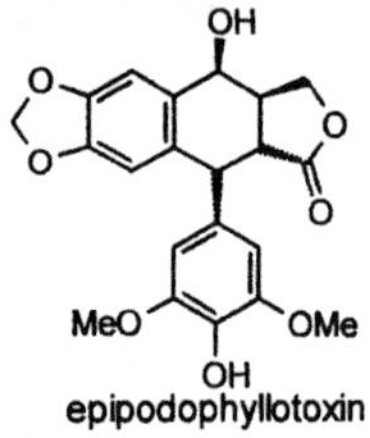
Anticancer drugs	irinotecan/alkal. deriv. 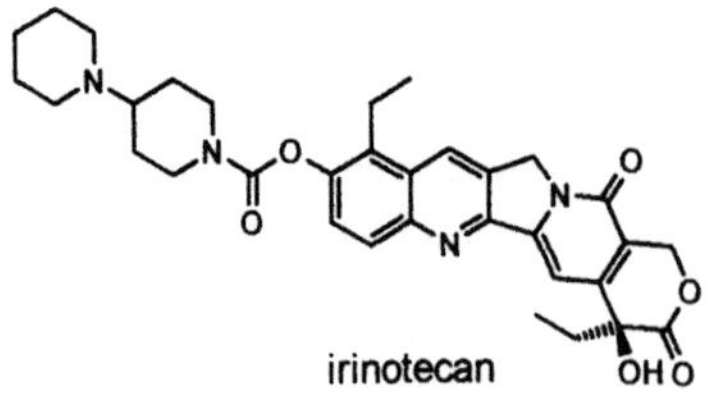	Camptosar/ Pharmacia & Upjohn	camptothecin from *Camptotheca acuminata* Decsne, Nyssaceae, Ang./alkal. 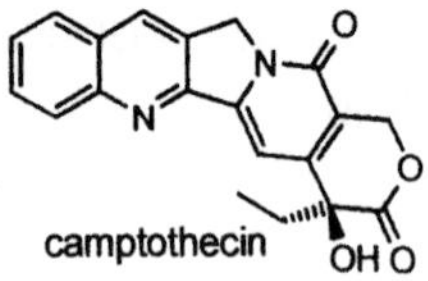

docetaxel

10-deactylbaccatin III

etoposide

epipodophyllotoxin

irinotecan

camptothecin

Table 15.II. Semisynthesis of drugs or technologically relevant agents from natural products

Area	Common or scientific name/Chemical class/Structure	Trade name/Company/phase	Precursor/Chemical class
Anticancer drugs	teniposide/shikim. polyket. glycoside 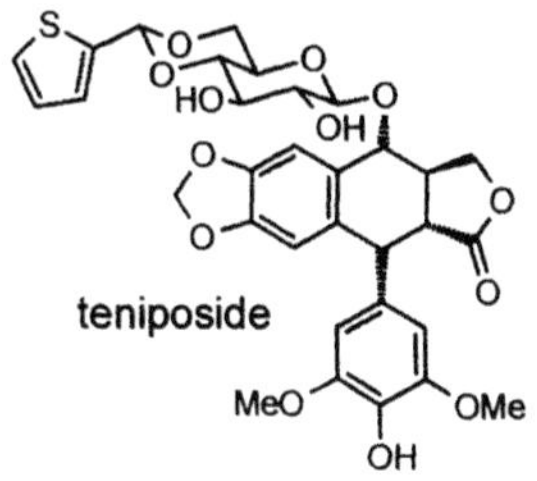 teniposide	Vumon/ Bristol- Myers Squibb	epipodophyllotoxin from N American *Podophyllum peltatum* L., Berberidaceae, Ang. 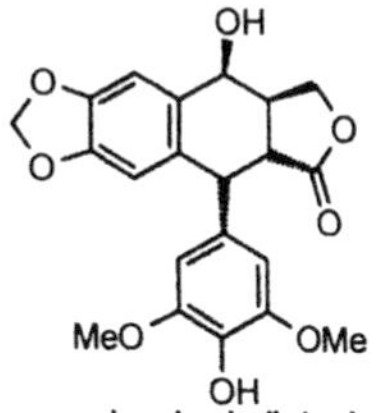 epipodophyllotoxin
Anticancer drugs (angio-genesis control)	TNP-470/terp. polyket. deriv. 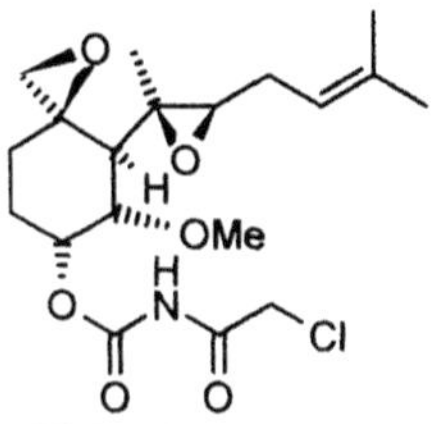 TNP-470	in development/ TAP Pharm (phase III)	fumagillin/terp. polyket. from *Aspergillus fumigatus* Fres, Deuterom., Eumyc. The cellular effects by this fungus were first discovered by Donald Ingber: Bailly 2000 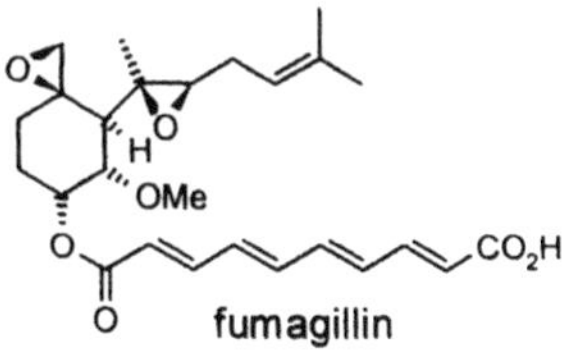 fumagillin
Anticancer drugs	topotecan/alkal. deriv. 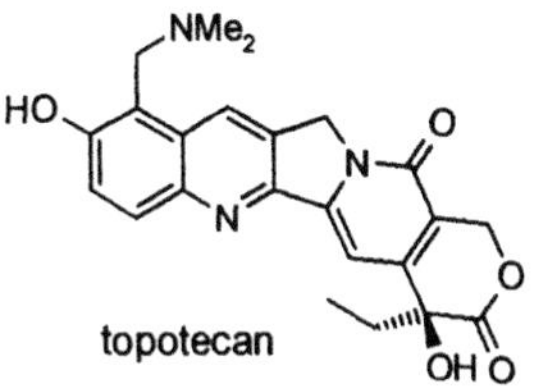 topotecan	Hycamptin/ SKB	camptothecin from the Chinese tree, *Camptotheca acuminata*, Nyssaceae/indolizine quinoline alkal. 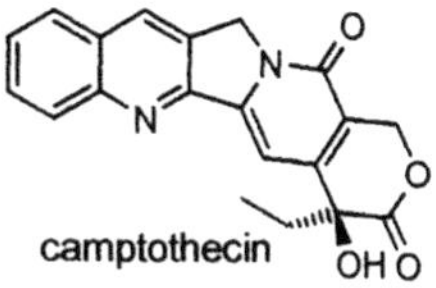 camptothecin

Table 15.II. Semisynthesis of drugs or technologically relevant agents from natural products

Area	Common or scientific name/Chemical class/Structure	Trade name/Company/phase	Precursor/Chemical class
Anticancer drugs (sensitizing cancer cells to apoptosis)	truncated apoptolidin/glycosidic macrolide that retains the activity of the precursor, with lessened toxicity 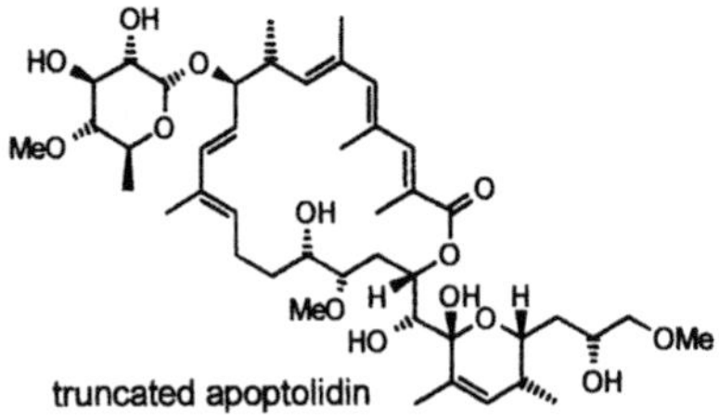 truncated apoptolidin	waiting for development	apoptolidin from *Nocardiopsis* sp., Actinom., Bact./polyglycosidic macrolide with a conserved motif: Salomon 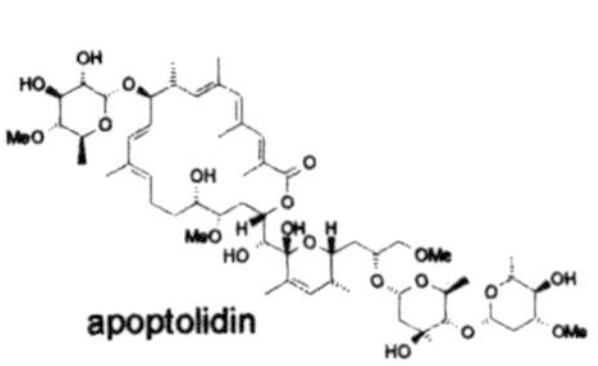 apoptolidin 2001
Anticancer drugs	vindesine and vinorelbine/alkal. derivs. 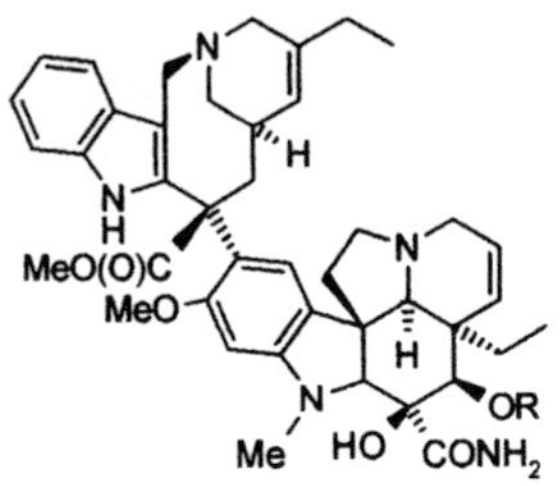 vindesine R=H vinorelbine R=Ac	vindesine (Eldisine); vinorelbine (Navelbine/ Fabre)	vinblastine from trop., cultivated, (*Vinca rosea* L. [= *Catharanthus roseus* (G. Don.)], Apocyanaceae, Ang./terp. alkal., approved in Europe and in clinical trials in the US 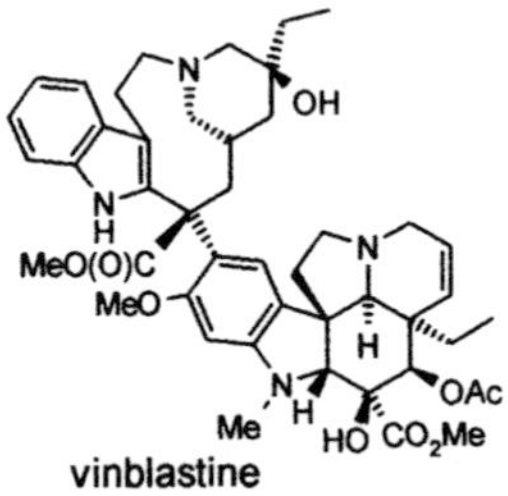 vinblastine
Anti-parasitic agents	dihydroartemisinin methyl ether = arteether dihydroartemisin ethyl ether = artheether dihydroartemisin hemihydrosuccinate = artesunate/sesquiterp. deriv.s 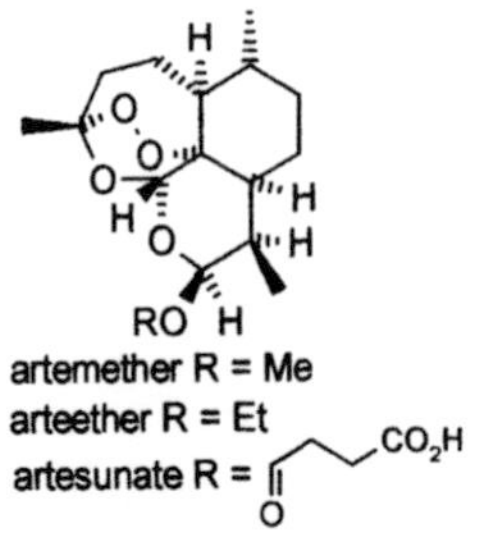 artemether R = Me arteether R = Et artesunate R = (CO₂H form)	artemether, artemether/ Kunning Pharm. Factory, China	artemisinin from the Chinese herb, *Artemisia annua* L., Compositae, Ang./sesquiterp. 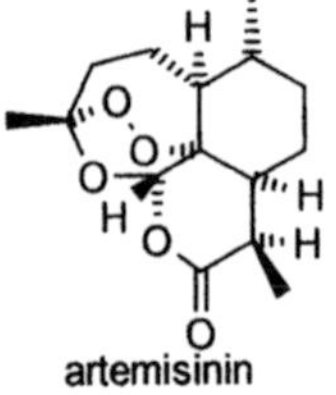 artemisinin

Table 15.II. Semisynthesis of drugs or technologically relevant agents from natural products

Area	Common or scientific name/Chemical class/Structure	Trade name/Company/phase	Precursor/Chemical class
Anti-parasitic agents (protecting dogs from deadly *Dirofilaria immitis* and *Dirofilaria repens*)	ivermectin/mixture of glycosidic macrolides 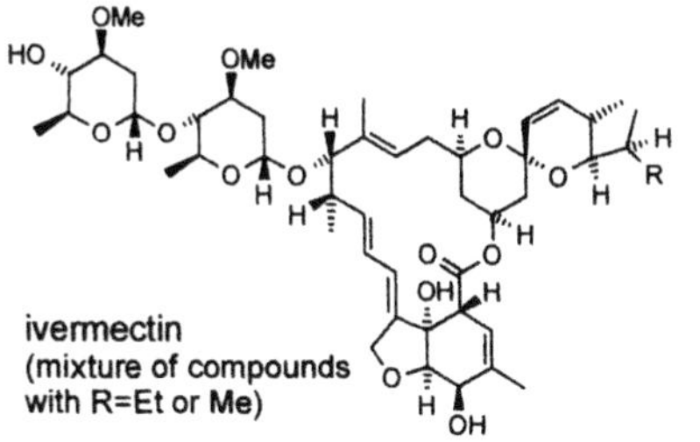 ivermectin (mixture of compounds with R=Et or Me)	Cardomec/ Merck & Co; Cardotek/ Merial	avermectin B_{2a} from *Streptomyces avermitilis*, Actinom., Bact./glycosidic macrolide
Antiviral agents	PEG-Intron/pegylated interferon	Schering-Plough-Enzon	Schering-Plough intron-A
Blood, inflammation and gout treating drugs	pravastatin sodium/polyket. deriv. 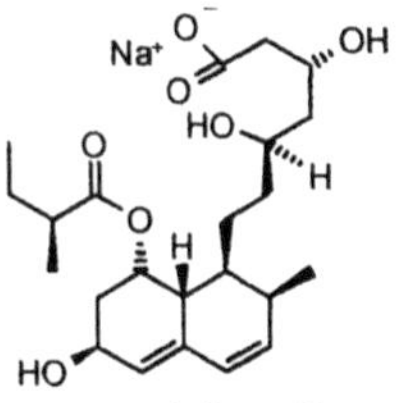 pravastatin sodium	Mevalotin/ Sankyo,	mevastatin (= compactin) from *Penicillium* and other Deuterom., Eumyc./polyket.: Endo 1997 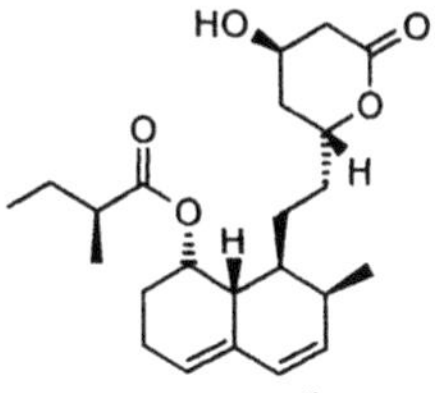 compactin
Blood, inflammation and gout treating drugs	Single-chain polypept. of 65 amino acids	Revasc/Ciba-Geigy	hirudin, anticoagulant protein from medical leech *Hirudo medicinalis* L., Annelida,

Table 15.II. Semisynthesis of drugs or technologically relevant agents from natural products

Area	Common or scientific name/Chemical class/Structure	Trade name/Company/phase	Precursor/Chemical class
Blood, inflammation and gout treating drugs	simvastatin/polyket. deriv. simvastatin	simvastatin/ Merck & Co	lovastatin (= mevinolin) from both *Monascus ruber* Tiegh. (1884) and *Aspergillus terreus* Thom (1918), Eumyc./polyket. lovastatin (=mevinolin)
CNS affecting drugs	bromocriptine/alkal. deriv. bromocriptine	Parlodel/ Sandoz	ergot alkal. (C 6.2.A2)
CNS affecting drugs (antitussive)	codeine/benzylisoquinoline alkal. codeine	galenic; restricted use in the US	morphine from opium/alkal. morphine
CNS affecting drugs (antitussive)	dextromethorphan/alkal. deriv. dextromethorphan	Benylin DM (Parke, Davis)	morphine from opium/alkal. morphine

Table 15.II. Semisynthesis of drugs or technologically relevant agents from natural products

Area	Common or scientific name/Chemical class/Structure	Trade name/Company/phase	Precursor/Chemical class
CNS affecting drugs	dihydrocodeinone and oxycodone/alkal. deriv. 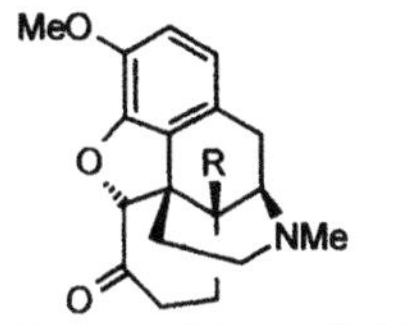 MeO O R NMe O dihydrocodeinone R=H oxycodone R=OH	dihydro-codeinone (Dicodid/ Knoll = Hydrocodone /Watson Labs); oxycodone (Dinarkon/ Pharcosar)	morphine from opium/alkal. 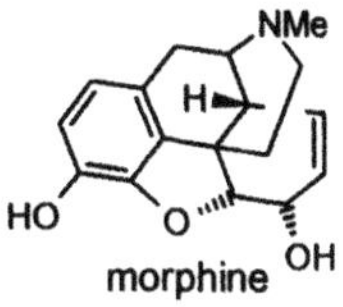 NMe H HO O OH morphine
Endocrine drugs	medroxyprogesterone/steroid deriv. 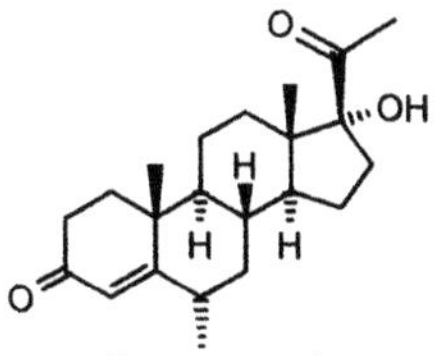 O OH H H H O medroxyprogesterone	Provera/ Upjohn; Cycrin/ Ayerst	pregnendione steroids
Endocrine drugs	norethindrone/steroid deriv. 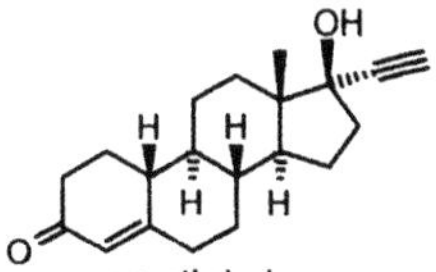 OH H H H H O norethrindone	Conludag Syntex; Norluten/ SKB	pregnenolone, via semisynthesis from androstenolone (the latter from the adrenal gland, Mamm.)/steroids 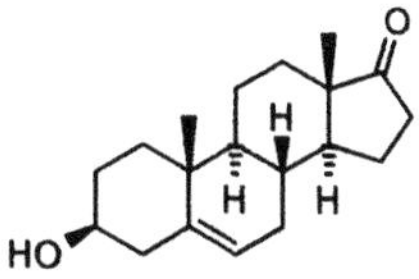 O H H H HO androstenolone
Endocrine drugs	(-)-norgestrel/steroid deriv. 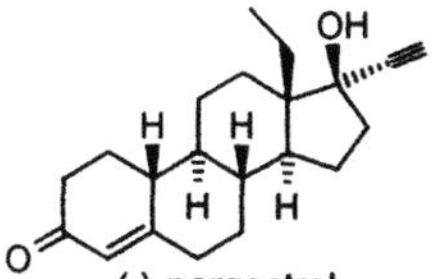 OH H H H H O (-)-norgestrel	LoOvral/ Wyeth-Ayest; Triphasil/ Wyeth; Microginon/ Schering	progesterone, from corpus luteum, Mamm./steroid 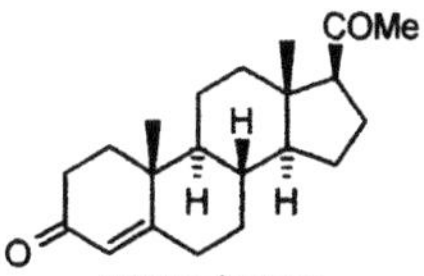 COMe H H H O progesterone

Table 15.II. Semisynthesis of drugs or technologically relevant agents from natural products

Area	Common or scientific name/Chemical class/Structure	Trade name/Company/phase	Precursor/Chemical class
Endocrine drugs	prednisone/steroid deriv. 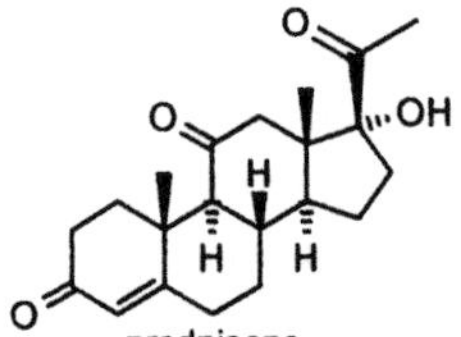 prednisone	Deltasone/ Upjohn; Deltacortone/ Merck & Co	corticosteroids
Environ- mental toxinology tools	nucleotide probes synthesized on the basis of identified sequences in D1 and D2 hypervariable regions of rRNA genes from *Alexandrium* spp. (toxic Dinofl.) are useful as genetic markers in the identification of species and bloom monitoring	Woods Hole Oceano- graphic Institution, Mass.	rRNA genes from *Alexandrium* spp., toxic Dinofl.: Anderson 1997
Oncology tools	salts prepared from agmatine; used as markers for tumor cells/alkal. deriv. 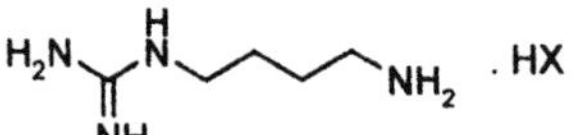 agmatine.HX		decarboxylated arginine (agmatine) from *Ambrosia artemisifolia* L., Compositae, Ang.; also **marine:** herring sperm or octopus muscle: MI/alkal.
Smooth muscle affecting drugs	beclomethasone/steroid deriv. 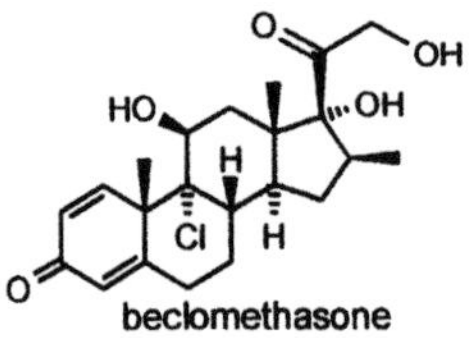 beclomethasone	Beconase AQ/Allen& Hanburys; Viarox/ Schering	corticosteroids
Smooth muscle affecting drugs	ipratropium bromide/alkal. deriv. 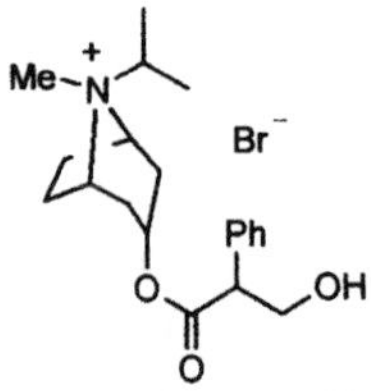 ipratropium bromide	Atrovent/ Boehringer Ing.	atropine from *Atropa belladonna* L., Ang./alkal. 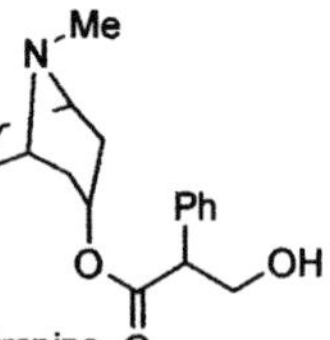 atropine

Table 15.II. Semisynthesis of drugs or technologically relevant agents from natural products

Area	Common or scientific name/Chemical class/Structure	Trade name/Company/phase	Precursor/Chemical class
Smooth muscle affecting drugs (vasodilator)	pentoxifylline/xanthine alkal. deriv. 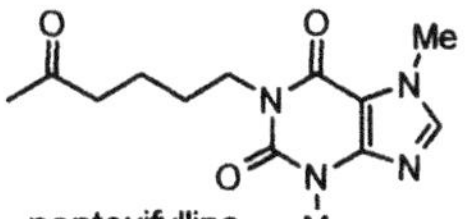 pentoxifylline	Torental/ Hoechst; Trental/ Hoechst Roussel Pharm. (off-patent)	theobromine, from cacao bean, *Theobroma cacao* L., Sterculariaceae, and other Ang./xanthine alkal. 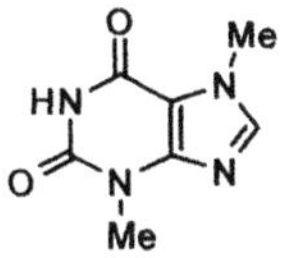 theobromine
Smooth muscle affecting drugs	triamcinolone/steroid deriv. 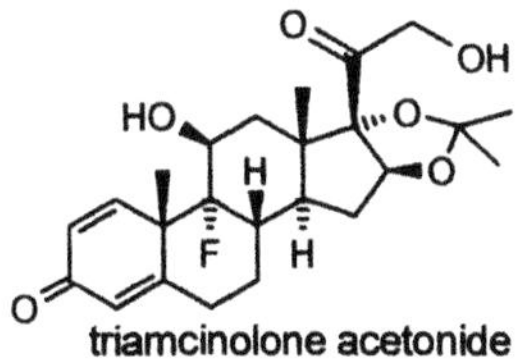 triamcinolone acetonide	Azmacort/ Rhône Poulenc Rorer; Adcortyl/ Bristol-Myers Squibb	corticosteroids

Table 15.III. Drugs and technologically relevant agents from synthesis patterned on natural products

Area	Name/class/structure	Trade name /company	Synthesis patterned on[a]
Adrenoceptor activating and antagonist drugs	timolol/alkal. deriv. 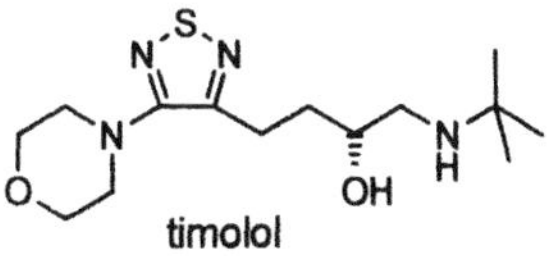 timolol	Timoptic/Merck	ephedrine, alkal. from *Ephedra* spp., Gnetaceae, Gymn. from China 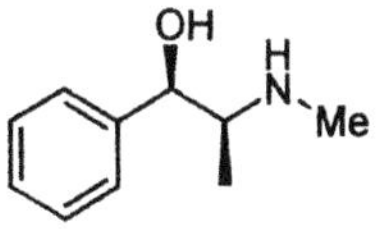 (-)-ephedrine

Table 15.III. Drugs and technologically relevant agents from synthesis patterned on natural products

Area	Name/class/structure	Trade name /company	Synthesis patterned on[a]
Analgesic drugs	ABT-594/alkal. deriv. 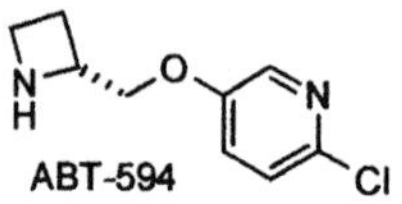 ABT-594	ABT-594/Abbott	epibatidine, alkal. from Dendrobatidae frogs: Daly 2000 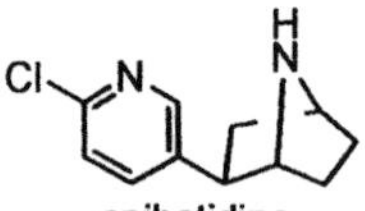 epibatidine
Antibacterial agents (against anthrax and other 14 infections)	ciprofloxacin hydrochloride/alkal. deriv. 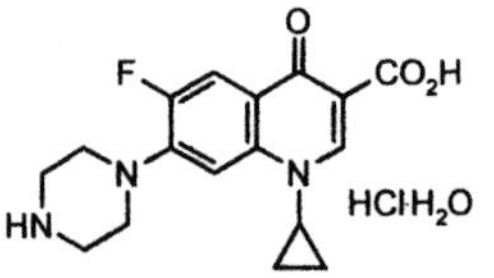 cyprofloxacin hydrochloride	Cipro/Bayer	antibacterial quinolones from bacteria
Antibacterial agents	trovafloxacin mesylate (oral) and alatrofloxacin mesylate (injection)/alkal. deriv. 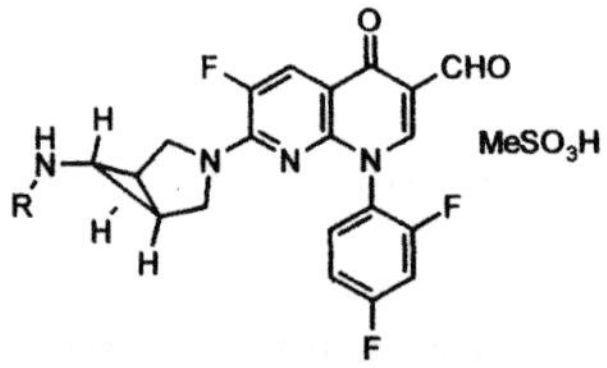 trovafloxacin mesylate R=H alatrofloxacin mesylate 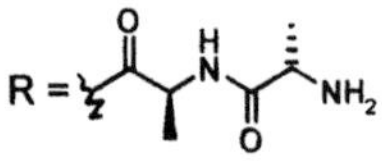 	Trovan and Trovan I.V., respect./Pfizer	antibacterial quinolones from bacteria; the L-alanyl-L-alanine deriv. is converted to trovafloxacin on intravenous administration; approved in the US in 1997 as a last resort against certain bacterial infections, because of serious side effects, particularly liver injury
Antidiabetic agents	insulin aspart (modified human insulin)	Novolog /Novo Nordisk	human insulin

Table 15.III. Drugs and technologically relevant agents from synthesis patterned on natural products

Area	Name/class/structure	Trade name /company	Synthesis patterned on[a]
Antiparasitic agents	chloroquine/alkal. deriv. 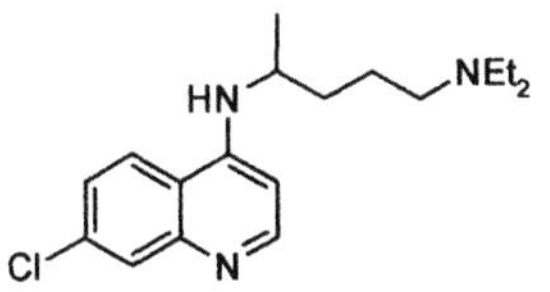 chloroquine	Tresochin/Bayer; Imagon/Astra (it was until recently the best antimalarial drug, now rendered useless in most areas by drug resistance acquired by the parasite)	quinine 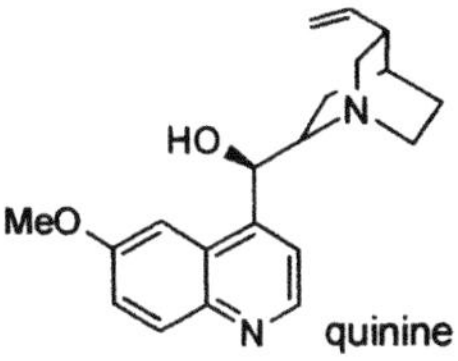 quinine
Antiparasitic agents	mefloquine/alkal. deriv. 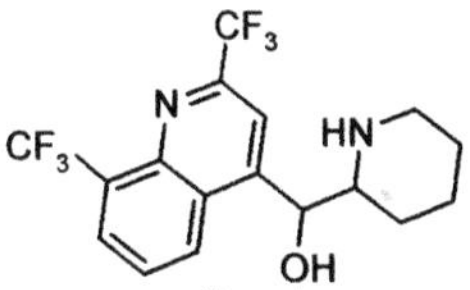 mefloquine	Lariam/Roche; Fansimef/Roche	quinine, alkal. from trop. S American *Cinchona* spp. Rubiales, Ang.
Antiviral agents (anti-HIV)	abacavir succinate/nucleoside analogue 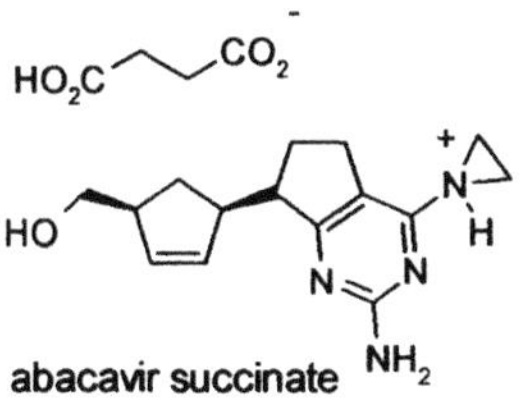 abacavir succinate	Ziagen/Glaxo-Wellcome	**marine:** nucleosides from Demosp., Porif.
Antiviral agents (anti-HIV)	acycloguanosine/nucleoside analogue 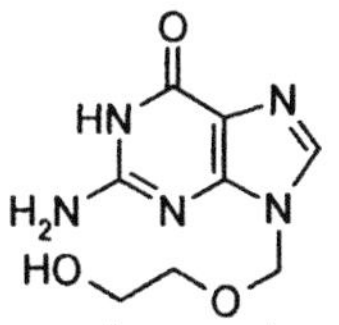 acycloguanosine	Acyclovir/Wellcome	**marine:** nucleosides from Demosp., Porif.

Table 15.III. Drugs and technologically relevant agents from synthesis patterned on natural products

Area	Name/class/structure	Trade name /company	Synthesis patterned on[a]
Antiviral agents (anti-HIV)	adenine arabinoside (Ara-A = vidarabine)/nucleoside analogue 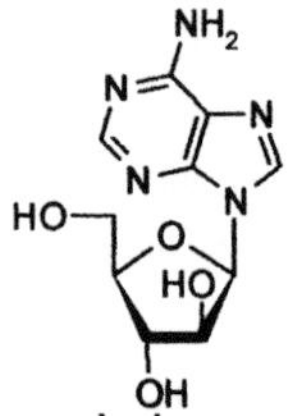	Vira-A/Parke-Davis	fermentation of *Streptomyces antibioticus* (Waksman and Woodruff 1941) Waksman and Henrici 1948, Actinom., Bact.; also **marine:** nucleosides from Demosp., Porif.
Antiviral agents (anti-HIV)	azidothymidine (AZT)/nucleoside analogue 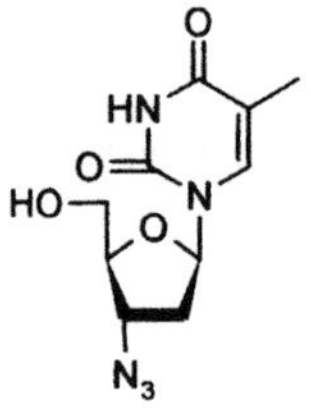	Retrovir/Glaxo-Wellcome	**marine:** nucleosides from Demosp., Porif.
Antiviral agents (anti-HIV)	2',3'-didehydro-3'-deoxythymidine (stavudine)/nucleoside analogue 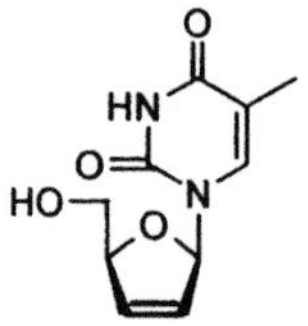	Zerit/Bristol-Myers Squibb	**marine:** nucleosides from Demosp., Porif.

adenine
arabinoside

azidothymidine

2',3'-didehydro-3'-
deoxythymidine

Table 15.III. Drugs and technologically relevant agents from synthesis patterned on natural products

Area	Name/class/structure	Trade name /company	Synthesis patterned on[a]
Antiviral agents (anti-HIV)	dideoxycytidine (zalcitabine)/nucleoside analogue 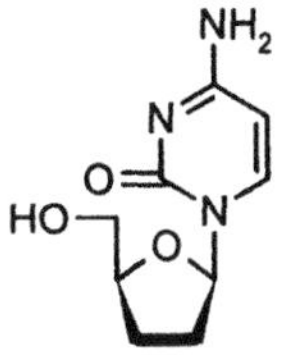	Hivid/Hoffmann-La Roche	**marine:** nucleosides from Demosp., Porif.

dideoxycytidine

Antiviral agents (anti-HIV)	dideoxyinosine (didanosine)/nucleoside analogue 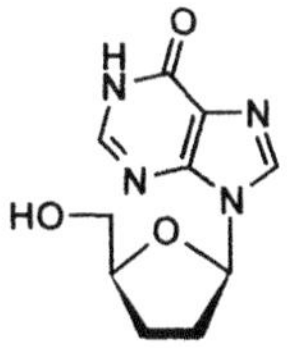	Videx/Bristol-Myers Squibb	**marine:** nucleosides from Demosp., Porif.

dideoxyinosine

Antiviral agents (anti-HIV)	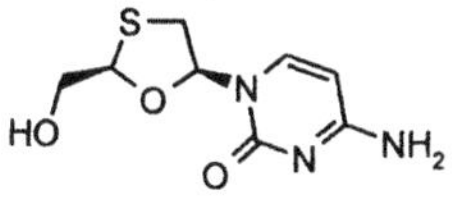 lamivudine/nucleoside analogue	Epivir/IAF Biochem. International and Glaxo-Wellcome	**marine:** nucleosides from Demosp., Porif.

lamivudine

Antiviral agents (anti-HIV protease inhibitors)	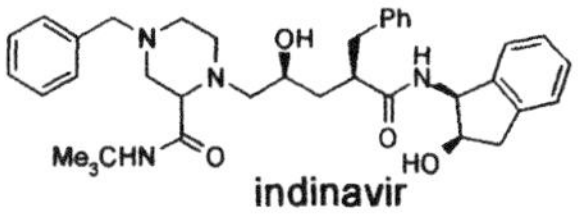 indinavir/pept.	Crixivan (MK-639)/Merck & Co	pepstatin (isovaleryl-Val-Val-AHMHA-Ala-AHMHA, where AHMHA= (3*S*, 4*S*)-4-amino-3-hydroxy-6-methyl-heptanoic acid) isolated from *Streptomyces testaceus* and *Streptomyces argenteolus* Tresner *et al.*, 1961, Actinom., Bact.

indinavir

Table 15.III. Drugs and technologically relevant agents from synthesis patterned on natural products

Area	Name/class/structure	Trade name /company	Synthesis patterned on[a]
Antiviral agents (anti-HIV protease inhibitors)	amprenavir/sulfonamide 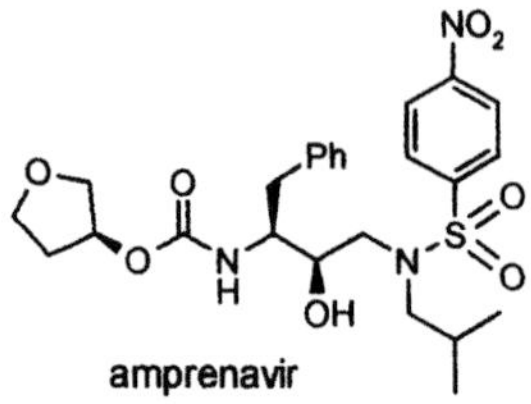 amprenavir	Agenerase/Vertex, Kissei, and Glaxo-Wellcome	synthetic, from rational drug design
Antiviral agents (anti-HIV protease inhibitors)	saquinavir/alkal. pept. 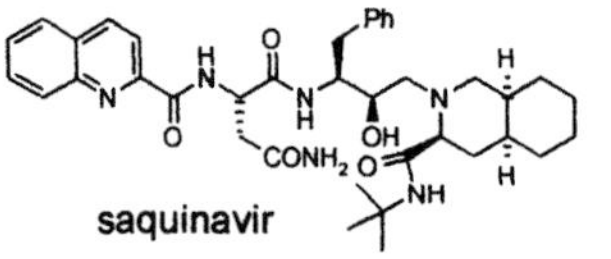 saquinavir	Fortovase/Roche	synthetic, transition state analogue
Anticancer drugs	curacin A analogue (identification via solution-phase combinatorial synthesis; mass production by individual synthesis: Wipf 2000B)/fatty acid deriv. 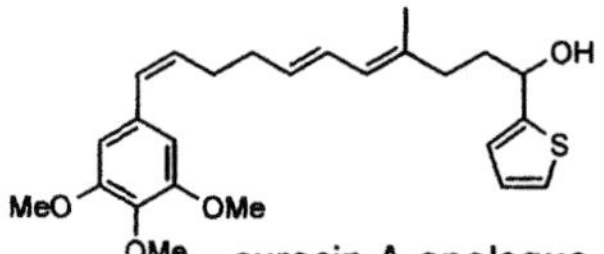 curacin A analogue	in development	curacin A 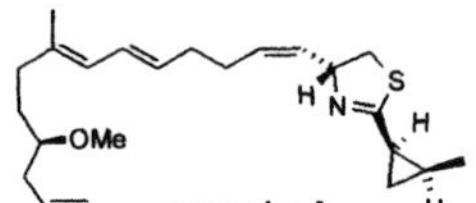 curacin A
Anticancer drugs	2-chlorodeoxyadenine (cladribine)/nucleoside analogue 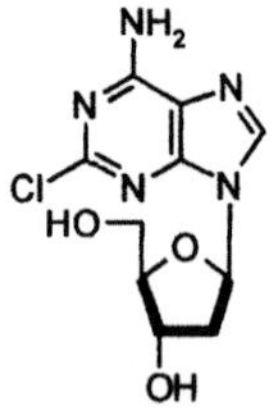 2-chlorodeoxy-adenine	Leustatin/Ortho Biotech	**marine:** nucleosides from Demosp., Porif.

Table 15.III. Drugs and technologically relevant agents from synthesis patterned on natural products

Area	Name/class/structure	Trade name /company	Synthesis patterned on[a]
Anticancer drugs	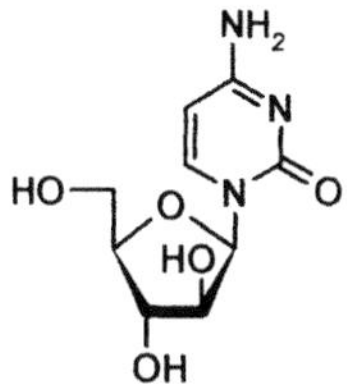 β-cytosine arabinoside (cytarabine)/nucleoside analogue	Ara-C/Alexan/Pfizer; Arabitin/Sankyo	**marine:** nucleosides from Demosp., Porif.

NH₂ structure labeled:
β-cytosine arabinoside

| Anticancer drugs | benzothiazole analogue of deoxyepothilone B with improved antiproliferative activity/macrolide | in development/Novartis Pharma AG: Altmann 2000 | deoxyepothilone B from *Sorangium cellulosum*, Myxobact. |

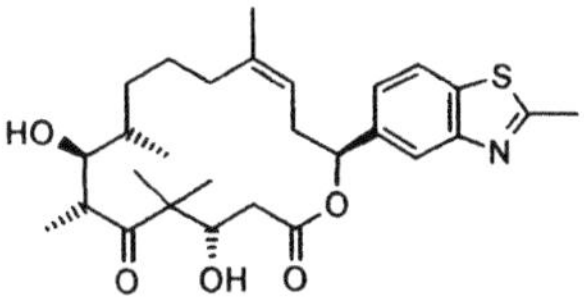

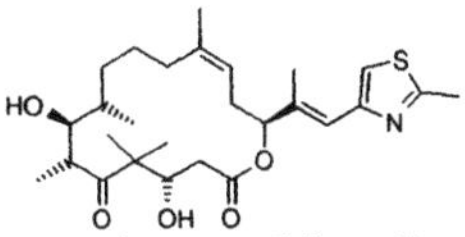

benzothiazole analogue of deoxyepothilone B

deoxyepothilone B

| Anticancer drugs | flavopiridol/shikim. deriv.s 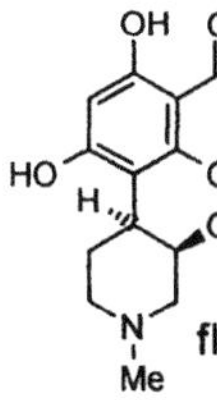 | clinical phase II: Newman 2000 | rohitukine, a flavonoid from tropical *Dysoxylum binectariferum*, Meliaceae, Ang. 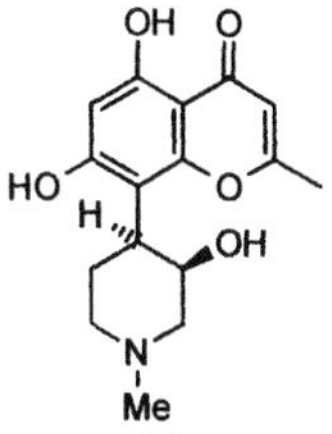 |

flavopiridol

rohitukine

Table 15.III. Drugs and technologically relevant agents from synthesis patterned on natural products

Area	Name/class/structure	Trade name /company	Synthesis patterned on[a]
Anticancer drugs	2-fluoro-9-β-D-arabinofuranosyladenine (fludarabine)/nucleoside analogues 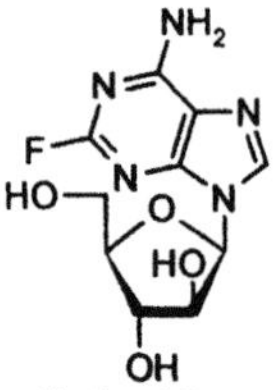	Fludara/Triton	**marine:** nucleosides from Demosp., Porif.

fludarabine

| Anticancer drugs | goserelin | Zoladex/ICI | luteinizing hormone - releasing pept., from Mamm. |

5-oxoProl-His-Try-Ser-Tyr-DSer(t-Bu)
|
H₂NOCHNHN-Prol-Arg-Leu
goserelin

| Anticancer drugs | halichondrin B right half diol/polyether polyket.: Stamos | waiting for development | **marine:** halichondrin B from Demosp., Porif. |

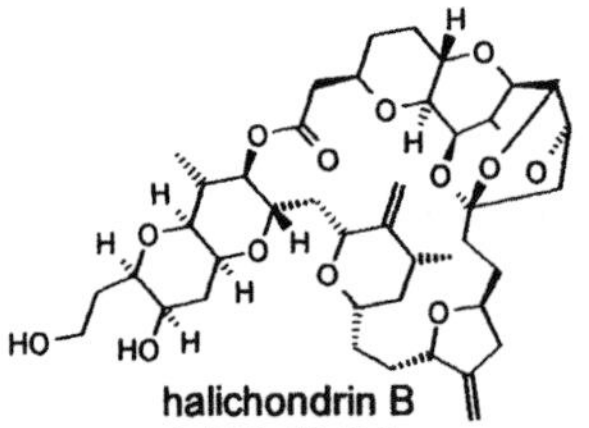

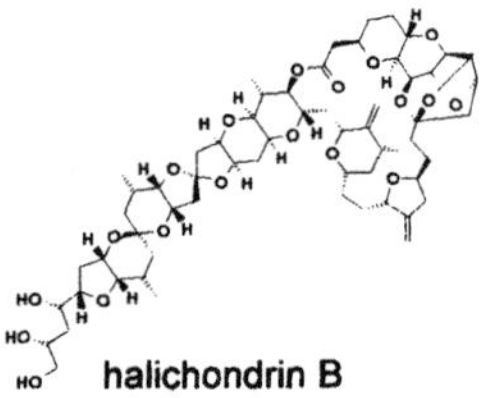

halichondrin B
right half diol

halichondrin B

1997

| Anticancer drugs | 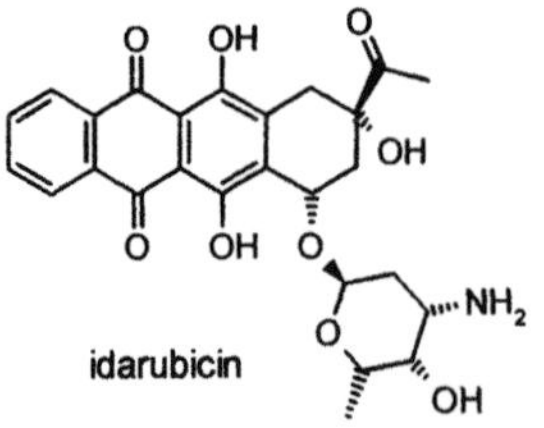 idarubicin/anthracycline glycoside | Zavedos/Farmitalia; Idamycin/Adria | daunomycin from *Streptomyces peucetius*, Actinom., Bact. 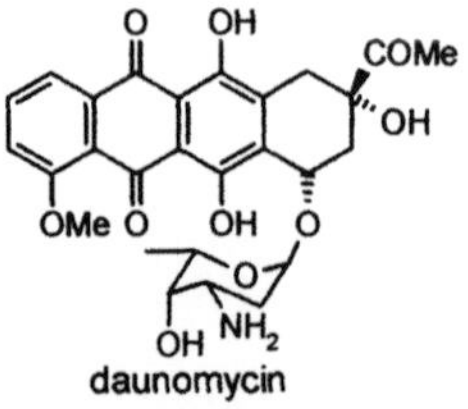 |

idarubicin

daunomycin

Table 15.III. Drugs and technologically relevant agents from synthesis patterned on natural products

Area	Name/class/structure	Trade name /company	Synthesis patterned on[a]
Anticancer drugs	phthalascidin/alkal. deriv.: Martinez 2000 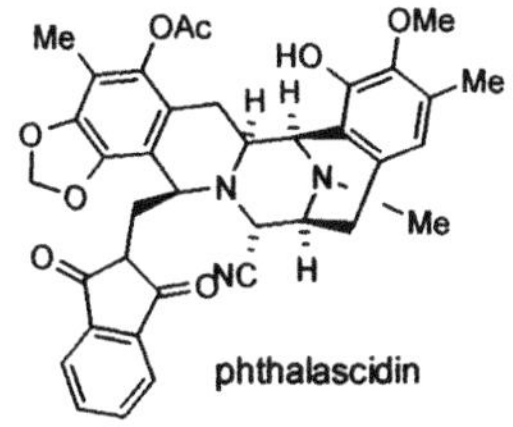 phthalascidin	waiting for development	**marine:** ecteinascidin 743 from Ascid. 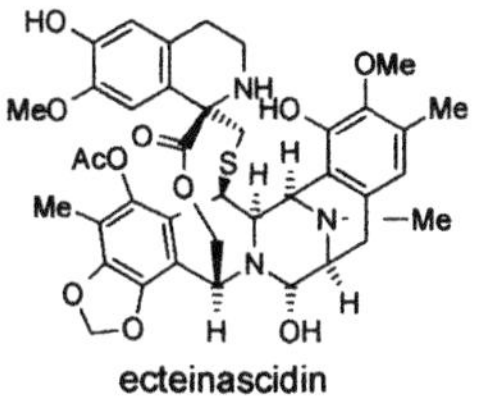 ecteinascidin
Anticancer drugs	sarcodictyin A analogue/diterp. 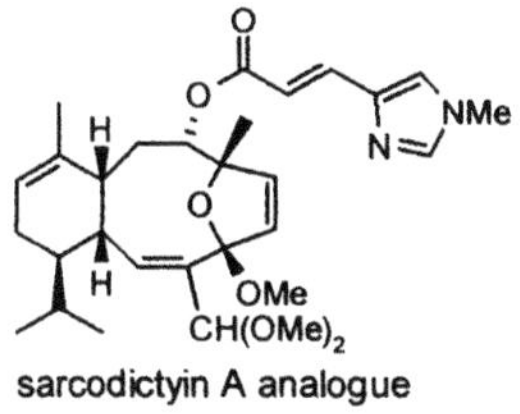 sarcodictyin A analogue	waiting for development	**marine:** combinatorial synthesis on sarcodictyin A from *Sarcodictyon roseum* (Philippi, 1842) and other Cnid.: Nicolaou 2000A 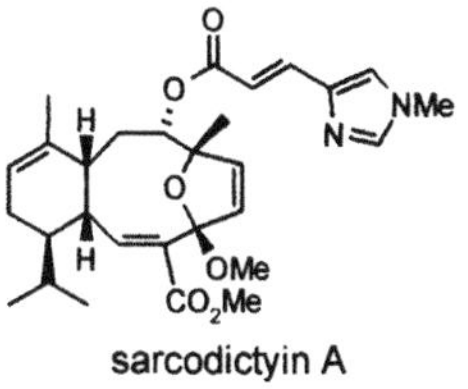 sarcodictyin A
Autonomic drugs	methylphenidate/alkal. deriv. 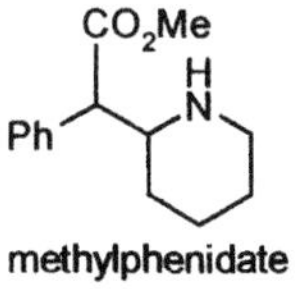 methylphenidate	Ritalin/Ciba; most prescribed in the US for attention deficit hyperactivity disorder in children (even under 6, though not approved for them by FDA)	ephedrine 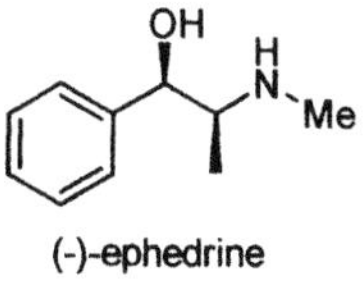 (-)-ephedrine
Blood, inflammation and gout treating drugs	tirofibian/sulfonamide 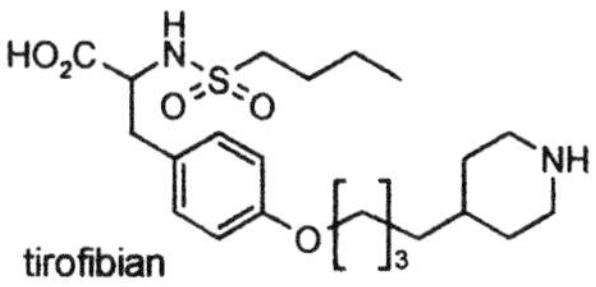 tirofibian	Aggrastat/Merck & Co (useful in conjunction with heparin and aspirin against acute coronary syndrome)	African saw-scaled viper venom

Table 15.III. Drugs and technologically relevant agents from synthesis patterned on natural products

Area	Name/class/structure	Trade name /company	Synthesis patterned on[a]
Blood, inflammation and gout treating drugs	eptifibatide/cyclic heptapept. 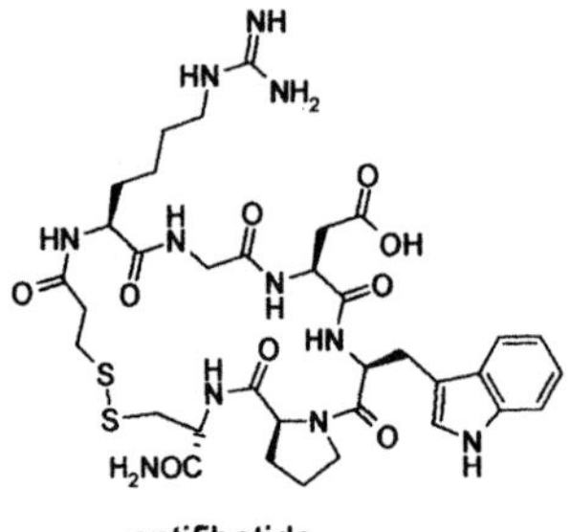 eptifibatide	Integrilin/COR Therapeutics and Schering Plough	SE pigmy rattlesnake venom
Cardio-vascular-renal drugs	captopril/proline deriv. 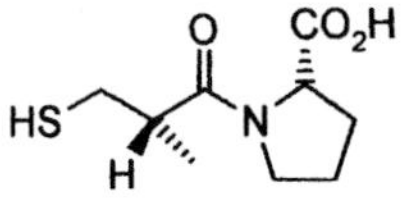 captopril	Captoten/Bristol-Myers Squibb; Acediur/Guidotti	C-terminal proline groups of teprotide, a nonapept. isolated from venom of the S American pit viper (*Bothrops jararaca* (Wied, 1824), Serpentes, Reptilia) 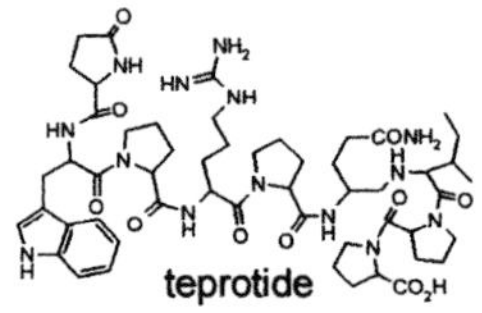 teprotide
Cardio-vascular-renal drugs	metoprolol tartrate/phenolic 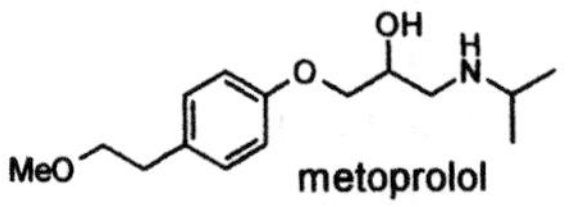 metoprolol	Lopressor/Ciba-Geigy; Beloc/Astra	ephedrine, from *Ephedra* spp., Gnetaceae, Gymn. from China 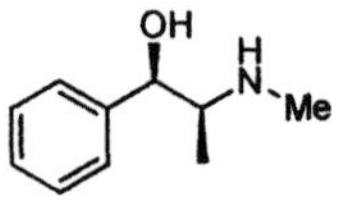 (-)-ephedrine
Cardio-vascular-renal drugs	nadolol/alkal. deriv. 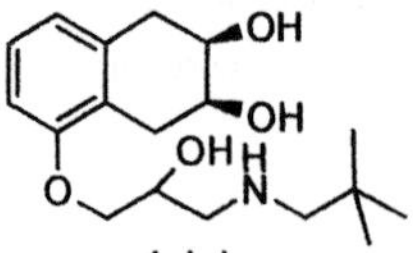 nadolol	Corgard (= Solgol)/Bristol-Myers Squibb	ephedrine

Table 15.III. Drugs and technologically relevant agents from synthesis patterned on natural products

Area	Name/class/structure	Trade name /company	Synthesis patterned on[a]
Cardio-vascular-renal drugs	propranolol/aminoalcohol deriv. 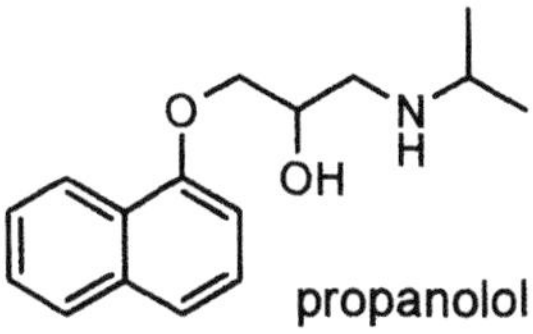	Angilol/DDSA	ephedrine
Cardio-vascular -renal drugs	verapamil/alkal. deriv. 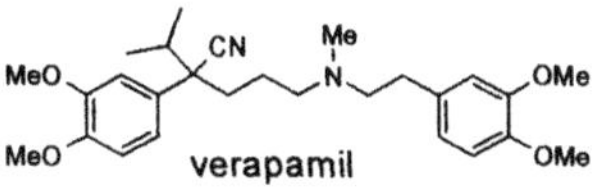	Verapamil, Calan/Searle; Verelan/Lederle	ephedrine
CNS drugs	(+)-amphetamine/alkal. deriv. 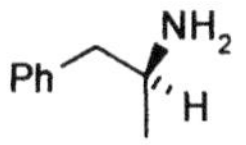	Alentol/SK & F	cathinone from *Catha edulis* Forsk., Celestraceae from E Africa 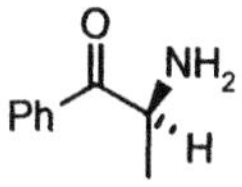
CNS drugs	cabergoline/alkal. deriv. 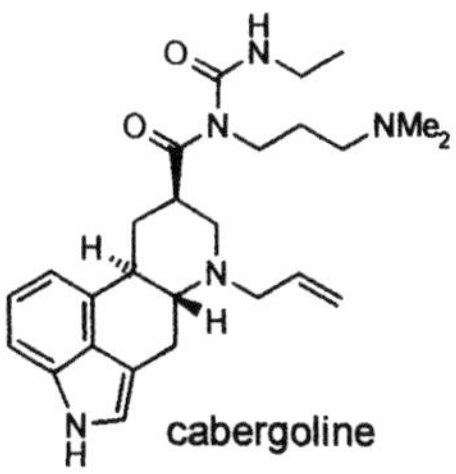	Cabaser/Pharmacia & Upjohn	ergot alkal. from *Clavices purpurea* (Fries) Tul., Eumyc./C 6.2.A2

propanolol

verapamil

(+)-amphetamine

cathinone

cabergoline

Table 15.III. Drugs and technologically relevant agents from synthesis patterned on natural products

Area	Name/class/structure	Trade name /company	Synthesis patterned on[a]
CNS drugs	[^{11}C]β-CPPIT for selective binding to the serotonin transporter for positron emission tomography/alkal. deriv. 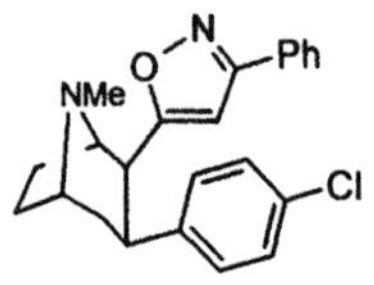 β-CPPIT	β-CPPIT/ETH Zürich: Ametamey 2000	cocaine from *Erythroxylum coca* Lamarck 1786, Erythroxylaceae, Ang. from hills in tropical SW America 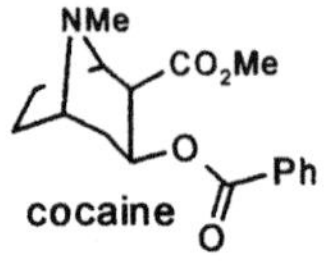 cocaine
CNS drugs	ω-conotoxins/polypept. forming a β-sheet made of three short anti-parallel strands; inside the folded structure are mainly found four of the disulfide-bonded cysteine residues: Kohno 1995, and its synthetic version SNX-111 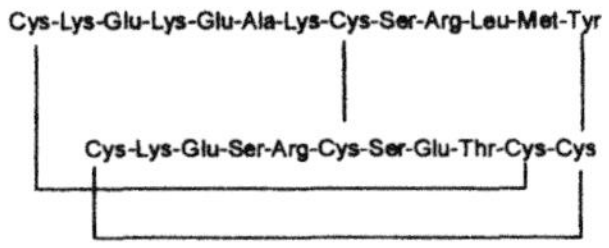 amino acid sequence and disulfide bridges of ω-conotoxin MVIIA	ω-conotoxins, such as ω-conotoxin MVIIA/Neurex (Elan Pharm.)	**marine:** polypept. from *Conus* spp., Moll. (phase III)
CNS drugs	neostigmine/alkal. derivative 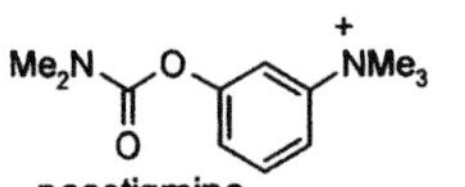 neostigmine	Hoffman-La Roche	physostigmine 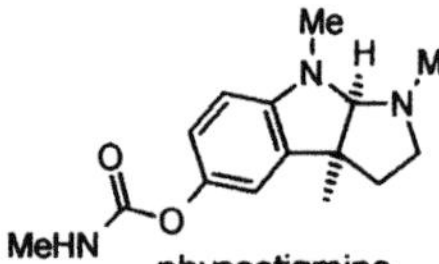 physostigmine
CNS drugs	NPS 1506/modified protein	NPS 1506/NPS Pharm (US)	neuroprotecting proteins from spider venoms, Arach. (clinical assays suspended in 1999 because of too high cost)

Table 15.III. Drugs and technologically relevant agents from synthesis patterned on natural products

Area	Name/class/structure	Trade name /company	Synthesis patterned on[a]
CNS drugs	rivastigmine/alkal. deriv. 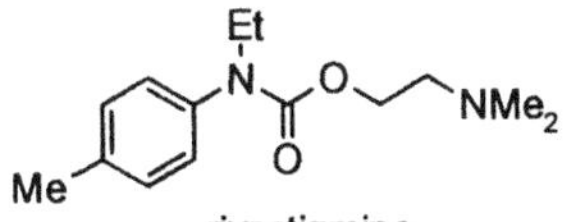 rivastigmine	Exelone/Sandoz (patent), Novartis (manufacturer)	physostigmine, from Nigerian *Physostigma venenosum* Balf., Leguminosae, Ang./indole alkal. 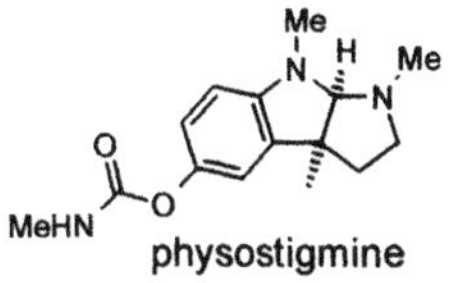 physostigmine
Cosmetics and fragrances	many companies are chiefly concerned with research in this area	hydroxypropyl cellulose, scent of the sea, and other base compounds for cosmetics and fragrances patterned on natural products	C 13.4
Drugs of abuse	3,4-methylenedioxy-methamphetamine 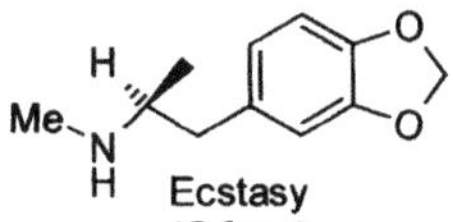 Ecstasy (*S* form)	Ecstasy (the *S* form is particularly active as a hallucinogen)	ultimately cathinone, from *Catha edulis* Forsk., Celestraceae from E Africa, which served as a template for the amphetamines 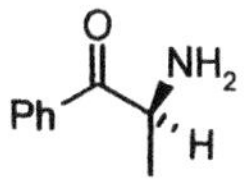 cathinone
Endocrine drugs (antidiabetic agents)	AC2993/pept. structurally correlated to the human hormone glucagon-like peptide-1	Amylin Pharm.	exendin-4, a 39-amino acid pept. from the lizard Gila monster, *Heloderma suspectum*, Sauria, Reptilia from Arizona and New Mexico: Harvey 2000; phase II clinical evaluation

Table 15.III. Drugs and technologically relevant agents from synthesis patterned on natural products

Area	Name/class/structure	Trade name /company	Synthesis patterned on[a]
Endocrine drugs	leuprolide/pept. deriv. 5-oxoProl-Hist-Try-DLeu-Leu │ H₅C₂HNProl-Arg leuprolide	Prostap/Lederle (= Carcinil/Abbott)	luteinizing hormone, Mamm.
Ocular disorders treating drugs	verteporphin/mixture of porphyrin deriv.  verteporphin R'=Me, R=H CO₂R' CO₂R R'=H, R'=Me	Visudyne/Ciba Vision-Novartis (phase III completed in 1999 for age-related macular degeneration; also in trial for psoriasis and leukemia) (see also timolol above, an antiglaucoma agent)	porphyrins
Orthopedics	synthetic polymer	Norian SRS/Norian	**marine:** synthesis patterned on hermatypic scleractinian skeletons: Constanz 1995
Smooth muscle affecting drugs	albuterol/alkal. deriv. 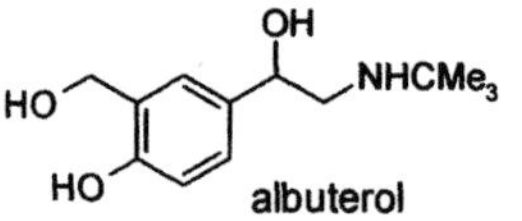 albuterol	Salbumol/Glaxo Wellcome; Proventil/Schering	ephedrine from *Ephedra* spp., Gnetaceae, Gymn. from China 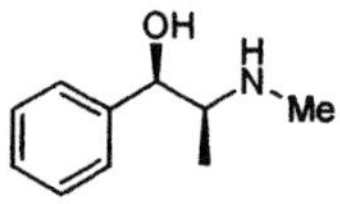 (-)-ephedrine
Smooth muscle affecting drugs	cromolyn sodium/flavonoid analogue 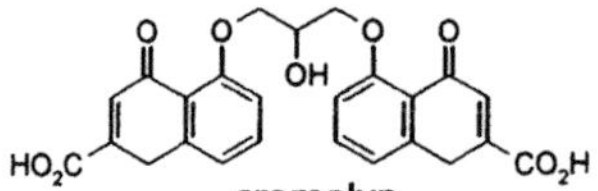 cromolyn	Intal/Fisons; Aarane/Synthex	flavonoids from Ang. (C 8.3.S)

Table 15.III. Drugs and technologically relevant agents from synthesis patterned on natural products

Area	Name/class/structure	Trade name /company	Synthesis patterned on[a]
Smooth muscle affecting drugs	guaifenesin/the shape only of the template alkal. was conserved OH, O, OH, OMe guaifenesin	Actifed-C/Burroughs Wellcome	ephedrine OH, H, N, Me (-)-ephedrine
Tire industry	copolymer of styrene and butadiene in 1:3 ratio	SBR Rubber	caoutchouc (isopr. polymer)

[a]Land origin is implied, unless marine origin is indicated in boldface.

Part VI. Threatening and management of natural product diversity

Quick are the mouths of Earth, and quick the teeth that fed upon this loveliness.

Thomas Wolfe

Chapter 16. Threatening natural product diversity

Evolution is accompanied by changes in biodiversity, as a result of nature imposing a limit on the existence of species. Abrupt disappearance of species and whole higher taxonomic groups also occurred during the history of Earth, as in the Alvarez's story.

These are processes where a certain set of biodiversity disappears, leaving space for another set. Abundant fossil remains speak for the turnover of species. What happened to natural products is more difficult to appreciate because fossil molecules are only rarely found, or have suffered from diagenetic transformations.

With the coming of man, the disappearance of species and metabolites took a steep course, characterized by the non recovery of the diversity. Early man was a hunter, particularly of large animals, but the greatest impact on biodiversity started with agricultural practices and the colonization of small islands; it took a non linear trend with the great expeditions of the Renaissance.

It is not only farming and industrialization that impoverishes the habitat. Forests are replaced by ponds in the Amazon region, fostering the spread of malaria (Walsh 1993). Substitution of humid land by dry cultures - fostered by the European Community - is against migratory aquatic birds. Evidence has also accumulated that loss of biodiversity in the European grassland turns out in a lower productivity (Hector 1999). Similarly, seed germination in tropical rain-forest fragments is less efficient than in the uninterrupted forest (Bruna 1999). The higher productivity of ecosystems rich in plant diversity has been linked to complementary patterns in the use of resources by the species (Loreau 2001). The Earth is becoming overpopulated and globalization is undergoing a major upheaval in our time. These are causes of an even redistribution of species and the disappearance of those that, like the aquatic migratory birds, need special habitats in close proximity to one another. The phenomenon has attained the size of mass extinction, the first man-made mass extinction.

If biodiversity decreases, natural product diversity also decreases, because of a link between the two (Part II). Modern times have seen the greatest natural product diversity. With the increase in plant species from 100,000 before the end-Cretaceous mass extinction to the 250,000 species of modern age, natural product diversity must also have increased. The estimate depends on how natural product diversity is evaluated, i.e. from the molecular skeletons or the actual metabolites. Metabolites of insects, fungi, and bacteria must be added to the list. The present man-made mass extinction may thus result in a larger numerical impact than any previous mass extinction, even if the loss of natural product diversity occurs at the same percentage of previous catastrophes.

Associated organisms have also to be considered. A large tree may host as many as 500 different species of insects, part of which go extinct with the tree. It is also believed that the diversity of symbionts correlates with that of the host, such as mycorrhizal fungi with terrestrial plants (van der Heijden 1998). Pathogens and herbivores may also control the plant species distribution, both in the tropics and temperate areas (Packer 2000).

Non-sustainable consumption of all available resources is the basis of threats to biodiversity and natural product diversity. Unfortunately, this gets *ad hoc* subsidies by governments, preventing recourse to alternative renewable energy (Myers 2000). It is the very concept of economic growth, as currently understood, that turns out against natural product diversity. Those who rule are often blurred by much misconception, such as in a proposal in my country that hounds and pointers are no more brought up, in order to preserve wild animals. This would endanger hounds, pointers, and an art that has ancient roots (Galloni 2000), while the wild animals, lacking the financial support of the hunters, would be threatened more than ever.

Starting from the far past (Chapter 16.1), until our days (Chapters 16.2.1-16.2.4), threats to natural product diversity are examined in the Tables 16.2.5.II-V, although disentangling the causes is seldom possible. A combination of various factors is usually responsible for the disappearance of metabolites. Threatening of natural product diversity has mostly political causes, first of all in permitting an indiscriminate use of the car. Therefore, the division of land into biomes, and the biological zonation of the oceans, have been abandoned in Tables 16.1.5.II-V in favor of the geopolitical subdivision.

16.1 Fossil molecules and past natural product diversity

Fossil remains as old as 3,000 My have been found of the precursors of modern cyanobacteria. The oldest remains of eukaryotes date to 2,500 My. Thus, together with many other landmark findings, the history of fossil remains is illustrative of the past events, in particular mass extinctions, which are dramatic episodes of biodiversity loss from unusual causes.

The first large mass mortality occurred during the end-Ordovician, 440 My ago (Fig. 16.1). A less extensive loss took place during the late Devonian, followed by the huge mass mortality of the end-Permian, 250 My ago, when the trilobites - marine arthropods that had survived the late Ordovician event - disappeared. Coming after a period of dry cold climate that saw the Appalachian mountains build up, the end-Permian mass mortality brought to extinction more than half the families of living species. It paved the way to the scleractinians, which formed the coral reefs (Stanley 2001).

After a warm period, in the Pangea/Panthalassa arrangement of Earth about 210 My ago, the end-Triassic mass mortality caused the disappearance of mammal-like reptiles leaving place to the dinosaurs, the first angiosperms, and the first mammals.

The end-Cretaceous mass mortality took place at a time, 65 My ago, that all major oceans were already in place, while India was still an island of the Indian Ocean. Perhaps caused by a great meteoric impact, this phenomenon brought extinction to the dinosaurs, leaving Earth fully open to the expansion and diversification of mammals.

All these were major causes of temporary loss of biodiversity, as arguably a unique class of phenomena or episodes of a continuum of physically induced transitions (Miller 1998) that had a great impact on the evolution of, and the competition among, the living things on Earth. That the competition was always serious is testified by the early Cambrian remains of arthropod-like predators in the genus *Anomalocaris*, which attained a staggering 2 meters (Briggs 1994).

Unraveling the history of natural products, how they arose and disappeared during the geological eras, is more difficult. The examination of the few fossil molecules that have been isolated allows us only scattered deductions that may be complemented by bold speculation (Pietra 1995). It is true that the analysis of ^{13}C enrichment aids the examination of fossil molecules, allowing us to discriminate C_3/C_4 biosynthetic routes of plants from those of microorganism (Hartgers 1994), but the pattern is blurred by diagenetic transformations that have occurred during the eras, catalyzed by clay. In the water column, steryl esters of pheophorbide-a were formed from chlorophyll-a, triggered by microbial enzymes. These esters have a wide distribution, in recent and ancient sediments, and in senescent

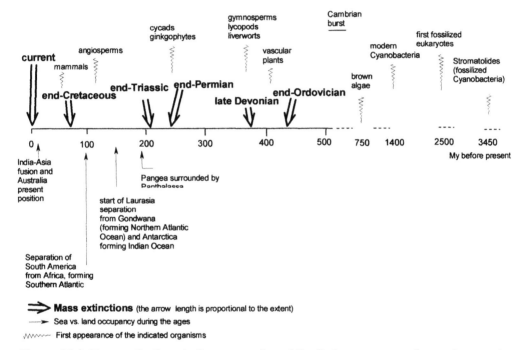

Figure 16.1. Mass extinctions in the perspective of the first appearance of organisms and sea vs. land occupancy

phytoplankton (Pearce 1993).

Representative examples of fossil molecules from the geological eras, from Precambrian to recent times, are shown in Chart 16.1. The molecular skeletons are displayed according to the strategy set forth in Part III; this is a minor change with respect to the actual fossil molecules, since most functionalization was lost in the diagenetic transformations. Blank spaces reflect our limited knowledge of fossil molecules.

Orphan skeletons are emphasized in boldface characters. Found both in the Tertiary and the Triassic eras, they have no equivalent in extant organisms and cannot be accounted for by diagenetic transformations alone. They comprise xanthenoxanthene quinones, quincyte pigments of likely polyketide origin (Prowse 1991), and three series of isoprenoids: polycyclic hydrocarbons from the

German Eocene (Schaeffer 1994A), tricyclopolyprenanes from Tasmanian tasmanite of Permian age (Schaeffer 1994B), and *S*-containing polycyclics from an Italian Triassic marl (Poinsot 1997). Judging from the structure, these isoprenoids may have originated from plants that probably do not exist anymore, or have undergone a dramatic biochemical change.

Most other fossil terpenoids in Chart 16.1 belong to skeleton classes known from extant plants, such as pre-Cambrian phytane (Blumer 1965), Permian arborane and fernane (Hauke 1992), Tertiary oleanane, ursane, and lupane, and their truncated derivatives (Trendel 1991), and recent botryococcane (Huang 1995B), rimuane (Huang 1997), and cheilantanes (Behrens 2000). Steranes, in particular cholestanes, isolated from 2,700 My old shales from the Pilbara Craton in Australia, suggest that eukaryotes existed before the dating from fossil remains (2,500 My, see above) (Brocks 1999). In general, molecular clock estimates suggest earlier colonization of Earth than was thought from fossil remains (Heckman 2001).

Hopanes, a most widespread and abundant class of triterpenes, and some diffused gammacerane triterpenes, deserve a special comment. It is believed that these terpenes came first from Cyanobacteria, in particular 2α-methylhopanes, found in the above said 2,700 My old Australian shales. This implies that oxygenic photosynthesis was already in place at those early times, slowly forming what much later became the dioxygen rich atmosphere (Brocks 1999). Hopanes came later also from protists (Ourisson 1992), which have also left tetrahydrophenanthrene derivatives of tetrahydrohymanol in German Jurassic sediments (Damsté 1999). Hopane and gammacerane triterpenes are also plant products (Ourisson 1992). Hopanoids found in an extant homoscleromorph sponge, *Plakortis simplex*, from the Caribbean most likely have a dietary origin (Costantino 2001).

The extant chlorophycean alga *Botryococcus braunii* is known to produce botryococcene in huge amounts (Huang 1995B). In this light, the isolation of 1,6,17,21-octahydrobotryococcene from sediments from the last ice age (1,262–40,465 years old) and sacredicene from interglacial sediments (20-10,280 years old) must reflect evolutionary changes undergone by these green algae (Huang 1995A). ^{13}C Fractionation data suggest that also carotenoids from Eocene sediments have algal rather than bacterial origin (Hartgers 1994).

Echinoderms have left fringelite pigments in Triassic sediments (Chart 16.1). Similar compounds have been found both in living fossil crinoids from seamounts of New Caledonia and extant angiosperms (Pietra 1995). Conservation of genes in extant echinoderms may be advocated to explain these findings, whereas convergence is the most likely explanation for fringelite pigments in plants. Chlorins from chlorophylls, and etioporphyrins, have also been found in abundance in sediments from the same period (Blumer 1965). Chlorins are also a proxy for Quaternary marine productivity (Harris 1996). Diatoms produce C_{25} and C_{30} isoprenoids which are also present in recent marine sediments (Chart 8.2.I) (Belt 2001), along with steroids of animal origin (Smallwood 1999).

Identities and similarities between terpenoids from plants and extant anthozoans suggest that cnidarians developed, during the Cambrian burst, the ability to make terpenoids, which was later acquired by the plants (Pietra 1995). Evidence from fossil molecules is missing, however.

Chart 16.1 Molecular-fossil skeletons (earliest occurrence in My from the present; names of orphan skeletons are displayed in boldface; dashed bonds may be present or not (A: *S* max=65, av=53; *S/H* max=0.68, av=0.57. I: *S* max=54, av=32; *S/H* max=0.81, av=0.39. FA/PO: *S* max=60, av=53; *S/H* max=0.81, av=0.79).

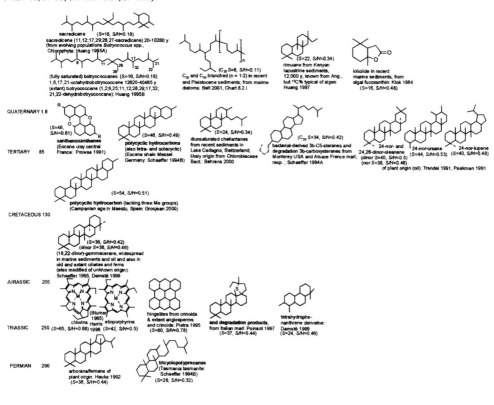

The bryozoans remain a mysterious group because of missing fossil molecules. The paleontologists have determined that the cheilostomes radiated about 100 My ago, when the cyclostomes, which where already in place, regressed, and that the cheilostomes recovered better than the cyclostomes from the End-Cretaceous mass extinction (McKinney 1998). Extant cheilostomes, especially in the genera *Bugula, Flustra,* and *Myriapora,* have given many pharmacologically promising natural products, while the cyclostomes are much less productive. Whether this explains the efficient radiation of cheilostomes is left to speculation.

16.2 Endangered natural products

Man has always exerted a heavy pressure on biodiversity. Evidence has accumulated that wherever the early human societies arrived, forest regressed, biodiversity was reduced, and the soil was eroded, while remedies were only sporadically applied (Moore 1998). Man appeared 100,000 years ago in his modern form, and was the main cause of the disappearance of the mammoth 11,000 years ago (Fig. 16.2). The practice of cutting and burning, introduced along with agriculture 7,000 years ago, resulted in extensive deforestation, which took a massive form about 300 years ago with the colonization of North America by the Europeans.

Locally, the most acute episodes of biodiversity threatening took place in isolated islands rich in endemism. An example is the colonization of Hawaii by the Polynesians, 1,600 years ago, which brought to extinction more than 1,000 species of endemic non-flying birds.

In our time, man's pressure on biodiversity has intensified because of rapid demographic expansion - with a staggering increase from 2.5 billion to 5.7 billion individuals in the last fifty years - and mechanization. Heavy incidence of malaria on newborns and spreading of AIDS notwithstanding, peaks of population growth are observed in Africa and Asia. China is an exception, having enforced drastic limits to human procreation. Anyway, it is expected that the world population will double in the next 50 years, posing dramatic problems of space and resources. All these affairs stimulate the rapid evolution of other species, in particular microorganisms subjected to control by antibiotics and insects treated with pesticides.

16.2.1 Threats from farming and urbanization

Farming and urbanization are main causes of habitat and biodiversity losses. A quarter of land's surface is cultivated and probably less than a fifth of the medieval forest is left in Europe. Forests are particularly endangered in dry periods because non-natural management has led to the replacement of grasses by small trees, favoring fires (MacNeil 2000). With the industrial dimension taken by agricultural practices, especially mechanized deforestation, breeding has intensified, with a reduction in the number of useful species, while arable ground is inhospitable to the wild fauna. Urbanization also occurs apace in emerging countries for the growing population.

The prevalent strategy in industrialized countries was building low and sparse, in the belief that houses are sheltered by the vegetation and made less visible. In the 1950s this strategy was vainly counteracted by the architect Bruno Zevi, who made clear that the only chance to leave large areas intact is building tall and concentrated. Detrimental effects on biodiversity by habitat loss are non linear; from a certain percentage on, the increase is exponential (Tilman 1994). Calculations based on the rule of thumb "species-area relationship" show that halving a wilderness area results in a loss of 15% of the species. This invokes the alarming prospect that spreading of humans will result in a loss of one-half the world's species during the next one hundred years.

Coral reefs are not exempt from urbanization, either directly to provide new space, or indirectly, like in Sri Lanka, where the traditional coral mining of lime for the construction industry dates to at least 400 years ago. It has taken a steep rise in the last three decades, endangering this particularly fragile ecosystem. In very dry regions, like the Red Sea coasts, simply discharging waters from hotel resources also heavily threatens the coral reef. Even eutrophication of the waters damages the coral

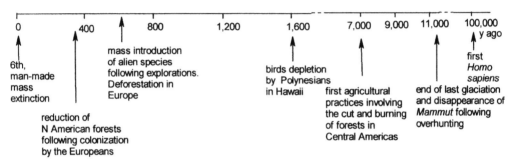

Figure 16.2. The role of man in mass extinctions

reef, because the spreading of the seaweeds prevents light from reaching the corals and their symbiotic photosynthetic zooxanthellae. Discharge of heavy metal ions, oil, detergents, and silt, either brought in by rivers or raised by mining or trawling, are other common threats to the coral reef.

For one reason or another, but always because of threatening human activities, Atlantic and western Indo-Pacific coral reefs are in poor condition, except the Lakshadweep archipelago, which has been banned from tourism. East Asian coral reefs are under stress by blast and cyanide fishing activities. Threatening constructions over the reefs were made during the last world war, and aircraft carriers and ships that sunk still pollute large areas, particularly in the Solomon archipelago. Plutonium remains from nuclear explosions at Johnston Atoll also make conditions unsuitable for life to exist.

Luckily, about 70% of the Pacific coral reefs are still in healthy condition, especially the Great Barrier Reef far away from the coast, Papua New Guinea's atolls, Cook Islands, and part of Micronesia. Global warming, however, is a standing menace. The New Caledonian lagoon, the largest in the planet, is unique in benefitting from upwelling of cold waters, which affords protection from global warming. However, the economic interests linked to nickel mining are an obstacle to inscribing on the list of the patrimony of humanity this sanctuary of marine invertebrates, with the myriad of other organisms that thrive on it, producing unique regulatory metabolites.

Any guess about the fate of the remaining tropical forests and coral reefs under such a pressure is far from optimistic. Pressure for space in developing countries resembles the situation in Europe centuries ago.

Deregulated hunting, and the availability of powerful rifle guns, has contributed in the nineteenth century to the rarefying of certain species of commercial interest, such as the bison in the US. Deregulated hunting is now rarer, limited, for the large mammals, to certain zones of Africa and tropical western Pacific islands because of wars and political instability, and, for migratory birds, to eastern and north-eastern European countries because this brings money from foreign hunters. Damage to the woodcock is immense. Even local traditions of deregulated hunting of migratory birds, such as in southern Mediterranean islands, are difficult to eradicate.

Overfishing has seen an escalation to the point that many fisheries have been closed because of lack of profit (Iudicello 1999). On the other hand, marine farming causes eutrophication and pollution of the waters. In many tropical areas, shrimp farming is also deleterious to the mangrove forest, which has been destroyed to make space for the farms. The size of the problem is made clear by a ban in India and Honduras to shrimp farming along the coasts.

Disappearance of predators may also unbalance the equilibrium, and the problem scales up, such as the disappearance of foxes, predators of the deer mouse, which has allowed spreading of the hantavirus in the US, carried by mice (Levins 1993). Similarly, Sabia virus has emerged in Brazil, Guaranito virus in Venezuela (Lisieux 1994), machupo virus in Bolivia, and Junin fever in Argentine (Garrett 1994). In contrast, in a robust ecosystem, elimination of a predator provides space for another predator, such as in the disappearance of the coyote, which has opened the control of field mice to snakes and owls. When both predator and prey are endangered, it may occur that the prey develops resistance. This is taken into account in Volterra's equation (Ehrlich 1986).

Under this situation of stress, natural products unique to endangered areas are the most threatened, while those of wide distribution are subjected to a lessened pressure. The analysis of natural product losses is made difficult also by the presence of microbial symbionts that may be responsible for the presence of unusual metabolites in macroorganisms.

16.2.2 Threats from the introduction of alien species

Alien species hinder isolation, and thus speciation, especially ecological speciation, and replace the indigenous species that regress and often disappear (Gibbons 1996; Tregenza 1999). If the introduction of alien species is massive, major problems may be posed to the diversity of natural products.

The introduction of alien species is not a new phenomenon. It began 50,000-60,000 years ago with the introduction in Australia of successful species by African populations, which brought to extinction the great mammals. Later, the Polynesian navigators introduced alien species into what is now New Zealand (Steadman 1995). Mixing of species took a global scale 500 years ago with the colonization of America by the Europeans, followed by Australia, New Zealand, and the smaller islands.

Introduction of many successful alien species in tropical islands by the European colonizers caused the disappearance of many endemic species, especially, two hundred years ago, in the Hawaiian archipelago, which had already been threatened 1,200 years before by the intrusion of the Polynesians. With the opening of the Panamanian and Suez streets and ballast water operations, the introduction of species has occurred to such a large extent that it has often become difficult distinguishing what is adapted from what is endemic.

The documented list of alien species is enormous, growing at no pace. It is updated by an international organization of ecologists, Marine-Pests, directed by Australian scientists of the CSIRO (CSIRO). Counting alien species is probably already beyond our resources, however. Governmental regulations appear slowly, such as a much controversial code of procedure for ballast waters recently issued by the US government.

The list of land invaders that have escaped control comprises plants, insects, invertebrates, and birds. Prominent in the plants are the Eurasian leafy spurge (*Euphorbia esula*, a perennial grass that threatens native grasses in northwestern US, favored by farm animals), the Eurasian cheatgrass (*Bromus tectorum*, which has also spread in the US), the tall fescue (*Festuca arundinacea*, the most abundant perennial grass introduced from Europe into eastern US, where it has been strengthened by the association with a fungus, *Neotyphodium coenophialum* (Clay 1999), and European and American pine trees (introduced since the year 1680 in South Africa, where they have become a major pest).

Insects out of control, which have become major pests, include the Argentine ant (responsible for alien iridoids, Table 16.2.2), the Asian longhorn beetle, the Eurasian weevil (originally introduced for biocontrol, it is now devouring Russian thistle and native North American thistles in the US) (Louda 1997), mites (in the genus *Varroa*, spreading in Europe, North America, and recently also in South Africa, where they parasitize honeybees), and the western corn rootworm, *Diabrotica vigifera* (Enserink 1999).

The zebra mussel, *Dreissena polymorpha*, is a most successful invertebrate invader. This Eurasian bivalve mollusk has recently entered the US, first in the Great Lakes region and then the Mississippi and Hudson rivers, where it is out of control.

The European house sparrow is the most successful alien bird. Sea birds, in the shortage of fish, have adapted to land. Particularly frequent is the sea-gull, adapted to land wastes. The cormorant is much more aggressive, fishing at trout farms, with great financial losses. The massive introduction of species of birds and mammals for hunting purposes is also at detriment of biodiversity, both directly for the globalization of the introduced species, and for the adverse impact on many other endemic species. Bad regulations favor this economic exploitation, particularly in my country.

More problematic is counting the species of alien microorganisms. With fungi, evidence has accumulated about the spreading of hybrid plant pathogens; hybridization is often detectable even at the phenotype level (Brasier 2000). Spreading of hybrid toxins from these hybrid pathogens has not been reported yet, though it is expected. Pathogenic filamentous fungi also seem to be responsible for the dramatic decline of frogs observed in our time.

The mechanisms involved in the spreading of alien species are investigated to devise methods of control. Even models are considered, suggesting that, at a local scale, high biodiversity hinders the spreading of alien species, whereas, at a community-wide scale, a favorable habitat favors indigenous and alien species equally (Levine 2000). Successful spreading of the Argentine ant, *Linepithema humile*, which has colonized large areas throughout the world, has been attributed to genetic globalization of the invaders, in contrast with the genetic variability of endemic populations. Globalization not only abolishes any fighting among the invading ants, but also enhances their cooperation (Tsutsui 2000).

Spreading of unusual metabolites by introduced alien species is accompanied by a decrease in the natural product diversity, and it may also aid spreading of the producer. This occurs at a global scale in the sea with alien toxic dinoflagellates; their lethal toxins and subtle tumor promoters are of major social concern. As shown in Table 16.2.2, a prominent position is occupied by lethal paralytic shellfish toxins, such as the saxitoxins, produced by dinoflagellates in the genera *Alexandrium*, *Gymnodinium*, and *Pyrodinium* (Committee O.R. 1999). Tumor promoters produced by dinoflagellates in the genera *Prorocentrum* and *Dinophysis*, like okadaic acid and dinophysistoxins (diarrheic shellfish toxins) are also shown in Table 16.2.2 (Committee O.R. 1999); they are accumulated in filter feeding mollusks, particularly edible mytilids, which have become a hazard. Ballast water procedures for oil transportation are the main cause of the spreading of toxic dinoflagellates (Committee O.R. 1999; Harvell 1999). Cars are the main triggers of these devastating procedures, adding to the problem of their gas emission from the combustion process and solid emission from the material of the catalytic converters.

Most other invaders do not raise immediate health problems but are of ecological concern. The tropical green seaweed *Caulerpa taxifolia*, which has entered the Mediterranean, replacing other species, is a special case. This alga has introduced in the Mediterranean alien metabolites that are distasteful to herbivorous species and toxic for many microorganisms (Table 16.2.2). These metabolites, in particular caulerpenyne, play a defensive role for *C. taxifolia* in the pressure for food of the tropics. Although in the Mediterranean such pressure is much lessened, the production of the distasteful metabolites by the seaweed is much higher than in the tropics, which further fosters its spreading (Académie 1997). Of its minor terpenoids, 10,11-epoxycaulerpenyne is toxic on the mice model, and caulerpenynol induces the proliferation of plant callus cultures because of satellite DNA amplification (Mancini 1998). Spreading of the Eurasian forb, *Centaurea diffusa*, in North America may be rationalized on similar grounds, although the structure of the metabolites involved has not been disclosed (Callaway 2000).

Mass mortality of fan gorgonians in the Caribbean is attributed to a pathogenic fungus in the genus *Aspergillus* (Geiser 1998). The disappearance of the monk seal in Mauritania may be due to either algal toxins or morbillivirus. The causative agent has not been identified for the death of *Acropora* spp. in many coral reefs, the sea urchin *Diadema antillarum* in the Caribbean, and turtle grass,

Thalassia testudinum, in Florida bay (Harvell 1999).

Even species that are not known to produce bioactive metabolites may have a large environmental impact, such as *Emiliania huxeleyi.* This haptomonad protist blooms periodically in the Bering Sea at the detriment of diatoms, disrupting the food chain with massive death of migratory birds and fish losses. *Mnemiopsis leidyi* is another major such problem; this comb jelly has invaded the Black Sea, subtracting zooplankton to fish.

Genetically engineered plants are special invaders. Modified crops of nutritional interest are believed to be more invasive that their natural counterparts (Wolfenbarger 2000), although these views have been challenged for oilseed rape, potato, maize, and sugar beet (Crawley 2001).

Table 16.2.2. Metabolites introduced with alien species

Invading species/Geographic origin/Invaded area	Metabolite introduced/Class of metabolite		Bioactivity of metabolites
Alexandrium, *Gymnodinium,* and *Pyrodinium,* Dinofl./now of worldwide distribution	saxitoxins, gonyautoxins/guanidinio alkal.	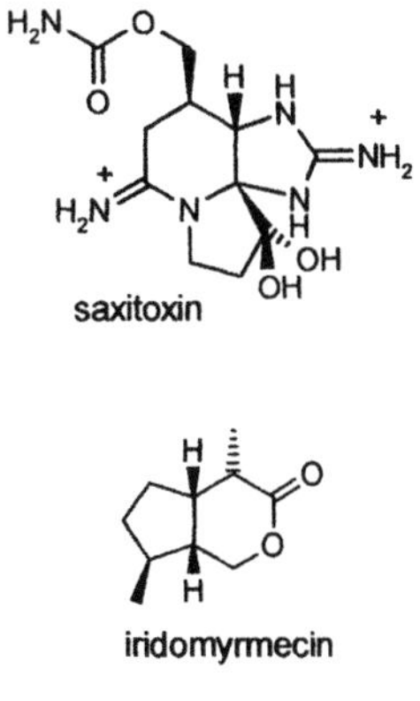 saxitoxin	toxins (paralytic shellfish poisoning) behaving as Na$^+$ channel blocking agents
Argentine ants, Ins./Argentina/worldwide	iridomyrmecin and other irodoids	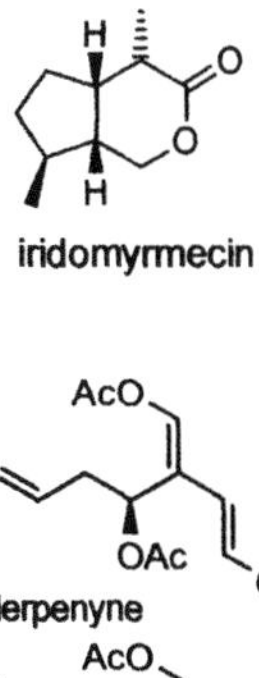 iridomyrmecin	antibacterial and insecticidal agents
Caulerpa taxifolia, Caulerpales, Ulvophyta/tropical/ Mediterranean basin	caulerpenyne, 10,11-epoxy-caulerpenyne, caulerpenynol/ sesquiterp.	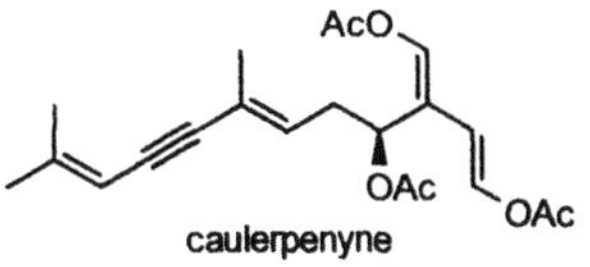 caulerpenyne 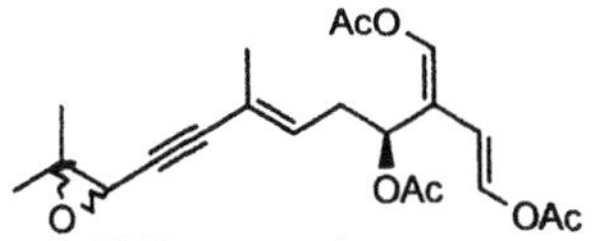 10,11-epoxycaulerpenyne 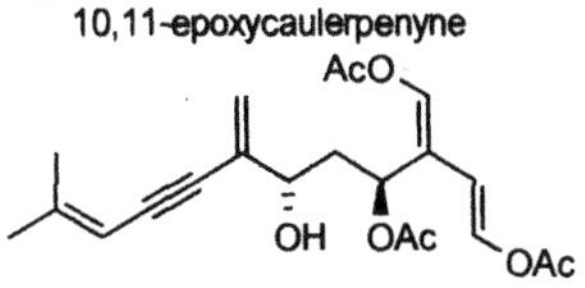 caulerpenynol	antimicrobial and cytotoxic; 10,11-caulerpenyne is a toxic agent on the mouse model

Prorocentrum,
Dinophysis/now of
worldwide distribution

tumor promoters

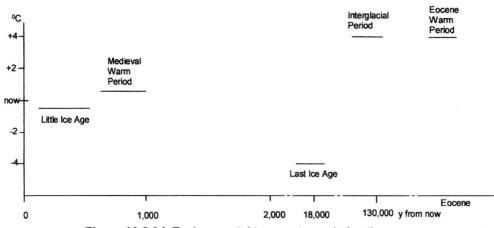

R = H   okadaic acid

R =   DTX-5a

okadaic acid, DTX-5a, and related polyket. polyethers

16.2.3 Threats from toxicity, ecotoxicity, and climate changes following industrialization

The present massive use of energetic resources, with the emission of greenhouse gases, is the cause of global warming and climate changes. This is of much concern since, according to models, biodiversity is threatened more by climate changes than local extirpation of species (Davis 2001). Whether this means that a species-poor community can be as stable as a species-rich community (Sankaran 1999) is contradicted by other observations, however.

In any event, forecasts about the extent and timing of global warming are subjected to frequent revisions, in contradiction with one another (Schneider 2001). Also, changes in biodiversity are hardly amenable to a quantitative evaluation because of the large number of factors involved. These are examined in the following.

First, the temperature changes. The mean environmental temperature is readily accessible, but data for the past are to be taken with prudence since the measurement of temperature is a recent acquisition. Indirect evidence (Fig. 16.2.3.I) suggests that the Little Ice Age, which lasted from the end of the Middle Ages until the mid 19th century, was colder than now (0.5 °C). Such conclusions derive mainly from the analysis of proxies, which are geographically scattered indicators, in the form of both natural (physical and biological) and documentary (written) archives for the last millennium

Figure 16.2.3.I. Environmental temperatures during the ages

(Jones 2001). According to sea level changes, the last Ice Age, centered at 18,000 years ago, was even colder than now (3-4 $^{\circ}$C) (Lambeck 2001). In contrast, the Medieval Warm Period was warmer than now (0.5 $^{\circ}$C), and both the Interglacial Period (130,000 y from now) and the Eocene Warm period (33-53 My from now) were even warmer (3-4 $^{\circ}$C) (Webb 1992).

Climate changes from the time of the disappearance of dinosaurs to our days have also been based up on deep-sea stable isotope data: oxygen isotope data provide an insight into the temperature variations, while carbon isotope data are informative on the kind of global carbon cycle perturbation (Zachos 2001). These data suggest that the present conditions of temperature are similar to the late Middle Age. However, the trend toward higher temperatures is now more difficult to contrast because of the drastic deforestation and emission of gases.

Oceanic temperatures and levels are also rising (Levitus 2000). According to models, greenhouse warming is expected to cause a collapse in the Labrador Sea convection, with the consequence of a further rise in the sea level (IPCC 1995).

Abrupt temperature changes are easier to appreciate than long-term changes. El Niño events, which are the best examples of the first type, occur in Indonesian forests in synergisms with threatening by human activities (Curran 1999). The same probably occurs on coral reefs too, where El Niño warming causes the death of seaweeds and invertebrates, in particular corals, which are extremely sensitive to temperature variations.

Carbon dioxide is needed for plant photosynthesis. Fast-growing trees take advantage of a high partial pressure of carbon dioxide in the atmosphere. However, this may cause damage to biodiversity since slow-growing plants respond inversely (Tangley 2001). Also, when the carbon dioxide level raises above normal limits in the atmosphere, a greenhouse effect is established. Therefore, the carbon dioxide level in the atmosphere is taken as a parameter of climate changes.

Estimating past levels of carbon dioxide in the Earth's atmosphere is not straightforward, however. According to ice-bubble data, the carbon dioxide level remained at a standing 270-280 ppm until the industrial era, when a gradual rise to the present 360 ppm level occurred. One has to be cautious about these estimates, since carbon dioxide is soluble in ice. However, looking at the problem from another side, the density of stomata (which decreases on increasing carbon dioxide availability) in fossil leaves and *Ginkgo* leaves, an agreement is found with isotope data from marine sediments: periods of low carbon dioxide concentration coincide with a cool climate on Earth (Kürschner 2001).

The emission of sulfur and nitrogen acidifying compounds from fossil oil burning has been put under control with recent regulations (Stoddard, J.L. 1999), but other emissions from cars are out of control for political reasons. Besides the problem of solid emissions from the material of catalytic converters, other subtle deleterious effects are expected from chemicals released into the environment that act as endocrine mimics. Formed during fossil fuel burning, together with acidic aerosols and substances that alter the immune system, they are a menace to biodiversity (Chivian 1997). Depression of the immune system goes hand in hand with spreading of viral illnesses, such as HIV or Ebola virus in Zaire, attained by losses in fertility and birth rate, and therefore in biodiversity. The risk is higher in dense human populations and uncontrolled large movements of humans, which occurs with the current emigration from Africa and the Middle East toward Europe, and in contacts between humans and animals. Activation of the expression of the HIV-1 gene by dioxin is also of much concern (Yao 1995).

Persistent synthetic toxins of major environmental concern include dioxins, furans, and chlorinated insecticides (DDT, Aldrin, Dieldrin, Chlordane, Heptachlor, hexachlorobenzene, polychlorinated biphenyls, and chlorinated camphenes). To be added to the list are the pesticide treadmill (which, in cotton plants in Central America, kill the natural predators, while the remaining arthropods become a pest), antifouling agents, and many other chemicals of extensive use. Oil spilling during oil exploration, transportation, and exploitation is a standing menace to biodiversity: forgetting the tragedy of the Amoco Cadiz in the English Channell is difficult. To this regard, we are little comforted by reading that *Lophelia pertusa*, a scleractinian coral, thrives healthily on oil platforms in the North Sea (Roberts 2000).

Drugs and drug metabolites, either expelled from human and animal bodies or used to control pathogens in agriculture, such as streptomycin, are another source of environmental toxicity. The damage is greatest to aquatic species. In addition, antitumor drugs and their metabolites are expelled in huge amounts from human and animal cells by p-glycoproteins, and antibiotics and their metabolites are released with the urine. Growth promotion of livestock by the addition to animal feed of antibiotics in sub-therapeutic doses has become a common practice. Totally banned in Sweden since 1986, it is largely accepted by the European Community, where the 1998 ban was limited to four classes of antibiotics in use for humans. From the drugs still allowed, exchange of genes may be expected that may foster the resistance of microbial pathogens. Direct menaces to farm animals may also be envisaged, because killing sensitive bacteria in the animal gut provides space and opportunity to resistant bacteria. Similar warnings was raised at a Division of Environmental Chemistry of the American Chemical Society meeting in San Francisco in March 2000 about synthetic products, like antiepileptic Carbamazepine, anti-inflammatory Diclofenac, the veterinary antibiotic sulfamethoxazole, clofibric acid (a drug metabolite), Iopromide (a X-ray contrasting agent), and even personal care products, like Galaxolide.

Certain inorganic toxic compounds are brought into circulation by natural causes, such as mercury ions from mount Amiata in Tuscany and arsenic at the mouth of Gange river. It is metal mining, however, that constitutes a major source of inorganic pollution. Serious damages have been produced since the late 1800 from copper mining at Ashio near Tokyo, in Montana, and in Andalusia. Likewise, nickel mining is threatening the New Caledonian lagoon.

The discharge of solvents and wastes during processes of industrial organic synthesis is another major problem the humanity is faced with. Pharmaceutical companies avoid national regulations by transferring plants to emerging countries, relying on loose environmental regulations. This shows that international treaties to limit pollution - including greenhouse gases - should include the emerging countries as strictly as the industrialized countries.

Biological and chemical welfare constitutes a hidden menace since the first world war. Monitoring of these agents is now made possible by sensitive, soft-ionization portable mass spectrometers.

On land, environmental pollution threatens especially the lichens. They are rarefying worldwide, which means that their particular pigments and other metabolites (Table 16.2.5.I) are also disappearing (Huneck 1999).

In the sea, coral reefs are particularly threatened areas because human intrusion adds to recurrent El Niño events. Bleaching of corals, that is loss of pigmented symbiotic zooxanthellae, following abnormal high water temperatures, is the cause of mass mortality of corals. Strong solar irradiation, even in the photosynthetically active wavelength range, also damages the coral reefs (Brown 2000). Thus, depletion of red seaweeds in Galápagos coastal waters (Wikelski 2000) results in a loss of ochtodane monoterpenes. Likewise, mass mortality of shallow-water gorgonians in the Caribbean (Goreau 1990) turns out in a depletion of unique cembranoids.

16.2.4 Threats from biotechnology

Biotechnology has promised to provide what is scarce or disappearing from nature and even to improve on nature. This is an alluring prospect at a time of great concern for food supply for a population increasing at a rate of 86 million individuals per year, mainly in underdeveloped countries (UNPD 1999). To meet this demand, an 80% increment in cereal yields over the 1990 average is needed by the first quarter of the 21st century (Serageldin 1997). The solution was seen in genetically modified crops, particularly corn and soy bean. These are mainly produced in the US, Canada, Brazil, and Argentina, but the way is full of obstacles (Chapter 14.1).

Herbicide-resistant and pesticide-resistant crops should avoid soil erosion and limit the spread of synthetic herbicides and pesticides. Modified crops may also provide heat-stable monounsaturated oleic acid, avoiding the problem of heat-unstable polyunsaturated fatty acids that give unhealthy *trans*-fatty acids as side products of industrial hydrogenation (Mazur 1999). On the longer term, biotechnology may also provide renewable fuel and raw chemicals that may replace petroleum.

Reducing crops to single species, shuffling genes, and all other biotechnological practices described above, tend to the uniformity, however, which is a menace to the conservation of biodiversity and natural product diversity (Serageldin 1999). Other problems may arise. Pollen from genetically engineered plants may reach neighboring non-target species, modifying their genes (Watkinson 2000). Do such cross-pollinated plant species pose a threat? Answering this question in general is difficult because studies of each genetically modified crop under every possible set of environmental conditions are an unrealistic goal (Crawley, 1999). However, the possibility that super-weeds arise cannot be dismissed (Freyssinet, 1988).

If plants have been genetically engineered with genetic material from toxic species, the toxins may be expressed in pollen, reaching other species. This is likely to occur with corn genetically engineered against the European corn borer by the insertion of genes for the insecticidal *Bt* toxin of the bacterium *Bacillus thuringiensis*. The larvae of the monarch butterfly, *Danaus plexippus*, which feed exclusively on milkweed, *Asclepias* spp., growing side by side to corn, are the threatened non-target species (Losey 1999). It is also a disadvantage that the *Bt* toxin is released with root exudates into the rhizosphere soil, where it becomes bound to humic substances and clays, inhibiting microbial degradation. Discriminating between the advantages of a continuing action of the toxin on pests and the disadvantages of the toxin affecting useful species is yet hardly possible (Saxena 1999). Further concern arises about new *Bt* toxins, used in the control of the American corn root-worm complex. The common agricultural practice of inserting antibiotic-resistant genes into crops is also regarded with suspicion because gene exchange may occur with microbes in the digestive apparatus of humans and animals, giving rise to super-bacteria.

It is also to be taken into account that pests, in response to genetically modified crops that bear genetic resistance, may rapidly become adapted, so that new genetically modified crops are needed to maintain high productivity. A gene associated with *Bt* resistance has been identified in the cotton pest, *Heliothis virescens* (Gahan 2001). The life span of a genetically modified crop is calculated to be a few years only (Plucknett 1986).

Once resistant pests attack a monoculture, all plants either survive or perish. Unfortunately, immediate economic interests push the farmer to see nothing else than the crop of interest. This turns against the conservation of natural product diversity: even side zones, as the field margins and roadsides, are no more left as a refugee for beneficial insects (Crawley 1999).

A fundamental shortcoming should also be borne in mind: transgenic organisms are made by altering one or a few genes, while it is the cell that has full attributes of autonomous replication. Further, in multicellular organisms this occurs in cooperation among differentiated cells.

Making an overall balance, benefits from replacing natural high diversity with genetically modified crops are doubtful. Recent extensive field experimentation with Chinese rice may be an indication: variety has increased production (Zhu 2000).

16.2.5 A tentative list of endangered natural products

Threats to biodiversity are threats to natural product diversity. Such menaces were analyzed and disentangled in Chapters 16.2.1-16.2.4. Usually, however, multiple causes concur to the loss of natural product diversity. Therefore, only a list of endangered metabolites can be reliably provided, as shown in Tables 16.2.5.I-V. The list is tentative because making predictions with nature   is difficult. Nature is flexible enough in seeking new equilibria. Organisms that are presently endangered on a worldwide basis, such as the lichens, may recover in a new equilibrium attained by nature, or the equilibrium may be displaced toward other species that will replace the lichens. This also means that planning conservation is difficult.

The examples of threatened metabolites in the tables that follow were selected from three main criteria. First, the focus was on endangered organisms that produce unusual secondary metabolites, like microbes, seaweeds, plants, insects, marine invertebrates, and certain vertebrates, in particular the amphibians. Second, particular attention was paid to threatened ecosystems. Third, rarity and slow growth of the producing organisms, and scarce productivity of unusual metabolites, were also taken into account.

Threats to orceins, orcinol, picrolichenic acid, usnic acid, and other metabolites not shown in Table 16.2.5.I, reflect the sensitivity of lichens to environmental pollution worldwide (Huneck 1999).

From Table 16.2.5.I one is struck by the small number of endangered plant metabolites, in contrast with the documented 7,300 globally endangered species of trees, which represent nine percent of the world's flora (Oldfield 1998). The reason is twofold. Firstly, most metabolites from endangered plants are also biosynthesized by other widely diffused plants, or can be provided by total synthesis or semisynthesis (Chapter 15), thus not contributing to Table 16.2.5.I. Catechin belongs to the first type, a flavonoid from endangered African mahogany, *Khaya senegalensis* and *Kaya madascarensis*, Meliaceae (MI), together with frankincense, polysaccharides and terpenes from endangered *Boswellia carterii*, Burseraceae from Ethiopia, Egypt, and Arabia (MI), all produced by many common plants too. Brazilin belongs to the second type, a polyphenol produced by *Caesalpina echinata*, Leguminosae, from Brazil (MI), a tree heavily exploited as dyewood since the discovery of the Americas; the dye was made available by chemical synthesis in 1875, albeit too late for an efficient recovery of the tree (Oldfield 1998).

Second, natural products from endangered plants are scarcely known just because of the rarity and protection of the species. Protection is extended towards commercial exploitation, which deters pharmaceutical companies from undertaking any study of the plants. The agreement between Merck and Costa Rica for examining the local flora is an exception, but the results remain under corporate secrecy.

As far as land animals are concerned, the amphibians are of most concern. They are rarefying worldwide for obscure reasons (Kiesecker 2001), depriving Earth of their unique alkaloids (Table 16.2.5.I).

It may be surprising that losses in marine natural product diversity (Table 16.2.5.II) are better documented than for natural products on land, in spite of the difficulty in collecting rare and low-yielding marine species. The paradox finds an explanation. Marine biodiversity has constituted in the past few years the largest untapped resource of novel natural products. The pharmaceutical industry entered the game early, but abandoned the scene soon; now it is back to marine natural products, albeit mainly for funding research at universities. In a time of powerful methodologies for the purification and structural elucidation of compounds, this has rapidly secured many unusual marine metabolites of a type never found on land.

In parallel, much has been learned about the biology and chemical ecology of the marine species discovered in these studies. In particular, the study of sponges, neglected for decades, has seen a resumption in the last few years. This is especially important for endangered species of the coral reefs because the rate of calcification by reef-building corals, coralline algae, and coccolithophorids, which together account for half the production of calcium carbonate on our planet, is decreasing with the present increase in the atmospheric level of carbon dioxide (Riebesell 2000). Many species are destined to disappear, and with them many unique metabolites. This is why many natural products from coral reefs appear in Tables 16.2.5.II-V.

Table 16.2.5.I. Endangered natural products on land

Name/chemical class	Structure	Origin	Role
	Worldwide		
orcein/alkal. ketide and orcinol/tetraketide	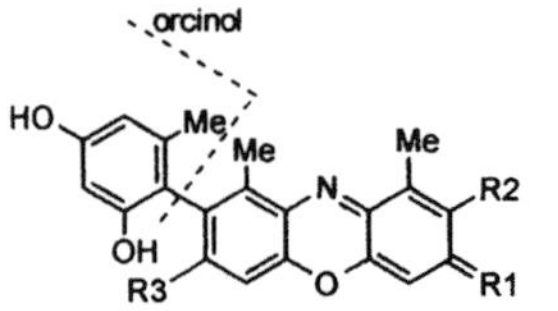 α-aminoorcein R1=O, R2=H, R3=NH₂ α-hydroxyorcein R1=O, R2=H, R3=OH β- and γ-aminoorcein R1=O, R2=orcinol, R3=NH₂	orcinol from *Roccella tinctoria* DC. (1805) and *Lecanora tartarea* (L.) Ach., lichens of the Scandinavian tundra; orcein, a mixture of dyes from oxidation of orcinol in the presence of ammonia	precursor of dyes: MI
picrolichenic acid/polyket.	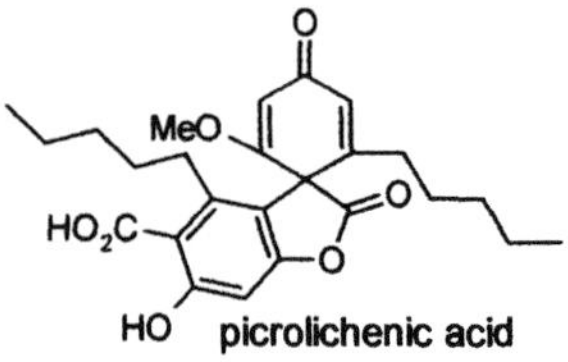 picrolichenic acid	*Pertusia amara* (Ach.) Nyl. (crustose lichen)	bitter principle: MI
usnic acid/polyket.	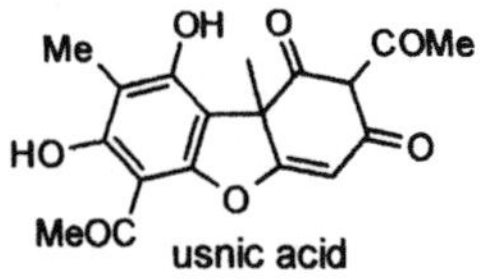 usnic acid	*Usnea barbata* (L.) Wigg., a lichen	milestone in classical biosynthetic studies: Herbert 1989

Table 16.2.5.I. Endangered natural products on land

Name/chemical class	Structure	Origin	Role
	Americas		
glabrescol/triterp.	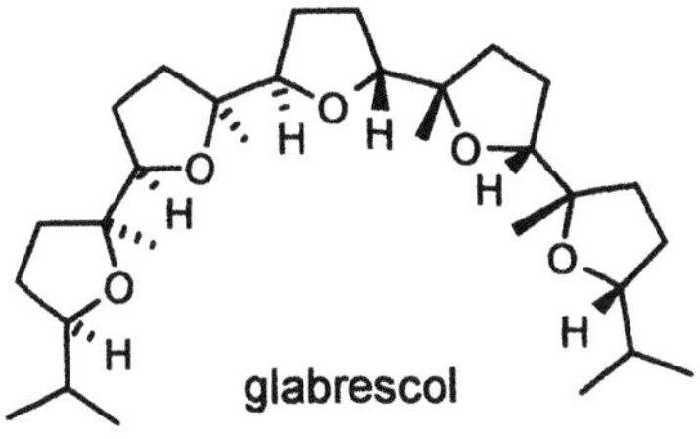 glabrescol	*Spathelia glabrescens*, Rutaceae, Ang. from Jamaica (a nearly threatened plant: Oldfield 1998)	ion binding, selective for K⁺: Bellenie 2001
indolizidine alkal., (orphans: pumiliotoxin B, 251D, 251H, 341H), and steroidal alkal. (batrachotoxin A), piperidine alkal. (histrionicotoxin), pyridine (epibatidine), and over 100 other undeterm. alkal.	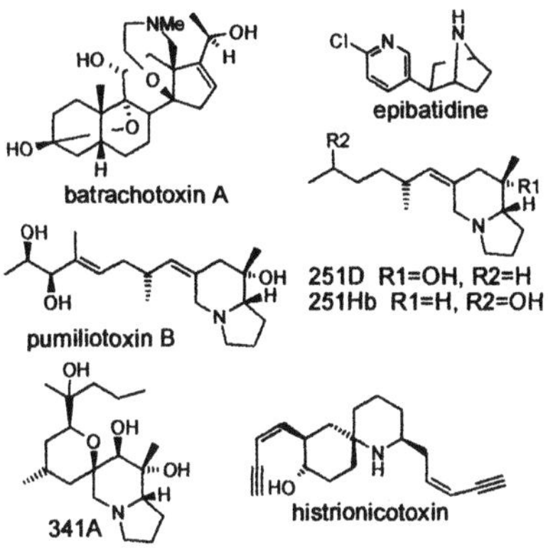 batrachotoxin A, epibatidine, 251D R1=OH, R2=H, 251Hb R1=H, R2=OH, pumiliotoxin B, 341A, histrionicotoxin	skin of Dendrobatidae frogs, Amph. from Colombia and Ecuador: Daly 2000; Braekman 1998	defensive toxins
	Asia		
ginkgolides/diterp.	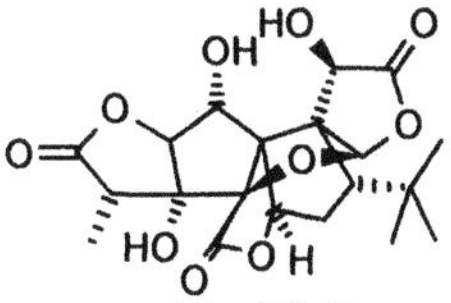 ginkgolide B	*Ginkgo biloba* L., Ginkgocaceae, Gymn. from Xitianmu Mountain, Zhejiang, China. Long living genus, first appeared in the Jurassic	specific platelet activating factors: MI; Oldfield 1998

Table 16.2.5.II. Endangered marine natural products in western Indo-Pacific coral reefs

Name/chemical class	Structure	Origin	Role
	Fiji		
citorellamine/ indole alkal.	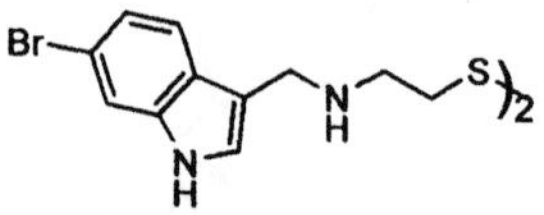 citorellamine	*Polycitorella mariae* Michaelsen, 1924, Polycitoridae, Ascid.	antibact., insecticidal: Moriarty 1987
dysidazirine/ unsat. aziridine alkal.	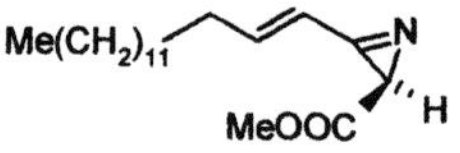 Me(CH$_2$)$_{11}$ / MeOOC / (-)-dysidazirine	*Dysidea fragilis* (Montagu, 1818), Dictyocer. Porif.	cytotoxic: Molinski 1988
fenestin A/cyclopept.	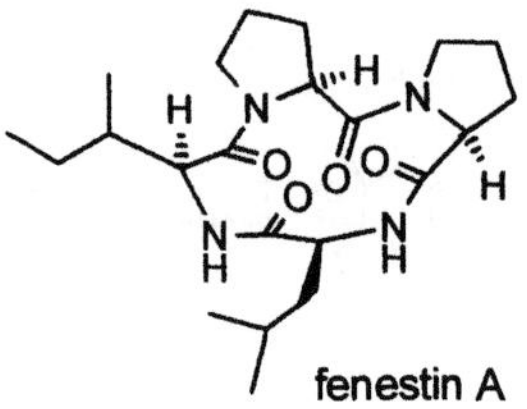 fenestin A	*Leucophloeus fenestrata,* Halichon., Porif.	not reported: Omar 1988
makaluvamine A/pyrrolo- iminoquinone alkal.	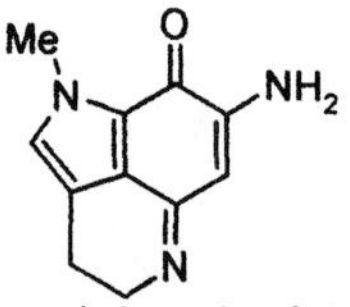 Me / NH$_2$ / makaluvamine A	*Zyzzya fuliginosa* Carter, 1879, Poecil. Porif. and *Didymium bahiense,* Mycetozoa: Ishibashi 2001	topo- isomerase II inhibitor: Radisky 1993
namenamicin/ calicheamicin- class enediyne polyket.	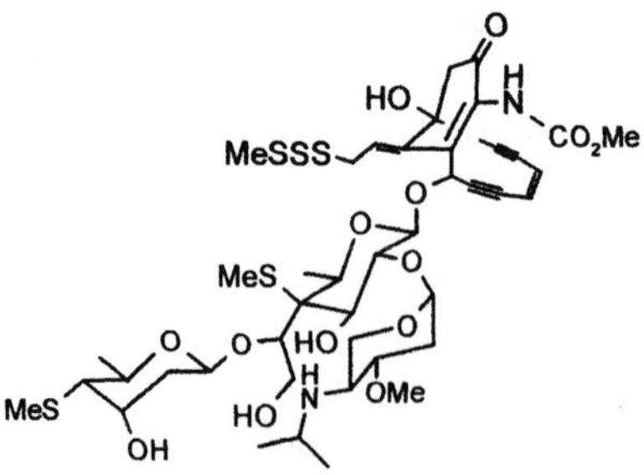 HO / MeSSS / CO$_2$Me / MeS / MeS / HO / OMe / HO / OH / namenamicin	*Polysyncraton lithostrotum,* Didemnidae, Ascid.	cytotox., antifungal: McDonald 1996

Table 16.2.5.II. Endangered marine natural products in western Indo-Pacific coral reefs

Name/chemical class	Structure	Origin	Role
varamine A and B/pyridoacridine alkal.	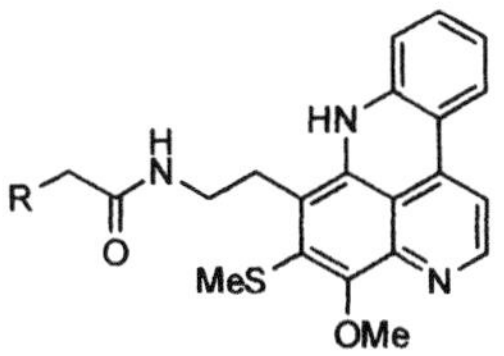 R=Me varamine A R=H varamine B	*Lissoclinum vareau* (Monniot and Monniot, 1987) and *Diplosoma* sp., Didemnidae, Ascid.	cytotoxic: Molinski 1989

New Caledonia

leucascandrolide A/macrolide	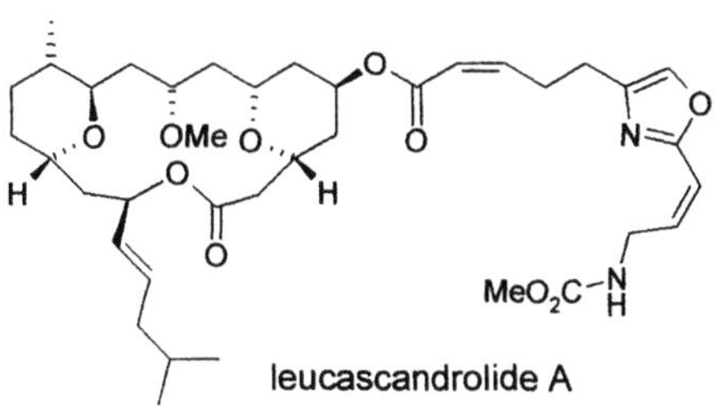 leucascandrolide A	*Leucascandra caveolata* Borojevic and Klautau, Calc., Porif.	antifungal, cytotoxic: D'Ambros io 1996B
lituarine A, B, and C/macrolides	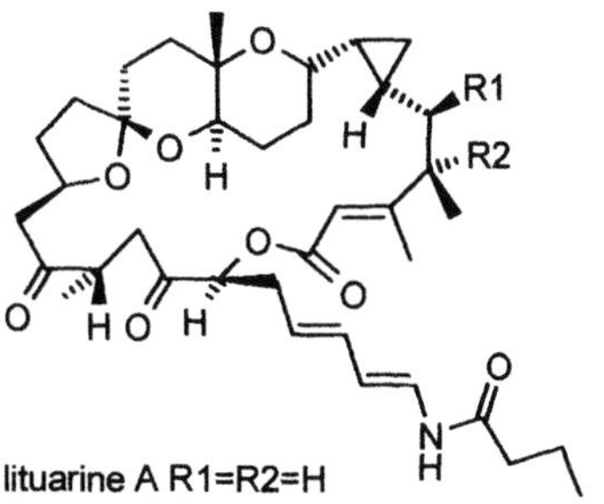 lituarine A R1=R2=H lituarine B R1=OCOMe, R2=OH lituarine C R1=R2=OH	*Lituaria australasiae* (Gray, 1970), Pennatulacea, Cnid.	cytotox., antifungal: Vidal 1992

Table 16.2.5.II. Endangered marine natural products in western Indo-Pacific coral reefs

Name/chemical class	Structure	Origin	Role
oceanapins/cera-mides C-branched at both chains	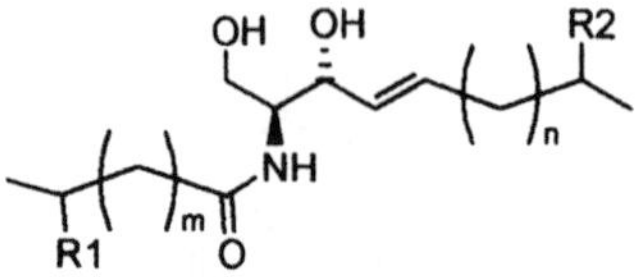 oceanapin A m=19, n=9, R1=Me, R2=H oceanapin B m=19, n=9, R1=R2=Me oceanapin C m=20, n=9, R1=H, R2=Me oceanapin D m=19, n=10, R1=R2=Me oceanapin E m=19, n=10, R1=H, R2=Et oceanapin F m=19, n=10, R1=Me, R2=Et	*Oceanapia* cf. *tenuis*, Haplosclerida, Porif.	not known: Mancini 1994B

Northern Marianas

Name/chemical class	Structure	Origin	Role
kumepaloxane/ rearr. trisnorsesquiterp.	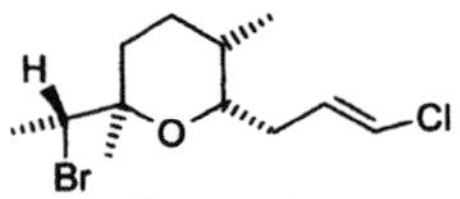 kumepaloxane	*Haminoea cymbalum* Quoy & Gaimard, 1835, Cephalaspidea, Opisthobr.	not reported: Poiner 1989
peroniatriol I/γ-pyrone polypropionate	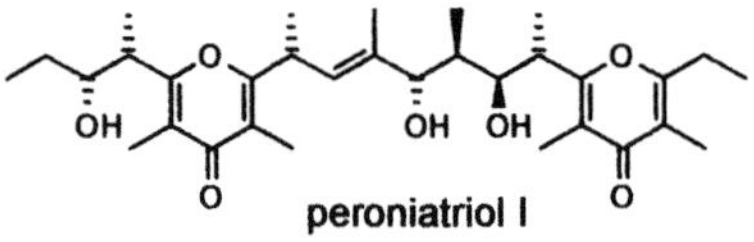 peroniatriol I	*Peronia peronii*, Pulmonata, Moll.	not reported: Arimoto 1995
(-)-polycaver-noside A/polyether macrolide	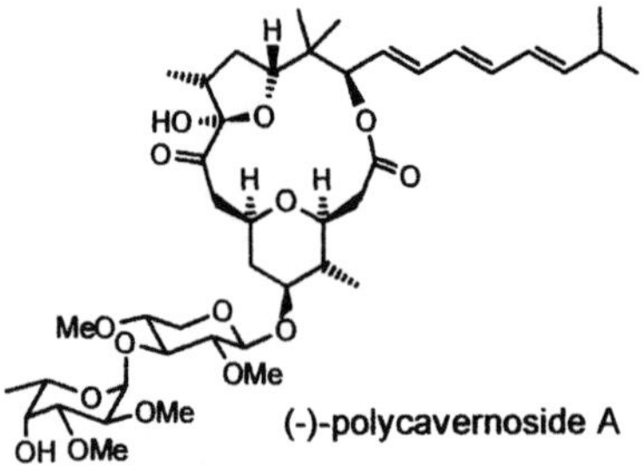 (-)-polycavernoside A	*Polycavernosa tsudai* [= *Gracilaria edulis*], Gigartinales, Rhodoph.	lethal toxin: Paquette 2000

Table 16.2.5.II. Endangered marine natural products in western Indo-Pacific coral reefs

Name/chemical class	Structure	Origin	Role
	French Polynesia		
maitotoxin/poly-ket. polyether	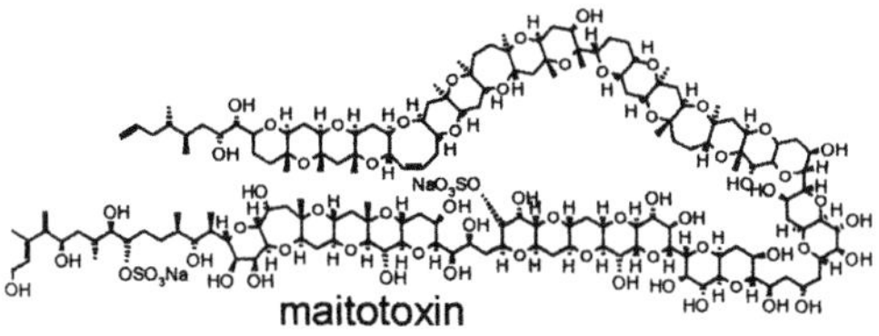	*Gambierdiscus toxicus* Adachi et Fukuio, 1979, Dinofl.	toxin: Murata 2000

maitotoxin

	Hawaii		
pulo'upone/pyri-dine alkal.	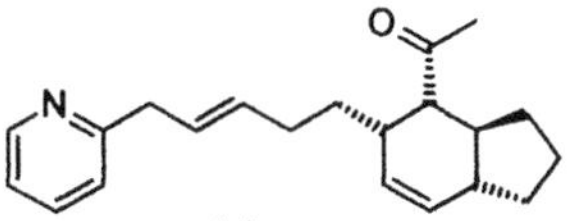	*Philinopsis speciosa* Pease, 1860, Cephalaspidea, Opisthobr.	not reported: Coval 1985

pulo'upone

punaglandins/fatty-acid deriv.	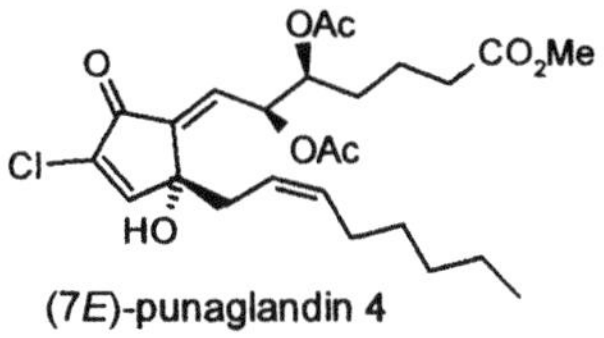	*Telesto riisei* Duchassaing and Michelotti, Telestacea, Cnid.	cytotoxic: Suzuki 1986

(7*E*)-punaglandin 4

ulapualide A/macrolide	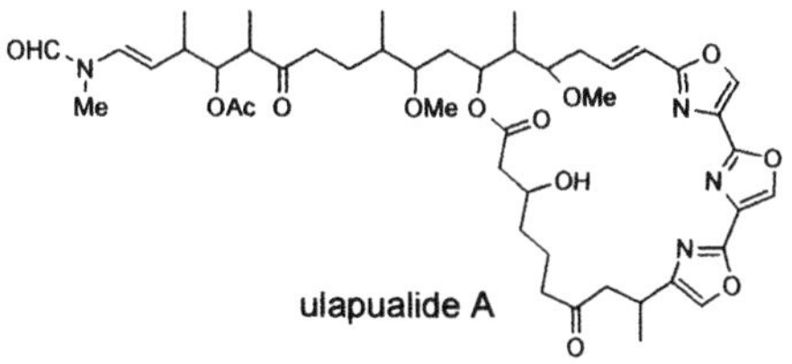	eggs of *Hexabranchus sanguineus* (Ruppell & Leuckart, 1828), Nudibr.	cytotoxic: Roesener 1986

ulapualide A

wailupemycin A/polyket.	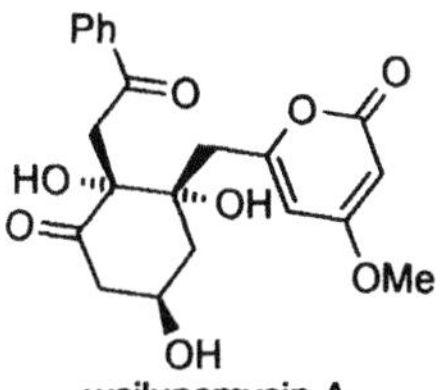	*Streptomyces* sp., Actinom., Bact.	antibact.: Sitachitta 1996

wailupemycin A

Table 16.2.5.III. Endangered marine natural products in Asiatic coral reefs

Name/chemical class	Structure	Origin	Role
	Vietnam		
callicladol/ squalene deriv. triterp. polyether	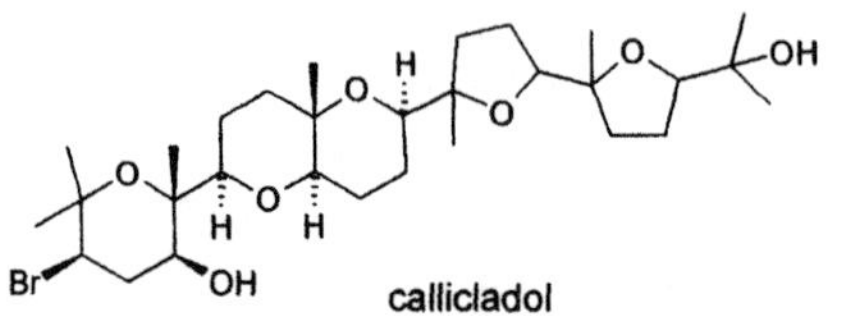 callicladol	*Laurencia calliclada* Masuda, Ceramiales, Rhodoph.	not reported: Suzuki 1995
	Indonesia		
kauluamine/ manzamine dimer, macrocyclic alkal.	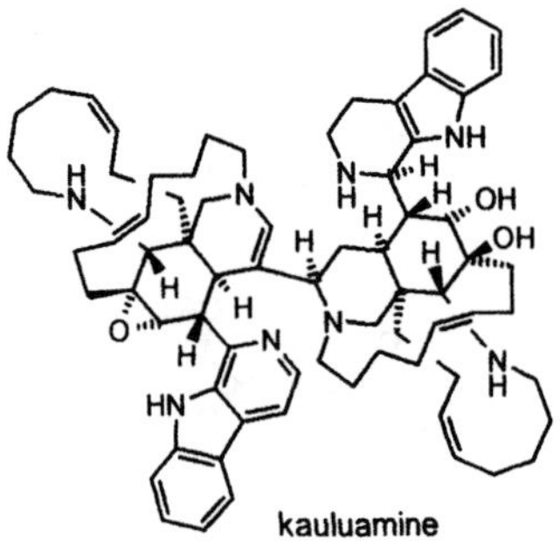 kauluamine	*Prianos* sp., Poecil., Porif.	mild immuno-suppres-sant: Ohtani 1995
vannusal A/non-squalene triterp. or sesquiterp. dimer	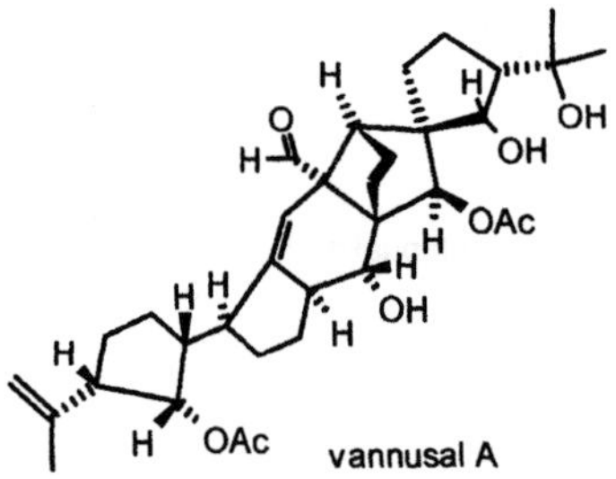 vannusal A	*Euplotes vannus*, Cilioph.	not reported: Guella 1999A

Table 16.2.5.III. Endangered marine natural products in Asiatic coral reefs

Name/chemical class	Structure	Origin	Role
	Philippines		
aciculitin A, B, and C/glycopept. with histidino-tyrosine bridge	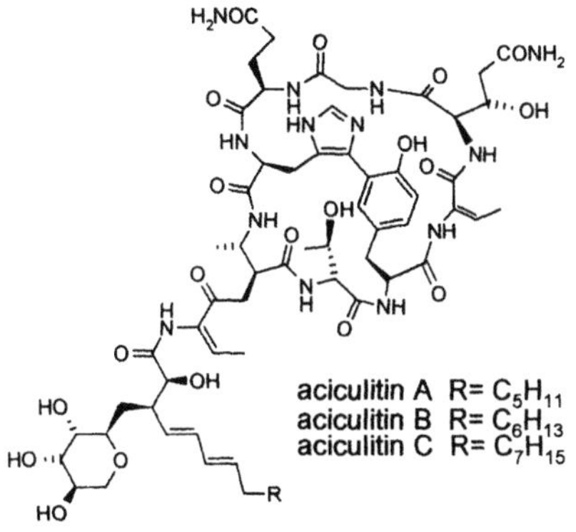	*Aciculites orientalis*, Lithistida, Porif.	cytotoxic, anti-fungal: Bewley 1996
2'-demethyl-aplysinopsin/ indole alkal.	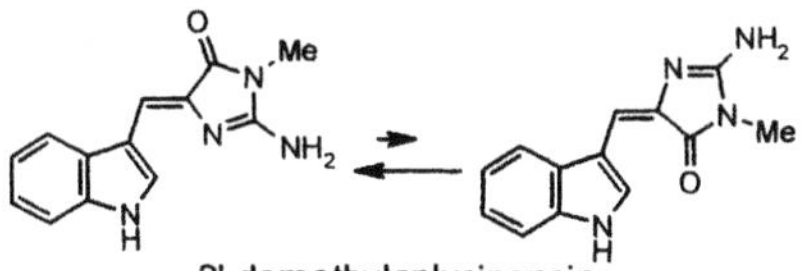	*Dendrophyllia* sp., Dendroph., Cnid.	light filter candidate: Guella 1989A
	Okinawa		
onnamide/pept. guanidine alkal.	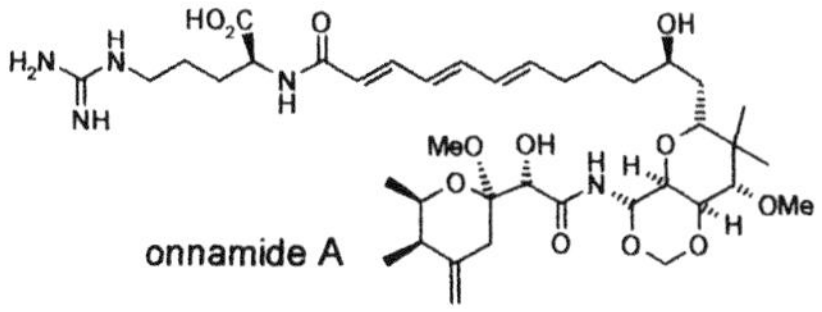	*Theonella* sp., Lithistida, Porif.	cytotoxic: Sakemi 1988
venustatriol/ squalene deriv. triterp. polyether	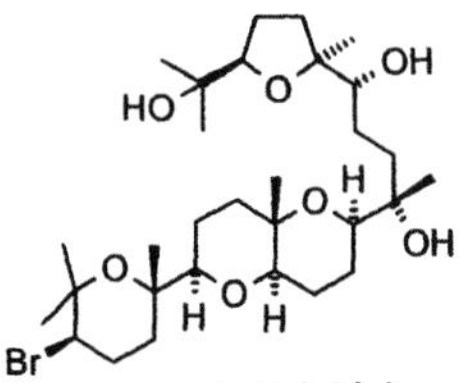	*Laurencia venusta* Yamada, Ceramiales, Rhodoph.	antiviral: Sakemi 1986

aciculitin A R= C_5H_{11}
aciculitin B R= C_6H_{13}
aciculitin C R= C_7H_{15}

2'-demethylaplysinopsin

onnamide A

venustatriol

Table 16.2.5.IV. Endangered unique natural products in Indian Ocean coral reefs

Name/chemical class	Structure	Origin	Role

Mozambique

thiocoraline/cyclopept.	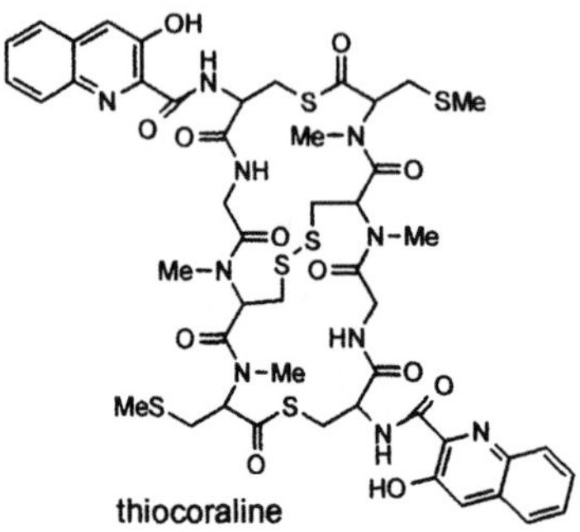	*Micromonospora* sp., Actinom., Bact.	RNA biosynthesis inhibitor: Baz 1997

thiocoraline

Comoros

mayotamid A and B/cyclopept.		*Didemnum molle* (Herdman, 1886), Didemnidae, Ascid.	cytotoxic: Rudi 1998

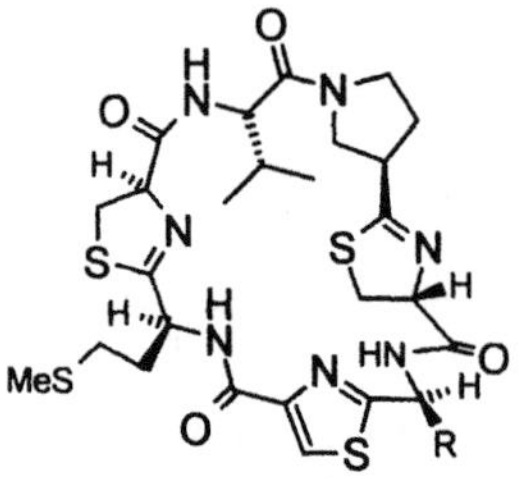

mayotamide A R= CH(Me)CH$_2$Me
mayotamide B R= CHMe$_2$

India

volvatellin/sesquiterp	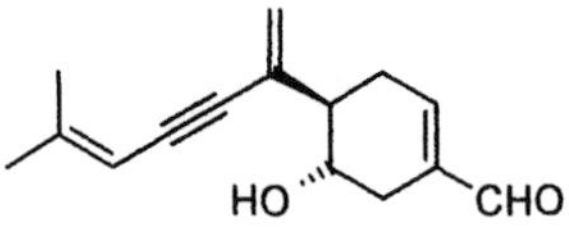	*Volvatella* sp., Sacoglossa, Moll.	not known: Mancini 2000

volvatellin

Table 16.2.5.V. Endangered marine natural products in American coral reefs

Name/chemical class	Structure	Origin	Role
	Bermuda		
clavepictine A/quinolizidine alkal.	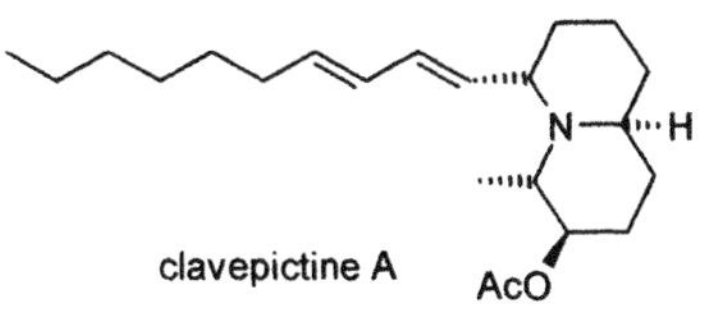 clavepictine A AcO	*Clavelina picta,* Aplousobranchia, Ascid.	fungicidal: Raub 1991
	Bahama		
discodermolide/ polyket.	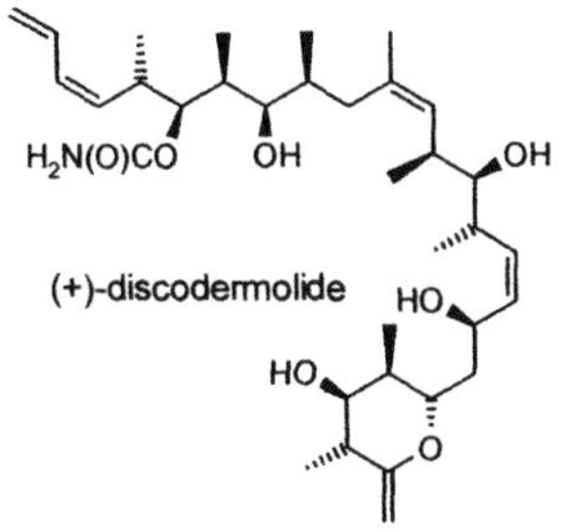 H₂N(O)CO OH OH (+)-discodermolide HO HO O	*Discodermia dissoluta,* Lithistida, Porif.	antitumor: Smith 2000; Paterson 2001
dysidiolide/rearr. sesterterp.	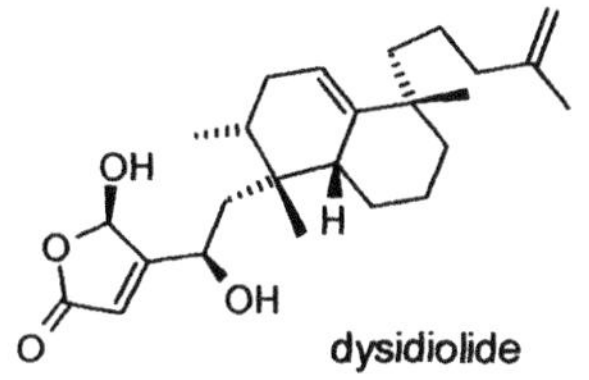 OH O OH O dysidiolide	*Dysidea etheria* de Laubenfels, Dictyocer., Porif.	cytotoxic: Corey 1997B
pseudopterosin C/diterp. glycoside	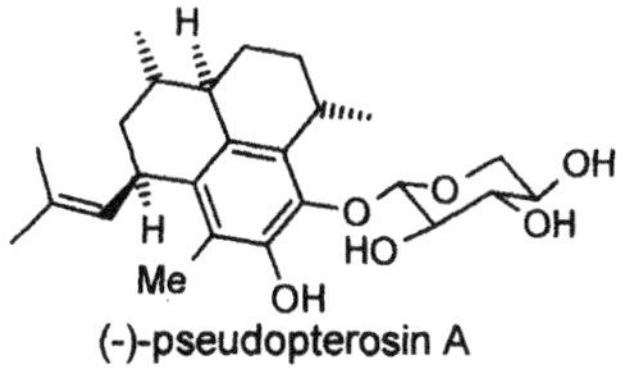 H H OH O O OH HO OH Me OH (-)-pseudopterosin A	*Pseudopterogorgia elisabethae* Bayer, Gorgonacea, Cnid.	anti-inflammatory : Corey 1998

Table 16.2.5.V. Endangered marine natural products in American coral reefs

Name/chemical class	Structure	Origin	Role

Florida

salinamide A and B/cyclopept.	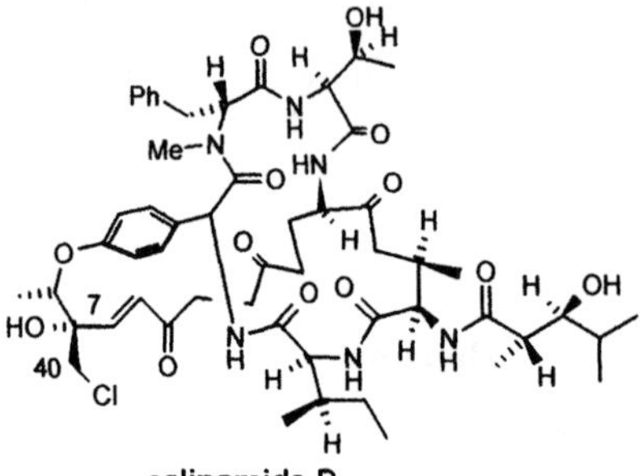 salinamide B salinamide A (7,40 epoxide)	*Streptomyces* sp., Actinom., Bact.	anti-inflammatory : Trischman 1994A

Other areas of the Gulf of Mexico

bryostatin 1 and 2/polyket. polyether macrolides	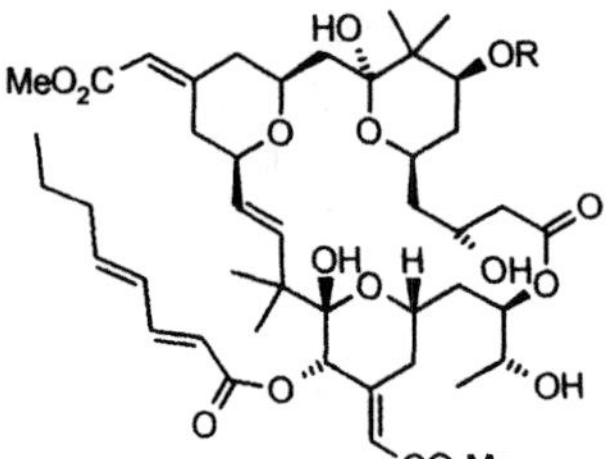 bryostatin 1 R = Ac bryostatin 2 R = H	*Bugula neritina* (L.), Bryoz.	antitumor: Evans 1999A

Venezuela

microcolin A and B/lipopept.	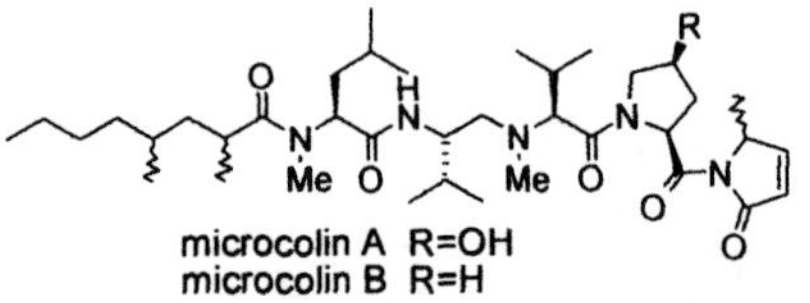 microcolin A R=OH microcolin B R=H	*Lyngbya majuscula* Gomont, Cyanobact.	immuno-suppressant: Koehn 1992

Table 16.2.5.V. Endangered marine natural products in American coral reefs

Name/chemical class	Structure	Origin	Role
	Netherlands Antilles		
curacin A/thiazoline-bearing lipid	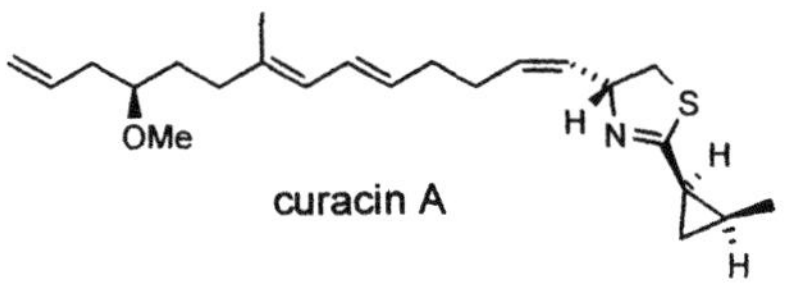 curacin A	*Lyngbya majuscula* Gomont, Cyanobact.	antimitotic: Wipf 2000B

16.3 Our biased view?

Long term forecasts about new equilibria attained in nature following man's intervention and natural causes, and their mutual interaction, indicate that the tropics and Mediterranean areas are at highest risk (Sala 2000). This, and all other menaces to the environment, were evaluated from our perspective, with species of immediate economic or aesthetic interest in mind. How could it be otherwise? Any complaint about the deteriorating environment is biased by our perspective. This is the limited view of a species that differs from all other ones not in the degree of organization, which is no better than with ants and bees, but in the syntactic ability to compose the various pieces (MacNeilage 2000) and to impose the rules of politics. Musil's "Politik ist Wille u nicht Warheit", with all the consequences (Musil 1996), was rarely contradicted after the great athenian period.

With complete disregard of this biased perspective, nature does not indulge man. A pollutant or adverse climate for man, may be an ideal condition for other forms of life in a new equilibrium of Earth. This is of no immediate concern, however, and it is easily forgotten.

Tables 16.2.5.I-V contain metabolites that future generations will probably never meet in nature. Such tables should also have many empty rows, for molecules that have not yet be seen. A question that may arise - actually frightening and fascinating at the same time - is which molecules are disappearing before they are discovered. Surely there are many, hidden in microorganisms that have not yet been investigated, or have resisted culture. They add to the books that have never been written.

Chapter 17. Management of natural product diversity

Biodiversity has never been as great as in modern times. The same is probably true for natural product diversity (Chapter 16). The trend is now to the opposite direction, however, because of the demographic expansion and over utilization of all available resources.

This is of much concern, since although it would be hopeless to preserve our planet as we know it - which never happened in the past - no advancement in technology or scientific knowledge will ever lessen the need of natural products as chemical messengers. Nature may attain a new equilibrium with different chemical messengers. The natural product may even disappear from the pharmacopeias once we have mastered the design of new drugs from scratch, but its value as a regulatory agent in nature cannot be substituted. It can only adapt to the course of the evolution.

In the forgiveness of any theological position, this implies that rapid extinction of natural-product-rich species goes to the detriment of the Earth's present equilibrium more than the disappearance of species that produce trivial secondary metabolites. From this viewpoint, the loss of microbes matters much more than the loss of the great mammals or man.

If these tenets are forgot, the answer to why natural product diversity should be preserved is biased by one's particular interest. The pharmaceutical industry has widely exploited natural product resources in the search of innovative drugs through the random screening of living species. This position has shifted, first toward educated screening of the living species and then to the synthetic product from combinatorial synthesis or design from scratch, based on receptor structures and gene data. Such changes were planned without forgetting that regulation of genes and receptors of recent discovery may require complex molecules that are best found in nature (Henkel 1999). To this concern, the 1992 Rio de Janeiro "Convention on Biological Diversity" has been redefined "Convention on Biotechnology Transfer" (Burk 1993).

The question is, how many novel valuable drugs, leads, or templates can still be expected from nature? The answer to this question is difficult. It is true that few plants have been thoroughly examined for natural products, but no plant family has been left out from the examination, so that compounds with entirely new skeletons can hardly be expected from the plants. The current phytochemical literature is replenished of variants on known themes. On the same vein, the exploration of marine organisms was started recently, but in an era of advanced technology. Therefore, the spectrum of entirely new marine natural products left for screening is also quite limited. Unculturable microorganisms (Chapter 14.1) have given no clear sign of interest so far.

However, the discovery of innovative natural drugs depends much on the new targets from the human genome. On the basis of the bare 500 targets of present pharmacopeias, it is difficult to foresee the discovery of many more natural drugs since the long past has provided only a few of them. The new gene sequences of the human genome have already led to novel laboratory tools, such as gene-based diagnostic assays for brest cancer and cystic fibrosis. This also means that the patent system is changing, putting in the hands of high technology companies the thread of life.

During past eras, mass mortalities were followed by a rapid recovery of biodiversity (Courtillot 1996). Now, however, mass mortality is man-made. For the first time the causes of biodiversity-loss are not temporary as in previous catastrophes. The demographic increase and escalation in the utilization of natural resources pose an increasing menace to biodiversity. Therefore, any prospect about natural product diversity in the far future is difficult to make. Views are pessimistic, unless we imagine a world populated by microbes that have survived the catastrophe or have mutated. Microbial life is typically rich in unusual secondary metabolites.

17.1 Preserving natural product diversity through the management of living species

From the above remarks, we cannot expect much help from industry as to the preservation of natural product diversity through the management of living species. On the other hand, governments may be short of means to sustain biodiversity, and emerging countries may also lack the knowledge (Terborg 1999).

Perhaps the only strategy to the conservation of natural product diversity through the management of living species is accepting private investments to achieve environmentally sustainable economies (Daily 2000). This is especially true for sparse oases of protection needed by migratory species. The disappearance of humid zones close to one another, and the consequent rarefying of the snipe (Devort 1997), are a fitting example. Preserving the humid zones - with hunters that pay the cost - is a method to save from extinction the snipe and other aquatic migratory birds, as well as all the macro- and micro-species associated to these zones. A different example is Costarica, for its agreement with the US pharmaceutical company Merck.

These conclusions do not contradict a recent analysis as to the cost of safeguarding the world's biodiversity through protected areas, which should be affordable by governments (James 1999). However, the dimensions of foreseeable protected areas are too small (Musters 2000) and separated by too great a distance. Another problem is the conflict between areas of high biodiversity and high density of human population, which often overlap, such as in the Nairobi National Park.

If it is agreed that accepting private investments can help preserving natural product diversity through the safeguarding of living species, the technical problem of species management arises. First, any indiscriminate saving of species is utopia in present times, and local saving may result in hollow victories. As in all scientific matter, general laws are of guidance to avoid harsh disappointment. A database tailored to biodiversity conservation may help (Smith, A.T. 2000), especially if integrated into the GBIF web of databases (Edwards 2000B).

Several points may then be considered. It is a dogma that refugees, camouflage, and the possibility of migration from one ecosystem to another one should be preserved to avoid elimination of both predators and preys (Levins 1970). The complexity of the relationships between biodiversity and body size of species was already considered in Chapter 9, taking into account also the internal biological scaling (West 1999). With plants, a scaling factor by which production declines uniformly with increasing size has been demonstrated (Enquist 1999). This may have a bearing on the conservation of plant diversity since those with more usable energy may become adapted to more numerous different niches. The importance of considering the energetic balance when dealing with biodiversity is thus emphasized. Control of invading species through parasites is also theoretically feasible for invaders with low genetic variability, such as in small islands (Epstein 1997). Finally, marginal zones are important from an evolutionary point of view (Guella 1997B) since the organisms living at border zones are the last to disappear. Exceptions, such as in Africa and the most heavily threatened Hawaii (Cox 1997), have received convincing explanation (Channell 2000).

Conservation of marine natural product diversity is even more difficult than on land. The sea is even less protected than land because what happens inside is sheltered by the medium, so that marine ecological problems are easily overlooked by both the layman and the authorities. Fisheries are an exception, because of the huge economies involved (Iudicello 1999). It is currently assumed that the total biomass of mature fish, measured by the spawner biomass, is related to the number of offspring produced (that is the recruitment) by the total number of eggs produced. This proposal is increasingly challenged, however, and more quantitative relationships are sought. Thus, measurements on Barents Sea cod stock, have shown that recruitment is proportional to the liver weights of spawners. This suggests that recruitment is limited by the amount of total lipid energy for mature females in the stock, which should allow a more rational planning of stock conservation (Marshall 1999). Management with whales may be possible through genetic tracking, which allows controlling the origin of whale meat on the market (Cipriano 1999). Fishery management in the tropics should take into account that coral-fish larvae stay near the spawning site (Swearer 1999). Whether this applies to other marine areas is not known. Anyway, there is an unbalance between insufficient primary production and too much fish caught. Not only, but the way fish is caught is threatening the environment: trawl beam and dredging extirpate all sizable living organisms, most of which are discarded, and the sea bottom is deprived of the necessary life.

This shows that the conservation of natural product diversity faces even greater difficulties in the sea than on land. Unusual marine natural products are mostly derived from slow-growing invertebrates, which are less amenable to culture and gene manipulation than plants. Marine invertebrates, except in the Great Barrier Reef, the New Caledonian seamounts, and deep waters in the Caribbean, thrive in coastal waters, which are most exposed to the human intrusion. In contrast, rain forest plants grow mainly inside isolated countries, which have always been less preferred than coastal zones by human populations. Moreover, marine life, even more than life on land, is sustained by a complex network of symbiosis, which is easily disrupted by human intrusion, because a defeat in one part is rapidly felt by the whole. Such network of symbiosis is particularly active in border zones, which contribute most to the biological evolution (Guella 1997B). These border zones are neglected by the so called environmentalists, who, while driving a car, perhaps with air conditioning, dictate environmental regulations, with traditionally high focus on the large mammals (Vogel 2001), which have scarce evolutionary significance.

Plantations and secondary growth deserve much attention for the preservation of forests and coral reefs. Ultimately, zoos remain living stores of animal biodiversity that is disappearing from the tropics. They may serve to reintroduce the species. Botanical gardens may also play a useful role as a store of a large variety of seeds. However, this may give rise to claims for germ plasm property by the countries of origin, which is a complex legal matter.

It has been observed that high biodiversity is tied to high cultural diversity, as measured by the number of different local languages. A bare hundred years ago there were about 6,000 spoken languages on Earth. No more than half those languages are still spoken, 80% of which by a few elderly persons only (Cox 1997). Signed languages are also threatened (Meier 2000), especially by the jargon of advertising and business, devised at attention getting. Should evolution be applicable to languages (MacNeilage 2000, Nowak 2000), slowing of the evolution of language is also expected.

Parsimony analysis, which in comparative linguistic is the last chance in reconstructing the history of disappearing languages (Gray 2000), might help unraveling past biosynthetic events (Chapter 16.1).

17.2 Preserving natural product diversity through collections and gene banks

The theory that a robust ecosystem is tied to high biodiversity (species redundancy) is gaining experimental support and wide consensus. This also holds for crop production, according to extensive experimentation in China with rice cultures, which achieve from diversity best protection against rice blast (Zhu 2000). Also decomposing microorganisms contribute to the species diversity, adding further complexity to the picture (Naeem 2000).

These ideas may be translated to natural product diversity. Ecosystems of greater natural product diversity are expected to be more robust (which, by analogy, may be termed natural product redundancy). Under conditions of high diversity, a semiochemical that disappears may be replaced by another one, in a fine tuning of the ecosystem. In contrast, compensation in an ecosystem of scarce natural product diversity needs drastic changes. Thus, the diversity of natural products may be taken as a measure of the contribution to biological evolution.

These ideas may have wide application, like in the resistance of marine ecosystems to invading species, which is believed to be correlated with species diversity, at least at local scale (Stachowicz 1999); in contrast, at community-wide scale a favorable habitat favors indigenous and alien species equally (Levine 2000). The bearing of natural product diversity on these affairs is as yet insufficiently known. Filling this gap will be an enormous task, requiring a detailed knowledge of the species productivity, as well as seasonal changes, bioactivity, stability, reactivity, and spreading ability of the metabolites. The presence of vectors that may favor the spreading of metabolites is also relevant.

Collections of living species may aid saving the natural product diversity. For several groups of productive microorganisms this is a long established strategy. Seeds, stored in a wide variety in botanical gardens, may serve to restoring plants and their metabolites. A bank of seeds, preserved at -20 ^{0}C at Kew Gardens, is representative of the British Islands; it is planned to represent 10% of Earth's flora in the next decade.

Preservation of natural products from the animals is more difficult. Zoos remain living stores of animals that, like the civet cat and the amphibians, give unusual metabolites. Aquaria might serve the same scope for marine invertebrates. However, many productive species, such as the sponges and the anthozoans, do not survive long in aquaria.

Saving the germ plasm to preserve all characters of the species, including their secondary metabolic production, in principle has no restrictions for microorganisms, plants, and animals. There are already examples of cloning endangered animals in the US, although the longest lived, a gaur, *Bibos gaurus* - which probably resulted from many unsuccessful attempts - survived only a couple of days.

Cloning is much easier with endangered plants than animals. Plants of nutritional value, or rich in unusual metabolites, are particularly attractive. Gene banks have already been set up for some of them, in particular rice. However, storing in gene banks subtracts from evolution. Germ plasm recovered from gene banks has no chance to code for evolved enzymes. The restriction is particularly severe if genes of a single species or strain are conserved.

Subtraction from evolution is total if virtual saving of germ plasm is conceived, by storing the gene structures in a computer data bank. Unraveling the whole story of genes, their mutual interaction, and the interaction with the environment, is difficult. It has yet to be done. Even when we know the whole story of genes, we should instruct our computer to simulate changes in germ plasm that would be too long to experience. The risk is that our virtual data bank of germ plasm is nothing more than a museum of extinct things.

References

Abas, S.; Hossain, M.B.; van der Helm, D.; Schmitz, F.J. (1996) Alkaloids from the tunicate *Polycarpa aurata* from Chuuk atoll. J. Org. Chem., *61*, 2709-12.

Abbenante, G.; Fairlie, D.P.; Gahan, L.R.; Hanson, G.R.; Pierens, G.K.; van den Brenk, A.L.(1996) Conformational control by thiazole and oxazoline rings in cyclooctapeptides of marine origin. Novel macrocyclic chair and boat conformations. J. Am. Chem. Soc. *118*, 10384-8.

Abbott, A. (2001) Genetic medicine gets real. Nature, *411*, 410-2.

Académie (1997) Dynamique d'espèces marine invasives: application à l'expansion de *Caulerpa taxifolia* en Méditerranée. TEC & DOC, Paris.

Acebal, C.; Alcazar, R.; Cañedo, L.M.; de la Calle, F.; Rodriguez, P.; Romero, F.; Fernández Puentes, J.L. (1998) Two marine *Agrobacterium* producers of sesbanimide antibiotics. J. Antibiot., *51*, 64-7.

Ahmad, V.U.; Zahid, M.; Shaig, A.M.; Ali, Z.; Jasshi, A.R.; Abbas, M.; Clardy, J.; Lobkovsky, E.; Tareen, R.B.; Iqbal, M.Z. (1999) Salvadiones A-B, two terpenoids having novel carbon skeleton from *Salvia bucharica*. J. Org. Chem., *64*, 8465-7.

Aiello, A.; Fattorusso, E.; Menna, M.; Pansini, M. (1996) The chemistry of the demosponge *Dysidea fragilis* from the lagoon of Venice. Biochem. System. Ecol., *24*, 37-42.

Aimi, N.; Odaka, H.; Sakat, S.; Fuliki, H.; Suganuma, M.; Moore, R.E.; Patterson, G.L. (1990) Lyngbyatoxins B and C, two new irritants from *Lyngbya majuscula*. J. Nat. Prod., *53*, 1593-6.

Albaugh, D.; Albert, G.; Bradford, P.; Cotter, V.; Froyd, J.; Gaughran, J.; Kirsch, D.R.; Lai, M.; Rehnig, A.; Sieverding, E.; Silverman, S. (1998) Cell wall active antifungal compounds produced by the marine fungus *Hypoxylon oceanicum* LL-15G256. J. Antibiot., *51*, 317-22.

Albizati, K.F.; Kim, F.; Pawlik, J.R.; Faulkner, D.J. (1985) Limatulone, a potent defensive metabolite of the intertidal limpet *Collisella limatula*. J. Org. Chem., *50*, 3428-30.

Alborn, H.T.; Turlings, T.C.J.; Jones, T.H.; Stenhagen, G.; Loughrin, J.H.; Tumlinson, J.H. (1997) An elicitor of plant volatiles from beet armyworm oral secretion. Science, *276*, 945-9.

Albrecht, M.; Takaichi, S.; Steiger, S.; Wang, Z.-Y.; Sandmann, G. (2000) Novel hydroxycarotenoids with improved antioxidative properties produced by gene combination in *Escherichia coli*. Nature Biotechnol., *18*, 843-6.

Albrizio, S.; Ciminiello, P.; Fattorusso, E.; Magno, S.; Pawlik, J.R. (1995) Amphitoxin, a new high molecular weight antifeedant pyridinium salt from the Caribbean sponge *Amphimedon compressa*. J. Nat. Prod., *58*, 647-52.

Altmann, K.-H.; Bold, G.; Caravatti, G.; End, N.; Flörsheimer, A.; Guagnano, V.; O'Relly, T.; Wartman, M. (2000) Epothilones and their analogs - Potential new weapons in the fight against cancer. Chimia, *54*, 612-21.

Alvarenga, N.L.; Ferro, E.A.; Ravelo, A.G.; Kennedy, M.L.; Maestro, M.A.; González, A.G. (2000) X-Ray analysis of volubilide, a new decacyclic Diels-Alder C_{20}-C_{30} adduct from *Hyppocratea volubilis* L. Tetrahedron, *56*, 3771-4.

Amagata, T.; Doi, M.; Tohgo, M.; Minoura, K.; Numata, A. (1999) Dankasterone, a new class of cytotoxic steroid produced by *Gymnascella dankaliensis*. Chem. Commun. 1321-2.

Ametamey, S.M.; Westera, G.; Gucker, P.; Schönbächler, R.; Honer, M.; Spang, J.E.; Schubiger, P.A. (2000) Functional brain receptor imaging with positron emission tomography. Chimia, *54*, 622-6.

Amico, V.; Oriente, G.; Piattelli, M.; Tringali, C.; Fattorusso, E.; Magno, S.; Mayol, L. (1979) Sesquiterpene based on the cadalane skeleton from the brown alga *Dilophus fasciola*. Experientia, *35*, 450-1.

Andersen, R.J.; Van Soest, R.W.M.; Kong, F. (1996) 3-Alkylpiperidine alkaloids isolated from marine sponges in the order Haplosclerida. In Alkaloids: Chemical and Biological Perspectives, Vol 10, Pelletier, S.W., ed., Pergamon Press, Oxford, pp. 301-55.

Anderson, D.M.; Scholin, C.A. (1997) Genetic markers derived from rRNAs and their use in detection of *Alexandrium* species of dinoflagellates associated with red tides. US Patent 5,582,983. Chem. Abstr., 1997, *126*, 101706k.

Anke, H. (1997) Zearalenone and zeranol. In Fungal Biotechnology. Anke, T., ed., Chapman & Hall, London, pp. 186-92.

Antonov, A.S.; Kalinovsky, A.J.; Stonik, V.A. (1998) Ulososide B, a new unusual norlanostane-triterpene glycoside and its genuine aglycone from the Madascar sponge *Ulosa* sp. Tetrahedron Lett., *39*, 3807-3808.

Arabidopsis Initiative (2000) Analysis of the genome sequence of the flowering plant *Arabidopsis thaliana*. Nature, *408*, 796-815.

Arimoto, H.; Arimoto, Y.; Okumura, S.N.; Shosuke, Y. (1995) Synthetic studies on fully substituted γ-pyrone-containing natural products: total synthesis and structural revision of onchitriol I. Tetrahedron Lett., *36*, 5357-8.

Asakawa, Y. (1995) Chemical constituents of the bryophytes. in Fortschr. Chem. Org. Naturst., Herz, W.; Kirby, G.W.; Moore, R.E.; Steglich, W.; Tamm, C., eds., *65*, 5-618.

Ashton, C. (1999) Reinventing drug development. Chem. Ind., 422-5.

AY = Ayensu, E.S. (1979) Plants for medicinal uses with special reference to arid zones. In Arid Land Plant Resources, Proc. Int. Arid Land Conf. on Plant Resources, Goodin, J.R; Northington, D.K., eds., Intern. Center for Arid and Semi-Arid Land Studies, Texas Tech University, Lubbock, TX, pp. 117-78.

Bae, M.-A.; Yamada, K.; Ijuin, Y.; Tsuji, T.; Yazawa, K.; Tomono, Y.; Uemura, D. (1996) Aburatubolactam A, a novel inhibitor of superoxide anion. Heterocycl. Commun., *2*, 315-8.

Bagley, M.C.; Bashford, K.E.; Hesketh, C.L.; Moody, C.J. (2000) Total synthesis of the thiopeptide promothiocin A. J. Am. Chem. Soc., *122*, 3301-13.

Bailey, J.E. (1999) Lessons from metabolic engineering for functional genomics and drug discovery. Nature Biotechnol., *17*, 616-8.

Bailly, C.; Lansiaux, A. (2000) Un inhibiteur de l'angiogenèse: le TNP-470. Bulletin du Cancer, *87*, 449-54.

Baker, J.T.; Wells, R.J. (1980) Biologically active substances from Australian marine organisms. In Natural Products as Medicinal Agents, Beal, J.L.; Reinhard, E., eds., Hippokrates Verlag, Stuttgart, pp. 281-318.

Bakes, M.J.; Nichols, P.D. (1995) Lipid, fatty acid and squalene composition of liver oil from six species of deep-sea sharks collected in southern Australian waters. Comp. Biochem. Physiol., *110B*, 267-75.

Baldwin, J.E.; Melnam, A.; Lee, V.; Firkin, C.; Whitehead, R.C. (1998) Biomimetic synthesis of (-)-xestospongin A, (+)-xestospongin C, (+)-araguspongine B and the correction of their absolute configuration. J. Am. Chem. Soc., *120*, 8559-60.

Baldwin, J.E.; Claridge, T.D.W.; Culshaw, A.J.; Heupel, F.A.; Lee, V.; Spring, D.R.; Whitehead, R.C. (1999) Studies on the biomimetic synthesis of the manzamine alkaloids. Chem. Eur. J., *5*, 3154-61.

Baldwin, J.E.; Fryer, A.M.; Pritchard, G.J. (2000) Novel C-4 heteroaromatic kainoid analogues: a parallel synthesis approach. Bioorg. Med. Chem. Lett., *10*, 309-11.

Ballantine, J.A.; Barrett, C.B.; Beer, R.J.S.; Eardley, S.; Robertson, A.; Shaw, B.L.; Simpson, T.H. (1958) The chemistry of bacteria. The structure of violacein. J. Chem. Soc., 755-60.

Ballantine, J.A.; Lavis, A.; Roberts, J.C.; Morris, R.J. (1977) Marine sterols. V. Sterol of some tunicata. The occurrence of saturated ring sterols in these filter-feeding organisms. J. Exp. Mar. Biol. Ecol., *30*, 29-44.

Ballio, A.; Brufani, M.; Casinovi, C.G.; Cerrini, S.; Fedeli, W.; Pellicciari, R.; Santurbano, B.; Vaciago, A. (1968) The structure of fusicoccin A. Experientia, *24*, 631-5.

Barco, A. (1997) D-(-)-Quinic acid: a chiron store for natural product synthesis. Tetrahedron Asymm., *8*, 3515-45.

Barrero, A.F.; Quílez del Moral, J.; Lara, A. (2000) Sesquiterpenes from *Juniperus thurifera* L. Stereochemistry in unusual cedrane and duprezianane series. Tetrahedron, *56*, 3717-23.

Barrett, M.D. (2000) www.quackwatch.com.

Barrow, P. (2000) An antibiotic alternative? Chem. Ind., 461-4.

Barrow, R.A.; Murray, L.M.; Lim, T.K.; Capon, R.J. (1996) Mirabilins (A-F): new alkaloids from a southern Australian marine sponge, *Arenochalina mirabilis*. Aust. J. Chem., *49*, 767-73.

Barton, D. H. R. (1994) The invention of chemical reactions of relevance to the chemistry of natural products. Pure Appl. Chem., *66*, 1943-54.

Basile, A.; Giordano, S.; Lopez-Saez, J.A.; Cobianchi, R.C. (1999) Antibacterial activity of pure flavonoids isolated from mosses. Phytochemistry, *52*, 1479-82.

Bauer, I.; Maranda, L.; Young, K.A.; Shimizu, Y. (1995) Isolation and structure of caribenolide I, a highly potent antitumor macrolide from a cultured free-swimming Caribbean dinoflagellate, *Amphidinium* sp. J. Org. Chem., *60*, 1084-6.

Bauer, S.M.; Armstrong, R.W. (1999) Total synthesis of motuporin (nodularin-V). J. Am. Chem. Soc., *121*, 6355-66.

Baumann, C.; Bröckelmann, M.; Fugmann, B.; Steffan, B.; Steglich, W.; Sheldrick, W.S. (1993) Haematopodin, an unusual pyrroloquinoline derivative isolated from the fungus *Mycena haematopus*, Agaricales. Angew. Chem. Int. Ed. Engl., *32*, 1087-9.

Baz, J.; Cañedo, L.M.; Fernández Puentes, J.L.; (1997) Thiocoraline, a new depsipeptide with antitumor activity produced by a marine *Micromonospora*. J. Antibiot., *50*, 738-41.

Beazley M. (2001) Hugh Johnson's Pocket Wine Book 2000. Octopus Publishing Group Ltd., London.

Beechan, C.M.; Djerassi, C.; Eggert, H. (1978) The sesquiterpenes from the soft coral *Sinularia mayi*. Tetrahedron, *34*, 2503-8.

Beetham, P.R.; Kipp, P.B.; Sawycky, X.L.; Arntzen, C.J.; May, G.D. (1999) A tool for functional plant genomics: chimeric RNA/DNA oligonucleotides cause *in vivo* gene-specific mutations. Proc. Natl. Acad. Sci. Usa, *96*, 8774-8.

Behrens, A.; Schaeffer, P.; Albrecht, P. (2000) Occurrence of precursors of regular tricyclopolyprenoids in recent sediments. Org. Lett., *2*, 361-3.

Bell, R.; Carmeli, S.; Sar, N. (1994) Vibrindole A, a metabolite of the marine bacterium *Vibrio parahaemolyticus*, isolated from the toxic mucus of the boxfish *Ostracion cubicus*. J. Nat. Prod., *57*, 1587-90.

Bellenie, B.R.; Goodman, J.M. (2001) The stereochemistry of glabrescol. Tetrahedron Lett., *42*, 7477-9.

Belt, S.T.; Allard, W.G.; Massé, G.; Robert, J.-M.; Owland, S.J. (2001) Structural characterisation of C_{30} highly branched isoprenoid alkenes (rhizenes) in the marine diatom *Rhizosolenia setigera*. Tetrahedron Lett., *42*, 5583-5.

Bensaude-Vincent, B. (2001) Chemical analysis. Language reform played an integral part in the development of a discipline. Nature, *410*, 415.

Bernan, V.S.; Greenstein, M.; Maiese, W.M. (1997) Marine microorganisms as a source of new natural products. Adv. Appl. Microbiol., *43*, 57; Chem. Abstr., 1997, *127*, 62907a.

Bertz, S.H. (1981) The first general index of molecular complexity. J. Am. Chem. Soc., *103*, 3599-601.

Bertz, S.H.; Sommer, T.J. (1993) Applications of graph theory to synthesis planning: complexity, reflexivity, and vulnerability. In Organic Synthesis. Theory and Applications. Hudlicky, T., ed., JAI Press, Greenwich, Conn., Vol. 2, pp. 67-92.

Bewley, C.A.; He, H.; Williams, D.H.; Faulkner, D.J. (1996) Aciculitins A-C: cytotoxic and antifungal cyclic peptides from the lithistid sponge *Aciculites orientalis*. J. Am. Chem. Soc., *118*, 4314-21.

Bhattacharya, D.; Friedl, T.; Schmidt, H.A. (1999) The phylogeny of thermophiles and hyperthermophiles and the three domains of life. In Enigmatic Microorganisms and Life in Extreme Environments, Seckbach, J., ed., Kluwer Academic Publishers, Dordrecht, Netherlands, pp. 291-304.

Biermann, U.; Friedt, W.; Lang, S.; Lühs, S.; Machmüller, G.; Metzger, J.O.; Rüsch, M.; Schäfer, H.J.; Schneider, M.P. (2000) New syntheses with oils and fats as renewable raw materials for the chemical industry. Angew. Chem. Int. Ed. Engl., *39*, 2207-24.

Bifulco, G.; Bruno, I.; Riccio, R.; Lavayre, J.; Bourdy, G. (1995A) Further brominated bis- and tris-indole alkaloids from the deep-water New Caledonian marine sponge *Orina* sp. J. Nat. Prod. *58*, 1254-60.

Bifulco, G.; Bruno, I.; Minale, L.; Riccio, R.; Debitus, C.; Bourdy, G.; Vassas, A.; Lavayre, J. (1995B) Bioactive prenylhydroquinone sulfates and a novel C_{31} furanoterpene alcohol sulfate from the marine sponge *Ircinia* sp. J. Nat. Prod., *58*, 1444-9.

Bisby, F.A. (2000) The quiet revolution: biodiversity informatics and the Internet. Science, *289*, 2309-12.

Blackman, A.J.; Hembley, T.W.; Picker, K.; Taylor, W.C.; Thirasasara (1987) Hinckdentine-A: a novel alkaloid from the marine bryozoan *Hincksinoflustra denticulata*. Tetrahedron Lett., *28*, 5561-2.

References

Blumer, M. (1965) Organic pigments: their long-term fate. Science, *149*, 722-6.

Bode, F.; Sachs, F.; Franz, M.R. (2001) Tarantula peptide inhibits atrial fibrillation. Nature, *409*, 35.

Boden, C.D.J.; Pattenden, G. (2000) Total syntheses and re-assignment of configurations of the cyclopeptides lissoclinamide and lissoclinamide 5 from *Lissoclinum patella*. J. Chem. Soc., Perkin Trans. 1, 875-82.

Boevink, P.; Martin, B.; Oparka, K.; Santa, S.; Hawes, C. (1999) Transport of virally expressed green fluorescent protein through the secretory pathway in tobacco leaves is inhibited by cold shock and brefeldin A. Planta, *208*, 392-400.

Bogenstätter, M.; Limberg, A.; Overman. L.E.; Tomasi, A.L. (1999) Enantioselective total synthesis of the kinesin motor protein inhibitor adociasulfate I. J. Am. Chem. Soc., *121*, 12206-7.

Boger, D.L.; Boyce, C.W.; Labroli, M.A.; Sehon, C.A.; Jin, Q. (1999A) Total synthesis of ningalin A, lamellarin O, lukianol A, and permethyl storniamide A utilizing heterocyclic azadiene Diels-Alder reactions. J. Am. Chem. Soc., *121*, 54-62.

Boger, D.L.; Ledeboer, M.W.; Kume, M. (1999B) Total synthesis of luzopeptins A-C. J. Am. Chem. Soc., *121*, 1098-9.

Boger, D.L.; Ichikawa, S. (2000) Total synthesis of thiocoralline and BE-22179: establishment of relative and absolute stereochemistry. J. Am. Chem. Soc., *122*, 2956-7.

Bongiorni, L.; Pietra F. (1996) Marine natural products for industrial application. Chem. Ind., 54-8.

Bonini, C.; Kinnel, R.B.; Li, M.; Scheuer, P.J.; Djerassi, C. (1983) Isolation, structure elucidation and partial synthesis of papakusterol, a new biosynthetically unusual marine sterol with a cyclopropyl-containing side chain. Tetrahedron Lett., *24*, 277-80.

Bonnett, R.; Head, E.J.; Herring, P.J. (1979) Porphyrin pigments of some deep-sea medusae. J. Mar. Biol. Ass. U.K., *59*, 565-73.

Bourguet-Kondracki, M.-L.; Guyoy, M.(1992) Smenoqualone, a novel sesquiterpenoid from the marine sponge *Smenospongia* sp. Tetrahedron Lett., *33*, 8079-80.

Bowden, B.F.; Clezy, P.S.; Coll, J.C.; Ravi, B.N.; Tapiolas, D.M. (1984) A new substituted pyrrole from a soft coral-sponge association. Aust. J. Chem., *37*, 227-30.

Bowden, B.F.; Coll, J.C.; Engelhardt, L.M.; Tapiolas, D.M.; White, A.H. (1986) The isolation and structural determination of calamenene-based sesquiterpenes from *Lemnalia cervicornis*. Aust. J. Chem., *39*, 103-21.

Bowden, B. F.; Coll, J.C.; Engelhardt, L.M.; Heaton, A.; White, A.H. (1987) Structure determination of a new sesquiterpene from *Xenia novae-britanniae*. Aust. J. Chem., *40*, 1483-9.

Brady, S.F.; Bondi, S.M.; Clardy, J. (2001) The guanacastepenes: a highly diverse family of secondary metabolites produced by an endophytic fungus. J. Am. Chem. Soc., *123*, 9900-1.

Braekman, J.C.; Daloze, D.; Pasteels, J.M. (1998) Alkaloids in animals. Alkaloids: Biochemistry Ecology and Medicinal Applications. Roberts, M.F.; Wink, M., eds., Plenum Press, pp. 349-78.

Brasier, C. (2000) The rise of the hybrid fungi. Nature, *405*, 134-5.

BR = Bruneton, J. (1995) Pharmacognosy and Phytochemistry of Medicinal Plants. Lavoisier TEC/DOC, Paris.

Briggs, D.E.G. (1994) Giant predators from the Cambrian of China. Science, *264*, 1283-4.

Brocks, J.J.; Logan, G.A.; Buick, R.; Summons R.E. (1999) Archean molecular fossils and the early rise of eukaryotes. Science, *285*, 1033-6.

Brookfield, J.F.Y. (2001) Predicting the future. Nature, *411*, 999.

Brown, B.E.; Dunne, R.P.; Goodson, M.S.; Douglas, A.E. (2000) Bleaching patterns in reef corals. Nature, *404*, 142-3.

Bruce, A.; Palfreyman, J.W., eds. (1998) Forest Products Biotechnology. Basingstoke: Taylor & Francis.

Bruna, E.M. (1999) Seed germination in rainforest fragments. Nature, *402*, 139.

Bublitz, G.B.; King, B.A.; Boxer, S.G. (1998) Electronic structure of the chromophore in green fluorescent protein (GFP). J. Am. Chem. Soc., *120*, 9370-1.

Burgoyne, D.L.; Miao, S.; Pathiran.; Andersen, R.J.; Ayer, W.A.; Singer, P.P.; Kokke, W.C.M.C.; Ross, D.M. (1990) The structure and partial synthesis of imbricatine, a benzyltetrahydroisoquinoline alkaloid from the starfish *Dermasterias imbricata*. Can. J. Chem., *69*, 20-7.

Burk, D.L.; Barovsky, K.; Monroy, G.H. (1993) Biodiversity and biotechnology. Science, *260*, 1900-1.

Butler, A.; Baldwin, A.H. (1997)Vanadium Bromoperoxidase and Functional Mimics. In Structure and Bonding: Metal Sites in Proteins and Models, Lewis Acids, and Vanadium, Sadler, P., Hill, H.A.O., Thompson, A., eds.; Springer-Verlag, New York, Volume 89, pp. 109-131.

Butler, M.S.; Capon, R.J.; Lu, C.C. (1992A) Psammopemmins (A-C), novel brominated 4-hydroxyindole alkaloids from an Antarctic sponge, *Psammopemma* sp. Aust. J. Chem., *45*, 1871-7.

Butler, M.S.; Capon, R.J. (1992B) The luffarins (A-Z), novel terpenes from an Australian marine sponge, *Luffariella geometrica*. Aust. J. Chem., *45*, 1705-43.

Caccamese, S.; Amico, V. (1990) Two new rearranged sesquiterpenoids from the red alga *Laurencia obtusa*. J. Nat. Prod., *53*, 1287-96.

Callaway, R.M.; Aschehough, E.T. (2000) Invasive plants versus their new and old neighbors: a mechanism for exotic invasion. Science, *290*, 521-3.

Cane, D.E.; Tan, W.; Ott, W.R. (1993) Nargenicin biosynthesis. J. Am. Chem. Soc., *115*, 527-35.

Cane, D.E.; Xue, Q.; van Epp, J.E.; Tsantrizos, Y.S. (1996) Enzymatic formation of isochamigrene, a novel sesquiterpene, by alteration of the aspartate-rich region of trichodiene synthase. J. Am. Chem. Soc., *118*, 8499-500.

Cantrell, C.L.; Groweiss, A.; Gustafson, K.R.; Boyd, M.R. (1999) A new staurosporine analog from the prosobranch mollusk *Coriocella nigra*. Nat. Prod. Lett., *14*, 39-46.

Capon, R.J.; Faulkner, D.J. (1985) Herbasterol, an ichthyotoxic 9,11-secosterol from the sponge *Dysidea herbacea*. J. Org. Chem., *50*, 4771-3.

Capon, R.J.; Elsbury, K.; Butler, M.S.; Lu, C.C.; Hooper, J.N.A.; Rostas, J.A.P.; O'Brien, K.J.; Mudge, L.M.; Sim, A.T.R. (1993) Extraordinary levels of cadmium and zinc in a marine sponge, *Tedania charcoti* Topsent: inorganic chemical defense agents. Experientia, *49*, 263-4.

Capon, R.J.; Rooney, F.; Murray, L.M.; Collins, E.; Sim, A.T.R.; Rostas, J.A.P.; Butler, M.S.; Carroll, A.R. (1998) Dragmacidins: new protein phosphatase inhibitors from a southern Australian deep-water marine sponge, *Spongosorites* sp. J. Nat. Prod., *61*, 660-2.

Capon, R.J.; Skene, C.; Lacey, E.; Gill, J.H.; Wadsworth, D.; Friedel, T. (1999) Geodin A magnesium salt: a novel nematocide from a southern Australian marine sponge, *Geodia*. J. Nat. Prod., *62*, 1256-9.

Caporale, L.H. (1995) Chemical ecology: a view from the pharmaceutical industry. Proc. Natl. Acad. Sci. USA, *92*, 75-82.

Carbonelli, S.; S.; Zampella, A.; Randazzo, A.; Debitus, C.; Gomez-Paloma, L. (1999) Sphinxolides E-G and reidiaspongiolide C: four new cytotoxic macrolides from the New Caledonian Lithistida sponges *N. superstes* and *R. coerulea*. Tetrahedron, *55*, 14665-74.

Carlson, T.J.; Cooper, R.; King, S.R.; Rozhon, E.J. (1997) Modern science and traditional healing. In Phytochemical Diversity. A Source of New Industrial Products. Wrigley, S.; Hayes, M.; Thomas, R.; Chrystal, E., eds., The Royal Society of Chemistry, Cambridge, UK., pp. 84-95.

Carroll, A.R.; Bowden, B.F.; Coll, J.C. (1993) Novel cytotoxic iodotyrosine-based alkaloids from colonial ascidians, *Aplidium* sp. Aust. J. Chem., *46*, 825-32.

Carroll, A.R.; Bowden, B.F.; Coll, J.C.; Hockless, D.C.R.; Skelton, B.W.; White, A.H. (1994) Mollamide, a cytotoxic cyclic heptapeptide from the compound ascidian *Didemnum molle*. Aust. J. Chem., *47*, 61-69.

Carroll, A.R.; Coll, J.C.; Bourne, D.J.; Macleod, J.K.; Zabriskie, T.M.; Ireland, C.M.; Bowden, B.F. (1996) Patellins 1-6 and trunkamide A: novel cyclic hexa-, hepta- and octa-peptides from colonial ascidians, *Lissoclinum* sp. Aust. J. Chem., *49*, 659-67.

Cassidy, M.P.; Ghisalberti, E.L. (1993) New terpene hydrocarbons from the sponge *Higginsia* sp. J. Nat. Prod., *56*, 1190-3.

Catalan, C.A.N.; Iglesias, D.I.; Retamar, J.A.; Iturraspe, J.B.; Dartayet, G.H.; Gros, E.G. (1992) 4,5-seco-African-4,5-dione from *Lippia integrifolia*. Phytochemistry, *31*, 4025-6.

CE (2001) = Columbia Encyclopedia (2001), 6th ed., Columbia University Press.

Chan, A.W.S.; Chong, K.Y.; Martinovich, C.; Simerly, C.; Schatten, G. (2001)Transgenic monkeys produced by retroviral gene transfer into mature oocytes. Science, *291*, 309-12 .

Chang, K.; Park, Y.; Chai, S.; kim, I.; Seo, Y.; Cho, K.; Shin, J. (1998) Anthraquinones and sterols from the Korean marine echiura *Urachis unicintus*. J. Korean Chem. Soc.,*42*, 64-8; Chem. Abstr., 1998, *129*, 2852k.

Channell, R.; Lomolino, M.V. (2000) Dynamic biogeography and conservation of endangered species. Nature, *403*, 84-6.

Chanon, M.; Barone, R.; Baralotto, C.; Julliard, M.; Hendrickson, J.B. (1998) Information theory description of synthetic strategies in the polyquinane series. The holosynthon concept. Synthesis, 1559-83.

Chbani, M.; Païs, M.; Delauneux, J.M.; Debitus, C. (1993) Brominated indole alkaloids from the marine tunicate *Pseudodistoma arborescens*. J. Nat. Prod., *56*, 99-104.

Chen, T.S.; Li, X.; Petuch, B.; Shafiee, A.; So, L. (1999) Structural modification of the immunosuppressant FK-506 and ascomycin using a biological approach. Chimia, *53*, 596-600.

Chicarelli-Robinson, M.I.; Gibbons, S.; McNicholas, C. (1997) Plants and microbes as complementary sources of chemical diversity for drug discovery. In Phytochemical Diversity. A Source of New Industrial Products. Wrigley, S.; Hayes, M.; Thomas, R.; Chrystal, E., eds., The Royal Society of Chemistry, Cambridge, UK., pp. 30-40.

Chivian, E. (1997) Global environmental degradation and biodiversity loss: implications for human health. In Biodiversity and Human Health. Grifo, F.; Rosenthal, J., eds., Island Press, Washington D.C., pp. 7-38.

Christel, P. Li, S. M.; Vert, M.; Patat, J.L. (1993) Material for bone prosthesis containing calcium carbonate particles dispersed in a bioabsorbable polymeric matrix. Eur. Pat. Appl. EP 564,369, 6 Oct to INOTEB. Chem. Abstr., 1993, *119*, 256592z.

Christie, M.J.; Vaughan, C.W. (2001) Cannabinoids act backwards. Nature, *410*, 527-30.

Christie, W.W.; Brechany, E.Y.; Marekov, I.N.; Stefanov, K.L.; Andreev (1994) The fatty acids of the sponge *Hymeniacidon sanguinea* from the Black Sea. Comp. Biochem. Physiol., *109B*, 245-52.

Christophersen, C. (1985) Secondary metabolites from marine bryozoans. Acta Chem. Scand., *B39*, 517-529.

Chumpolkulwong, N.; Kakizono, T.; Handa, T.; Nishio, N. (1997) Isolation and characterization of compactin resistant mutants of an astaxanthin synthesizing green alga *Haematococcus pluvialis*. Biotechnol. Lett., *19*, 299-302.

Chung, B.G.; Lee, S.J. (1997) *Aspergillus fumigatus* mutant for manufacture of chitosan oligosaccharide from chitosan. Jpn. Kokai Tokkyo Koho JP 08,322,554 of 10 Dec. 1996. Chem. Abstr., 1997, *126*, 101706k.

Ciminiello, P.; Fattorusso, E.; Magno, S.; Mangoni, A.; Pansini, M. (1990) Incisterols, a new class of highly degraded sterols from the marine sponge *Dictyonella incisa*. J. Am. Chem. Soc., *112*, 3505-9.

Cimino, G.; Spinella, A.; Sodano, G. (1989) Potential alarm pheromones from the Mediterranean opisthobranch *Scaphander lignarius*. Tetrahedron Lett., *30*, 5003-4.

Cimino, G.; Crispino, A.; Di Marzo, V.; Spinella, A.; Sodano, G. (1991) Prostaglandin 1,15-lactones of the F series from the nudibranch mollusc *Tethys fimbria*. J. Org. Chem., *56*, 2907-11.

Cimino, G.; Crispino, A.; Madaio, A.; Trivellone, E.; Uriz, M. (1993) Raspacionin B, a further triterpenoid from the Mediterranean sponge *Raspaciona aculeta*. J. Nat. Prod., *56*, 534-8.

Cinel, B.; Patrick, B.O.; Roberge, M.; Andersen, R.J. (2000) Solid-state and solution conformations of eleutherobin from X-ray diffraction analysis and solution NOE data. Tetrahedron Lett., *41*, 2811-5.

Ciomei, M.; Albanese, C.; Pastori, W.; Grandi, M.; Pietra, F.; D'Ambrosio, M.; Guerriero, A.; Battistini, C. (1997) Sarcodictyins: a new class of marine derivatives with mode of action similar to Taxol. American Association for Cancer Research, Eighty-eighth Annual Meeting, April 12-16, 1997, San Diego, CA, Volume 38, Abstract Nr. 30, Section Pharmacology/Therapeutics (Preclinical and Clinical) .

Cipriano, F.; Palumbi, S.R. (1999) Genetic tracking of a protected whale. Nature, *397*, 307-8.

Ciufolini, M.A.; Shen, Y.-C.; Bishop, M.J. (1995) A unified strategy for the synthesis of sulfur-containing pyridoacridine alkaloids: antitumor agents of marine origin. J. Am. Chem. Soc., *117*, 12460-9.

Clark, R.J.; Field, K.L.; Charan, R.D.; Garson, M.J.; Brereton, I.M.; Willis, A.C. (1998) The haliclonacyclamines, cytotoxic tertiary alkaloids from the tropical marine sponge *Haliclona* sp. Tetrahedron, *54*, 8811-26.

Clay, K.; Holah, J. (1999) Fungal endophyte symbiosis and plant diversity in successional fields. Science, *285*, 1742-4.

Coffey, D.S.; McDonald, A.I.; Overman, L.E.; Rabinowitz, M.H.; Renhowe, P.A. (2000) A practical entry to the crambescidin family of guanidine alkaloids. Enantioselective total syntheses of ptilomycalin A, crambescidin 657 and its methyl ester, and crambescidin 800. J. Am. Chem. Soc., *122*, 4893-903.

Cohen E., ed. (1979) Biomedical Applications of the Horseshoe Crab, Alan R. Riss , Inc. New York.

Coleman, J.E.; de Silva, E.D.; Kong, F.; Andersen, R.J.; Allen, T.M. (1995) Cytotoxic peptides from the marine sponge *Cymbastela* sp. Tetrahedron, *51*, 10653-62.

Colin M.; Guenard, D.; Gueritte-Voegelein, F.; Potier, P (1988) Preparation of Taxol derivatives as antitumor agents. US Patent 4,814,470; Eur. Pat. Appl. (1988).

Coll, J.; Skelton, B.W.; White, A.H.; Wright, A.D. (1989) The structure determination of two novel sesquiterpenes from the red alga *Laurencia tenera*. Aust. J. Chem., *42*, 1695-703.

Coll, J.; Kearns, P.S.; Rideout, J.A.; Hooper, J. (1997) Ircianin sulfate from the marine sponge *Ircinia (Psammocinia) wistari*. J. Nat. Prod., *60*, 1178-9.

Collum, D.B.; McDonald III, J.H.; Still, W.C. (1980) Synthesis of the polyether antibiotic monensin. J. Am. Chem. Soc., *102*, 2120-1.

Comin, J.; Gonçalves de Lima, O.; Grant, H.N.; Jackman, L.M.; Keller-Schierlein, W.; Prelog, V. (1963) Über die Konstitution des Biflorins, eines *o*-Chinons der Diterpene-Reihe. Helv. Chim. Acta, *46*, 409-15.

Committee O.R. (1999) From Monsoons to Microbes. Understanding the Ocean's Role in Human Health. National Academy Press. Washington D.C.

Conde-Petit, B.; Nüssli, J.; Arrigoni, E.; Escher, F.; Amadò, R. (2001) Perspectives of starch in food science. Chimia, *55*, 201-5.

Conference (2001) 8th Conference on Retroviruses and Opportunistic Infections, Chicago, Ill., 4-8 February.

Connolly, J.D.; Hill, R.A. (1997) Triterpenoids. Nat. Prod. Rep., *14*, 661-79.

Constanz, B.R.; Ison, I.C.; Fulmer, M.T.; Poser, R.D.; Smith, S.T.; Van Wagoner, M.; Ross, J.; Goldstein, S.A.; Jupiter, J.B.; Rosenthal, D.I. (1995) Skeletal repair by in situ formation of the mineral phase of bone. Science, *267*, 1796-8.

Contag, C.H.; Contag, P.R. (1999) Illuminating drug discovery. Chem. Ind., 664-6.

Corey, E.J.; Ohno, M.; Vatachencherry, P.A.; Mitra, R.B. (1961) Total synthesis of *d,l*-longifolene. J. Am. Chem. Soc., *83*, 1251-3.

Corey, E.J.; Gosh, A.K. (1988) Enantioselective route to a key intermediate in the total synthesis of ginkgolide B. Tetrahedron Lett., *29*, 3201-4.

Corey, E.J.; Che, X.-M. (1989) The Logic of Chemical Synthesis. Wiley, New York.

Corey, E.J.; Luo, G.; Lin, L.S. (1997A) A simple enantioselective synthesis of the biologically active tetracyclic marine sesterterpene scalarenedial. J. Am. Chem. Soc., *119*, 9927-8.

Corey, E.J.; Roberts, B.E. (1997B) Total synthesis of dysidiolide. J. Am. Chem. Soc., *119*, 12425-31.

Corey, E.J.; Lazerwith, S.E. (1998) A direct and efficient streocontrolled synthetic route to the pseudopterosins, potent marine anti-inflammatory agents. J. Am. Chem. Soc., *120*, 12777-82.

Cormier, M.J.; Lorenz, W.W. (1994) Cloning and expressions of the gene for luciferase of *Renilla*. US Patent 5,292,658, 29 Dec. to Univ of Georgia Res Found. Chem. Abstr., 1994, *120*, 318319x.

Costantino, V.; Fattorusso, E.; Imperatore, C.; Mangoni, A. (2001) A biosynthetically significant bacteriohopanoid present in large amounts in the Caribbean sponge *Plakortis simplex*. Tetrahedron, *57*, 4045-8.

Courtillot. V.; Gaudemer, Y. (1996) Effects of mass extinctions on biodiversity. Nature, *381*, 146-8.

Coval, S.J.; Scheuer, P.J. (1985) An intriguing C_{16}-alkadienone-substituted 2-pyridine from a marine mollusk. J. Org. Chem., *50*, 3024-5.

Coval, S.J.; Puar, M.S.; Phife, D.W.; Terracciano, J.S.; Patel, M. (1995) SCH57404, an antifungal agent possessing the rare sordaricin skeleton and a tricyclic sugar moiety. J. Antibiot., *48*, 1171-2.

Cox, P.A. (1997) Indigenous people and conservation. In Biodiversity and Human Health. Grifo, F.; Rosenthal, J., eds., Island Press, Washington D.C., pp. 207-21.

Cox, P.A. (2000) Will tribal knowledge survive the millennium? Science, *287*, 44-5.

Cragg, G.M.; Newman, D.J.; Sue Yang, S. (1998) Bioprospecting for drugs. Nature, *393*, 301.

Crameri, A. (1999) Enhanced enzyme performance by DNA shuffling. Chimia, *53*, 617-20.

Crawley, M.J. (1999) Bollworm, genes and ecologists. Nature, *400*, 501-2.

Crawley, M.J.; Brown, S.L.; Hails, R.S.; Kohn, D.D.; Rees, M. (2001) Transgenic crops in natural habitats. Nature, *409*, 682-3.

Crews, P.; Myers, B.L.; Naylor, S.; Clason, E.L.; Jacobs, R.S.; Staal, G.B. (1984) Bio-active monoterpenes from red seaweeds. Phytochemistry, *23*, 1449-51.

Crews, P.; Farias, J.J.; Emrich, R.; Keifer, P.A. (1994) Milnamide A, an unusual cytotoxic tripeptide from the marine sponge *Auletta* cf. *constricta*. J. Org. Chem., *59*, 2932-4.

Crimmins, M.T.; Pace, J.M.; Nantermet, P.G.; Kim-Meade, A.S.; Thomas, J.B.; Watterson, S.H.; Wagman, A.S. (1999) Total synthesis of (±)-ginkgolide B. J. Am. Chem. Soc., *121*, 10249-50.

Crist, B.V.; Li, X.; Bergquist, P.R.; Djerassi, C. (1983) Isolation, partial synthesis, and determination of absolute configuration of pulchrasterol. The first example of double bioalkylation of the sterol side chain at position 26. J. Org. Chem., *48*, 4472-9.

Cronan,, J.M., Jr.; Davidson, T.R.; Singleton, F.L.; Colwell, R.R.; Cardellina II, J.H. (1998) Plant growth promoters isolated from a marine bacterium associated with *Palythoa* sp. Nat. Prod. Lett., *11*, 271-8.

CSIRO majordomo@ml.csiro.au.

Curran, L.M.; Caniago, I.; Paoli, G.D.; Astianti, D.; Kusneti, M.; Leighton, M.; Nirarita, C.E.; Haeruman, H. (1999) Impact of El Niño and logging on canopy tree recruitment in Borneo. Science, *286*, 2184-8.

Currie, C.R.; Scott, J.A.; Summerbell, R.C.; Malloch, D. (1999) Fungus-growing ants use antibiotic-producing bacteria to control garden parasites. Nature, *398*, 701-4.

Dabrah, T.T.; Harwood, H.J., Jr.; Huang, L.H.; Jankovich, N.D.; Kaneko, T.; Li, J.-C.; Lindsey, S.; Moshier, P.M.; Subashi, T.A.; Therrien, M.; Watts, P.C. (1997) CP-225,917 and CP-263,114, novel RAS farnesylation inhibitors from an unidentified fungus. J. Antibiot., *50*, 1-7.

Dai, J.-R.; Hallock, F.; Cardellina II, J.H.; Boyd, M.R. (1996) Vasculyne, a new cytotoxic acetylenic alcohol from the marine sponge *Cribrochalina vasculum*. J. Nat. Prod., *59*, 88-9.

Daily, G.C.; Walker, B.H. (2000) Seeking the great transition. Nature, *403*, 243-5.

Daly, J.W.; Garraffo, H.M.; Spande, T.F.; Decker, M.W.; Sullivan, J.P.; Williams, M. (2000) Alkaloids from frog skin: the discovery of epibatidine and the potential for developing novel non-opioid analgesics. Nat. Prod. Rep., *17*, 131-5.

D'Ambrosio, M.; Guerriero, A.; Pietra, F. (1984) Arboxeniolide-1, a new, naturally occurring xeniolide diterpenoid from the gorgonian *Paragorgia arborea* of the Crozet Is. (S. Indian Ocean). Zeit. Naturforsch. C, *39*, 1180-3.

D'Ambrosio, M.; Guerriero, A.; Fabbri, D.; Pietra, F. (1987) Coralloidolide A and coralloidolide B, the first cembranoids from a Mediterranean organisms, the alcyonacean *Alcyonium coralloides*. Helv. Chim. Acta, *70*, 63-70.

D'Ambrosio, M.; Guerriero, A.; Pietra, F. (1990) Coralloidolide F, the first example of a 2,6-cyclized cembranolide: isolation from the Mediterranean alcyonacean *Alcyonium coralloides*. Helv. Chim. Acta, *73*, 804-7.

D'Ambrosio, M.; Guerriero, A.; Debitus, C.; Ribes, O.; Pietra, F.; (1993) On the novel free porphyrins corallistin B, C, D, and E. Isolation from the demosponge *Corallistes* sp. of the Coral Sea and reactivity of their nickel(II) complexes toward formylating agents. Helv. Chim. Acta, *76*, 1489-96.

D'Ambrosio, M.; Guerriero, A.; Chiasera, G.; Pietra, F. (1994) Conformational preferences and absolute configuration of agelastatin A, a cytotoxic alkaloid of the axinellid sponge *Agelas dendromorpha* from the Coral Sea, *via* combined molecular modeling, NMR and exciton splitting for diamides and hydroxamide derivatives. Helv. Chim. Acta, *77*, 1895-902.

D'Ambrosio, M.; Guerriero, A.; Chiasera, G.; Pietra, F.; Tatò, M. (1996A) Epinardins A-C, new pyrroloiminoquinone alkaloids of undetermined deep-water green demosponges from pre-Antarctic Indian Ocean. Tetrahedron, *52*, 8889-906.

D'Ambrosio. M.; Guerriero, A.; Debitus, C.; Pietra, F. (1996B) Leucascandrolide A, a new type of macrolide: the first powerfully bioactive metabolite of calcareous sponges (*Leucascandra caveolata*, a new genus from the Coral Sea) Helv. Chim. Acta, *79*, 51-60.

D'Ambrosio, M.; Guerriero, A.; Pietra, F. (1996C) Glycerol enol ethers of the brachiopod *Gryphus vitreus* from the Tuscan archipelago. Experientia, *52*, 624-7.

D'Ambrosio, M.; Guerriero, A.; Deharo, E.; Debitus, C.; Munoz, V.; Pietra, F. (1998) New types of potentially antimalarial agents: norditerpene and norsesterterpene epidioxides from the marine sponge *Diacarnus levii*. Helv. Chim. Acta, *81*, 1285-92.

Damsté, J.S.S.; Köster, J.; Baas, M.; Ossebaar, J.; (1999) A sedimentary tetrahydrophenanthrene derivative of tetrahymanol. Dekker, M.; Pool, W.; Geenevasen, J.A.J. Tetrahedron Lett., *40*, 3949-52.

Darling, K.F.; Wade, C.M.; Stewart, I.A.; Kroon, D.; Dingle, R.; Leigh Brown, A.J. (2000) Molecular evidence for genetic mixing of Arctic and Antarctic subpolar populations of planktonic foraminifers. Nature, *405*, 43-7.

D'Auria, M.V.; Gomez Paloma, L.; Minale, L.; Riccio, R.; Debitus, C. (1992) Structure characterization of two marine triterpene oligoglycosides from a Pacific sponge of the genus *Erylus*. Tetrahedron, *48*, 491-8.

D'Auria M.V. ; Gomez Paloma, L.; Minale, L.; Riccio, R.; Zampella, A.; Debitus, C. (1993) Metabolites of the New Caledonian sponge *Cladocroce incurvata*. J. Nat. Prod., *56*, 418-23.

D'Auria M.V.; Gomez Paloma, L.; Minale, L.; Debitus, C. (1994) A novel cytotoxic macrolide, superstolide B. J. Nat. Prod., *57*, 1595-7.

Davidson, B.S.; Schumacher, R.W. (1993) Isolation and synthesis of caprolactins A and B, new caprolactams from a marine bacterium. Tetrahedron, *49*, 6569-74.

Davies-Coleman, M.T.; Faulkner, D.J.; Dubowchik, G.M.; Roth, G.P.; Polson, C.; Fairchild, C. (1993) A new EGF-active polymeric pyridinium alkaloid from the sponge *Callyspongia fibrosa*. J. Org. Chem., *58*, 5925-30.

Davies-Coleman, M.T.; Garson, M.J. (1998) Marine polypropionates. Nat. Prod. Rep., *15*, 477-93.

Davis, B. (2000) Hand in glove. Chem. Ind., 135-8.

Davis, M.B.; Shaw, R.G. (2001) Range shifts and adaptive responses to quaternary climate changes. Science, *292*, 673-9.

Davis, R.A.; Carroll, A.R.; Quinn, R.J. (1998) Eudistomin V, a new β-carboline from the Australian ascidian *Pseudodistoma aureum*. J. Nat. Prod., *61*, 959-60.

Davis, R.A.; Carroll, A.R.; Pierens, G.K.; Quinn, R.J. (1999A) New lamellarin alkaloids from the Australian ascidian, *Didemnum chartaceum*. J. Nat. Prod., *62*, 419-24.

Davis, R.A.; Carroll, A.R.; Quinn, R.J. (1999B) Longithorones J and K. J. Nat. Prod., *62*, 158-60.

Debitus, C.; Cesario, M.; Guilhem, J.; Pascard, C.; Païs, M. (1989) Tetrahedron Lett., (1989) Corallistine, a new polynitrogen compound from the sponge *Corallistes fulvodesmus*. Tetrahedron Lett., *30*, 1535-8.

Debono. M.; Turner. W.W.; Lagrandeur, L.; Burkhardt, F.J.; Nissen, J.S.; Nichols, K.K.; Rodriguez, M.J., Zweifel, M.J; Zeckner, D.J.; Gordee, R.S.; Tang, J.; Parr, T.R. (1995) Semisynthetic chemical modification of the antifungal lipopeptide echinocandin-b (ecb): structure-activity studies of the lipophilic and geometric parameters of polyarylated acyl analogs of ecb. J. Med. Chem., *38*, 3271-81.

de Guzman, F.S.; Carte, B.; Troupe, N.; Faulkner, D.J.; Harper, M.K.; Concepcion, G.P.; Mangalindan, G.C.; Matsumoto, S.S.; Barrows, L.R.; Ireland, C.M. (1999) Neoamphimedine: a new pyridoacridine topoisomerase II inhibitor which catenates DNA. J. Org. Chem., *64*, 1400-2.

Delbert, L.H.; Cascarano, G.; Pettit, G.R.; Srirangam, J.K. (1997) Crystal conformation of the cyclic decapeptide phakellistatin 8: comparison with antamanide. J. Am. Chem. Soc., *119*, 6962-73.

Dellar, J.E.; Cole, M.D.; Waterman, P.G. (1996) Antifungal polyoxygenated fatty acids from *Aeollanthus parvifolius*. J. Chem. Ecol., *22*, 897-906.

DeLong, E.F.; Wu, K.Y.; Prézelin, B.B.; Jovine, R.V.M. (1994) High abundance of Archaea in Antarctic marine picoplankton. Nature, *371*, 695-7.

Dembisky, V.M.; Řezanka, T.; Kashin, A.G. (1994) Comparative study of the endemic freshwater fauna of Lake Baikal - VI. Unusual fatty acid and lipid composition of the endemic sponge *Lubomirskia baicalensis* and its amphipod crustacean parasite *Brandtia (Spinacanthus) parasitica*. Comp. Biochem. Physiol., *109B*, 415-26.

Deml, R.; Huth, A. (2000) Benzoquinones and hydroquinones in defensive secretions of tropical millepedes. Naturwissenshaften, *87*, 80-2.

Deng, J.; Hamada, Y.; Shioiri, T. (1995) Total synthesis of alterobactin A, a super siderophore from an open-ocean bacterium. J. Am. Chem. Soc., *117*, 7824-5.

de Nys, R.; Coll, J.C.; Carroll, A.R.; Bowden, B.F. (1993) Isolaurefucin methyl ether, a new lauroxane derivative from the red alga *Dasyphila plumarioides*. Aust. J. Chem., *46*, 1073-7.

de Riccardis, F.; Iorizzi, M.; Minale, L.; Riccio, R.; Richer de Forges, B.; Debitus, C. (1991) The gymnochromes: novel marine brominated phenanthroperylene-quinone pigments from the stalked crinoid *Gymnocrinus richeri*. J. Org. Chem., *56*, 6781-7.

de Rosa, M.; de Rosa, S.; Gambacorta, A.; Minale, L.; Bu'Lock, J.D. (1977) Chemical structure of the ether lipids of thermophilic acidiphilic bacteria of the *Caldariella* group. Phytochemistry, *16*, 1961-5.

de Solla Price, D. (1963) Little Science, Big Science. Columbia University Press., N.Y.

Devort, M. (1997) La Bécassine des Marais. Éléments pour un Plain d'Action. Éditions Confluences CICB-OMPO, Paris.

Distel, D.L.; Baco, A.R.; Chuang, E.; Morrill, W.; Cavanaugh, C.; Smith, C.R. (2000) Marine ecology: do mussel take wooden steps to deep-sea vents? Nature, *403*, 725-6.

Djerassi, C.; Silva, C.J. (1991) Sponge sterols: origin and biosynthesis. Acc. Chem. Res., *24*, 371-8.

Doi, M.; Ishida, T.; Kobayashi, M.; Kitagawa, I. (1991) Molecular conformation of swinholide A, a potent cytotoxic dimeric macrolide from the Okinawan marine sponge *Theonella swinhoei*. J. Org. Chem., *56*, 3629-32.

Domoto, N.; Kasutoori, U.; Tanaka, H.; Miki, W. (1994) Protease, its manufacture with *Alteromonas*, and *Alteromonas* microorganisms. Jpn. Kokai Tokkyo Koho JP Patent 06 38,749, 15 Feb. to Kaiyo Bio Tech Lab. Chem. Abstr., 1994, *120*, 318322t.

Doty, M.S. (1988) Prodromus ad systematica Eucheumatoideorum: a tribe of commercial seaweeds related to *Eucheuma* (Solieriaceae, Gigartinales). In Taxonomy of Economic Seaweeds with Special Reference to Pacific and Caribbean Species. Volume 11. Abbott, I. A. (ed.), California Sea Grant College Program Rep. No. T-CSGCP-018, pp. 159-207.

Dounay, A.B.; Urbanek, R.A.; Frydrychowski, V.A.; Forsyth, C.J. (2001) Expedient access to the okadaic acid architecture. J. Org. Chem., *66*, 925-38.

Draths, K.M.; Knop, D.R.; Frost, J.W. (1999) Shikimic acid and quinic acid: replacing isolation from plant sources with recombinant microbial biocatalysis. J. Am. Chem. Soc., *121*, 1603-4.

Drews J. (2000) Drug discovery: a historical perspective. Science, *287*, 1960-4.

DSII (Devon, T.K.; Scott, A.I., 1972) Handbook of Naturally Occurring Compounds. Vol II, Terpenes. Academic Press, NY.

Duan, M.; Paquette, L. (2001) Enantioselective total synthesis of the cyclophilin-binding immunosuppressive agent sanglifehrin A. Angew. Chem. Int. Ed. Engl., *40*, 3632-5.

Dumdei, E.J.; Blunt, J.W.; Munro, M.H.G.; Pannell, L.K. (1997) Isolation of calyculins, calyculinamides, and swinholide H from the New Zealand deep-water marine sponge *Lamellomorpha strongylata*. J. Org. Chem., *62*, 2636-9.

Dunlop, R.W.; Kazlauskas, R.; March, G.; Murphy, P.T.; Wells, R.J. (1982) New furano-sesquiterpenes from the sponge *Dysidea herbacea*. Aust. J. Chem., *35*, 95-103.

Edwards, J.L.; Lane, M.A.; Nielsen, E.S. (2000A) Interoperability of biodiversity databases: biodiversity information on every desktop. Science, *289*, 2312-4.

Edwards. J.L.; Lane, M.A.; Nielsen, E.S. (2000B) A response to Smith 2000. Science, *290*, 2073-4.

Ehrlich, P.R. (1986) The Machinery of Nature. Simon and Schuster, New York.

Eichberg, M.J.; Dorta, R.L.; Grotjahn, D.B.; Lamottke, K.; Schmidt, M.; Vollhardt, K.P.C. (2001) Approaches to the synthesis of (±)-strychnine via the cobalt-mediated [2 + 2 + 2] cycloaddition: rapid assembly of a classical framework. J. Am. Chem. Soc., *123*, 9324-37.

Endo, A.; Hasumi, K. (1997) Mevinic acids. In Fungal Biotechnology. Anke, T. ed., Chapman & Hall, London, pp. 162-72.

Engberts, J.B.F.N.; Blandamer, M.J. (2001) Understanding organic reactions in water: from hydrophobic encounters to surfactant aggregates. Chem. Commun. 1701-8.

Enoki, N.; Furusaki, A.; Suehiro, K.; Ishida, R.; Matsumoto, T. (1983) Epoxydictymene, a new diterpene from the brown alga *Dictyota dichotoma*. Tetrahedron Lett., *24*, 4341-2.

Enquist, B.J.; West, G.B.; Charnov, E.L.; Brown, J.H. (1999) Allometric scaling of production and life-history variation in vascular plants. Nature, *401*, 907-11.

Enserink, M. (1999) Biological invaders sweep in. Science, *285*, 1834-6.

Epstein, P.R.; Dobson, A.; Vandermeer, J. (1997) Biodiversity and emerging infectious diseases: integrating health and ecosystems monitoring. In Biodiversity and Human Health. Grifo, F.; Rosenthal, J., eds. Island Press, Washington D.C., pp. 60-86.

Erkel, G. (1997) Fusidic acid, griseofulvin and pleuromutilin. In Fungal Biotechnology. Anke, T. ed., Chapman & Hall, London, pp. 128-35.

Eschenmoser, A.; Wintner, C.E. (1977) Natural product synthesis and vitamin B_{12}. Science, *196*, 1410-20.

Eschenmoser, A. (1999) Chemical etiology of nuclei acid structure. Science, *284*, 2118-24.

Espada, A.; Jiménez, C.; Debitus, C.; Riguera, R. (1993) Villagorgin A and B. New type of indole alkaloids. Tetrahedron Lett., *34*, 7773-6.

Evans, D.A.; Carter, P.H.; Carreira, E.M; Charette, A.B.; Prunet, J.A.; Lautens, M. (1999A) Total synthesis of bryostatin 2. J. Am. Chem. Soc., *121*, 7540-52.

Evans, D.A.; Trotter, B.W.; Coleman, P.J.; Cote, B.; Dias, L.C.; Rajapakse, H.A.; Tyler, A.N. (1999B) Enantioselective total synthesis of altohyrtin C (spongistatin 2). Tetrahedron, *55*, 8671-726.

Evidente, A.; Capasso, R.; Andolfi, A.; Vurro, M.; Zonno, M.C. (1998). Putaminoxins D and E from *Phoma putaminum*. Phytochemistry, *48*, 941-5.

Ewart, K.V. (2000) In from the cold. Chem. Ind., 366-9.

Facciotti, D.; Metz, J.G.; Lassner, M. (1998) Polyketide synthesis genes of marine microbes and production of polyunsaturated fatty acids and PUFA-containing plant oils with transgenic plants. PCT Int. Appl. 10 Dec. to Calgene LLC Co); Chem. Abstr., 1999, *130*, 62050.

Fedorov, S.N.; Levina, E.V.; Kalinovsky, A.I.; Dmitrenok, P.S.; Stonik, V.A. (1994) Sulfated steroids from Pacific brittle stars. J. Nat. Prod., *57*, 1631-7.

Fedorov, S.N.; Shubina, L.K.; Kalinovsky, A.I.; Lyakhova, E.G.; Stonik, V.A. (2000) Structure and absolute configuration of a new rearranged chamigrane-type sesquiterpenoid from the sea hare *Aplysia* sp. Tetrahedron Lett., *41*, 1979-82.

Fernández, R.; Dherbomez, M.; Letourneux, Y.; Nabil, M.; Verbist, J.F.; Biard, J.F. (1999) Antifungal metabolites from the marine sponge *Pachastrissa* sp.: new bengamides and bengazole derivatives. J. Nat. Prod., *62*, 678-89.

Finlay, B.J.; Corliss, J.O.; Esteban, G.; Fenchel, T. (1996) Biodiversity at the microbial level: the number of free-living ciliates in the biosphere. Quart Rev. Biol., *71*, 221-37.

Finlay, B.J.; Clarke, K.J. (1999) Ubiquitous dispersal of microbial species. Nature, *400*, 828.

Fisher, H.; Kansy, M.; Bur, D. (2000) CAFCA: a novel tool for the calculation of amphiphilic properties of charged drug molecules. Chimia, *54*, 640-5.

Flamm, F. (1994) The chemistry of life at the margins. Science, *265*, 471-2.

Fleishmann, R.D.; Adams, M.D.; White, O.; Clayton, R.A.; Kirkness, E.F.; Kerlavage, A.R.; Bult, C.J.; Tomb, J.F.; Dougherty, B.A.; Merrick, J.M. (1995) Whole-genome random sequencing and assembly of *Haemophilus influenzae*. Science, *269*, 496-512.

Flowers, A.E.; Garson, M.J.; Webb, R.I.; Dumdei, E.J.; Charan, R.D. (1998) Cellular origin of chlorinated diketopiperazines in the dictyoceratid sponge *Dysidea herbacea* (Keller). Cell Tissue Res., *292*, 597-607.

Fontana, A.; Muniaín, C.; Cimino, G. (1998) First chemical study of Patagonian nudibranchs. J. Nat. Prod., *61*, 1027-9.

Forsyth, C.J.; Hao, J.; Aiguade, J. (2001) Synthesis of the (+)-C26-C40 domain of the azaspiric acids by a novel double intramolecular hetero-Michael addition strategy. Angew. Che. Int. Ed. Engl., *40*, 3663-7.

Fostel, J.M.; Lartey, P.A. (2000) Emerging novel antifungal agents. DDT (Elsevier), *5*, 25-32.

Foster, M.P.; Conceptión, G.P.; Caraan, G.B.; Ireland, C.M. (1992) Bistratamides C and D. Two new oxazole-containing cyclic hexapeptides isolated from a Philippine *Lissoclinum bistratum* ascidian. J. Org. Chem., *57*, 6671-5.

Franco, L.H.; Joffé, E.B.; Puricelli, L.; Tatian, M.; Seldes, A.M.; Palermo, J.A. (1998) Indole alkaloids from the tunicate *Aplidium meridianum*. J. Nat. Prod., *61*, 1130-2.

Franklin, M.A.;Penn, S.G.; Lebrilla, C.B.; Lam, T.H.; Pessah, I.N.; Molinski, T.F. (1996) Bastadin 20 and bastadin *O*-sulfate esters from *Ianthella basta*: novel modulators of the Ry_1R FKBP12 receptor complex. J. Nat. Prod., *59*, 1121-7.

Frey, M.; Chomet, P.; Glawischnig, E.; Stetter, C.; Grün, S.; Winklmair, A.; Eisenreich, W.; Bacher, A.; Meeley, R.B.; Briggs, S.P.; Simcox, K.; Gierl, A. (1997) Analysis of a chemical plant defense mechanism in grasses. Science, *277*, 696-9.

Freyssinet, T.T.; Thomas, T. (1988) Plants as a factory to produce molecules. Pure Appl. Chem. *70*, 61-6.

Frisvad, J.C.; Thrane, U.; Filtenborg, O. (1998) Role and use of secondary metabolites in fungal taxonomy. In Chemical Fungal Taxonomy. Frisvad, J.C.; Bridge, P.D.; Arora, D.K., eds., M. Dekker, New York, N.Y., pp. 289-319.

Fu, X.; Hossain, M.B.; Schmitz, F.J.; van der Helm, D. (1997) Longithorones, unique prenylated para- and metacyclophane type quinones from the tunicate *Aplidium longithorax*. J. Org. Chem., *62*, 3810-19.

Fu, X.; Schmitz, F.J.; Kelly-Borges, M.; McCready, T.L.; Holmes, C.F.B. (1998A) Clavosines A-C from the marine sponge *Myriastra clavosa*. Potent cytotoxins and inhibitors of protein phosphatases 1 and 2A. J. Nat. Prod., *63*, 7957-63.

Fu, X.; Ferreira, M.L.G.; Schmitz, F.J.; Kelly-Borges, M. (1998B) New diketopiperazines from the sponge *Dysidea chlorea*. J. Nat. Prod., *61*, 1226-31.

Fu, X.; Schmitz, F.J.; Kelly, M. (1999) Swinholides and new acetylenic compounds from an undescribed species of *Theonella* sponge. J. Nat. Prod., *62*, 1336-8.

Fuhrman, J.A. (1999) Marine viruses and their biogeochemical and ecological effects. Nature, *399*, 541-8.

Fujimura, T.; Kawai, T.; Shiga, M.; Kajiwara, T.; Hatanaka, A. (1990) Long-chain aldehyde production in thalli culture of the marine green alga *Ulva pertusa*. Phytochemistry, *29*, 745-7.

Fujiwara, Y.; Naithou, K.; Miyazaki, T.; Hashimoto, K.; Mori, K.; Yamamoto, Y. (2001) Two new alkaloids, pipercyclobutanamides A and B, from *Piper nigrum*. Tetrahedron Lett., *42*, 2497-9.

Fukuyama, T.; Akasaka, K.; Karanewsky, D.S.; Wang, C.-L.; Schmid, G.; Kishi, Y. (1979) Stereocontrolled total synthesis of monensin. J. Am. Chem. Soc., *101*, 262-3.

Fukuyama, Y.; Kodama, M.; Kinzyo, Z.; Mori, H.; Nakayama, Y.; Takahashi, M. (1989) Structures of novel dimeric eckols isolated from the brown alga *Ecklonia kurome* Okamura. Chem. Pharm. Bull. *37*, 2438-40.

Funa, N.; Ohnishi, Y.; Fujii, I.; Shibuya, M.; Ebizuka, Y.; Horinouchi, S. (1999) A new pathway for polyketide synthesis in microorganisms. Nature, *400*, 897-9.

Fusetani, N.; Ejima, D.; Matsunaga, S.; Hashimoto, K.; Itagaki, K.; Akagi, Y.; Taga, N.; Suzuki, K. (1987) 3-Amino-3-deoxy-D-glucose: an antibiotic produced by a deep-sea bacterium. Experientia, *43*, 464-5.

Fusetani, N.; Yasumuro, K.; Matsunaga, S.; Hirota, H. (1989) Haliclamines A and B, cytotoxic macrocyclic alkaloids from a sponge of the genus *Haliclona*. Tetrahedron Lett., *30*, 6891-4.

Fusetani, N.; Asai, N.; Matsunaga, S.; Honda, K.; Yasumuro, K. (1994) Cyclostellettamines A-F, pyridine alkaloids which inhibit binding of methyl quinuclidinyl benzilate to muscarinic acetylcholine receptors, from the marine sponge, *Stelletta maxima*. Tetrahedron Lett., *35*, 3967-70.

Gabius, H.-J. (2000) Biological information beyond the genetic code: the sugar code. Naturwissenschaften, *87*, 108-121.

Gábor, L.L. (1997) Global change through invasion. Nature, *388*, 627-8.

Gahan, L.J.; Gould, F.; Heckel, D.G. (2001) Identification of a gene associated with *Bt* resistance in *Heliothis virescens*. Science, *293*, 857-64.

Galloni, P. (2000) Storia e Cultura della Caccia. Dalla Preistoria a Oggi. Laterza, Bari.

Garrett, L. (1994) The Coming Plague: Newly Emerging Diseases in a World Out of Balance. Farrar, Strauss, and Giroux, New York.

Geiser, D.M. (1998) Cause of sea fan death in the West Indies. Nature, *394*, 137-8.

George, J.D.; George, J.J. (1979) Marine Life. An Illustrated Encyclopedia of Invertebrates in the Sea. J. Wiley & Sons, N.Y.

George R.Y (1981) Functional adaptation of deep-sea organisms. in Functional Adaptation of Marine Organisms, Vernberg, F.J.; Vernberg. W.B., eds., Academic Press, N.Y., pp. 280-332.

Gerard, J.; Lloyd, R.; Barsby, T.; Haden, P.; Kelly, M.T.; Andersen, R.J. (1997) Massetolides A-H, antimycobacterial cyclic depsipeptides produced by two pseudomonads isolated from marine habitats. J. Nat. Prod., *60*, 223-9.

Gerard, J.; Haden, P.; Kelly, M.T.; Andersen, R.J. (1999) Loloatins A-D, cyclic decapeptide antibiotics produced in culture by a tropical marine bacterium. J. Nat. Prod., *62*, 80-5.

Gerber, N. (1983) Cycloprodigiosin from *Beneckea gazogenes*. Tetrahedron Lett., *24*, 2797-8.

Gerwick, W.H.; Fenical, W. (1983) Spatane diterpenes from the tropical algae *Spatoglossum schmittii* and *Spatoglossum holeii*. J. Org. Chem., *48*, 3325-9.

Gibbons, A. (1996) On the many origins of species. Science, *273*, 1496-9.

Giner, J.-L.; Djerassi, C. (1991) Biosynthesis of dinosterol, peridinosterol, and gorgosterol. J. Org. Chem., *56*, 2357-63.

Glombitza, K.-W.; Li, S.-M. (1991) Fucophlorethrols from the brown alga *Carpophyllum maschalocarpum*. Phytochemistry, *30*, 3423-7.

Gonthier, I.; Rager, M.-N.; Metzger, P.; Guezennec, J.; Largeau, C. (2001) A di-*O*-dihydrogeranylgeranyl glycerol from *Thermococcus* S 557, a novel ether lipid, and likely intermediate in the biosynthesis of diethers in Archaea. Tetrahedron Lett., *42*, 2795-7.

Goodwin, T.E.; Rasmussen, E.L.; Guinn, A.C.; McKelvey, S.S.; Gunawardena, R.; Riddle, S.W.; Riddle, H.S. (1999) African elephant sesquiterpenes. J. Nat. Prod., *62*, 1570-2.

Goreau, T.J.; Macfarlane, A.H. (1990) Coral bleaching in Jamaica. Nature, *343*, 417.

Gorinsky, C. (1998) Cunaniol polyacetylene derivatives as heart blocking agents. US Patent application 96-6444894 19960510; Eur. Pat. Appl. 94-300725.

Gottlieb, O.R.; de M. B. Borin, M.R. (1998) Natural products and evolutionary ecology. Pure Appl. Chem., *70*, 299-302.

Govindan, M.; Abbas, S.A.; Schmitz, F.J.; Lee, R.H.; Parkoff, J.S.; Slate, D.L. (1994) New cycloartanol sulfates from the alga *Tydemania expeditionitis*. J. Nat. Prod., *57*, 74-8.

Gray, R.; Jordan, F.M. (2000) Language trees support the express-train sequence of Austronesian expansion.. Nature, *405*, 1052-5.

Grayner, R.J.; Chase, M.W.; Simmonds, M.S.J. (1999) A comparison between chemical and molecular characters for the determination of phylogenetic relationships among plant families: an appreciation of Hegnauer's Chemotaxonomie der Pflanzen. Biochem. Syst. Ecol., *27*, 369-93.

Graziani, E.I.; Allen, T.M.; Andersen, R.J. (1995) Lovenone, a cytotoxic degraded triterpenoid isolated from skin extracts of the North Sea dorid nudibranch *Adalaria loveni*. Tetrahedron Lett., *36*, 1763-6.

Griffin, H.L.; Greene, R.; V.; Cotta, M. A. (1992) Industrial alkaline protease from shipworm bacteria. US Pat. Appl. US 880,912, 15 Sept. to US Dept Agriculture. Chem. Abstr., 1993, *118*, 142526e.

Groom. 1992 (Groombridge B., ed., 1992) Global Biodiversity. Status of the Earth's Living Resources. Chapman & Hall, London.

Grosjean, E.; Poinsot, J.; Charrié-Duhaut, A.; Tabuteau, S.; Adam, P.; Trendel, J.; Schaeffer, J.; Connan, J.; Dessort, D.; Albrecht, P. (2000) A heptacyclic polyprenoid hydrocarbon in sediments: a clue to unprecedented biological lipids. Chem. Commun., 923-4.

Guella, G.; Guerriero, A.; Pietra, F. (1985) Sesquiterpenoids of the sponge *Dysidea fragilis* of the North-Brittany sea. Helv. Chim. Acta, *68*, 39-48.

Guella, G.; Mancini, I; Pietra, F. (1987) Oxidative breaking of long-chain acetylenic enol ethers of glycerol of the marine sponges *Raspailia pumila* and *R. ramosa* and of model compounds with aerial oxygen. Helv. Chim. Acta, *70*, 1400-11.

Guella, G.; Mancini, I; Pietra, F. (1988) Isolation of ergosta-4,24(28)-dien-9-one from both Astrophorida demosponges and subantarctic hexactinellides. Comp. Biochem. Physiol., *90B*, 113-5.

Guella, G.; Mancini, I.; Zibrowius, H.; Pietra, F. (1989A) Aplysinopsin-type alkaloids from *Dendrophyllia* sp., a scleractinian coral of the family Dendrophylliidae of the Philippines. Facile photochemical *Z/E* photoisomerization and thermal reversal. Helv. Chim. Acta, *72*, 1444-50.

Guella, G.; Mancini, I.; Duhet, D.; Richer de Forges, B.; Pietra, F. (1989B) Ethyl 6-bromo-3-indolcarboxylate and 3-hydroxyacetal-6-bromoindole, novel bromoindoles from the sponge *Pleroma menoui* of the Coral Sea. Z. Naturforsch., C: Biosci., *44c*, 914-6.

Guella, G.; Mancini, I.; Chiasera, G.; Pietra, F. (1989C) Sphinxolide, a 26-membered antitumoral macrolide isolated from an unidentified Pacific nudibranch. Helv. Chim. Acta, *72*, 237-46.

Guella, G.; Pietra, F (1991) Rogiolenyne A, B, and C: the first branched marine C_{15} acetogenins. Isolation from the red seaweed *Laurencia microcladia* or the sponge *Spongia zimocca* of Il Rogiolo. Helv. Chim. Acta, *74*, 47-54.

Guella, G.; Mancini, I; Pietra, F. (1992) C_{15} acetogenins and terpenes of the dictyoceratid sponge *Spongia zimocca* of Il Rogiolo: a case of seaweed-metabolite transfer to, and elaboration within, a sponge? Comp. Biochem. Physiol. B, *103B*, 1019-23.

Guella, G.; Öztunç, A.; Mancini, I.; Pietra, F. (1997A) Stereochemical features of sesquiterpene metabolites as a distinctive trait of red seaweeds in the genus *Laurencia*. Tetrahedron Lett., *38*, 8261-4; Francisco M.E.Y.; Turnbull, M.M.; Erickson, K.L. Cartilagineol, the fourth lineage of *Laurencia*-derived polyhalogenated chamigrene. Ibid., *39*, 5289-92.

Guella, G.; Öztunç, A.; Chiasera, G.; Mancini, I.; Pietra, F. (1997B) Slowly interconverting conformers of twelve-membered, *O*-bridged, cyclic ethers of red seaweeds in the genus *Laurencia*. Chem. Eur. J., *3*, 1223-31.

Guella, G.; Pietra, F. (1998) Antipodal pathways to secondary metabolites in the same eukaryotic organism. Chem. Eur. J., *4*, 1692-7.

Guella, G.; Dini, F.; Pietra, F. (1999A) Metabolites with a novel C_{30} backbone from marine ciliates. Angew. Chem. Int. Ed. Engl., *38*, 1134-6.

Guella, G.; Mancini, I.; Öztunç, A.; Pietra, F. (1999B) Conformational bias in macrocyclic ethers and observation of high solvolytic reactivity at a masked furfuryl (= 2-furylmethyl) C-atom. Helv. Chim. Acta, *82*, 336-48.

Guella, G.; Skropeta, D.; Breuils, S.; Mancini, I.; Pietra, F. (2001A) Calenzanol, the first member of a new class of sesquiterpene with a novel skeleton isolated from the red seaweed *Laurencia microcladia* from the Bay of Calenzana, Elba Island. Tetrahedron Lett., *42*, 723-5.

Guella, G.; Pietra, F. (2001B) unpublished results on sesquiterpenes from *Euplotes vannus*.

Guerriero, A.; D'Ambrosio, M.; Pietra, F. (1988) Leiopathic acid, a novel optically active hydroxydocosapentaenoic acid, and related compounds, from the black coral *Leiopathes* sp. of Saint Paul Island (S. Indian Ocean). Helv. Chim. Acta, *71*, 1094-100.

Guerriero, A.; Debitus, C.; Pietra, F. (1991) On the first stigmastane sterols and sterones having 24,25-double bond. Isolation from the sponge *Stelletta* sp. of deep Coral Sea. Helv. Chim. Acta, *74*, 487-94.

Guerriero, A.; D'Ambrosio, M.; Pietra, F.; Debitus, C.; Ribes, O. (1993) Pteridines, sterols, and indole derivatives from the lithistid sponge *Corallistes undulatus* of the Coral Sea. J. Nat. Prod., *56*, 1962-70.

Guerriero, A.; D'Ambrosio, M.; Pietra, F. (1995) Bis-allylic reactivity of the funicolides, 5,8(17)-diunsaturated briarane diterpenes of the sea pen *Funiculina quadrangularis* from the Tuscan archipelago, leading to 16-nortaxane derivatives. Helv Chim. Acta, *78*, 1465-78.

Guerriero, A.; D'Ambrosio, M.; Zibrowius, H.; Pietra, F. (1996) Novel cholic-acid-type sterones of *Deltocyathus magnificus*, a deep-water scleractinian coral from the Loyalty Islands, SW Pacific. Helv. Chim. Acta, *79*, 982-8.

Guerriero, A.; Debitus, C.; Laurent, D.; D'Ambrosio, M.; Pietra, F. (1998) Aztèquynol A, the first clearly defined, C-branched polyacetylene and the analogue aztèquynol B. Isolation from the tropical marine sponge *Petrosia* sp. Tetrahedron Lett., *39*, 6395-8.

Gulavita, N.K.; Gunasekera, S.P.; Pomponi, S.A.; Robinson, E.V. (1992) Polydiscamide A: a new bioactive depsipeptide from the marine sponge *Discodermia* sp. J. Org. Chem., *57*, 1767-72.

Gulavita, N.K.; Wright, A.E.; Kelly-Borges, M.; Longley, R.E; Yarwood, D.; Sills, M.A. (1994) Eryloside E from an Atlantic spoge *Erylus goffrilleri*. Tetrahedron Lett., *35*, 4299-302.

Gulavita, N.K.; Pomponi, S.A.; Wright, A.E.; Garay, M.; Sills, M.A. (1995) Aplysillin A, a thrombin receptor antagonist from the marine sponge *Aplysina fistularis fulva*. J. Nat. Prod., *58*, 954-7.

Gunasekera, S.P.; Pomponi, S.A.; McCarthy, P.J. (1994A) Discobahamins A and B, new peptides from the Bahamian deep water marine sponge *Discodermia* sp. J. Nat. Prod., *57*, 79-83.

Gunasekera, S.P.; McCarthy, P.J.; Kelly-Borges, M (1994B) Hamacanthins A and B, new antifungal bis indole alkaloids from the deep-water marine sponge, *Hamacantha* sp. J. Nat. Prod., *57*, 1437-41.

Gunasekera, S.P.; McCarthy, P.J.; Kelly-Borges, M.; Lobkovsky, E.; Clardy, J. (1996A) Dysidiolide: a novel protein phosphatase inhibitor from the Caribbean sponge *Dysidea etheria*. J. Am. Chem. Soc., *118*, 8759-60.

Gunasekera, S.P.; Kelly-Borges, M.; Longley, R.E. (1996B) A new cytotoxic sterol methoxymethyl ether from a deep water marine sponge *Scleritonema* sp. cf. *paccardi*. J. Nat. Prod., *59*, 161-2.

Gunasekera, S.P.; McCarthy, P.J.; Longley, R.E.; Pomponi, S.A.; Wright, A.E. (1999) Secobatzellines A and B, two new enzyme inhibitors from a deep-water Caribbean sponge of the genus *Batzella*. J. Nat. Prod., *62*, 1208-11.

Guo, Y.; Trivellone, E.; Scognamiglio, G.; Cimino, G. (1998A) Absolute stereochemistry of isosaraine-1 and -2. Tetrahedron Lett., *39*, 463-6.

Guo, Y.; Trivellone, E.; Scognamiglio, G.; Cimino, G. (1998B) Misenine, a novel macrocyclic alkaloid with an unusual skeleton from the Mediterranean sponge *Reniera* sp. Tetrahedron, *54*, 533-50.

Gustafson, K.R.; Cardellina II, J.H.; McMahon, J.B.; Gulakowski, R.J.; Ishitoya, J.; Szallasi, Z.; Lewin, N.E.; Blumberg, P.M.; Weislow, O.S.; Beutler, J.A.; Buckheit, R.W.; Cragg, G.M.; Cox, P.A.; Bader, J.P.; Boyd, M.R. (1992) A nonpromoting phorbol from Samoan medicinal plant *Homalanthus nutans* inhibits cell killing by HIV-1. J. Med. Chem., *35*, 1978-86.

Güven, K.C.; Bora, A.; Sunam, G. (1970) Hordenine from the alga *Phyllophora nervosa*. Phytochemistry, *9*, 1893.

Guz, N.R.; Stermitz, F.R. (2000) Synthesis and structures of regioisomeric hydnocarpin-type flavolignans. J. Nat. Prod., *63*, 1140-5.

Häberli, A.; Bircher, C.; Pfander, H. (2000) Isolation of a new carotenoid and two new carotenoid glycosides from *Curtobacterium flaccumfaciens pvar poinsettiae*. Helv. Chim. Acta, *83*, 328-35.

Haddad, J.; Vakulenko, S.; Mobashery, S. (1999) An antibiotic cloaked by its own resistance enzyme. J. Am. Chem. Soc., *121*, 11922-3.

Hadjieva, P.; Popov, S.; Budevska, B.; Dyulgerov, A.; Andreev, S. (1987) Terpenoids from a Black Sea bryozoan *Conopeum seuratum*. Z. Naturforsch., C: Biosci., *42c*, 1019-1022.

Haggarty, S.J.; Mayer, T.U.; Miyamoto, D.T.; Fathi, R.; King, R.W.; Mitchison, T.J.; Schreiber, S.L. (2000) Dissecting cellular processes using small molecules: identification of colchicine-like, taxol-like and other small molecules that perturb mitosis. Chem. Biol., *7*, 275-86.

Hagiwara, H.; Uda, H. (1988) Total synthesis of (+)-dysideapalaunic acid. J. Chem. Soc., Chem. Commun., 815-6.

Hagiwara, H.; Nakano, T.; Kon-no, M.; Uda, H. (1995) Total synthesis of (+)-compactin by a double Michael protocol. J. Chem. Soc., Perkin Trans. 1, 777-83.

Hamada, Y.; Kondo, Y.; Shibata, M.; Shioiri, T. (1989) Efficient total synthesis of didemnins A and B. J. Am. Chem. Soc., *111*, 669-73.

Hamberg, M.; Gardner, H.W. (1992) Oxylipin pathway to jasmonates: biochemistry and biological significance. Biochim. Biophys. Acta, 1165(1), 1-18.

Hanson, J.R. (1996) The sesterterpenoids. Nat Prod. Rep., *13*, 529-35.

Harada, K.; Nagai, H.; Kimura, Y.; Suzuki, M.; Park, H.-D. *et al.* (1993) Liquid chromatography/mass spectrometry detection of anatoxin-a, a neurotoxin from cyanobacteria. Tetrahedron, *49*, 9251-60.

Harborne, J.B. (1999A) Recent advances in chemical ecology. Nat. Prod. Rep., *16*, 509-23.

Harborne, J.B. (1999B) The comparative biochemistry of phytoalexin induction in plants. Biochem. System. Ecol., *27*, 335-67 (unfortunately this otherwise useful paper is replete of wrong or incomplete structural representations and no reference is given to the original chemical publications).

Harbour, G.C.; Tymiak, A.A.; Rinehart, K.L., Jr.; Shaw, P.D.; Hughes, R.G., Jr. (1981) Ptilocaulin and isoptilocaulin, antimicrobial and cytotoxic cyclic guanidines from the Caribbean sponge *Ptilocaulis* aff. *P. spiculifer*. J. Am. Chem. Soc., *103*, 5604-6.

Harrigan, G.G.; Harrigan, B.L.; Davidson, B.S. (1997) Kailuins A-D, new cyclic acyldepsipeptides from cultures of a marine-derived bacterium. Tetrahedron, *53*, 1577-82.

Harrigan, G.G.; Luesch, H.; Yoshida, W.Y.; Moore, R.E.; Nagle, D.G.; Paul, V.J. (1999) Symplostatin 2: a dolastatin 13 analogue from the marine cyanobacterium *Symploca hydnoides*. J. Nat. Prod., *62*, 655-8.

Harris, P.G.; Zhao, M.; Rosell-Melé, A.; Tiedemann, R.; Sarnthein, M.; Maxwell, J.R. (1996) Chlorin accumulation rate as a proxy for Quaternary marine primary productivity. Nature, *383*, 63-5.

Hartgers, W.A.; Damsté, J.S.S.; Requejo, A.G.; Allan, J.; Hayes, J.M.; de Leeuw, J.W. (1994) Evidence for only minor contributions from bacteria to sedimentary organic carbon. Nature, *369*, 224-7.

Harvell, C.D.; Kim, K.; Burkholder, J.M.; Colwell, R.R.; Epstein, P.R.; Grimes, D.J.; Hofmann, E.E.; Lipp, E.K.; Osterhaus, A.D.M.E.; Overstreet, R.M.; Porter, J.W.; Smith, G.W.; Vasta, G.R. (1999) Emerging marine diseases - Climate links and anthropogenic factors. Science, *285*, 1505-10.

Harvey, A.L. (2000) In search of venomous cures. Chem. Ind., 174-6.

Harvey, H.R.; Ederington, M.C.; McManus, G.B. (1997) Lipid composition of the marine ciliates *Pleuronema* sp. and *Fabrea salina*. J. Euk. Microbiol., *44*, 189-93.

Hauer, B.; Breuer, M.; Ditrich, K.; Matuschek, M.; Ress-Löschke, M.; Stürmer, R. (1999) The development of enzymes for the preparation of chemicals. Chimia, *53*, 613-6.

Hauke, V.; Graff, R.; Wehrung, P.; Trendel, J.M.; Albrecht, L.; Keely, B.J.; Peakman, T.M. (1992) Novel triterpene-derived hydrocarbons of arborane/fernane series in sediments. Tetrahedron, *48*, 3915-24.

Hay, M.E. (1992) The role of seaweed chemical defenses in the evolution of feeding specialization and in the mediation of complex interactions. In Paul, V.J., ed., Ecological Roles of Marine Natural Products. Comstock Publishing Associates, Ithaca and London, pp. 93-118.

Hay, M.E.; Fenical, W. (1996) Chemical ecology and marine biodiversity: insights and products from the sea. Oceanography, *9*, 10-20 (the structure of pachydictyol A at p. 13 is incorrect).

Head, J.F.; Inouye, S.; Teranishi, K.; Shimomura, O. (2000) The crystal structure of the photoprotein aequorin at 2.3 Å resolution. Nature, *405*, 372-6.

Heckman, D.S.; Geiser, D.M.; Eidell, B.R.; Stauffer, R.L.; Kardos, N.L.; Hedges, S.B. (2001) Molecular evidence for the early colonization of land by fungi and plants. Science, *293*, 1129-1133.

Hector, A.; Schmid, B.; Beierkuhnlein, C.; Caldeira, M.C.; Diemer, M. (1999) Plant diversity and productivity experiments in European grassland. Science, *286*, 1123-7.

Hegde, V.R.; Puar, M.S.; Dai, P.; Patel, M.; Gullo, V.P.; Das, P.R.; Bond, R.W.; McPhail, A.T. (2000) A novel microbial metabolite, activator of low density lipoprotein receptor promoter. Tetrahedron Lett., *41*, 1351-4.

Henkel, T.; Brunne, R.M.; Müller, H.; Reichel, F. (1999) Statistical investigation into the structural complementarity of natural products and synthetic products. Angew. Chem. Int. Ed. Engl., *38*, 643-7.

Henry, C.M. (2001) Pharmacogenomics. Chem. Eng. News, 13 August, 37-42.

Hepworth, D. (2000) Solving a synthetic puzzle. Chem. Ind., 59-65.

Herald, D.L.; Cascarano, G.L.; Pettit, G.R.; Srirangam, J.K. (1997) Crystal conformation of the cyclic decapeptide phakellistatin 8: comparison with antamanide. J. Am. Chem. Soc., *119*, 6962-73.

Herb, R.; Carroll, A.R.; Yoshida, W.Y.; Scheuer, P.J.; Paul, V.J. (1990) Polyalkylated cyclopentindoles: cytotoxic fish antifeedants from a sponge: *Axinella* sp. Tetrahedron, *46*, 3089-92.

Herbert, R.B. (1989) The Biosynthesis of Secondary Metabolites. Chapman & Hall, London.

Herbert, R.B.; Venter, H.; Pos, S. (2000) Do mammals make their own morphine? Nat. Prod. Rep., *17*, 317-22.

Hermann, T. (2000) Strategies for the design of drugs targeting RNA and RNA-protein complexes. Angew. Chem. Int. Ed. Engl., *39*, 1890-905.

Herold, S.; Koppenol, W.H. (2001) Living in an oxygen atmosphere - NO problem? Chimia, *55*, 870-4.

Hess, B.A., Jr.; Smentek, L.; Brash, A.R.; Cha, J.K. (1999) Mechanism of the rearrangement of vinyl allene oxide to 2-cyclopenten-1-one. J. Am. Chem. Soc., *121*, 5603-4.

Higa, T.; Scheuer, P.J. (1974) Thelepin, a new metabolite from the marine annelid *Thelepus setosus*. J. Am. Chem. Soc., *96*, 2246-8.

Hijri, M.; Hosny, M.; van Tuinen, D.; Dulieu, H. (1999) Intraspecific ITS polymorphism in *Scutellospora castanea* (Glomales, Zygomycota) is structured within multinucleate spores. Fungal Genet. Biol., *26*, 141-51.

Hinrichs K.-U.; Hayes, J.M.; Sylva, S.P.; Brewer, P.G.; DeLong, E.F. (1999) Methane-consuming archaebacteria in marine sediments. Nature, *398*, 802-5.

Hinterding, K.; Alonso-Díaz, D.; Walmann, H. (1998) Organic synthesis and biological signal transduction. Angew. Chem. Int. Ed. Engl. *37*, 688-749.

Hirayae, K.; Hirata, A.; Akutsu, K.; Hara, S.; Havukkala, I.; Nishizawa, Y.; Hibi, T. (1996) *In vitro* growth inhibition of plant pathogenic fungi, *Botrytis* spp., by *Escherichia coli* transformed with a chitinolytic enzyme gene from a marine bacterium, *Alteromonas* sp. Strain 79401. Ann. Phytopathol. Soc. Jpn., *62*, 30-6.

Hoffmüller, U.; Schneider-Mergener, J. (1998) *In vitro* evolution and selection of proteins: ribosome display for larger libraries. Angew. Chem. Int. Ed. Engl., *37*, 3141-3.

Höltzel, A.; Gänzle, M.G.; Nicholson, G.J.; Hammes, W.P.; Jung, G. (2000) The first low molecular weight antibiotic from lactic acid bacteria: reutericyclin, a new tetramic acid. Angew. Chem. Int. Ed. Engl., *39*, 2766-8.

Hong, T.W.; Jimenez, D.R.; Molinski, T.F. (1998) Agelastatins C and D, new pentacyclic bromopyrroles from the sponge *Cymbastela* sp., and potent toxicity of (-)-agelastatin A. J. Nat. Prod. *61*, 158-61.

Hopwood, D.A. (1997) Genetic contributions to understanding polyketide synthases. Chem. Rev., *97*, 2465-97.

Hornberger, K.R.; Hamblett, C.L.; Leighton, J.L. (2001) Total synthesis of leucascandrolide A. J. Am. Chem. Soc., *122*, 12894-5.

Horton, P.A.; Longley, R.E.; Kelly-Borges, M.; McConnell, O.J.; Ballas, L.M. (1994) New cytotoxic peroxylactones from the marine sponge, *Plakinastrella onkodes*. J. Nat. Prod., *57*, 1374-81.

Hosny, M.; Hijri, M.; Passerieux, E.; Dulieu, H. (1999) rDNA Units are highly polymorphic in *Scutellospora castanea* (Glomales, Zygomycetes). Gene, *226*, 61-71.

Houghton, P.J. (1999) Roots of remedies: plants, people and pharmaceuticals. Chem. Ind., 15-9.

Hoye, T.R.; Jenkins, S.A. (1987) Asymmetric synthesis of achiral molecules: *meso* selectivity. J. Am. Chem. Soc., *109*, 6196-8.

Hoye, T.R.; North, J.T.; Yao, L.J. (1994) Conformational considerations in 1-oxaquinolizidines related to the xestospongin/araguspongine family. J. Org. Chem., *59*, 6904-10.

Hoye, T.R.; Humpal, P.E.; Moon, B. (2000) Total synthesis of (-)-cylindrocyclophane A via double Horner-Emmons macrocyclic dimerization event. J. Am. Chem. Soc., *122*, 4982-3.

Hu, T.; Curtis, J.M.; Walter, J.A.; Wright, J.L.C. (1999) Hoffmanniolide: a novel macrolide from *Prorocentrum hoffmannianum*. Tetrahedron Lett., *40*, 3977-80.

Huang, Y.; Murray, M.; Eglinton, G.; Metzger, P. (1995A) Sacredicene, a novel monocyclic C_{33} hydrocarbon from sediment of Sacret lake, a tropical freshwater lake, Mount Kenya. Tetrahedron Lett., *36*, 5973-5976.

Huang, Y.; Murray, M. (1995B) Identification of 1,6,17,21-octahydrobotryococcene in a sediment. J. Chem. Soc., Chem. Commun., 335-6.

Huang, Y.; Peakman, T.M.; Murray, M. (1997) 8β,9α,10β-Rimuane: a novel, optically active, tricyclic hydrocarbon of algal origin. Tetrahedron Lett., *38*, 5363-6.

Hubbs, J.L.; Heathcock, C.H. (1999) Total synthesis of (±)-isoschizogamine. Org. Lett., *1*, 1315-7.

Huber, R.; Kurr, M.; Jannasch, H.W.; Stetter, K.O. (1989) A novel group of abyssal methanogenic archaebacteria (*Methanopyrus*) growing at 110 °C. Nature, *342*, 833-4.

Huelsenbeck, J.P; Rannala, B.; Masly, J.P. (2000) Accommodating phylogenetic uncertainty in evolutionary studies. Science, *288*, 2349-50.

Huisman, J.; Weissing, F.J. (1999) Biodiversity of plankton by species oscillations and chaos. Nature, *402*, 407-10.

Huneck, S. (1999) The significance of lichens and their metabolites. Naturwissenschaften, *86*, 559-70.

Igarashi, T.; Satake, M.; Yasumoto, T. (1999) Structures and partial stereochemical assignments for prymnesin-1 and prymnesin-2: potent hemolytic and ichthyotoxic glycosides isolated from the red tide alga *Prymnesium parvum*. J. Am. Chem. Soc., *121*, 8499-511.

Ihara, M.; Fukumoto, K. (1997) Recent progress in the chemistry of non-monoterpenoid indole alkaloids. Nat. Prod. Rep., *14*, 413-29.

Imamura, N.; Nishijima, M.; Adachi, K.; Sano, H. (1993) Novel antimycin antibiotics, urauchimycins A and B, produced by a marine actinomycete. J. Antibiot., *46*, 241-6.

Imamura, N.; Adachi, K.; Sano, H. (1994) Magnesidin A, a component of marine antibiotic magnesidin, produced by *Vibrio gazogenes*. J. Antibiot., *47*, 257-61.

Imre, S.; Öztung, A.; Çelik, T.; Wagner, H. (1987) Isolation of caffeine from the gorgonian *Paramuricea chamaeleon*. J. Nat. Prod., *50*, 1187.

Iorizzi, M.; de Riccardis, F.; Minale, L.; Palagiano, E.; Riccio, R.; Debitus, C.; Duhet, D. (1994) Polyoxygenated marine steroids from the deep water starfish *Styracaster caroli*. J. Nat. Prod., *57*, 1361-73.

IPCC Intergovernmental Panel on Climate Change (1995) Summary for Policy Makers: Impacts, Adaptation and Mitigation Options. World Meteorological Organization and United Nations Environment Programme.

Ireland, R.E. (1969) Organic Synthesis. Prentice-Hall, Englewood Cliffs, N.J.

Ireland, R.E.; Gleason, J.L.; Gegnas, L.D.; Highsmith, T.K. (1996) A total synthesis of FK-506. J. Org. Chem., *61*, 6856-72.

Isaacs, S.; Hizi, A.; Kashman, Y. (1993) Toxicols A-C and toxiusol, hexapreoid hydroquinones from *Toxiclona toxius*. Tetrahedron, *49*, 4275-82.

Ishibashi, M.; Iwasaki, T.; Imai, S.; Sakamoto, S.; Yamaguchi, K.; Ito, A. (2001) Laboratory culture of the Myxomycetes: formation of fruiting bodies of *Didymium bahiense* and its plasmodial production of makaluvamine A. J. Nat. Prod., *64*, 108-10.

Ishihara, K.; Nakamura, S.; Yamamoto, H. (1999) The first enantioselective biomimetic cyclization of polyprenoids. J. Am. Chem. Soc., *121*, 4906-7.

Iudicello, S.; Weber, M.; Wieland, R. (1999) Fish, Markets, and Fishermen: the Economics of Overfishing. Earthscan/Island Press, Washington D.C.

Iwashima, M.; Okamoto, K.; Miyai, Y.; Iguchi, K. (1999) 4-Epiclavulones, new marine prostanoids from the Okinawan soft coral, *Clavularia viridis*. Chem. Pharm. Bull., *47*, 884-6.

Izac, R.R.; Poet, S.; Fenical, W.; van Engen, D.; Clardy, J. (1982) The structure of pacifigorgiol, an ichthyotoxic sesquiterpenoid from the Pacific gorgonian coral *Pacifigorgia* cf. *adamsi*. Tetrahedron Lett., *23*, 3743-6.

Jablonski, D. (1993) The tropics as a source of evolutionary novelty through geological times. Nature, *364*, 142-4.

Jacobi, P.A.; Lee, K. (2000) Total synthesis of (±)- and (-)-stemoamide. J. Am. Chem. Soc., *122*, 4295-303.

Jacobs, M. (1981) The Tropical Rain Forest. A First Encounter. Springer-Verlag, Berlin.

Jacobson, M. (1967) The structure of echinacein, the insecticidal component of American coneflower roots. J. Org. Chem., *32*, 1646-7.

Jalal, M.A.F.; Hossain, M.B.; van der Helm, D.; Sanders-Loehr, J.; Actis, L.A.; Crosa, J.H. (1989) Structure of anguibactin, a unique plasmid-related bacterial siderophore from the fish pathogen *Vibrio anguillarum*. J. Am. Chem. Soc., *111*, 292-6.

James, A.N.; Gaston, K.J.; Balmford, A. (1999) Balancing the Eart's accounts. Nature, *401*, 323-4.

Jiménez-Estrada, M.; Reyes-Chilpa, R.; Hernández-Ortega, S.; Cristobal-Telésforo, E.; Torres-Colín, L.; Jankowski, C.K.; Aumelas, A.; Van Calsteren, M.R. (2000) Two novel Diels-Alder adducts from *Hippocratea celastroides* roots and their insecticidal activity. Can. J. Chem., *78*, 248-54.

Jayasuriya, H.; Zink, D.L.; Singh, S.B.; Borris, R.P.; Nanakorn, W.; Beck, H.T.; Balik, M.J.; Goetz, M.A.; Slayton, L.; Gregory, L.; Zakson-Aiken, M.; Shoop, W.; Singh, S.B. (2000) Structure and stereochemistry of rediocide A, a highly modified daphnane from *Trigonostemon reidioides* exibiting potent insecticidal activity. J. Am. Chem. Soc., *122*, 4998-9.

Jenkins, K.M.; Jensen, P.R.; Fenical, W. (1999) Thraustochytrosides A-C: new glycosphingolipids from a unique marine protist, *Thraustochytrium globosum*. Tetrahedron Lett., *40*, 7637-40.

Jiang, Z.-D.; Jensen, P.R.; Fenical, W. (1997) Actinoflavoside, a novel flavonoid-like glycoside produced by a marine bacterium of the genus *Streptomyces*. Tetrahedron Lett., *38*, 5065-8.

Jones, P.D.; Osborn, T.J.; Briffa, K.R. (2001) The evolution of climate over the last millennium. Science, *292*, 662-7.

Jurek, J.; Scheuer, P.J. (1993) Sesquiterpenoids and norsesquiterpenoids from the soft coral *Lemnalia africana*. J. Nat. Prod., *56*, 508-13.

Kageyama, M.; Tamura, T.; Nantz, M.H.; Roberts, J.C.; Somfai, P.; Whritenour, D.C.; Masamune, S. (1990) Synthesis of bryostatin 7. J. Am. Chem. Soc., *112*, 7407-8.

Kahn, Y.; Uemura, D.; Hirata, Y.; Ishiguro, M.; Iwashita, T. (2001) Complete NMR signal assignment of palytoxin and *N*-acetylpalytoxin. Tetrahedron Lett., *42*, 3197-202.

Kaiser, R. (2000) Scents from rain forests. Chimia, *54*, 346-63.

Kakinuma, K.; Yamagishi, M.; Fujimoto, Y.; Ikckawa, N.; Oshima, T. (1988) Stereochemistry of the biosynthesis of *sn*-2,3-*O*-diphytanyl glycerol, membrane lipid of Archaebacteria *Halobacterium halobium*. J. Am. Chem. Soc., *110*, 4861-3.

Kalinovskaya. N.I.; Kuznetsova, T.A.; Rashkes, Ya.V.; Mil'grom, Yu.M.; Mil'grom, E.G. *et al.* (1995) Surfactin-like structures of five cyclic depsipeptides from the marine isolate of *Bacillus pumilus*. Russ. Chem. Bull., *44*, 951-5.

Kameyama, T.; Takahashi, A.; Kurasawa, S.; Ishizuka, M.; Okami, Y.; Takeuchi, T.; Umezawa, H. (1987) Bisucaberin, a new siderophore, sensiting tumor cells to macrophage-mediated cytolysis. J. Antibiot., *40*, 1671-6.

Karner, M.B.; DeLong, E.F.; Karl, D.M. (2001) Archaeal dominance in the mesopelagic zone of the Pacific Ocean. Nature, *409*, 507-10.

Kashman, Y.; Hirsch, S. (1991) Isoreiswigin, a new diterpene from a marine sponge. J. Nat. Prod., *54*, 1430-2.

Kassen, R.; Bockiln, A.; Bell, G.; Rainey, P.B. (2000) Diversity peaks at intermediate productivity in a laboratory microcosm. Nature, *406*, 508-12.

Kates, M. (1993) Biology of halophilic bacteria, II. Membrane lipids of extreme halophiles: biosynthesis, function and evolutionary significance. Experientia, *49*, 1027-36.

Kato, M.; Mizuno, K.; Crozier, A.; Fujimura, T.; Ashihara, H. (2000) Caffeine synthase gene from tea leaves. Nature, *406*, 956-7.

Kavouras, I.G.; Mihalopoulos, N.; Stephanou, E.G. (1998) Formation of atmospheric particles from organic acids produced by forests. Nature, *395*, 683-6.

Kaya, K.; Sano, T.; Beattie, K.A.; Codd, G.A. (1996) Nostocyclin, a novel 3-amino-6-hydroxy-2-piperidone containing cyclic depsipeptide from the cyanobacterium *Nostoc* sp. Tetrahedron Lett., *37*, 6725-8.

Kearns, P.S.; Coll, J.C.; Rideout, J.A. (1995) A β-carboline dimer from an ascidian, *Didemnum* sp. J. Nat. Prod., *58*, 1075–6.

Keeley, J.E. (1998) Diel acid fluctuation in C_4 amphibious grasses. Photosynthetica, *35*, 273-7.

Kem, W.R.; Abbott, B.C.; Coates, R.M. (1971) Isolation and structure of a hoplonemertine toxin. Toxicon, *9*, 15-22.

Kendall, J.M.; Buckling, A.; Bell, G.; Rainey, P.B. (1998) *Aequorea victoria* bioluminescence moves into an exciting new era. Trends in Biotechnol., *16*, 216-24.

Kende, A.S.; Smalley, T.L., Jr.; Huang, H. (1999) Total synthesis of (±)-isostemopholine. J. Am. Chem. Soc., *121*, 7431-2.

Kessler, B.; Ren, Q.; de Roo, G.; Prieto, M.A.; Witholt, B. (2001) Engineering of biological systems for the synthesis of tailor-made polyhydroxyalkanoates, a class of versatile polymers. Chimia, *55*, 119-22.

Khosla, C. (1997) Harnessing the biosynthetic potential of modular polyketide synthases. Chem. Rev., *97*, 2577-90.

Kicha, A.A.; Kalinovsky, A.I.; Ivanchina, N.V.; Stonik, V.A. (1999) New steroid glycosides from the deep-water starfish *Mediaster murrayi*. J. Nat. Prod., *62*, 279-82.

Kiesecker, J.M.; Blaustein, A.R.; Belden, L.K. (2001) Complex causes of amphibian population declines. Nature, *410*, 681-4.

Kiguchi, T.; Yuumoto, Y.; Ninomiya, I.; Naito, T. (1997) Synthesis of (±)-pseudodistomins A and B acetates. Chem. Pharm. Bull., *45*, 1212-5.

Killday, K.B.; Longley, R.; McCarthy, P.J.; Pomponi,S.A.; Wright, A.E.; Neale, R.F.; Sills, M.A. (1993) Sesquiterpene-derived metabolites from the deep water marine sponge *Poecillastra sollasi*. J. Nat. Prod., *56*, 500-7.

Kishi, Y. (1989) Natural product synthesis: palytoxin. Pure Appl. Chem., *61*, 313-24.

Kita, Y.; Tohma, H.; Inagaki, M.; Hatanaka, K.; Yakura, T. (1992) Total synthesis of discorhabdin C: a general aza spiro dienone formation from *O*-silylated phenol derivatives using a hypervalent iodine reagent. J. Am. Chem. Soc., *114*, 2175-80.

Kittelman, M.; Oberer, L.; Blum, W.; Ghisalba, O. (1999) Microbial hydroxylation and simultaneous formation of the 4"-*O*-methylglucoside of the tyrosine-kinase inhibitor CGP 62706. Chimia, *53*, 594-6.

Kiviranta, J.; Namikoshi, M.; Sivonen, K.; Evans, W.R.; Carmichael, W.W.; Rinehart, K.L., Jr. (1992) Structure determination and toxicity of a new microcystin from *Microcystis aeruginosa* . Toxicon, *30*, 1093-8.

Klein, D.; Braekman, J.-C.; Daloze, D.; Hoffman, L.; Castillo, G.; Demoulin, V. (1999) Lyngbyapeptin A, a modified tetrapeptide from *Lyngbya bouillonii*. Tetrahedron Lett., *40*, 695-6.

Kleinkauf, H.; von Döhren, H. (1997) Biosynthesis of cyclosporins and related peptides. In Fungal Biotechnology. Anke, T., ed., Chapman & Hall, London, pp. 147-61.

Klok, J.; Baas, M.; Cox, H.C.; de Leeuw, J.W.; Schenck, P.A. (1984) Loliolide and dihydroactinidiolide in a recent marine sediment probably indicate a major transformation pathway of carotenoids. Tetrahedron Lett., *25*, 5577-80.

Knowlton, N. (1993) Sibling species in the sea. Annu. Rev. Ecol. Syst., *24*, 189-216.

Kobayashi, J.; Mikami, S.; Shigemori, H.; Takao, T.; Shimonishi, Y.; Izuta, S.; Yoshida, S. (1995) Flavocristamides A and B, new DNA polymerase α inhibitors from a marine bacterium, *Flavobacterium* sp. Tetrahedron, *51*, 10487-90.

Kobayashi, J.; Yuasa, K.; Kobayashi, T.; Tsuda, M.; Sasaki, T. (1996A) Jaspiferals A-G, new cytotoxic isomalabaricane-type nortriterpenoids from Okinawan marine sponge *Jaspis stellifera*. Tetrahedron, *52*, 5745-50.

Kobayashi, J.; Kawasaki, N.; Tsuda, M. (1996B) Absolute stereochemistry of keramaphidin B. Tetrahedron Lett., *37*, 8203-4.

Kobayashi, M.; Yasuzawa, T.; Kobayashi, Y.; Kyogoku, Y.; Kitagawa, I. (1981) Alcyonolide, a novel diterpenoid from a soft coral. Tetrahedron Lett., *22*, 4445-8.

Kobayashi, M.; Aoki, S.; Gato, K.; Matsunami, K.; Kurosu, M.; Kitagawa, I. (1994) Trisindoline, a new antibiotic indole trimer, produced by a bacterium of *Vibrio* sp. separated from the marine sponge *Hyrtios altum*. Chem. Pharm. Bull., *42*, 2449-51.

Kobayashi, Y.; Nakayama, Y.; Yoshida, S. (2000) Determination of the stereoisomer of korormicin from eight possible stereoisomers by total synthesis. Tetrahedron Lett., *41*, 1465-8.

Kodama, M.; Sato, S.; Sakamoto, S.; Ogata, T. (1996) Occurrence of tetrodotoxin in *Alexandrium tamarense*, a causative dinoflagellate of paralytic shellfish poisoning. Toxicon, *34*, 1101-5.

Koehn, F.E.; Longley, R.E.; Reed, J.K. (1992) Microcolins A and B, new immunosuppressive peptides from the blue-green alga *Lyngbya majuscula*. J. Nat. Prod., *55*, 613-9.

Kohmoto, S.; McConnell, O.J.; Wright, A.; Cross, S. (1987) Isospongiadiol, a cytotoxic and antiviral diterpene from a Caribbean deep water marine sponge, *Spongia* sp. Chem. Lett., 1687-90.

Kohmoto, S.; Kashman, Y.; McConnell, O.J.; Rinehart, K.L., Jr.; Wright, A.; Koehn, F. (1988) Dragmacidin, a new cytotoxic bis(indole) alkaloid from a deep water marine sponge, *Dragmacidon* sp. J. Org. Chem., *53*, 3116-8.

Kohno, T.; Kim, J.I.; Kobayashi, K.; Kodera, Y.; Maeda, T.; Sato, K. (1995) Three-dimensional structure in solution of the calcium channel blocker ω-conotoxin MVIIA. Biochemistry, *34*, 10256-65.

Koljak, R.; Lopp, A.; Penk, T.; Varvas, K.; Mueuerisepp, A.-M. Müürisepp *et al.* (1998) New cytotoxic sterols from the soft coral *Gersemia fruticosa*. Tetrahedron, *54*, 179-86.

Kong, F.; Burgoyne, D.L.; Andersen, R.J.; Allen, T.M. (1992) Pseudoaxinellin, a cyclic heptapeptide isolated from the Papua New Guinea sponge *Pseudoaxinella massa*. Tetrahedron Lett., *33*, 3269-72.

König, G.M.; Wright, A.D. (1993) A new caryophyllene-based diterpene from the soft coral, *Cespitularia* sp. J. Nat. Prod., *56*, 2198-200.

König, G.M.; Wright, A.D. (1997) New and unusual sesquiterpenes from the tropical sponge *Cymbastela hooperi*. J. Org. Chem., *62*, 837-40.

Konoki, K.; Sugiyama, N.; Murata, M.; Tachibana, K. (1999) Direct observation of binding between biotinylated okadaic acid and protein phosphatase 2A monitored by surface plasmon resonance. Tetrahedron Lett., *40*, 887-90.

Kopp, A.; Duncan, I.; Carroll, S.B. (2000) Genetic control and evolution of sexually dimorphic characters in *Drosophila*. Nature, *408*, 553-9.

Kosuge, T.; Tsuji, K.; Hirai, K.; Fukuyama, T. (1985) First evidence of toxin production by bacteria in a marine organism. Chem. Pharm. Bull., *33*, 3059-61.

Kouno, I.; Hirai, A.; Fukushige, A.; Jiang, Z.-H.; Tanaka, T. (1999) A novel rearranged type of secoeudesmane from the roots of *Lindera strychnifolia*, Chem. Pharm. Bull., *47*, 1056-7.

Kourany-Lefoll, E.; Laprevote, O.; Sevenet, T.; Montagnac, A.; Païs, M.; Debitus, C. (1994) Phloeodictines A1-A7 and C1-C2, Antibiotic and Cytotoxic Guanidine Alkaloids from the New Caledonian Sponge, *Phloeodictyon* sp. Tetrahedron, *50*, 3415-26.

Kozikowski, A.P.; Tückmantel, W. (1999) Chemistry, pharmacology, and clinical efficacy of the Chinese nootropic agent huperzine A. Acc. Chem. Res., *32*, 641-50.

Kröger, N.; Deutzmann, R.; Sumper, M. (1999) Polycationic peptides from diatom biosilica that direct silica nanosphere formation. Science, *286*, 1129-32.

Ksebati, M.B.; Schmitz, F.J.; Gunasekera, S.P. (1988) Puosides A-E, novel triterpene galactosides from a marine sponge, *Asteropus* sp. J. Org. Chem., *53*, 3917-21.

Kubanek, J.; Andersen, R.J. (1999) Evidence for *de novo* biosynthesis of the polyketide fragment of diaululusterol A by the northeastern Pacific nudibranch *Diaululua sandigensis*. J. Nat. Prod., *62*, 777-9.

Kubo, M.; Minami, H.; Hayashi, E.; Kodama, M.; Kawazu, K.; Fukuyama, Y. (1999) Neovibsamin C, a macrocyclic peroxide-containing neovibsane-type diterpene from *Viburnum awabuki*. Tetrahedron Lett., *40*, 6261-5.

Kulowski, K.; Wendt-Pienkowski, E.; Han, L.; Yang, K.; Vining, L.C.; Hutchinson, C. R. (1999) Functional characterization of the *jadl* gene as a cyclase forming angucyclinones. J. Am. Chem. Soc., *121*, 1786-94.

Kurbanov, I.; Gullyev, M.N.; Tesler, I.D.; Onov, A.; Khekimov, Y.K. (1997) Composition and structure of biologically active fatty acids of Caspian Sea algae. Zdravookhr. Turkm. (3), 17-20; Chem. Abstr., 1998, *128*, 59227u.

Kürschner, W.M. (2001) Leaf sensor for CO_2 in deep time. Nature, *411*, 247-8.

Kushiro, T.; Shibuya, M.; Ebizuka, Y. (1999) Chimeric triterpene synthase. A possible model for multifunctional triterpene synthase. J. Am. Chem. Soc., *121*, 1208-16.

Laatsch, H.; Thomson, R.H.; Cox, P.J. (1984) Spectroscopic properties of violacein and related compounds: crystal structure of tetramethyl violacein. J. Chem. Soc., Perkin Trans. 2, 1331-9.

LaCour, T.G.; Guo, C.; Bhandaru, S.; Boyd, M.R.; Fuchs, P.L. (1998) Interphylal product splicing: the first total syntheses of cephalostatin 1, the North hemisphere of ritterazine G, and the highly active hybrid analogue, ritterostatin GN1$_N$. J. Am. Chem. Soc., *120*, 692-707.

Lagos, N.; Onodera, H.; Zagatto, P.A.; Andrinolo, D.; Azevedo, S.M.F.Q.; Oshima Y. (1999) The first evidence of paralytic shellfish toxins in the freshwater cyanobacterium *Cylindrospermopsis raciborskii*, isolated from Brazil. Toxicon, *37*, 1359-73.

Lambeck, K.; Chappell, J. (2001) Sea level change through the last glacial cycle. Science, *292*, 679-86.

Lander, E.S.; Linton, L.M.; Birren, B.; Nusbaum, C.; Zody, M.C.; Baldwin, J.; Devon. K *et al.* (2001) Initial sequencing and analysis of the human genome. Nature, *409*, 860-921.

Lauffenburger, D.A. (1998) From a lecture at the symposium Challenges for the Chemical Sciences in the 21st Century. Chem. Eng. News, 30 March 1998, 38.

Laurent, D.; Guella.; Roquebert, M.-F.; Farinole, F.; Mancini, I.; Pietra, F. (2000) Cytotoxins, mycotoxins and drugs from a new deuteromycete, *Acremonium neo-caledoniae*, from the southwestern lagoon of New Caledonia. Planta Medica, *66*, 63-6.

Lebrun, B.; Braekman, J.-C.; Daloze, D.; Kalushkov, P.; Pasteels, J.M. (1999) Isopsylloborine A, a new dimeric azaphenalene alkaloid from ladybird beetles (Coleoptera: Coccinellidae) Tetrahedron Lett., *40*, 8115-6.

Lee, C.Y.; Chiappinelli, V.A.; Takasaki, C.; Yanagisawa, M.; Kimura, S.; Goto, K.; Masaki, T. (1988) Similarity of endothelin to snake venom toxin. Nature, *335*, 303.

Lehn, J.-M.; Eliseev, A.V. (2001) Dynamic combinatorial chemistry. Science, *291*, 2331-2.

Leinders-Zufall, F. (2000) Ultrasensitive pheromone detection by mammalian vomeronasal neurons. Nature, *405*, 792-6.

Lenis, L.A.; Ferreiro, M.J.; Debitus, C.; Jimenez, C.; Quinoa, E.; Riguera, R. (1998) The unusual presence of hydroxylated furanosesquiterpenes in the deep ocean tunicate *Ritterella rete*. Tetrahedron, *54*, 5385-406.

L'Epplatenier, F. (2000) New enterprises emerging from industry and universities: their importance for the Swiss economy. Chimia, *54*, 133-5, and related contributions in this issue, pp. 135-248.

Levin, S. (1999) Fragile Dominion. Complexity and the Common. Helix (Perseus), Reading, MA.

Levina, E.V.; Andriyashchenko, P.V.; Kalinovsky, A.I.; Stonik, V.A. (1998) New ophiuroid-type steroids from the starfish *Pteraster tesselatus*. J. Nat. Prod., *61*, 1423-6.

Levine, J. M. (2000) Species diversity and biological invasions: relating local processes to community patterns. Science, *288*, 852-4.

Levins, R. (1970) Extinction. Lectures on Mathematics in the Life Sciences, *2*, 75-107.

Levins, R. *et al.* (1993) Hantavirus disease emerging. Lancet, *342*, 1292.

Levitus, S.; Antonov, J.I.; Boyer, T.P.; Stephens, C. (2000) Warming of the world ocean. Science, *287*, 2225-9.

Lewin, R.A. (2001) Why rename things? Nature, *410*, 637.

Ley, S.V.; Humphries, A.C.; Erick, H.; Downham, R.; Ross, A.R. *et al.* (1998) Total synthesis of the protein phosphatase inhibitor okadaic acid. J. Chem. Soc., Perkin Trans. 1, 3907-11.

Li, C.-J.; Schmitz, F.J.; Kelly-Borges, M. (1999) Six new spongian diterpenes from the sponge *Spongia matamata*. J. Nat. Prod., *62*, 287-90.

Li, M.K.W.; Scheuer, P.J. (1984) A guaianolide pigment from a deep sea gorgonian. Tetrahedron Lett., *25*, 2109-10.

Liaaen-Jensen, S. (1990) Marine carotenoids. New J. Chem., *14*, 747-59.

Lindquist, N.; Fenical, W.; Sesin, D.F.; Ireland, C.M.; van Duyne, G.D.; Forsyth, C.J.; Clardy, J. (1988) Isolation and structure determination of the didemnenones, novel cytotoxic metabolites from tunicates. J. Am. Chem. Soc., *110*, 1308-9.

Lisieux, T.; Coimbra, M.; Nassar, E.S.; Burattini, M.N.; de Souza, L.T.; Ferreira, I.; Rocco, I.M. da Rosa, A.P.; Vasconcelos, P.F.; Pinheiro, F.P. (1994) New arenavirus isolated in Brazil. Lancet, *343*, 391-2.

Liu, D.R.; Schultz, P.G.(1999) Generating new molecular function: a lesson from nature. Angew. Chem. Int. Ed. Engl., *38*, 36-54.

Liu, L.; Thamchaipenet, A.; Fu, H.; Betlach, M.; Ashley, G. (1997) Biosynthesis of 2-nor-6-deoxyerythronolide B by rationally designed domain substitution. J. Am. Chem. Soc., *119*, 10553-4.

Liu, W.; Shen, B. (2000) Genes for production of the enediyne antitumor antibiotic C-1027 in *Streptomyces globisporum* are clustered with *cagA* gene that encodes the C-1027 apoprotein. Antimicr. Agents Chemother., *44*, 382-92.

Lohmeyer, M.; Tudzynski, P. (1997) Claviceps alkaloids. In Fungal Biotechnology. Anke, T., ed., Chapman & Hall, London, pp. 173-85.

Loreau, M.; Hector, A. (2001) Partitioning selection and complementarity in biodiversity experiments. Nature, *412*, 72-6.

Losey, J.E.; Rayor, L.S.; Carter, M.E. (1999) Transgenic pollen harms monarch larvae. Nature, *399*, 214.

Louda, S.M.; Kendall, D.; Connor, J.; Simberloff, D. (1997) Ecological effects of an insect introduced for the biological control of weeds. Science, *277*, 1088-90.

Luesch, H.; Yoshida, W.Y.; Moore, R.E.; Paul, V.J. (2000) Isolation and structure of the cytotoxin lyngbyabellin B and absolute configuration of lyngbyapeptin A from the marine cyanobacterium *Lyngbya majuscula*. J. Nat. Prod., *63*, 1437-9.

Lumbsch, H.T. (2000) Phylogeny of filamentous ascomycetes. Naturwissenschaften, *87*, 335-42.

Luo, Z. (2000) In search of the whales' sisters. Nature, *404*, 235-7.

MacNeil, J. (2000) Forest fire plan kindles debates. Science, *289*, 1448-9.

MacNeilage, P.F.; Davis, B.L. (2000) On the origin of internal structure of word forms. Science, *288*, 527-31.

Magnus, P.; Mansley, T.E. (1999) Synthesis of the ABCD-rings of the insecticidal indole alkaloid nodulisporic acid. Tetrahedron Lett., *40*, 6909-12.

Magnuson, S.R.; Sepp-Lorenzino, L.; Rosen, N.; Danishefsky, S.J. (1998) A concise total synthesis of (racemic) dysidiolide through application of dioxolenium-mediated Diels-Alder reaction. J. Am. Chem. Soc., *120*, 1615-6.

Makarieva, T.N.; Stonik, V.A.; Dmitrenok, A.S.; Grebnev, B.B.; Isakov, V.V. *et al.* (1995A) Varacin and three new marine antimicrobial polysulfides from the far-eastern ascidian *Polycitor* sp. J. Nat. Prod., *58*, 254-8.

Makarieva, T.N.; Stonik, V.A.; D'yachuk, O.G.; Dmitrenok, A.S. (1995B) Annasterol sulfate, a novel marine sulfated steroid, inhibitor of glucanase activity from a deep water sponge, *Poecillastra laminaris*. Tetrahedron Lett., *36*, 129-32.

Makarieva, T.N.; Ilyin, S.G.; Stonik, V.A.; Lyssenko, K.A.; Denisenko, V.A. (1999) Pibocin, the first ergoline alkaloid from the far-eastern ascidian *Eudistoma* sp. Tetrahedron Lett., *40*, 1591-4.

Mancini, I.; Guella, G.; Debitus, C.; Duhet, D.; Pietra, F. (1994A) Imidazolone and imidazolidinone artifacts of a pivotal imidazolthione, zyzzin, from the poecilosclerid sponge *Zyzza massalis* from the Coral Sea. The first thermochromic systems of marine origin. Helv. Chim. Acta, *77*, 1886-94.

Mancini, I.; Guella, G.; Debitus, C.; Pietra, F. (1994B) Oceanapins A-F, unique branched ceramides isolated from the haplosclerid Sponge *Oceanapia* cf. *tenuis* of the Coral Sea. Helv. Chim. Acta, *76*, 51-8.

Mancini, I.; Guella, G.; Debitus, C.; Pietra, F. (1995) Novel naamidine-type alkaloids and mixed-ligand zinc(II) complexes from a calcareous sponge, *Leucetta* sp., of the Coral Sea. Helv. Chim. Acta, *78*, 1178-84.

Mancini, I.; Guella, G.; Debitus, C.; Waikedre, J.; Pietra, F. (1996) From inactive nortopsentin D, a novel bis(indole) alkaloid isolated from the axinellid sponge *Dragmacidon* sp. from deep waters south of New Caledonia , to a strongly cytotoxic derivative. Helv. Chim. Acta, *79*, 2075-82.

Mancini, I.; Guella, G.; Defant, A.; Candenas, M.L.; Armesto, C.P.; Pietra, F. (1998) Polar metabolites of the tropical green seaweed *Caulerpa taxifolia* which is spreading in the Mediterranean Sea: glycoglycerolipids and stable enols. Helv. Chim. Acta., *81*, 1681-91.

Mancini, I.; Guella, G.; Zibrowius, H.; Laurent, D.; Pietra, F. (1999A) A novel type of second epoxy bridge in eunicellane diterpenes: isolation and characterization of massileunicellins A-C from the gorgonian *Eunicella cavolinii*. Helv. Chim. Acta, *82*, 1681-9.

Mancini, I.; Guerriero, A.; Guella, G.; Bakken, T.; Zibrowius, H.; Pietra, F. (1999B) Novel 10-hydroxydocosapolyenoic acids from deep-water scleractinian corals. Helv. Chim. Acta, *82*, 677-84.

Mancini, I.; Guella, G.; Pietra, F. (2000) Highly diastereoselective, biogenetically-patterned synthesis of (+)(1S,6R)-volvatellin (= (+)-(4R,5S)-5-hydroxy-4-(5-methyl-1-methylenehex-4-en-2-ynyl)cyclohex-1-ene-1-carboxaldehyde. Helv. Chim. Acta, *83*, 694-701.

Mann, V.; Harker, M.; Pecker, I.; Hirschberg, J. (2000) Metabolic engineering of astaxanthin production in tobacco flowers. Nat. Biotechnol., *18*, 888-92.

Marshall. C.T.; Yaragina, N.A.; Lambert, Y.; Kjesbu, O.S. (1999) Total lipid energy as a proxy for total egg production by fish stocks. Nature, *402*, 288-90.

Martin, G.E.; Sanduja, R.; Alam, M. (1986) Isolation of isopteropodine from the marine mollusk *Nerita albicilla*. J. Nat. Prod., *49*, 406-11.

Martin, J.F.; Gutiérrez, S.; Demain, A.L. (1997) β-Lactams. In Fungal Biotechnology. Anke, T., ed., Chapman & Hall, London, pp. 91-127.

Martin, S.F.; Humphrey, J.M; Ali, A.; Hillier, M.H. (1999) Enantioselective total synthesis of ircinal A and related manzamine alkaloids. J. Am. Chem. Soc., *121*, 866-7.

Martinez, E.J.; Corey, E.J. (2000) A new, more efficient, and effective process for the synthesis of a key pentacyclic intermediate for production of ecteinascidin and phthalascidin antitumor agents. Org. Lett., *2*, 993-6.

Masaki, H.; Maeyama, J.; Kamada, K.; Esumi, T.; Iwabuchi, Y.;Hatakeyama, S.; (2000) Total synthesis of (-)-dysiherbaine. J. Am. Chem. Soc., *122*, 5216-7.

Mata de Urquiza, A.; Liu, S.; Sjöberg, M.; Zetterström, R.H., Griffiths, W., Sjövall, J.; Perlmann, T. (2001) Docosahexaenoic acid, a ligand for the retinoid X receptor in mouse brain. Science, *290*, 2140-4.

Matcham, G.; Bhatia, M.; Lang, W.; Lewis, C.; Nelson, R.; Wang, A.; Wu, W. (1999) Enzyme and reaction engineering in biocatalysis: synthesis of (*S*)-methoxyisopropylamine (= (*S*)-1-methoxypropan-2-amine). Chimia, *53*, 584-9.

Matsumura K. (1995) Tetrodotoxin as a pheromone. Nature, *378*, 563-4.

Matsunaga, S.; Yamashita, T.; Tsukamoto, S.; Fusetani, N. (1999) Three new antibacterial alkaloids from a marine sponge *Stelletta* species. J. Nat. Prod., *62*, 1202-4.

Matthée. G.F.; König, G.M.; Wright, A.D. (1998) Three new diterpenes from the marine soft coral *Lobophytum crassum*. J. Nat. Prod., *61*, 237-40.

Maugh, K.J.; Anderson, D.M.; Strausberg, R.; Strausberg, S.L.; Mccandliss, R.; Wei, T.; Filpula, D. (1998) Recombinant bioadhesive proteins of marine animals and their use in adhesive compositions. PCT Int. Appl. WO 88 03,953 2 Jun to Genex Corporation. Chem. Abstr., 1989, *110*, 2190a.

May, R.M. (1994) Biological diversity: differences between land and sea. Phil. Trans. R. Soc. London B, *343*, 105-11.

Mayr, E. (1942) Systematics and the Origin of Species. Columbia University Press, N.Y.

Mayr, E. (1969) Principles of Systematic Zoology. McGraw-Hill, N.Y.

Mazur, B.; Krebbers, E.; Tingey, S. (1999) Gene discovery and product development for grain quality traits. Science, *285*, 372-5.

McClintock, J.B.; Baker, B.J. (1997) A review of the chemical ecology of antarctic marine invertebrates. Amer. Zool., *37*, 329-42.

McCoy, M. (2000) Antibiotic restructuring follows pricing woes. Chem. Eng. News, 24 April, 21-5.

McDonald, L.A.; Foster, M.P.; Phillips, D.R.; Ireland, C.M.; Lee, A.Y.; Clardy, J. (1992) Tawicyclamides A and B, new cyclic peptides from the ascidian *Lissoclinum patella*. J. Org. Chem., *57*, 4616-24.

McDonald, L.A.; Capson, T.L.; Krishnamurthy, G.; Ding, W.-D.; Ellestad, G.A. *et al.* (1996) Namenamicin, a new enediyne antitumor antibiotic from the marine ascidian *Polysyncraton lithostrotum*. J. Am. Chem. Soc., *118*, 10898-9.

McKinney, F.K.; Lidgard, S.; Sepkoski, J.J., Jr.; Taylor P.D. (1998) Decoupled temporal pattern of evolution and ecology in two post-Paleozoic clades. Science, *281*, 807-9.

McMillan, J. (1997) Biosynthesis of the gibberellin plant hormones. Nat. Prod. Rep., *14*, 221-43.

Meier, R.P. (2000) Diminishing diversity of signed languages. Science, *288*, 1965.

Menzel, D.; Kazlauskas, R.; Reichelt, J. (1983) Coumarins in the siphonalean green algal family Dasycladaceae. Bot. Mar., *26*, 23-9.

Mesguiche, V.; Valls, R.; Piovetti, L.; Peiffer, G. (1999) Characterization and synthesis of (-)-7-methoxydodec-4(*E*)-enoic acid, a novel fatty acid isolated from *Lyngbya majuscula*. Tetrahedron Lett., *40*, 7473-6.

MI (The Merck Index, 1998, 12:2, on disk), Chapman & Hall, London.

Mierzwa, R.; King, A.; Conover, M.A.; Tozzi, S.; Puar, M.S.; Patel, M.; Coval, S.J.; Pomponi, S.A. (1994) Verongamine, a novel bromotyrosine-derived histamine H_4-antagonist from the marine sponge *Verongula gigantea*. J. Nat. Prod., *57*, 175-177.

Miles, D.W.; Chittawong, V.; Lho, D.-S.; Payne, A.M.; de la Cruz, A.; Gomez, E.D.; Weeks, J.A.; Atwood, J.L. (1991) Toxicants from mangrove plants, VII. Vallapin and vallapianin, novel sesquiterpene lactones from the mangrove plant *Heritiera littoralis*. J. Nat. Prod., *54*, 286-9.

Milkova, T.; Talev, G.; Christov, R.; Dimitrova-Konaklieva, S.; Popov, S. (1997) Sterols and volatiles in *Cystoseira barbata* and *Cystoseira crinita* from the Black Sea. Phytochemistry, *45*, 93-5.

Miller, A.I. (1998) Biotic transitions in global marine diversity. Science, *281*, 1157-60.

Miller, S.L.; Tinto, W.F.; Yang, J.-P.; McLean, S.; Reynolds, W.F. (1995) 9,11-seco-24-Hydroxydinosterol from *Pseudopterogorgia americana*. Tetrahedron Lett., *36*, 1227-8.

Misra, T.N.; Singh, R.S.; Pandey, H.S.; Prasad, C.; Singh, B.P. (1993) Two long chain compounds from *Achyranthes aspera*. Phytochemistry, *33*, 221-3.

Miyaoka, H.; Nakano, M.; Iguchi, K.; Yamada, Y.(1999) Three new xenicane diterpenoids from Okinawan soft coral of the genus, *Xenia*. Tetrahedron, *55*, 12977-82.

Miyazaki, K.; Kobayashi, M.; Natsume, T.; Gondo, M.; Mikami, T. *et al.* (1995) Synthesis and antitumor activity of novel dolastatin 10 analogs. Chem. Pharm. Bull., *43*, 1706-18.

Mochizuki, H.; Yamakawa, T. (1998) Preparation of aequorin derivatives with improved stability. Jpn. Kokai Tokkyo Koho JP 09,278,796, 28 Oct. to Mochida Pharmaceutical Co. Chem. Abstr., 1998, *128*, 32705m.

Mojzsis, S.J.; Arrhenius G.; McKeegan, K.D.; Harrison, T.M.; Nutman, A.P.; Friend, C.R.L. (1996) Evidence for life on Earth before 3,800 million years ago. Nature, *384*, 55-69.

Molinski, T.F.; Ireland, C.M. (1988) Dysiazirine, a cytotoxic azacyclopropene from the marine sponge *Dysidea fragilis*. J. Org. Chem., *53*, 2103-5.

Molinski, T.F.; Ireland, C.M. (1989) Varamines A and B, new cytotoxic thioalkaloids from *Lissoclinum vareau*. J. Org. Chem., *54*, 4256-9.

Monteagudo, E.; Cicero, D.O.; Cornett, B.; Myles, D.C.; Snyder, J.P. (2001) The conformation of discodermolide in DMSO. J. Am. Chem. Soc., *123*, 6929-30, and references to the TB complex there.

Moon-van der Staay, S.Y.; De Wachter, R.; Vaulot, D. (2001) Oceanic 18S rDNA sequences from picoplankton reveal unsuspected eukaryotic diversity. Nature, *409*, 607-10.

Moore, K.S.; Wehrli, S.; Rogers, M.; McCrimmin, D.; Zasloff, M. (1993) Squalamine: an aminosterol antibiotic from the shark. Proc. Natl. Acad. Sci. USA, *90*, 1354-8.

Moore, P.D. (1998) Incas and alders. Nature, *394*, 224-5.

Moore, R.E.; Entzeroth, M. (1988) Majusculamide D and deoxymajusculamide D, two cytotoxins from *Lyngbya majuscula*. Phytochemistry, *27*, 3101-3.

Moreau, R.; Dabrowski, K. (1998) Body pool and synthesis of ascorbic acid in adult sea lamprey (*Petromyzon marinus*): an agnatan fish with gulonolactone oxidase activity. Proc. Natl. Acad. Sci. USA, *95*, 10279-82.

Morel, A.F.; Flach, A.; Zanatta, N.; Ethur, E.M.; Mostardeiro, M.A.; Gehrke, I.T.S. (1999) A new cyclopeptide alkaloid from the bark of *Waltheria douradinha*. Tetrahedron Lett., *40*, 9205-9.

Mori, K. (2000) Organic synthesis and chemical ecology. Acc. Chem. Res., *33*, 102-10.

Moriarty, R.M.; Roll, D.M.; Ku, Y.-Y.; Nelson, C.; Ireland, C.M. (1987) A revised structure for the marine bromoindole derivative citorellamine. Tetrahedron Lett. *28*, 749-52.

Morita, H.; Yoshida, N.; Kobayashi, J. (1999) Daphnezomines C-E, new alkaloids with an *N*-oxide moiety from *Daphniphyllum humile*. Tetrahedron, *55*, 12549-56.

Morris, L.A.; Jaspars, M.; Adamson, K.; Woods, S.; Wallace, H.M. (1998) The capnellenes revisited: new structures and new biological activity. Tetrahedron, *54*, 12953-8.

Morris, S.A.; Andersen, R.J. (1990) Brominated bis(indole) alkaloids from the marine sponge *Hexadella* sp. Tetrahedron, *46*, 715-20.

Morris, S.A.; Northcote, P.T.; Andersen, R.J. (1991) Triterpenoid glycosides from the northeastern Pacific marine sponge *Xestospongia vanilla*. Can. J. Chem., *69*, 1352-64.

Morse, D.E.; Hooker, N.; Duncan, H.; Jensen, L. (1979) γ-Aminobutyric acid, a neurotransmitter, induces planktonic larvae to settle and begin metamorphosis. Science, *204*, 407-10.

Mueller-Dombois, D. (1999) Biodiversity and environmental gradients across the tropical Pacific islands: a new strategy for research and conservation. Naturwissenschaften, *86*, 253-61.

Müller, W.E.G.; Schröder, H.C. (1997) Bioaktive Substanzen aus Schwämmen: Gene weisen den Weg bei der Suche nach neuen Arzneimitteln. Biologie in unserer Zeit, *27*, 389-98.

Mulzer, J.; Waldmann, H, eds. (1998) Organic Synthesis Highlights III. Wiley-VCH.

Munda, I.M. (1987) Preliminary information on the ascorbic acid content in some Adriatic seaweeds. Hydrobiologia, *151/152*, 477-81.

Munson, M.C.; Barany, G. (1993) Synthesis of α-conotoxin SI, a bicyclic tridecapeptide amide with two disulfide bridges: illustration of novel protection schemes and oxidation strategies. J. Am. Chem. Soc., *115*, 10203-10.

Murakami, M.; Okita, Y.; Matsuda, H.; Okino, T.; Yamaguchi, K. (1998) From the dinoflagellate *Alexandrium hiranoi*. Phytochemistry, *48*, 85-8.

Murata, K.; Satake, M.; Naoki, H.; Kaspar, H.F.; Yasumoto, T. (1998) Isolation and structure of a new brevetoxin analog, brevetoxin B2, from greenshell mussels from New Zealand. Tetrahedron, *54*, 735-42.

Murata, M.; Shigeru M.; Nobuaki M.; Gopal K.P.; Kazuo T. (1999) Absolute configuration of amphidinol 3. J. Am. Chem. Soc., *121*, 870-1.

Murata, M.; Yasumoto, T. (2000) The structure elucidation and biological activities of high molecular weight algal toxins: maitotoxin, prymnesins and zooxanthellatoxins. Nat. Prod. Rep., *17*, 293-314.

Murray, L.; Currie, G.; Capon, R.J. (1995) A new macrocyclic γ-pyrone from a southern Australian marine red alga. Aust. J. Chem., *48*, 1485-9.

Musil, R. (1996) Allerhand Fragliches. Aphorismen. Rowolt Verlag, Hamburg, 102-103.

Musmar, M.J.; Weinheimer,A.J.; Martin, G.E. (1983) Assignment of the high-field resonances of a gorgosterol derivative. J. Org. Chem., *48*, 3580-1.

Musters, C.J.M.; de Graaf, H.J.; ter Keurs, W.J. (2000) Can protected areas be expanded in Africa? Science, *287*, 1759-60.

Myers, N. (2000) Sustainable consumption. Science, *287*, 2419.

Naeem, S.; Hahn, D.R.; Schuurman, G. (2000) Producer-decomposer co-dependency influences biodiversity effects. Nature, *403*, 762-4.

Nagai, H.; Yasumoto, T.; Hokama, Y. (1997) Manauealides, some of the causative agents of a red alga *Gracilaria coronopifoliua* poisoning in Hawaii. J. Nat. Prod., *60*, 925-8.

Nagle, D.G.; Gerwick, W.H. (1995) Nakienone A-C and nakitriol, new cytotoxic cyclic C_{11} metabolites from an Okinawan cyanobacterial (*Synechocystis* sp.) overgrowth of coral. Tetrahedron Lett., *36*, 849-52.

Nagle, D.G.; Paul, V.J.; Roberts, M.A. (1996) Ypaoamide, a new broadly acting feeding deterrent from the marine cyanobacterium *Lyngbya majuscula*. Tetrahedron Lett., *37*, 6263-6.

Nakada, T.; Yamamura, S. (2000) Three new metabolites of hybrid strain KO 0231, derived from *Penicillium citreo-viride* IFO 6200 and 4692. Tetrahedron, *56*, 2595-602.

Nakamura, H.; Kishi, Y.; Shimomura, O.; Morse, D.; Hastings, J.W. (1989) Structure of dinoflagellate luciferin and its enzymatic and non enzymatic air-oxidation products. J. Am. Chem. Soc., *111*, 7607-11.

Nakayama, T.; Yonekura-Sakakibara, K.; Sato, T.; Kikuchi, S.; Fukui, Y.; Fukuchi-Mizutani, M.; Nakao, M.; Tanaka, Y.; Kusumi, T.; Nishino, T. (2000) Aureusidin synthase: polyphenol oxidase homolog responsible for flower coloration. Science, *290*, 1163-6.

Needham, J.; Kelly, M.T.; Ishige, M.; Andersen, R.J. (1994) Andrimid and moiramides A-C, metabolites produced in culture by a marine isolate of the bacterium *Pseudomonas fluorescens*. J. Org. Chem., *59*, 2058-63.

Nemoto, T.; Yoshino, G.; Ojika, M.; Sakagami, Y. (1997) Amphimic acids, novel unsaturated C_{28} fatty acids as DNA topoisomerase I inhibitors. Tetrahedron Lett., *38*, 5667-70.

Neuschütz, K.; Velker, J.; Neier, R. (1998) Tandem reactions combining Diels-Alder reactions with sigmatropic rearrangement processes and their use in synthesis. Synthesis, *3*, 227-55.

Newman, D.J.; Cragg, G.M.; Snader, K.M. (2000) The influence of natural products upon drug discovery. Nat. Prod. Rep., *17*, 215-33.

Nicolaou, K.C.; Wissinger, N.; Vourloumis, D.; Ohshima, T.; Kim, S.; Pfefferkorn, J.; Xu, J.-Y.; Li, T. (1998) Solid and solution phase synthesis and biological evaluation of combinatorial sarcodictyin libraries. J. Am. Chem. Soc., *120*, 10814-26.

Nicolaou, K.C.; Mitchell, H.J.; Suzuki, H.; Rodríguez, R.M.; Baudoin, O.; Fylaktakidou, K.C. (1999) Total synthesis of everninomicin: A₁B(A)C fragment. Angew. Chem. Int. Ed. Engl., *38*, 3334-9.

Nicolaou, K.C.; Vourloumis, D.; Wissinger, N.; Baran, P.S. (2000A) The art and science of total synthesis at the dawn of the twenty-first century. Angew. Chem. Int. Ed. Engl., *39*, 45-122.

Nicolaou, K.C.; Vassilikogiannakis, G.; Simonsen, K.B.; Baran, P.S.; Zhong, Y.-L.; Vidali, V.P.; Pitsinos, E.N.; Couladouros, E.A. (2000B) Biomimetic total synthesis of bisorbicillinol, bisorbibutenolide, trichodimerol, and designed analogues of the bisorbicillinoids. J. Am. Chem. Soc., *122*, 3071-9.

Nicolaou, K.C.; Pfefferkorn, J.A.; Barluenga, S.; Roecker, A.J.; Cao, G.-Q. (2000C) Natural product-like combinatorial libraries based on privileged structures. 3. The "libraries from libraries" principle for diversity enhancement of benzopyran libraries. J. Am. Chem. Soc., *122*, 968-76.

Nicolaou, K.C.; Hughes, R.; Pfefferkorn, J.A.; Barluenga, S. (2001) Optimization and mechanistic studies of psammaplin A type antibacterial agents active against methicillin-resistant *Staphylococcus aureus*. Angew. Chem. Int. Ed. Engl., *40*, 4296-310.

Nikki Paden, A.; Dillon, V.M.; John, P.; Edmonds, J.; Collins, M.D.; Alvarez, N. (1998) *Clostridium* used in mediaeval dyeing. Nature, *396*, 225.

Northcote, P.T.; Blunt, J.W.; Munro, M.H.G. (1991) Pateamine: a potent cytotoxin from the New Zealand marine sponge *Mycale* sp. Tetrahedron Lett., *32*, 6411-4.

Nowak, M.A.; Plotkin, J.B.; Jansen, V.A.A. (2000) The evolution of syntactic communication. Nature, *404*, 495-8.

Numata, A.; Takahashi, C.; Ito, Y.; Minoura, K.; Yamada, T. *et al.* (1996) Penochalasins, a novel class of cytotoxic cytochalasans from a *Penicillium* species separated from a marine alga. J. Chem. Soc., Perkin Trans. 1, 239-45.

Nuzillard, J.-M.; Connolly, J.D.; Dclaude, C.; Richard, B.; Zèches-Hanrot, M.; Le Men-Olivier, L. (1999) Alkaloids with a novel diaza-adamantane skeleton from the seeds of *Acosmium panamense* (Fabaceae). Tetrahedron, *55*, 11511-8.

Ochi, M.; Yamada, K.; Kotsuki, H.; Shibata, K. (1991) Calicoferols A and B, two novel secosterols possessing brine-shrimp lethality from the gorgonian *Calicogorgia* sp. Chem. Lett., 427-30.

Odum, E.P. (1983) Basic Ecology. Saunders College Publishing, Philadelphia.

Ogawa, A.K.; Armstrong, R.W. (1998) Total synthesis of calyculin C. J. Am. Chem. Soc., *120*, 12435-42.

O'Hagan, D. (1995) Biosynthesis of fatty acids and polyketide metabolites. Nat. Prod. Rep., *12*, 1-32.

Ohmoto, T. (1969) Triterpene methyl ethers from Gramineae plants. Chem. Commun., 601.

Ohta, T.; Uwai, K.; Kikuchi, R.; Nozoe, S.; Oshima, Y.; Sasaki, K.; Yoshizaki, F. (1999) Absolute stereochemistry of cicutoxin and related toxic polyacetylenic alcohols from *Cicuta virosa*. Tetrahedron, *55*, 12087-98.

Ohtani, I.; Kusumi, T.; Kashman, Y.; Kakisawa, H. (1991) High-field FT NMR applications of Mosher's method. The absolute configuration of marine terpenoids. J. Am. Chem. Soc., *113*, 4092-6.

Ohtani, I.; Ichiba, T.; Isobe, M.; Kelly-Borges, M.; Scheuer, P.J. (1995) Kauluamine: an unprecedented manzamine dimer from an Indonesian marine sponge, *Prianos* sp. J. Am. Chem. Soc., *117*, 10743-4.

Okamoto, Y.; Ojika, M.; Sakagami, Y. (1999) Iantheran A, a dimeric polybrominated benzofuran as a Na,K-ATPase inhibitor from a marine sponge, *Ianthella* sp. Tetrahedron Lett., *40*, 507-10.

Okino T.; Yoshimura, E.; Hirota, H.; Fusetani, N. (1996) New antifouling sesquiterpenes from four nudibranchs of the family Phyllidiidae. Tetrahedron, *52*, 9447-54.

Oldfield, S.; Lusty, C.; MacKinven, A. (1998) The World List of Threatened Trees. World Conservation Press, Cambridge, UK.

Oldham, N.J.; Boland, W. (1996) Chemical ecology: multifunctional compounds and multitrophic interaction. Naturwissenschaften, *83*, 248-54.

Omar, S.; Tenenbaum, L.; Manes, L.V.; Crews, P. (1988) Novel marine sponge derived amino acids 7. The fenestins. Tetrahedron Lett., *29*, 5489-92.

Ormund, R.F.G.; Gage, J.D.; Angel, M.V., eds. (1997) Marine Biodiversity: Patterns and Processes. Cambridge University Press.

Orsat, B.; Wirz, B.; Bishof, S. (1999) A continuous lipase-catalyzed acylation process for the large-scale production of vitamin A precursors. Chimia, *53*, 579-84.

Oscarson, S.; Sehgelmeble, F. (2000) A novel-directing fructofuranosyl donor concept. Stereospecific synthesis of sucrose. J. Am. Chem. Soc., *122*, 8869-72.

Ourisson, G.; Albrecht, P. (1992) Geohopanoids: the most abundant natural products on Earth? Acc. Chem. Res., *25*, 398-402.

Ovenden, S.P.B.; Capon, R.J. (1998) Trunculins G-I: norsesterterpene cyclic peroxides from a southern Australian marine sponge, *Latrunculia* sp. Aust. J. Chem., *51*, 573-9.

Ovenden, S.P.B.; Capon, R.J. (1999) Nuapapuin A and sigmosceptrellins D and E: new norterpene cyclic peroxides from a southern Australian marine sponge, *Sigmosceptrella* sp. J. Nat. Prod., *62*, 214-8.

Overman, L.E.; Rogers, B.; Tellew, J.; Trenk, W.C. (1997) Stereocontrolled synthesis of the tetracyclic core of the bisguanidine alkaloids palau'amine and styloguanidine. J. Am. Chem. Soc., *119*, 7159-60.

Pachlatko, J.P (1999) Industrial biocatalysis. Chimia, *53*, 577.

Packer, A.; Clay, K. (2000) Soil pathogens and spatial patterns of seedling mortality in a temperate tree. Nature, *404*, 278-81.

Païs, M.; Fontaine, C.; Laurent, D.; Barre, S.L.; Guittet, E. (1987) Stylotelline, a new sesquiterpene isocyanide from the sponge *Stylotella* sp. Tetrahedron Lett., *28*, 1409-12.

Palagiano, E.; De Marino, S.; Minale, L.; Riccio, R.; Zollo, F.; Iorizzi, M.; Carre, J.B.; Debitus, C.; Lucarain, L.; Provost, J. (1995) Ptilomycalin A, crambescidin 800 and related new highly cytotoxic guanidine alkaloids from the starfishes *Fromia monilis* and *Celerina heffernani*. Tetrahedron, *51*, 3675-82.

Palagiano, E.; Zollo, F.; Minale, L.; Iorizzi, M.; Bryan, P.; McClintock, J.; Hopkins, T. (1996) Isolation of 20 glycosides from the starfish *Henricia downeyae*, collected in the Gulf of Mexico. J. Nat. Prod., *59*, 348-54.

Palermo, J.A.; Rodríguez Brascoa,, M.F.; Hughes, E.A.; Seldes, A.M.; Balzaretti, V.T.; Cabezas, E. (1996) Short side chain sterols from the tunicate *Polizoa opuntia*. Steroids, *61*, 1-6.

Palermo, J.A.; Brasco, M.F.R.; Seldes, A.M. (1998) Storniamides A-D: alkaloids from a Patagonian sponge *Cliona* sp. Tetrahedron, *52*, 2727-34.

Panek, J.S.; Liu, P. (2000) Total synthesis of the actin-depolymerizing agent (-)-mycalolide A: application of chiral silane-sased bond construction methodology. J. Am. Chem. Soc., *122*, 11090-7.

Pani, A.K.; Anctil, M. (1994) Evidence for biosynthesis and catabolism of monoamines in the sea pansy *Renilla koellikeri*. Neurochem. Int., *25*, 465-74.

Paquette, L.A.; Barriault, L.; Pissarnitski, D.; Johnston, J.N. (2000) Stereocontrolled elaboration of natural (-)-polycavernoside A, a powerfully toxic metabolite of the red alga *Polycavernosa tsudai*. J. Am. Chem. Soc., *122*, 619-31.

Paterson, I.; Florence, G.J.; Gerlach, K.; Scott, J.P.; Sereinig, N. (2001) A practical synthesis of (+)-discodermolide and analogues: fragment union by complex aldol reactions. J. Am. Chem. Soc., *123*, 9535-44.

Pathirana, C.; Jensen, P.R.; Dwight, R.; Fenical, W. (1992) Rare phenazine L-quinovose esters from a marine actinomycete. J. Org. Chem., *57*, 740-2.

Patil, A.D.; Freyer, A.J.; Reichwein, R.; Bean, M.F.; Faucette, L.; Johnson, R.K.; Haltiwanger, R.C.; Eggleston, D.S. (1997) Two new nitrogenous sesquiterpenes from the sponge *Axinyssa aplysinoides*. J. Nat. Prod., *60*, 507-10.

Paul, V.J. (1992) Ecological Roles of Marine Natural Products. Comstock Publishing Associates, Ithaca and London.

Peakman, T.M.; Lo ten Haven, H.; Rullkötter, J.; Curiale, J.A. (1991) Characterisation of 24-*nor*-triterpenoids occurring in sediments and crude oils by comparison with synthesized standards. Tetrahedron, *47*, 3779-86.

Pearce, G.; Moura, D.S.; Stratmann, J.; Ryan, C.A. (2001) Production of multiple plant hormones from a single polyprotein precursor. Nature, *411*, 817-20.

Pearce, G.E.S.; Keely, B.J.; Harradine, P.J.; Eckardt, C.B.; Maxwell, J.R. (1993) Characterization of naturally occurring steryl esters derived from chlorophyll a. Tetrahedron Lett., *34*, 2989-92.

Pei, D.; Schultz, P.G. (1991) Engineering protein specificity: gene manipulation with semisynthetic nucleases. J. Am. Chem. Soc., *113*, 9398-400.

Perry, N.B.; Blunt, J.W.; Munro, M.H.G. (1988) Cytotoxic pigments from New Zealand sponges of the genus *Latrunculia*: discorhabdins A, B and C. Tetrahedron, *44*, 1727-34.

Petersen, M.; Sauter, M. (1999) Biotechnology in the fine-chemicals industry: cyclic amino acids by enantioselective catalysis. Chimia, *53*, 608-12.

Peterson, C.J.; Cossé, A.; Coats, J.R. (2000) Insecticidal components in the meal of *Crambe abyssinica*. J. Agric. Urban Entomol., *17*, 27-36.

Pettit, G.R.; Herald, D.L.; Gao, F.; Sengupta, D.; Herald, C.L. (1991) Absolute configuration of the bryostatins. J. Org. Chem., *56*, 1337-40.

Pettit, G.R.; Tan, R.; Gao, F.; Williams, M.D. Boyd, M.R. *et al.* (1993) Isolation and structure of halistatin 1 from the Eastern Indian Ocean marine sponge *Phakellia carteri*. J. Org. Chem., *58*, 2538-43.

Pettit, G.R.; Tan, R.; Herald, D.L.; cerny, R.L.; Williams, M.D. (1994) Isolation and structure of phakellistatin 3 and isophakellistatin 3 from a Republic of Comoros marine sponge. J. Org. Chem., *59*, 1593-5.

Pettit, G.R.; Butler, M.S.; Williams, M.D.; Filiatrault, M.J.; Pettit, R.K. (1996) Isolation and structure of hemibastadinols 1-3 from the Papua New Guinea marine sponge *Ianthella basta*. J. Nat. Prod., *59*, 927-34.

Pfaar, U.; Gygax, D.; Gertsch, W.; Winkler, T.; Ghisalba, O. (1999) Enzymatic synthesis of β-D-glucuronides in an enzyme membrane reactor. Chimia, *53*, 590-3.

Pfeifer, B.A.; Admiraal, S.J.; Gramajo, H.; Cane, D.E.; Khosla, C. (2001) Biosynthesis of complex polyketides in a metabolically engineered strain of *E. coli*. Science, *291*, 1790-2.

Pfizenmayer, A.J.; Ramanjulu, J.M.; Vera, M.D.; Ding, X.; Dong, X.; Chen, W.-C.; Joullie, M.M. (1999) Synthesis and biological activities of [*N*-MeLeu⁵]- and [*N*-MePhe⁵]-didemnin B. Tetrahedron, *55*, 313-34.

Phoon, C.W.; Abell, C. (1999) Use of quinic acid as template in solid-phase combinatorial synthesis. J. Comb. Chem., *1*, 485-92.

Pietra, F. (1995) Structurally similar natural products in phylogenetically distant marine organisms, and a com parison with terrestrial species. Chem. Soc. Rev., *24*, 65-71.

Pietra, F. (1997) Secondary metabolites from marine microorganisms: bacteria, protozoa, algae and fungi. Achievements and prospects. Nat. Prod. Rep., *14*, 453-64.

Pika, J.; Faulkner, D.J. (1995) Unusual chlorinated homo-diterpenes from the South African nudibranch *Chromodoris hamiltoni*. Tetrahedron, *51*, 8189-98.

Piscitelli, S.; Burstein, A.H.; Chaitt, D.; Alfaro, R.M.; Falloon, J. (2000) Indinavir concentrations and St John's wort. Lancet, 547-8.

Plante, O.J.; Palmacci, E.R.; Seeberger, P.H. (2001) Automated solid-phase synthesis of oligosaccharides. Science, *291*, 1523-7.

Plasman, V.; Daloze, D.; Braekman, J.-C.; Connétable, S.; Robert, A.; Bordereau, C. (1999) New macrolactones from the defensive salivary secretions of soldiers of the African termite *Pseudacanthotermes spiniger*. Tetrahedron Lett., *40*, 9229-32.

Plucknett, D.L.; Smith, N.J.H. (1986) Sustaining agricultural yields. BioScience, *36*, 40-5.

Poch, G.G.; Gloer, J.B. (1989) Helicascolides A and B: new lactones from the marine fungus *Helicascus kanaloanus*. J. Nat. Prod., *52*, 257-60.

Poiner, A.; Valerie, P.J.; Scheuer, P.J. (1989) Kumepaloxane, a rearranged trisnor sesquiterpene from the bubble shell *Haminoea cymbalum*. Tetrahedron, *45*, 617-22.

Poinsot, J.; Schneckenburger, P.; Adam, P.; Schaeffer, P.; Trendel, J.; Albrecht, P. (1997) Novel polycyclic polyprenoid sulfides in sediments. Chem. Commun., 2191-2.

Polne-Fuller, M. (1988) Microorganisms and methods for degrading plant cell walls and complex hydrocarbons. Eur. Pat. Appl. EP 289,908; Chem. Abstr., 1989, *110*, 151224s.

Ponomarenko, L.P.; Makarieva, T.N.; Stonik, V.A.; Dmitrenok, A.S.; Dmitrenok, P.S. (1995) Sterol composition of *Linneus torquatus* (Nemertini). Comp. Biochem. Physiol., *111B*, 575-7.

Porzel, A.; Lien, T.P.; Schmidt, J.; Drosihn, S.; Wagner, C.; Merzweiler, K.; van Sung, T. (2000) Fissistigmatis A-D: novel type natural products with flavonoid-sesquiterpene hybrid structure from *Fissistigma bracteolatum*. Tetrahedron, *56*, 865-72.

Prasad, A.V.K.; Shimizu, Y. (1989) The structure of hemibrevetoxin-B: a new type of toxin in the Gulf of Mexico red tide organism. J. Am. Chem. Soc., *111*, 6476-7.

Prowse, W.G.; Arnot, K.I.; Recka, J.A.; Thomson, R.H.; Maxwell, J.R. (1991) The quincyte pigments: fossil quinones in an Eocene clay mineral. Tetrahedron, *47*, 1095-108.

Qureshi, A.; Colin, P.L.; Faulkner, D.J. (2000) Microsclerodermins F-I, antitumor and antifungal cyclic peptides from the lithistid sponge *Microscleroderma* sp. Tetrahedron, *56*, 3679-85.

Rademacher, W. (1997) Gibberellins. In Fungal Biotechnology. Anke, T., ed., Chapman & Hall, London, pp. 193-205.

Radisky, D.C.; Radisky, E.S.; Barrows, L.R.; Copp, B.R.; Kramer, R.A.; Ireland, C.M. (1993) Novel cytotoxic topoisomerase II inhibiting pyrroloiminoquinones from Fijian sponges of the genus *Zyzzya*. J. Am. Chem. Soc., *115*, 1632-8.

Radman, M. (2001) Fidelity and infidelity. Nature, *413*, 115.

Rahbæk L.; Carsten, C. (1997) Three new alkaloids, securamines E-G, from the marine bryozoan *Securiflustra securifrons*. J. Nat. Prod., *60*, 175-7.

Ramesh, P.; Reddy, N.S.; Rao, T.P.; Venkateswarlu, Y. (1999) New oxygenated africanenes from the soft coral *Sinularia dissecta*. J. Nat. Prod., *62*, 1019-21.

Rasmussen, T.; Jensen, J.; Anthoni, U.; Christoffersen, C.; Nielsen, P. (1993) Structure and synthesis of bromoindoles from the marine sponge *Pseudosuberites hyalinus*. J. Nat. Prod., *56*, 1553-8.

Raub, M.F.; Cardellina, II, J.H.; Choudhary, M.I.; Ni, C.Z.; Clardy, J. (1991) Clavepictines A and B: cytotoxic quinolizidines from the tunicate *Clavelina picta*. J. Am. Chem. Soc., *113*, 3178-80.

Raub, M.F.; Cardellina, II, J.H.; Spande, T.F. (1992) The piclavines, antimicrobial indolizidines from the tunicate *Clavelina picta*. Tetrahedron Lett., *33*, 2257-60.

Reid, R.T.; Live, D.H.; Faulkner, D.J.; Butler, A. (1993) A siderophore from a marine bacterium with an exceptional ferric ion affinity constant. Nature, *366*, 455-8.

Rein, K.S.; Gawley, R.E.; Baden, D.G. (1994) Conformational analysis of the sodium channel modulator, brevetoxin A. J. Org. Chem., *59*, 2101-6.

Renner, M.K.; Shen, Y.-C.; Cheng, X.-C.; Jensen, P.R.; Frankmoelle, W.; Kauffman, C.A.; Fenical, W.; Lobkovsky, E.; Clardy, J. (1999) Cyclomarins A-C, new antiinflammatory cyclic peptides produced by a marine bacterium (*Streptomyces* sp.). J. Am. Chem. Soc., *121*, 11273-6.

Repa, J.J.; Turley, S.D.; Lobaccaro, J.-M.A.; Medina, J.; Li, L.; Lustig, K.; Shan, B.; Heyman, R.A.; Dietschy, J.M.; Mangelsdorf, D.J. (2000) Regulation of absorption and ABC1-mediated efflux of cholesterol by RXR heterodimers. Science, *289*, 1524-9.

Rheinheimer, G. (1998) Pollution in the Baltic Sea. Naturwissenschaften, *85*, 318-29.

Rhew, R.C.; Miller, B.R.; Weiss, R.F. (2000) Natural methyl bromide and methyl chloride emissions from coastal salt marshes. Nature, *403*, 292-8.

Richer de Forges, B.; Koslow, J.A.; Poore, G.C.B. (2000) Diversity and endemism of the benthic seamount fauna in the southwest Pacific. Nature, *405*, 944-7.

Riebesell, U.; Zondervan, I.; Rost, B.; Tortell, B.D.; Zeebe, R.E.; Morel, F.M.M. (2000) Reduced calcification of marine plankton in response to increased atmospheric CO_2. Nature, *407*, 364-7.

Rinehart, K.L., Jr.; Kishore, V.; Bible, K.C.; Sakai, R.; Sullins, D.; Li, K.-M. (1988) Didemnins and tunichlorin: novel natural products from the marine tunicate *Trididemnum solidum*. J. Nat. Prod., *51*, 1-21.

Rinehart, K.L., Jr. (2000) Antitumor compounds from tunicates. Medic. Res. Rev., *20*, 1-27.

Roberts, J.M. (2000) Full effects of oil rigs on corals are not yet known. Nature, *403*, 242.

Robin, J.-P.; Dhal, R.; Dujardin, G.; Girodier, L.; Mevellec, L.; Poutot, S. (2000) The first semi-synthesis of enantiopure homoharringtonine *via* anhydrohomoharringtonine from a preformed chiral acyl moiety. Tetrahedron Lett., *40*, 2931-4.

Robinson, R. (1917) A synthesis of tropinone. J. Chem. Soc., 762-8.

Rochfort, S.J.; Atkin, D.; Hobbs, L.; Capon, R.J. (1996) Hippospongins A-F: new furanoterpenes from a southern Australian marine sponge *Hippospongia* sp. J. Nat. Prod., *59*, 1024-8.

Rodríguez, A.D.; Ramirez, C.; Rodríguez, I.I. (1999A) Sandresolides A and B: novel nor-diterpenes from the sea whip *Pseudopterogorgia elisabethae*. Tetrahedron Lett., *40*, 7627-31.

Rodríguez, A.D.; Ramírez, C.; Rodríguez, I.I.; González, E. (1999B) Novel antimycobacterial benzoxazole alkaloids, from the West Indian sea whip *Pseudopterogorgia elisabethae*. Org. Lett., *1*, 527-30.

Rodríguez, J.P.; Ashenfelter, G.; Rojas-Suárez, F.; García, Fernández, J.J.; Suárez, L.; Dobson, A.P. (2000) Local data are vital to worldwide conservation. Nature, *403*, 241.

Roenneberg, T.; Nakamura, H.; Cranmer, L.D.; Ryan, K.; Kishi, Y.; Hastings, J.W. (1991) Gonyauline: a novel endogenous substance shortening the period of the circadian clock of a unicellular alga. Experientia, *47*, 103-6.

Roesener, J.A.; Scheuer, P.J. (1986) Ulapualide A and B, extraordinary antitumor macrolides from nudibranch eggmasses. J. Am. Chem. Soc., *108*, 846-7.

Rohmer, M.; Sutter, B.; Sahm, H. (1989) Bacterial sterol surrogates. Biosynthesis of the side-chain of bacteriohopanetetrol. J.C.S., Chem. Commun., 1471-2.

Romo, D.; Rzasa, R.M.; Shea, H.A.; Park, K.; Langenhan, J.M.; Sun, L.; Akhiezer, A.; Liu, J.O. (1998) Total synthesis and immunosuppressive activity of (-)-pateamine A and related compounds: implementation of a β-lactam-based macrocyclization. J. Am. Chem. Soc., *120*, 12237-54.

Rose, H.; Rose, S., eds. (2000) Alas, Poor Darwin: Arguments Against Evolutionary Psychology. Jonathan Cape, London.

Rougeaux, H.; Pichon, R.; Kervarec, N.; Raguénès, G.H.C.; Guezennec, J.G. (1996) Novel bacterial exopolysaccharides from deep-sea hydrothermal vents. Carbohydr. Polym., *31*, 237-42.

Rudi, A.; Aknin, M.; Gaydou, E.M.; Kashman, Y. (1998) Four new cytotoxic cyclic hexa- and heptapeptides from the marine ascidian *Didemnum molle*. Tetrahedron, *54*, 13203-10.

Rudi, A.; Yosief, T.; Schleyer, M.; Kashman, Y. (1999A) Bilosespenes A and B: two novel cytotoxic sesterterpenes from the marine sponge *Dysidea cinerea*. Org. Lett., *1*, 471-2.

Rudi, A.; Yosief, T.; Schleyer, M.; Kashman, Y. (1999B) Several new isoprenoids from two marine sponges of the family Axinellidae. Tetrahedron, *55*, 5555-66.

Ryan, C.A. (2001) Night moves the pregnant moths. Nature, *410*, 530-1.

Rychnovsky, S.D.; Skalitzky, D.J.; Pathirana, C.; Jensen, P.J.; Fenical, W. (1992) Stereochemistry of the macrolactins. J. Am. Chem. Soc., *114*, 671-7.

Saito, K.; Yamazaki, M.; Murakoschi, I. (1992) Transgenic medicinal plants: *Agrobacterium*-mediated foreign gene transfer and production of secondary metabolites. J. Nat. Prod., *55*, 149-62.

Sakai, R.; Oiwa, C.; Takaishi, K.; Kamiya, H.; Tagawa, M. (1999) Dysibetaine: a new α,α-disubstituted α-amino acid derivative from the marine sponge *Dysidea herbacea*. Tetrahedron Lett., *40*, 6941-4.

Sakemi, S.; Higa, T.; Jefford, C.W.; Bernardinelli, G. (1986) Venustatriol. A new, anti-viral, triterpene tetracyclic ether from *Laurencia venusta*. Tetrahedron Lett., *27*, 4287-90.

Sakemi, S.; Higa, T.; Jefford, C.W.; Bernardinelli, G. (1988) Isolation and structure elucidation of onnamide A, a new bioactive metabolite of a marine sponge, *Theonella* sp. J. Am. Chem. Soc., *110*, 4851-3.

Sakowicz, R.; Berdelis, M.S.; Ray, K.; Blackburn, C.L.; Hopmann, C.; Faulkner, D.J.; Goldstein, L.S. (1998) A marine natural product inhibitor of kinesin motors. Science, *280*, 292-5.

Sala, O.E.; Chapin III, F.S.; Armesto, J.J.; Berlow, E.; Bloomfield, J.; Dirzo, R.; Huber-Sanwald, E.; Huenneke, L.F.; Jackson, R.B.; Kinzig, A.; Leemans, R.; Lodge, D.M.; Mooney, H.A.; Oesteheld, M.; LeRoy Poff, N.; Sykes, M.T.; Walker, B.H.; Walker, M.; Wall, D.H. (2000) Global biodiversity scenarios for the year 2100. Science, *287*, 1770-4.

Salomon, A.R.; Zhang, Y.; Khosla, C. (2001) Structure-activity relationships within a family of selectively cytotoxic macrolide natural products. Org. Lett., *3*, 57-9.

Sánchez-Ferrando, F.; San-Martín, A. (1995) Epitaondiol: the first polycyclic meroditerpenoid containing two fused six-membered rings forced into the twist-boat conformation. J. Org. Chem., *60*, 1475-8.

Sanduja R.; Weinheimer, A.J.; Alam, M.; Hossain, M.B.; van der Helm, D. (1984) Echinolactone A and B: isolation and structures of two novel lactones from the marine coelenterate *Echinopora lamellosa*. J. Chem. Soc., Chem. Commun., 1091-2.

Sanduja, R.; Weinheimer, A.J.; Alam, M. (1985) Albizoin: isolation and structure of a deoxybenzoin from the marin mollusk *Nerita albicilla*. J. Chem. Res. (S), 56-7.

Sanglier, J.-J.; Quesniaux, V.; Fehr, T.; Hofmann, H.; Mahnke, M.; Memmert, K.; Schuler, W.; Zenke, G.; Gschwind, L.; Maurer, C.; Schilling, W. (1999) Sanglifehrins A-D, novel cyclophilin binding compounds isolated from *Streptomyces* sp. J. Antibiot., *52*, 466-73.

Sankaran, M.; McNaughton, S.J. (1999) Determinants of biodiversity regulate compositional stability of communities. Nature, *401*, 691-3.

San-Martín, A.; Darias, J.; Soto, H.; Contreras, C.; Herrera, J.S.; Rovirosa, J. (1997) A new C_{15} acetogenin from the marine alga *Laurencia claviformis*. Nat. Prod. Lett., *10*, 303-11.

Santos, G.A.; Doty, M.S. (1971) Constituents of the green alga *Caulerpa lamourouxii*. Lloydia, *34*, 88-90.

Sarma, N.S.; Rambabu, M.; Anjaneyulu, A.S.R.;Rao, C.B.S. (1987) Occurrence of picrotoxins in marine sponge *Spirastrella incostans*. Indian J. Chem., B, *26*, 189-90.

Sata, N.U.; Wada, S.I.; Matsunaga, S.; Watabe, S.; van Soest, R.W.M.; Fusetani, N. (1999) Rubrosides A-H, new bioactive tetramic acid glycosides from the marine sponge *Siliquariaspongia japonica*. J. Org. Chem., *64*, 2331-9.

Satake, M.; Ishibashi,Y.; Legrand, A.M.;, Yasumoto, T. (1997) Isolation and structure of ciguatoxin-4A, a new ciguatoxin precursor, from cultures of dinoflagellate *Gambierdiscus toxicus* and parrotfish *Scarus gibbus*. Biosci. Biotech. Biochem., *60*, 2103-5.

Sato, A.; Morishita, T.; Shiraki, T.; Yoshioka, S.; Horikoshi, H. *et al.* (1993) Aldose reductase inhibitors from a marine sponge, *Dictyodendrilla* sp. J. Org. Chem., *58*, 7632-4.

Saxena, D.; Flores, S.; Stotzky, G. (1999) Insecticidal toxin in root exudates from *Bt* corn. Nature, *402*, 480.

Schaeffer, P.; Fache-Dany, F.; Trifilieff, S.; Trendel, J.M.; Albrecht, P. (1994A) Characterisation of novel 3-carboxyalkyl-steranes occurring inn geological samples. Tetrahedron, *50*, 12633-42.

Schaeffer, P.; Poinsot, J.; Hauke, V.; Adam, P.; Wehrung, P.; Trendel, J.-M.; Albrecht, P.; Dessort, D.; Connan, J. (1994B) Novel optically active hydrocarbons in sediments: evidence for an extensive biological cyclization of higher regular polyprenols. Angew. Chem. Int. Ed. Engl., *33*, 1166-9.

Schaeffer, P.; Trendel, J.-M.; Albrecht, P. (1995) Structure and origin of two triterpene-derived aromatic hydrocarbons in Messel shale. J. Chem. Soc., Chem. Commun., 1275-6.

Schatz, A.; Bugie, E.; Waksman, S.A. (1944) Streptomycin, a substance exhibiting antibiotic activity against Gram-positive and Gram-negative bacteria. Proc. Spc. Exp. Biol. Med., *55*, 66-9.

Scheuer, P.J. (1973) Chemistry of Marine Natural Products. Academic Press, New York and London, 67.

Scheuer, P.J. (1996) Marine toxin research - An historical overview. J. Nat. Toxins, *5*, 181-9.

Schkeryantz, J.M.; Woo, J.C.G.; Siliphaivanh, P.; Depew, K.M.; Danishefsky, S.J. (1999) Total synthesis of gypsetin, deoxybrevianamide E, brevianamide E, and tryprostatin B. J. Am. Chem. Soc., *121*, 11964-75.

Schlama, T.; Baati, R.; Gouverneur, V.; Valleix, A.; Falk, J.R.; Mioskowski, C. (1998) Total synthesis of (±)-halomon by a Johnson-Claisen rearrangement. Angew. Chem. Int. Ed. Engl., *37*, 2085-7.

Schmid, A.; Dordick, J.S.; Hauer, B.; Kiener, A.; Wubbolts, M.; Witholt, B. (2001) Industrial biocatalysis today and tomorrow. Nature, *409*, 258-68.

Schmid, P.E.; Tokeshi, M.; Schmid-Araya, J.M. (2000) Relation between population density and body size in stream communities. Science, *289*, 1557-60.

Schmidt-Dannert, C.; Umeno, D.; Arnold, F.H. (2000) Molecular breeding of carotenoid biosynthetic pathways. Nat. Biotechnol., *18*, 750-3.

Schneider, S.H. (2001) What is "dangerous" climate change? Nature, *411*, 17-19.

Schopf, J.W. (1993) Microfossils of the early Archean chert: new evidence of the antiquity of life. Science, *260*, 640-6.

Schroeder, F.C.; Farmer, J.J.; Smedley, S.R.; Attygalle, A.B.; Eisner, T.; Meinwald, J. (2000) A combinatorial library of macrocyclic polyamines produced by a ladybird beetle. J. Am. Chem. Soc., *122*, 3628-34.

Schroeder, J.L.; Kwak, J.M.; Allen, G.J. (2001) Guard cell abscisic acid signalling and engineering drought hardiness in plants. Nature, *410*, 327-30.

Schumacher, R.W.; Davidson, B.S.; Montenegro, D.A.; Bernan, V.S. (1995) γ-Indomycinone, a new pluramycin metabolite from a deep-sea derived actinomycete. J. Nat. Prod., *58*, 613-7.

Schupp, P.; Eder, C.; Proksch, P.; Wray, V.; Schneider, B.; Herderich, M.; Paul, V. (1999) Staurosporine derivatives from the ascidian *Eudistoma toealensis* and its predatory flatworm *Pseudoceros* sp. J. Nat. Prod., *62*, 959-62.

Schwartz, R.E.; Yunker, M.B.; Scheuer, P.J.; Ottersen, T. (1979) Pseudozoanthoxanthins from gold coral. Can. J. Chem., *57*, 1707-11.

Schwartz, R.E.; Scheuer, P.J.; Zabel, V.; Watson, W.H. (1981) The coraxeniolides, constituents of pink coral, *Corallium* sp. Tetrahedron, *37*, 2725-33.

Science whaling (2000) A debate on scientific/illegal whaling, *290*, 1695-8.

Scott, A.I. (1992) Genetically engineered synthesis of complex natural products. Tetrahedron, *48*, 2559-78.

Scott, A.I. (1994) Towards a total, genetically engineered synthesis of vitamin B_{12}. Synlett, 871-84.

Seaman, F.C.; Hurley, L.H. (1998) Molecular basis for the DNA sequence selectivity of ecteinascidin 736 and 743: evidence for the dominant role of direct readout via hydrogen bonding. J. Am. Chem. Soc., *120*, 13028-41.

Searle, P.A.; Jamal, N.M.; Lee, G.M.; Molinsky, T.F. (1994A) Configurational analysis of new furanosesquiterpenes from *Dysidea herbacea*. Tetrahedron, *50*, 3879-88.

Searle, P.A.; Molinsky, T.F. (1994B) Five new alkaloids from the tropical ascidian, *Lissoclinum* sp. J. Org. Chem., *59*, 6600-5.

Sebahar, P.R.; Williams, R.M. (2000) The asymmetric total synthesis of (+)- and (-)-spirotryprostatin B. J. Am. Chem. Soc., *122*, 5666-7.

Seebach, D.; Beck, A.K.; Rueping, M.; Schreiber, J.V.; Sellner, H. (2001) Excursions of synthetic organic chemists to the world of oligomers and polymers. Chimia, *55*, 98-103; Seebach, D.; Albert, M.; Arvidsson, P.I.; Rueping, M.; Schreiber, J.V. (2001) From the biopolymer PHB to biological investigations of unnatural β- and γ-peptides. Chimia, *55*, 345-53.

Seemann, M.; Zhai, G.; Umezawa, K.; Cane, D. (1999) Pentalenene synthase. Histidine-309 is not required for catalytic activity. J. Am. Chem. Soc., *121*, 591-2.

Segawa, M.; Enoki, N.; Ikura, M.; Hikichi, K.; Ishida, R.; Shirahama, H.; Matsumoto, T. (1987) Dictymal, a new seco-fusicoccin type diterpene from the brown alga *Dictyota dichotoma*. Tetrahedron Lett., *28*, 3703-4.

Sellman, B.R.; Mourez, M.; Collier, R.J. (2001) Dominant-negative mutants of a toxin subunit: an approach to therapy of anthrax. Science, *292*, 695-7.

Seo, Y.; Rho, J.R.; Geum, N.; Yoon, J.B.; Shin, J. (1996) Isolation of guaianoid pigments from the gorgonian *Calicogorgia granulosa*. J. Nat. Prod., *59*, 985-6.

Sepčić, K.; Guella, G.; Mancini, I.; Pietra, F.; Dalla Serra, M.; Menestrina, G.; Tubbs, K.; Maček, P.; Turk, T. (1997) Characterization of anticholinesterase-active 3-alkylpyridinium polymers from the marine sponge *Reniera sarai* in aqueous solution. J. Nat. Prod., *60*, 991-6.

Serageldin, I. (1997) Rural Well-Being: From Vision to Action. Proceedings of the 4th Annual World Bank Conference on Environmentally Sustainable Development. World Bank, Washington, DC, 43.

Serageldin, I. (1999) Biotechnology and food security in the 21st century. Science, *285*, 387-9.

Sgró, C.M.; Partridge, L. (1999) A delayed wave of death from reproduction in *Drosophila*. Science, *286*, 2521-4.

Shen, Y.-C.; Lin, T.T.; Sheu, J.-H.; Duh, C.-Y. (1999) Structures and cytotoxicity relationships of isoaaptamine and aaptamine. J. Nat. Prod., *62*, 1264-7.

Sheu, J.-H.; Sung, P.-J.; Su, J.-H.; Wang, G.-H.; Duh, C.-Y.; Shen, Y.-C.; Chiang, M.Y.; Chen, I.-T. (1999) Excavatolides U-Z, new briarane diterpenes from the gorgonian *Briareum excavatum*. J. Nat. Prod., *62*, 1415-20.

Shibuya H.; Bohgaki, T.; Ohashi, K. (1999) Two novel migrated pimarane-type diterpenes from the leaves of *Orthosiphon aristatus*. Chem. Pharm. Bull., *47*, 911-2.

Shigemori, H.; Bae, M.-A.; Yazawa, K.; Sasaki, T.; Kobayashi, J. (1992) Alteramide A, a new tetracyclic alkaloid from a bacterium *Alteromonas* sp. associated with the marine sponge *Halichondria okadai*. J. Org. Chem., *57*, 4317-20.

Shigemori, H.; Tenma, M.; Shimazaki, K.; Kobayashi, J. (1998) Three new metabolites from the marine yeast *Aureobasidium pullulans*. J. Nat. Prod., *61*, 696-8.

Shimizu, K.; Cha, J.; Stucky, G.D.; Morse, D.E. (1998) Silicatein α: cathepsin L-like protein in sponge biosilica. Proc. Natl. Acad. Sci. USA, *95*, 6234-8.

Shimizu, Y. (1996) Microalgal metabolites: a new perspective. Annu. Rev. Microbiol., *50*, 431-65.

Shin, J.; Seo, Y.; Rho, J.-R.; Baek, E.; Kwon, B.-M.; Jeong, T.-S.; Bok, S.-H. (1995) Suberitones A and B: sesterterpenoids of an unprecedented skeletal class from the Antarctic sponge *Suberites* sp. J. Org. Chem., *60*, 7582-8.

Shintani, D.; DellaPenna, D. (1998) Elevating the vitamin E content of plants through metabolic engineering. Science, *282*, 2098-100.

Siemann, E.; Tilman, D.; Haarstad, J. (1996) Insect species diversity, abundance, and body size relationships. Nature, *380*, 704-6.

Sigma (2000) Biochemical Reagents. Sigma 2000/2001.

Simpson, A.J.G.; Reinach, F.C.; Arruda, P.; Abreu, F.A.; Acencio, M. *et al.* (2000) The genome sequencing of the plant pathogen *Xylella fastidiosa*. Nature, *406*, 151-7.

Sitachitta, N.; Gadepalli, M; Davidson, B.S. (1996) New α-pyrone-containing metabolites from a marine-derived actinomycete. Tetrahedron, *52*, 8073-80.

Slater, G.P.; Blok, V.C. (1983) Isolation and identification of odorous compounds from a lake subject to cyanobacterial blooms. Water Sci. Technol., *15*, 229-40.

Smallwood, B.J.; Wolff, G.A.; Bett, B.J.; Smith, C.R.; Hoover, D.; Gage, J.D.; Patience, A. (1999) Megafauna can control the quality of organic matter in marine sediments. Naturwissenschaften, *86*, 320-4.

Smith, A.B., III; Friestad, G.K.; Barbosa, J.; Bertounesque, E.; Duan, J.J.-W.; Hull, K.G.; Iwashima, M.; Qiu, Y.; Spoors, P.G.; Salvatore, B.A. (1998) Total synthesis of (+)-calyculin A and (-)-calyculin B. J. Org. Chem., *63*, 7596-7.

Smith, A.B., III; Beauchamp, T.J.; LaMarche, M.J.; Kaufman, M.D.; Qiu, Y.; Arimoto, H.; Jones, D.R.; Kobayashi, K. (2000) Evolution of a gram-scale synthesis of (+)-discodermolide. J. Am. Chem. Soc., *122*, 8654-64.

Smith, A.B., III; Doughty, V.A; Lin, Q.; Zhuang, L.; McBriar, M.D.; Boldi, A.M.; Moser, W.H.; Murase, N.; Nakayama, K.; Sobukawa, M. (2001) The spongistatins: architecturally complex natural products. Synthesis of the C(29-51) subunit, fragment assembly, and final elaboration to (+)-spongistatin 2. Angew. Chem. Int. Ed. Engl., *40*, 196-9.

Smith, A.T.; Boitani, L.; Bibby, C.; Brackett, D.; Corsi, F.; da Fonseca, G.A.B.; Gascon, C.; Gimenez, M.; Hilton-Taylor, C.; Mace, G.; Mittermeier, R.A.; Rabinovich, J.; Richardson, B.J.; Rylands, A.; Stein, B.; Stuart, S.; Thomsen, J.; Wilson, C. (2000) Databases tailored for biodiversity conservation. Science, *290*, 2073.

Somerville, C.; Dangl, J. (2000) Plant biology in 2010. Science, *290*, 20-1.

Sommaruga, R.; Garchia-Pichel, F. (1999) UV-absorbing mycosporine-like compounds in planktonic and benthic organisms from a high-mountain lake. Arch. Hydrobiol., *144(3)*, 255-69.

Species (2000) global database through Internet at http://www.sp2000.org. (a few animals only are included).

Speight, M.R.; Hunter, M.D.; Watt, A.D. (1999) Ecology of Insects: Concepts and Applications. Blackwell Science.

Spinella, A.; Mollo, E.; Trivellone, E.; Cimino, G. (1997) Testudinariol A and B, two unusual triterpenoids from the skin and the mucus of the marine mollusc *Pleurobranchus testudinarius*. Tetrahedron, *53*, 16891-6.

Stachowicz, J.J.; Whitatch, R.B.; Osman, R.W. (1999) Species diversity and invasion in a marine ecosystem. Science, *286*, 1577-9.

Stamos, D.P.; Chen, S.S.; Kishi, Y. (1997) New synthetic route to the C.14-C.38 segment of halichondrins. J. Org. Chem., *62*, 7552.

Stanley, G.D., Jr.; Fautin, D.G. (2001) The origins of modern coral. Science, *291*, 1913-4.

Staunton, J.; Wilkinson, B. (1997) Biosynthesis of erythromycin and rapamycin. Chem. Rev., *97*, 2611-29.

Staunton, J.; Weissman, K.J. (2001) Polyketide biosynthesis: a millennium review. Nat. Prod. Rep., *18*, 380-416.

Steadman, D.W. (1995) Prehistoric extinctions of Pacific island birds: biodiversity meets zooarcheology. Science, *267*, 1123-31.

Stern, K.; McClintock, M.K. (1998) Regulation of ovulation by human pheromones. Nature, *392*, 177-9.

Stetter, K.O.; Huber, R.; Blöchl, E.; Kurr, M.; Eden, R.D.; Flelder, M.; Cash, H.; Vance, I. (1993) Hyperthermophilic archaea are thriving in deep North Sea and Alaskan oil reservoirs. Nature, *365*, 743-5.

Steward, M.; Blunt, J. W.; Munro, M.H G.; Robinson, W.T.; Hannah, D.J. (1997) The absolute stereochemistry of the New Zealand shellfish toxin gymnodimine. Tetrahedron Lett., *38*, 4889-90.

Stien, D.; Anderson, G.T.; Chase, C.E.; Koh, Y.-H.; Weinreb, SM. (1999) Total synthesis of the antitumor marine sponge alkaloid agelastatin A. J. Am. Chem. Soc., *121*, 9574-9.

Stierle, A.; Strobel, G.; Stierle, D. (1993) Taxol and taxane production by *Taxomyces andreanae*, an endophyte fungus of Pacific Yew. Science, *260*, 214-6.

Stierle, D.B.; Strobel, G.; Stierle, D. (1991) Two new pyrroloquinoline alkaloids from the sponge *Damiria* sp. J. Nat. Prod., *54*, 1131-3.

Stierle, D.B.; Stierle, A.A. (1992) Pseudomonic acid derivatives from a marine bacterium. Experientia, *48*, 1165-9.

Stocking, E.M.; Sanz-Cervera, J.F.; Williams, R.M. (2000) Total synthesis of VM55599. Utilization of an intramolecular Diels-Alder cycloaddition of potential biogenetic relevance. J. Am. Chem. Soc., *122*, 1675-83.

Stoddard, J.L.; Jeffries, D.S.; Lükewille, A.; Clair, T.A.; Dillon, P.J. (1999) Regional trends in aquatic recovery from acidification in North America and Europe. Nature, *401*, 575-8.

Stoddard, P.K. (1999) Predation enhances complexity in the evolution of electric fish signals. Nature, *400*, 254-6.

Stonik, V.A.; Makarieva, T.N.; Dmitrenok, A.S. (1992) Sarcochromenol sulfates A-C and sarcohydroquinone sulfates A-C, new natural products from the sponge *Sarcotragus spinulosus*. J. Nat. Prod., *55*, 1256-60.

Stout, T.J.; Clardy, J.; Pathirana, I.C.; Fenical, W. (1991) Aplasmomycin C: structural studies of a marine antibiotic. Tetrahedron, *47*, 3511-20.

Su, B.-N.; Takaishi, Y.; Tori, M.; Takaoka, S.; Honda, G.; Itoh, M.; Takeda, Y.; Kodzhimatov, O.K.; Ashurmetov, O. (2000) Macrophyllidimers A and B, two novel sesquiterpene dimers from the bark of *Inula macrophylla*. Tetrahedron Lett., *41*, 1475-9.

Su, J.-Y.; Zhong, Y.-L.; Zeng, L.-M. (1993) Two new sesquiterpenoids from the soft coral *Paralemnalia thyrsoides*. J. Nat. Prod., *56*, 288-91.

Sugawara, K.; Nishiyama, Y.; Toda, S.; Komiyama, N.; Hatori, M.; Moriyama, T.; Sawada, Y.; Kamei, H.; Konishi, M.; Oki, T. (1992) Lactimidomycin, a new glutarimide group antibiotic. J. Antibiot., *45*, 1433-41.

Sugawara, K.; Toda, S.; Moriyama, T.; Konishi, M.; Oki, T. (1993) Verucopeptin, a new antitumor antibiotic active against B16 melanoma. J. Antibiot., *46*, 928-35.

Summons, R.E.; Janke, L.L.; Hope, J.M.; Logan, G.A. (1999) 2-Methylhopanoids as biomarkers for cyanobacterial oxygenic photosynthesis. Nature, *400*, 554-7.

Sun, H.H.; Fenical, W. (1979) Diterpenoids from the brown seaweed *Glossophora galapagensis*. Phytochemistry, *18*, 340-1.

Sun, H.H.; Sakemi, S. (1991) A brominated (aminoimidazolinyl) indole from the sponge *Discodermia polydiscus*. J. Org. Chem., *56*, 4307-8.

Sunazuka, T.; Hirose, T.; Shirahata, T.; Harigaya, Y.; Hayashi, M.; Komiyama, K.; Ōmura, S. (2000) Total synthesis of (+)-madindoline A and (-)-madindoline B, potent, selective inhibitors of interleukin 6. J. Am. Chem. Soc., *122*, 2122-3.

Sutton, A.E.; Clardy, J. (2001) Synthesis and biological evaluation of analogues of the antibiotic pantocin B. J. Am. Chem. Soc., *123*, 9935-46.

Suzuki, M.; Morita, Y.; Yanagisawa, A.; Baker, B.J.; Scheuer, P.J.; Noyori, R.(1986) Synthesis of (*7E*)- and (*7Z*)-punaglandin A. Structural revision. J. Am. Chem. Soc., *108*, 5021-2.

Suzuki, M.; Matsuo, Y.; Takahashi, Y.; Masuda, M. (1995) Callicladol, a novel cytotoxic bromotriterpene polyether from a Vietnamese species of the red algal genus *Laurencia*. Chem. Lett., 1045-6.

Suzuki, T.; Takeda, S.; Suzuki, M.; Kurosawa, E.; Kato, A.; Imanaka, Y. (1987) Cytotoxic squalene-derived polyethers from the marine red alga *Laurencia obtusa*. Chem. Lett., 361-4.

Swearer, S.E.; Caselle, J.E.; Lea, D.W.; Warner, R.R. (1999) Larval retention and recruitment in an island population of a coral-reef fish. Nature, *402*, 799-802.

Swersey, J.C.; Ireland, C.M.; Cornell, L.M.; Peterson, R.W. (1994) Eusynstyelamide, a highly modified dimer peptide from the ascidian *Eusynstyela misakiensis*. J. Nat. Prod., *57*, 842-5.

Swift, M.J.; Vandermeer, J.; Ramakrishnan, P.S.; Anderson, J.M.; Ong, C.K.; Hawkins, B.A. (1996) Biodiversity and agroecosystem function. In Functional Roles of Biodiversity: a Global Perspective, Mooney, H.A. ed, J. Wiley, New York, pp. 261-98.

Takada, N.; Umemura, N.; Suenaga, K.; Uemura, D. (2001) Structural determination of pteriatoxins A, B and C, extremely potent toxins from the bivalve *Pteria penguin*. Tetrahedron Lett., *42*, 3495-7.

Takahashi, A.; Kurasawa, S.; Ikeda, D.; Okami, Y.; Takeuchi, T. (1989) Altemicidin, a new acaricidal and antitumor substance. J. Antibiot., *42*, 1556-61.

Takahashi, C.; Takada, T.; Yamada, T.; Minoura, K.; Uchida, K. *et al.* (1994) Halichomycin, a new class of potent cytotoxic macrolactams produced by an actinomycete from a marine fish. Tetrahedron Lett., *35*, 5013-4.

Takahashi, M.; Dodo, K.; Hashimoto, Y.; Shirai, R. (2000) Concise asymmetric synthesis of dysidiolide. Tetrahedron. Lett., *41*, 2111-4.

Takeda, S.; Iimura, Y.; Tanaka, K.; Kurosawa, E.; Suzuki, T. (1990) A new naturally occurring racemic compound from the marine red alga *Laurencia obtusa*. Chem. Lett., 155-6.

Takeshi, O.; Teruo, H.; Aiya, S.; Kohei, F. (1987) Antibiotic zofimarin manufacture by *Zofiela marina* and its fungal activity. Jpn. Kokai Tokkyo Koho JP 62 40,292; Chem. Abstr., 1987, *107*, 5745j.

Takizawa, P.A.; Yucel, J.K.; Veit, B.; Faulkner, D.J.; Deerinck, T.; Soto, G.; Ellisman, M.; Malhotra, V. (1993) Complete vesiculation of Golgi membranes and inhibition of protein transport by a novel sea sponge metabolite. Cell, *73*, 1079-90.

Talpir, R.; Benayahu, Y.; Kashman, Y.; Panell, L.; Schleyer, M. (1994) Hemiasterlin and geodiamolide TA; two new cytotoxic peptides from the marine sponge *Hemiasterella minor*. Tetrahedron Lett. *35*, 4453-6.

Tan, R.X.; Zou, W.X. (2001) Endophytes: a rich source of functional metabolites. Nat. Prod. Rep., *18*, 448-59.

Tanaka, R.; Kumagai, T.; In, Y.; Ishida, T.; Nishino, H.; Matsunaga, S. (1999) Piceanols A and B, triterpenoids bearing a novel skeletal system isolated from the bark of *Picea jezoensis* var. *hondoensis*. Tetrahedron Lett., *40*, 6415-8.

Tangley, L. (2001) High CO_2 levels may give fast-growing trees an edge. Science, *292*, 36-7.

Tanimoto, H.; Oritani, T. (1997) Synthesis of (+)-ambrein. Tetrahedron, *53*, 3527-36.

Terborg, J. (1999) Requiem for Nature. Island Press, Washington D.C.

Testa, B.; Caron, G.; Crivori, P.; Rey, S.; Reist, M.; Carrupt, P.A. (2000) Lipophilicity and related molecular properties as determinants of pharmacokinetic behaviour. Chimia, *54*, 672-7.

Thompson, A.M.; Blunt, J.W.; Munro, M.H.G.; Perry, N.B.; Pannell, L.K. (1994) Chemistry of the mycalamides. J. Chem. Soc., Perkin Trans. 1, 1025-31.

Tilman, D.; May, R.M.; Lehman, C.L.; Nowak, M.A. (1994) Habitat destruction and the extinction debt. Nature, *371*, 65-6.

Todd, J.S.; Proteau, P.J.; Gerwick, W.H. (1994) The absolute configuration of ecklonialactones A, B, and E, novel oxylipins from brown algae of the genera *Ecklonia* and *Egregia*. J. Nat. Prod., *57*, 171-4.

Toró, A.; Nowak, P.; Deslongchamps, P. (2000) Transannular Diels-Alder entry into stemodanes: first asymmetric total synthesis of (+)-maritimol. J. Am. Chem. Soc., *122*, 4526-7.

Tregenza, T.; Butlin, R.K. (1999) Speciation without isolation. Nature, *400*, 311-2.

Trendel, J.M.; Graff, R.; Albrecht, P.; Riva, A. (1991) 24,28-Dinor-18α-oleanane, a novel demethylated higher plant derived triterpene hydrocarbon in petroleum. Tetrahedron Lett., *32*, 2959-62.

Trischman, J.A.; Tapiolas, D.M.; Jensen, P.R.; Dwight, R.; Fenical, W.; McKee, T.C.; Ireland, C.M.; Stout, T.J.; Clardy, J. (1994A) Salinamides A and B: anti-inflammatory depsipeptides from a marine streptomycete. J. Am. Chem. Soc., *116*, 757-8.

Trischman, J.A.; Jensen, P.R.; Fenical, W. (1994B) Halobacillin b: a cytotoxic cyclic acylpeptide of the iturin class produced by a marine *Bacillus*. Tetrahedron Lett., *35*, 5571-4.

Tsuda, M.; Hirano, K.; Kubota, T.; Kobayashi, J. (1999A) Pyrinodermin A, a cytotoxic pyridine alkaloid with an isoxazolidine moiety from sponge *Amphimedon* sp. Tetrahedron Lett., *40*, 4819-20.

Tsuda, M.; Shimbo, K.; Kubota, T.; Mikami, Y.; Kobayashi, J. (1999B) Two theonellapeptolide congeners from marine sponge *Theonella* sp. Tetrahedron, *55*, 10305-14.

Tsuda, M.; Endo, T.; Kobayashi, J. (2000) Amphidinolide T, novel 19-membered macrolide from marine dinoflagellate *Amphidinium* sp. J. Org. Chem., *65*, 1349-52.

Tsujii, S.; Rinehart, K.L., Jr.; Gunasekera, S.P.; Kashman, Y.; Cross, S.S. *et al.* (1988) Topsentin, bromotopsentin, and dihydrodeoxybromotopsentin: antiviral and antitumor bis(indolyl)imidazoles from Caribbean deep-sea sponges of the family Halichondriidae. J. Org. Chem., *53*, 5446-53.

Tsukamoto, S.; Hirota, H.; Kato, H.; Fusetani, N. (1993) Urochordamines A and B: larval settlement/metamorphosis promoting pteridine-containing physostigmine alkaloids from the tunicate *Ciona savignyi*. Tetrahedron Lett., *34*, 4819-22.

Tsutsui, N.D.; Suarez, A.V.; Holway, D.A.; Case, T.J. (2000) Reduced genetic variation and the success of an invasive species. Proc. Natl. Acad. Sci., *97*, 5948-53.

Tumlinson, J.H.; Lewis, W.J.;Vet, L.E.M. (1993) How parasitic wasp find their hosts. Sci. Am., March, 100-6.

Turner, W.B.; Aldridge, D.C. (1983) Fungal metabolites II. Academic Press. London.

Ueda, M.; Sohtome, Y.; Ueda, K.; Yamamura, S. (2001) Potassium 2,3,4-trihydroxy-2-methylbutanoate, a leaf-closing substance of *Leucaena leucocephalum*. Tetrahedron Lett., *42*, 3109-11.

Uehara, H.; Oishi, T.; Yoshikawa, K.; Mochida, K.; Hirama, M. (1999) The absolute configuration and total synthesis of korormicin. Tetrahedron Lett., *40*, 8641-5.

Umeyama, A.; Machida, M.; Nozaki, M.; Arihara, S. (1998) Peroxypolasol and mugipolasol: two novel diterpenes from the marine sponge *Epipolasis* sp. J. Nat. Prod., *61*, 1435-6.

UNPD (United Nations Population Division) World Population Prospects: The 1998 Revision. United Nations, N.Y.

Unson, M.D.; Rose, C.B.; Faulkner, D.J.; Brinen, L.S.; Steiner, J.R.; Clardy, J. (1993) New polychlorinated amino acid derivatives from the marine sponge *Dysidea herbacea*. J. Org. Chem., *58*, 6336-43.

Urban, S.; Capon, R.J. (1992) Cometins (A-C), new furanosesterterpenes from an Australian marine sponge, *Spongia* sp. Aust. J. Chem., *45*, 1255-63.

van den Ent, F.; Amos, L.A.; Löwe, J. (2001) Prokaryotic origin of the actin cytoskeleton. Nature, *413*, 39-44.

van der Heijden, M.G.A.; Klironomos, J.N.; Ursic, M.; Moutoglis, P.; Streitwolf-Engel, R.; Boller, T.; Wiemken, A.; Sanders, I.R. (1998) Mycorrhizal fungal diversity determines plant biodiversity, ecosystem variability and productivity. Nature, *396*, 69-72.

van Dover, C.L. (2000) The Ecology of Deep-Sea Hydrothermal Vents. Princeton University Press, Princeton, NJ.

Vankar, Y.D.; Schmidt, R.R. (2000) Chemistry of glycosphingolipds - carbohydrate molecules of biological significance. Chem. Soc. Rev., *29*, 201-16.

Varvas, K.; Koljak, R.; Jaerving, I.; Pehk, T.; Samel, N. (1994) Endoperoxide pathway in prostaglandin biosynthesis in the soft coral *Gersemia fruticosa*. Tetrahedron Lett., *35*, 8267-70.

Venkataraman, G.; Shriver, Z.; Raman, R.; Sasisekharan, R. (1999) Sequencing complex polysaccharides. Science, *286*, 537-42.

Verberne, M.C.; Verpoorte, R.; Bol, J.F.; Mercado-Blanco, J.; Linthorst, H.J.M. (2000) Overproduction of salicylic acid in plants by bacterial transgenes enhances pathogen resistance. Nature Biotechnol., *18*, 779-83.

Vernberg, F.J.; Vernberg, W.B. (1981) Functional Adaptation of Marine Organisms. Academic Press, New York.

Vervoort, H.C.; Fenical, W.; Keifer, P.A. (1999) A cyclized didemnimide alkaloid from the Caribbean ascidian *Didemnum conchyliatum*. J. Nat. Prod., *62*, 389-91.

Vidal, J.-P.; Escale, R.; Girard, J.-P.; Rossi, J.-C.; Chantraine, J.-M.; Aumelas, A. (1992) Lituarines A, B, and C: a new class of mavrocyclic lactones from the New Caledonian sea pen *Lituaria australasiae*. J. Org. Chem., *57*, 5857-60.

Voelker, T.A.; Worrell, A.C.; Anderson, L.; Bleibaum, J.; Fan, J.; Hawkins, D.J.; Radke, S.E.; Davies, H.M. (1992) Fatty acid biosynthesis redirected to medium chains in transgenic oilseed plants. Science, *257*, 72-4.

Vogel, G. (2001) No easy answers for biodiversity in Africa. Science, *291*, 2529-30.

von Nussbaum, F.; Danishefsky, S.J. (2000) A rapid total synthesis of spirotryprostatin B: proof of its relative and absolute stereochemistry. Angew. Chem. Int. Ed. Engl., *39*, 2175-8.

Vuong, D.; Capon, R.J. (2000) Phorbasin A: a novel skeleton diterpene from a southern Australian marine sponge, *Phorbas* species. J. Nat. Prod., *63*, 1684-5.

Wabnitz, P.A.; Bowie, J.H.; Tyler, M.J.; Wallace, J.C.; Smith, B.P. (2000) Differences in the skin peptides of the male and female Australian tree frog *Litoria splendida*. Eur. J. Biochem., *267*, 1-8.

Wahlberg, I.; Eklung, A.-M. (1992A) Cembranoids, pseudopteranoids, and cubitanoids of natural occurrence. In Fortschr. Chem. Org. Naturst., Herz, W.; Kirby, G.W.; Moore, R.E.; Steglich, W.; Tamm, C., eds., *59*, pp. 141-294.

Wahlberg, I.; Eklung, A.-M. (1992B) Cyclized cembranoids of natural occurrence. In Fortschr. Chem. Org. Naturst., Herz, W.; Kirby, G.W.; Moore, R.E.; Steglich, W.; Tamm, C., eds., *60*, pp. 1-142.

Waibel, R.; Novak-Hofer, I.; Schibli, R.; Bläuenstein, P.; Garcia-Garayoa, E.; Schwarzbach, R.; Zimmermann, Pellikka, R.; Gasser, O.; Blanc, A.; Brühlmeier, M.; Schubiger, P.A. (2000) Radiopharmaceuticals for targeted tumor diagnosis and therapy. Chimia, *54*, 683-9.

Waite, H.; Qin, X.X.; Coyne, K.J. (1998) The peculiar collagens of mussel byssus. Matrix Biol., *17*, 93-106.

Walker, F. A. (1999) Novel nitric oxide-liberating heme proteins from the saliva of bloodsucking insects. Met. Ions Biol. Syst., *36*, 621-63; Chem. Abstr., 1999, *131*, 56600.

Walker, R.P.; Faulkner, D.J.; van Engen, D.; Clardy, J. (1981) Sceptrin, an antimicrobial agent from the sponge *Agelas sceptrum*. J. Am. Chem. Soc., *103*, 6772-3.

Walsh, J.F. (1993) Deforestation: effects on vector-borne disease. Parasitology, *106*, 55-75.

Wang, C.; Wang, M.; Su, J. (1998) Research on the chemical constituents of *Acanthophora spicifera* in the South Cina Sea. Bopuxue Zazhi, *15*, 237-42; Chem. Abstr., 1998, *129*, 158921t.

Wang, H.; Ganesam, A. (2000) Total synthesis of the fumiquinazoline alkaloids: solution-phase studies. J. Org. Chem., *65*, 1022-30.

Watanabe, A.; Fujii, I.; Sankawa, U.; Mayorga, M.E.; Timberlake, W.E.; Ebizuka, Y. (1999) Re-identification of *Aspergillus nidulans wA* gene to code for a polyketide synthase of naphthopyrone. Tetrahedron Lett., *40*, 91-4.

Watanabe, K.; Miyakado, M.; Ohno, N.; Okada, A.; Yanagi, K.; Moriguchi, K. (1989) A polyhalogenated insecticidal monoterpene from the red alga, *Plocamium telfairiae*. Phytochemistry, *28*, 77-8.

Watkinson, A.R.; Freckleton, R.P.; Robinson, R.A.; Sutherland, W.J. (2000) Predictions of biodiversity response to genetically modified herbicide-tolerant crops. Science, *289*, 1555-7.

Webb, T. (1992) Past changes in vegetation and climate: lessons for the future. In Global Warming and Biological Diversity. Peters, R.T.; Lovejoy, T.E., eds., Yale University Press, pp. 59-75.

Wegner, C.; Hamburger, M.; Kunert, O.; Haslinger, E. (2000) Tensioactive compounds from the aquatic plant *Ranunculus fluitans* L. Helv. Chim. Acta, *83*, 1454-64.

Weiner, R.M.; Colwell, R.R.; Bonar, D.B.; Coon, S.L.; Walsh, M. (1991) Induction of settlement and metamorphosis in *Crassostrea virginica* by melanin-synthesizing bacteria and ammonia, and metabolic products of said bacteria. US 5,047,344, 30 Dec. to Research Corp. Techniques. Chem. Abstr., 1992, *116*, 18334k.

Welzel, P.; Röhrig, S.; Milkova, Z. (1999) Strigol-type germination stimulants. Chem. Commun., 2017-8.

Wendt, K.U.; Schulz, G.E.; Corey, E.J.; Liu, D.R. (2000) Enzyme mechanisms for polycyclic triterpene formation. Angew. Chem. Int. Ed. Engl., *39*, 2812-33.

West, G.B.; Brown, J.H.; Enquist, B.J. (1999) The fourth dimension of life: fractal geometry and allometric scaling of organisms. Science, *284*, 1677-9.

White, J.D.; Bolton, G.L.; Dantanarayana, A.P.; Fox, C.M.J.; Hiner, R.N.; Jackson, R.W.; Sakuma, K.; Warrier, U.S. (1995) Total synthesis of the antiparasitic agent avermectin B_{1a}. J. Am. Chem. Soc., *117*, 1908-39.

White, J.D.; Carter, R.G.; Sundermann, K.F.; Wartmann, M. (2001) Total synthesis of epothilone B, epothilone D, and cis- and trans-9,10-dehydroepothilone D. J. Am. Chem. Soc., *123*, 5407-13.

White, R.H. (1988) Structural diversity among methanofurans from different methanogenic bacteria. J. Bacteriol., 4594-7.

Whitlock, H.W. (1998) On the structure of total synthesis of complex natural products. J. Org. Chem., *63*, 7982-9.

Whitwam, R.E.; Ahlert, J.; Holman, T.R.; Ruppen, M.; Thorson, J.S. (2000) The gene *calC* encodes for a non-heme iron metalloprotein responsible for calicheamicin self-resistance in *Micromonospora*. J. Am. Chem. Soc., *122*, 1556-7.

Wikelski, M.; Thom, C. (2000) Marine iguanas shrink to survive El Niño. Nature, *403*, 37.

Willard, H.F. (2000) Artificial chromosomes coming to life. Science, *290*, 1308-9.

Williams, D.E.; Andersen, R.J.; van Duyne, G.; Clardy, J. (1987) Gersemolide, a diterpenoid with a new rearranged carbon skeleton from the soft coral *Gersemia rubiformis*. Tetrahedron Lett., *28*, 5079-80.

Williams, D.E.; Lassota, P.; Andersen, R.J. (1998) Motuporamines A-C, cytotoxic alkaloids isolated from the marine sponge *Xestospongia exigua*. J. Org. Chem., *63*, 4838-41.

Williams, D.R.; Brooks, D.A; Berliner, M.A. (1999) Total synthesis of (-)-hennoxazole A. J. Am. Chem. Soc., *121*, 4924-5.

Winston, J.E. (2000) Describing Species: Practical Taxonomy Procedure for Biologists. Columbia University Press.

Wipf, P.; Lim, S. (1996) Total synthesis and structural studies of the antiviral natural product hennoxazole A. Chimia, *50*, 157-67.

Wipf, P.; Yoshikazu, U. (2000A) Total synthesis and revision of stereochemistry of the marine metabolite trunkamide A. J. Org. Chem., *65*, 1037-49.

Wipf, P.; Reeves, J.T.; Balachandran, R.; Giuliano, K.A.; Hamel, E.; Day, B.W. (2000B) Synthesis and biological evaluation of a focused mixture library of analogues of the antimitotic marine natural product curacin A. J. Am. Chem. Soc., *122*, 9391-5.

Wolfenbarger, L.L.; Phifer, P.R. (2000) The ecological risk and benefits of genetically engineered plants. Science, *290*, 2088-93.

Woolfson, A. (2000) Life Without Genes. Harper Collins.

Wratten, S.J.; Faulkner, D.J. (1977) Metabolites of the red alga *Laurencia suboppposita*. J. Org. Chem., *42*, 3343-9.

Wratten, S.J.; Faulkner, D.J. (1978) Antimicrobial metabolites from the marine sponge *Ulosa* sp. Tetrahedron Lett., 961-4.

Wratten, S.J.; Meinwald, J. (1981) Antimicrobial metabolites from the marine sponge *Axinella polycapella*. Experientia, *37*, 13.

Wright, A.E.; Pomponi, S.A.; McConnell, O.J.; Kohmoto, S.; McCarthy, P.J. (1987) (+)-Curcuphenol and (+)-curcudiol, sesquiterpene phenols from shallow and deep water collections of the marine sponge *Didiscus flavus*. J. Nat. Prod., *50*, 976-8.

Wright, A.E.; McCarthy, P.J.; Schulte, G.K. (1989) Sulfircin: a new sesterterpene sulfate from a deep-water sponge of the genus *Ircinia*. J. Org. Chem., *54*, 3472-4.

Wright, A.E.; Rueth, S.A.; Cross, S.S. (1991) An antiviral sesquiterpene hydroquinone from the marine sponge *Strongylophora hartmani*. J. Nat. Prod., *54*, 1108-11.

Wright, A.E.; Pomponi, S.A.; Cross, S.S.; McCarthy, P. (1992) A new bis(indole) alkaloid from a deep-water marine sponge of the genus *Spongosorites*. J. Org. Chem., *57*, 4772-5.

Wu, M.; Milligan, K.E.; Gerwick, W.H. (1997) Three new malyngamides from the marine cyanobacterium *Lyngbya majuscula*. Tetrahedron, *53*, 15983-90.

Wu, N.; Kudo, F.; Cane, D.E.; Khosla, C. (2000) Analysis of the molecular recognition features of individual modules derived from the erythromycin polyketide synthase. J. Am. Chem. Soc., *122*, 4847-52.

Wu Won, J.J.; Chalker, B.E. and Rideout, JA. (1997) Two new UV-absorbing compounds from *Stylophora pistillata*: sulfate esters of mycosporine-like amino acids. Tetrahedron Lett., *38*, 2525-6.

www.IBISmedical.com (2000) for a list of possible interactions of hypericin with drugs.

Xia, Q.; Ganem, B. (2001) Asymmetric total synthesis of (-)-α-kainic acid using an enantioselective, metal-promoted ene cyclization. Org. Lett., *3*, 485-7.

Xu, X.; de Guzman, F.; Gloer, J.B.; Shearer, C.A. (1992) Stachybotrins A and B: novel bioactive metabolites from a brackish water isolate of the fungus *Stachybotrys* sp. J. Org. Chem., *57*, 6700-3.

Yaalon, D.H. (2000) Down to the Earth. Why soil - and soil science - matters. Nature, *407*, 301.

Yamamoto, Y. (1995) Adhesive proteins of pearl oyster for medical use. Jpn. Kokai Tokkyo Koho JP 07,188,640, 25 Jul. to Hitachi Chemical Co. Chem. Abstr., 1995, *123*, 266073c.

Yamano, N.; Higashida, N.; Endo, C.; Sakata, N.; Fujishima, S.; Maruyama, A.; Higashihara, T. (2000) Purification and characterization of *N*-acetylglucosamine-6-phosphate deacetylase from a psychrotrophic marine bacterium, *Alteromonas* species. Mar. Biotechnol., *2*, 57-64.

Yamashita, F.; Nakatsuka, T. (1985) New antibiotic production generated by protoplasm fusion treatment between *Streptomyces griseus* and *S. tenjimariensis*. J. Antibiot., *38*, 58-63.

Yao, Y.; Hoffer, A.; Ching-yi, C.; Puga, A. (1995) Dioxin activates HIV gene expression by an oxidative stress pathway requiring a functional cytochrome *P*450 CYP1A1 enzyme. Envir. Health Perspect., *103*, 366-71.

Yayli, N.; Findlay, J.A. (1999) A triterpenoid saponin from *Cucumaria frondosa*. Phytochemistry, *50*, 135-8.

Yokokawa, F.; Fujiwara, H.; Shioiri, T. (2000) Total synthesis and revision of absolute stereochemistry of antillatoxin. Tetrahedron, *56*, 1759-75.

Yokoyama, A.; Shizuri, Y.; Misawa, N. (1998) Production of new carotenoids, astaxanthin glucosides, by *Escherichia coli* transformants carrying carotenoid biosynthetic genes. Tetrahedron Lett., *39*, 3709-12.

Yukimune, Y.; Tabata, H.; Higashi, Y.; Hara, Y. (1996) Methyl jasmonate-induced overproduction of paclitaxel and baccatin III in *Taxus* cell suspension cultures. Nature Biotechnol., *14*, 1129-32.

Yunker, M.B.; Scheuer, P.J. (1978) Mokupalides, three novel hexaprenoids from a marine sponge. J. Am. Chem. Soc., *100*, 307-9.

Zachos, J.; Pagani, M.; Sloan, L.; Thomas, E.; Billups, K. (2001) Trends, rhythms, and aberrations in global climate 65 Ma to present. Science, *292*, 686-93.

Zhang, X.; Zhang, X.; Rao, M.N.; Jones, S.R.; Shao, B.; Feibush, P.; McGuigan, M.; Tzodikov, N.; Feibush, B.; Sharkansky, I.; Snyder, B.; Mallis, L.M.; Sarkahian, A.; Wilder, S.; Turse, J.E.; Kinney, W.A. (1998) Synthesis of squalamine utilizing a readily accessible spermidine equivalent. J. Org. Chem., *63*, 8599-603.

Zhao, L.; Ahlert, J.; Xue, Y.; Thorson, J.S.; Sherman, D.H.; Liu, H.-w. (1999) Engineering a methymycin/picromycin-calicheamicin hybrid: construction of two new macrolides carrying a designed sugar moiety. J. Am. Chem. Soc., *121*, 9881-2.

Zheng, N.; Shimizu, Y. (1997) The isolation and structure of bacillariolide III, an extracellular metabolite of the diatom, *Pseudo-nitzschia multiseries*. Chem. Commun. 399-400.

Zhu, T.; Peterson, D.J.; Tagliani, L.; St. Clair, G.; Baszczynski, C.L.; Bowen, B. (1999) Targeted manipulation of maize genes *in vivo* using chimeric RNA/DNA oligonucleotides. Proc. Natl. Acad. Sci. USA, *96*, 8768-73.

Zhu, Y.; Chen, H.; Fan, J.; Wang, Y.; Li, Y.; Chen, J.; Fan, J.; Yang, S.; Hu, L.; Leung, H.; Mew, T.W.; Teng, P.S.; Wang, Z.; Mundt, C.C. (2000) Genetic diversity and disease control in rice. Nature, *406*, 718-22.

Zimmerman, M.P. (1989) The distribution of lipids and sterols in cell types from the marine sponge *Pseudaxinyssa* sp. Lipids, *24*, 210-6.

Index

Entries in italics, boldface, and underlined characters refer to charts, tables, and figures, respectively

Index

Printed and bound by CPI Group (UK) Ltd, Croydon, CR0 4YY

03/10/2024

01040311-0005